INTRODUCTION TO NEURAL ENGINEERING FOR MOTOR REHABILITATION

INTRODUCTION TO NEURAL ENGINEERING FOR MOTOR REHABILITATION

EDITED BY

DARIO FARINA
Department of Neurorehabilitation Engineering, Bernstein Focus Neurotechnology Göttingen, Bernstein Center for Computational Neuroscience, University Medical Center Göttingen, Georg-August University, Göttingen, Germany

WINNIE JENSEN
Center for Sensory–Motor Interaction, Department of Health Science and Technology, Aalborg University, Aalborg, Denmark

METIN AKAY
Department of Biomedical Engineering, Cullen College of Engineering, University of Houston, Houston, TX, USA

EMB IEEE Engineering in Medicine and Biology Society, *Sponsor*

IEEE Press Series in Biomedical Engineering
Metin Akay, *Series Editor*

IEEE
IEEE PRESS

WILEY

Published by John Wiley & Sons, Inc., Hoboken, New Jersey.
Published simultaneously in Canada.

Library of Congress Cataloging-in-Publication Data:

Introduction to neural engineering for motor rehabilitation / edited by Dario Farina, Winnie Jensen, Metin Akay.
p. ; cm.
ISBN 978-0-470-91673-5 (cloth)
I. Farina, Dario. II. Jensen, Winnie. III. Akay, Metin.
[DNLM: 1. Biomedical Engineering. 2. Nervous System. 3. Neural Networks (Computer) 4. Signal Processing, Computer-Assisted. 5. Trauma, Nervous System–rehabilitation. 6. User-Computer Interface. QT 36]

610.28–dc23

2012041247

10 9 8 7 6 5 4 3 2 1

CONTENTS

CONTRIBUTORS

Ole K. Andersen, Integrative Neuroscience Group, Center for Sensory–Motor Interaction, Aalborg University, Aalborg, Denmark

Felix Biessmann, Machine Learning Group, Technische Universität Berlin, Berlin, Germany

Alberto Botter, Laboratory for Engineering of the Neuromuscular System (LISiN), Politecnico di Torino, Torino, Italy

Germana Cappellini, Laboratory of Neuromotor Physiology, Santa Lucia Foundation, Rome, Italy

Jacopo Carpaneto, Biorobotics Institute, Scuola Superiore Sant'Anna, Pisa, Italy

Maura Casadio, Department of Informatics, Systems and Telematics, University of Genoa, Italy; Sensory Motor Performance Program, Rehabilitation Institute of Chicago and Department of Physiology, Northwestern University Medical School, Chicago, IL, USA

Luca Citi, Department of Anesthesia, Critical Care and Pain Medicine, Massachusetts General Hospital, Harvard Medical School, Boston, MA, USA; Department of Brain and Cognitive Sciences, Massachusetts Institute of Technology, Cambridge, MA, USA; Biorobotics Institute, Scuola Superiore Sant'Anna, Pisa, Italy

Elaine A. Corbett, Department of Biomedical Engineering, Northwestern University, Evanston, IL, USA

Janis J. Daly, Director, Brain Rehabilitation Research Center of Excellence, MR Gainesville DVA Medical Center; Research Career Scientist, DVA; Professor, Department of Neurology, College of Medicine, University of Florida, Gainesville, FL, USA; Director, Brain Rehabilitation Research Program, McKnight Brain Institute, University of Florida, Gainesville, FL, USA

Peter Detemple, Institut für Mikrotechnik Mainz GmbH, Mainz, Germany

Hans Dietl, Otto Bock HealthCare GmbH, Duderstadt, Germany; Otto Bock HealthCare Products GmbH, Vienna, Austria

Omar Feix do Nascimento, Center for Sensory–Motor Interaction, Department of Health Science and Technology, Aalborg University, Aalborg, Denmark

Strahinja Došen, Department of Neurorehabilitation Engineering, University Medical Center, Georg-August University, Göttingen, Germany

Kim Dremstrup, Center for Sensory–Motor Interaction, Department of Health Science and Technology, Aalborg University, Aalborg, Denmark

Günter Edlinger, g.tec medical engineering GmbH/Guger Technologies OG Schiedlberg, Austria

Kevin Englehart, Institute of Biomedical Engineering, University of New Brunswick, Fredericton, NB, Canada

Christian Ethier, Department of Physiology, Feinberg School of Medicine, Northwestern University, Chicago, IL, USA

Dario Farina, Department of Neurorehabilitation Engineering, Bernstein Focus Neurotechnology Göttingen, Bernstein Center for Computational Neuroscience, University Medical Center Göttingen, Georg-August University, Göttingen, Germany

Marco Gazzoni, Laboratory for Engineering of the Neuromuscular System (LISiN), Politecnico di Torino, Torino, Italy

Di Ge, Glaizer Groupe, Malakoff, France

Bernhard Graimann, Otto Bock HealthCare GmbH, Duderstadt, Germany

Ying Gu, Center for Sensory–Motor Interaction, Department of Health Science and Technology, Aalborg University, Aalborg, Denmark

Christoph Guger, g.tec medical engineering GmbH/Guger Technologies OG, Schiedlberg, Austria

Levi Hargrove, Center for Bionic Medicine, Rehabilitation Institute of Chicago, Chicago, IL, USA; Department of Physical Medicine and Rehabilitation, Northwestern University Feinberg School of Medicine, Chicago, IL, USA

Kristian Rauhe Harreby, Center for Sensory–Motor Interaction, Department of Health Science and Technology, Aalborg University, Aalborg, Denmark

Ulrich G. Hofmann, Neuroelectronic Systems, Department for Neurosurgery, University Hospital Freiburg, Freiburg, Germany; Graduate School for Computing in Medicine and Life Sciences, University of Lübeck, Lübeck, Germany

Yuri P. Ivanenko, Laboratory of Neuromotor Physiology, Santa Lucia Foundation, Rome, Italy

Winnie Jensen, Center for Sensory–Motor Interaction, Department of Health Science and Technology, Aalborg University, Aalborg, Denmark

Konrad Kording, Department of Physical Medicine and Rehabilitation, Feinberg School of Medicine, Northwestern University, Chicago, IL, USA; Department of Physiology, Feinberg School of Medicine, Northwestern University, Chicago, IL, USA

Francesco Lacquaniti, Department of Neuroscience and Centre of Space Bio-Medicine, University of Rome Tor Vergata, Rome, Italy; Laboratory of Neuromotor Physiology, Santa Lucia Foundation, Rome, Italy

Birgit Larsen, Neurodan A/S, Aalborg, Denmark

Lorenzo Masia, Department of Robotics, Brain and Cognitive Sciences, Italian Institute of Technology, Genoa, Italy

Kevin A. Mauser, Biomedical Engineering Department, Indiana University–Purdue University Indianapolis, Indianapolis, IN, USA

Frank C. Meinecke, Machine Learning Group, Technische Universität Berlin, Berlin, Germany

Roberto Merletti, Laboratory for Engineering of the Neuromuscular System (LISiN), Politecnico di Torino, Torino, Italy

Silvestro Micera, Biorobotics Institute, Scuola Superiore Sant'Anna, Pisa, Italy; Center for Neuroprosthetics and Institute of Bioengineering, School of Engineering, Ecole Polytechnique Fédérale de Lausanne (EPFL), Lausanne, Switzerland

José del R. Millán, Center for Neuroprosthetics, School of Engineering, Ecole Polytechnique Fédérale de Lausanne (EPFL), Lausanne, Switzerland

Lee E. Miller, Department of Physical Medicine and Rehabilitation, Feinberg School of Medicine, Northwestern University, Chicago, IL, USA; Department of Physiology, Feinberg School of Medicine, Northwestern University, Chicago, IL, USA; Department of Biomedical Engineering, Northwestern University, Evanston, IL, USA

Marco Molinari, Spinal Cord Unit, Santa Lucia Foundation, Rome, Italy

Pietro G. Morasso, Department of Robotics, Brain and Cognitive Sciences, Italian Institute of Technology, Genoa, Italy

Juan C. Moreno, Bioengineering Group, Spanish Research Council, CSIC, Arganda del Rey, Spain

Natalie Mrachacz-Kersting, Center for Sensory–Motor Interaction (SMI), Department of Health Science and Technology, Aalborg University, Aalborg, Denmark

Klaus-Robert Müller, Machine Learning Group, Technische Universität Berlin, Berlin, Germany

Xavier Navarro, Institute of Neurosciences and Department of Cell Biology, Physiology and Immunology, Universitat Autònoma de Barcelona; Centro de Investigación en Red sobre Enfermedades Neurodegenerativas (CIBERNED), Spain

Emily R. Oby, Department of Physiology, Feinberg School of Medicine, Northwestern University, Chicago, IL, USA

António R.C. Paiva, Scientific Computing and Imaging Institute, University of Utah, Salt Lake City, UT, USA; presently at ExxonMobil Upstream Research Company, Houston, TX, USA

Il Park, Center for Perceptual Systems and Institute for Neuroscience, University of Texas, Austin, TX, USA

Andrei Patriciu, Neurodan A/S, Aalborg, Denmark

Eric J. Perreault, Department of Physical Medicine and Rehabilitation, Feinberg School of Medicine, Northwestern University, Chicago, IL, USA; Department of Biomedical Engineering, Northwestern University, Evanston, IL, USA

José Luis Pons, Bioengineering Group, Spanish Research Council, CSIC, Arganda del Rey, Spain

Dejan B. Popović, Center for Sensory–Motor Interaction, Aalborg University, Aalborg, Denmark; University of Belgrade, Faculty of Electrical Engineering, Belgrade, Serbia

Mirjana B. Popović, University of Belgrade, Faculty of Electrical Engineering, Belgrade, Serbia; Aalborg University, Department for Health Science and Engineering, Aalborg, Denmark; University of Belgrade, Institute for Multidisciplinary Research, Belgrade, Serbia

José C. Príncipe, Department of Electrical and Computer Engineering, University of Florida, Gainesville, FL, USA

Shaoyu Qiao, Weldon School of Biomedical Engineering, Purdue University, West Lafayette, IN, USA

Stanisa Raspopovic, Biorobotics Institute, Scuola Superiore Sant'Anna, Pisa, Italy; Center for Neuroprosthetics and Institute of Bioengineering, School of Engineering, Ecole Polytechnique Fédérale de Lausanne (EPFL), Lausanne, Switzerland

Jacopo Rigosa, Biorobotics Institute, Scuola Superiore Sant'Anna, Pisa, Italy

Eduardo Rocon, Bioengineering Group, Spanish Research Council, CSIC, Arganda del Rey, Spain

Justin C. Sanchez, Department of Biomedical Engineering, University of Miami, Coral Gables, FL, USA

Vittorio Sanguineti, Department of Informatics, Systems and Telematics, University of Genoa, Italy; Department of Robotics, Brain and Cognitive Sciences, Italian Institute of Technology, Genoa, Italy

Erik Scheme, Institute of Biomedical Engineering, University of New Brunswick, Fredericton, NB, Canada

Thomas Sinkjær, Center for Sensory–Motor Interaction, Aalborg University, Aalborg, Denmark

Erika G. Spaich, Integrative Neuroscience Group, Center for Sensory–Motor Interaction, Aalborg University, Aalborg, Denmark

Valentina Squeri, Department of Robotics, Brain and Cognitive Sciences, Italian Institute of Technology, Genoa, Italy

Aiko K. Thompson, Program for Translational Neurological Research, Helen Hayes Hospital and the Wadsworth Center, New York State Department of Health, New York, NY, USA

Jonathan R. Wolpaw, Program for Translational Neurological Research, Helen Hayes Hospital and the Wadsworth Center, New York State Department of Health, New York, NY, USA

Yijing Xie, Graduate School for Computing in Medicine and Life Sciences, University of Lübeck, Lübeck, Germany

Ken Yoshida, Biomedical Engineering Department, Indiana University–Purdue University Indianapolis, Indianapolis, IN, USA; Weldon School of Biomedical Engineering, Purdue University, West Lafayette, IN, USA; Center for Sensory–Motor Interaction, Department of Health Science and Technology, Aalborg University, Aalborg, Denmark

PREFACE

Neural engineering is an interdisciplinary research area that brings to bear methods from neuroscience and engineering to analyze neurological functions and to design solutions to problems associated with neurological limitations and dysfunctions (definition by the Editorial Board of the *Journal of Neural Engineering* [Durand, 2007]). Despite neural engineering's being a relatively new research area, the field is developing rapidly. This development requires continuously updated didactic material for the increasing number of undergraduate, graduate, and Ph.D. courses on the topic. The applications of neural engineering to rehabilitation of movement cover a broad range of engineering challenges, from electrode design to signal processing and from the neurophysiology of movement to robotics.

The three main approaches of neural engineering used for rehabilitation of impaired motor functions are restoration, replacement, and neuromodulation. Restoration consists in retaining existing neural and anatomical structures and in controlling them for reestablishing a motor function. An example of such an approach is functional electrical stimulation (FES). Replacement consists in substituting the impaired motor apparatus with an artificial one, controlled by residual, but still functional, neural or muscular structures. An example of these methods is the control of artificial limbs (active prostheses). The aim of neuromodulation is (re)training the central nervous system to induce plasticity through artificial stimulation of afferent pathways and/or by artificial enhancement of efferent neural and muscular signals provided as feedback. Examples of such an approach are the application of patterned peripheral electrical neuromuscular stimulation (e.g., transcutaneous electrical nerve stimulation, TENS), mechanical stimulation using robots, or repetitive transcranial magnetic stimulation for retraining the diseased central nervous system.

The aim of this book is to present the state of the art in technologies for motor neurorehabilitation and to give an overview of the current challenges and recent advances within neural rehabilitation technology. The book is intended for undergraduate, graduate, and Ph.D. students as well as senior researchers who work in the field of biomedical engineering, and it is organized in five parts. Part I reviews aspects related to injuries of the nervous system that determine motor impairments. It is considered as a prerequisite that the reader is familiar with the physiology of the neuromuscular system, which is not included in this book. Part II reviews engineering methods for interfacing the neuromuscular system and for conditioning and processing neural and muscular signals. The methods described in Part II are also used in the last three parts of the book, which describe examples of neurotechnologies within the areas of restoration, replacement, and neuromodulation. The topics in each part are collected with the focus on the application (e.g., replacement of function) rather than on the principle on which such application is exploited. Therefore, for example, the principle of brain-interfacing is used in applications described in both Parts III (replacement) and V (neuromodulation), according to the different uses of brain-interfacing in these two sections. Each part begins with a short introduction that serves to put into perspective the topics addressed in that part and to guide the reader to the research areas detailed there. The book's parts comprise introductory chapters, which provide a broad perspective (review chapters), and chapters with a strong focus on more specialized topics (focused chapters), as indicated at the beginning of each chapter.

The book is intended to provide a broad perspective within the field of motor neurorehabilitation engineering by including several topics that in most other books are treated separately. At the same time, the book does not intend to provide an exhaustive treatment of all methods and approaches for motor neurorehabilitation. Rather, the topics presented have been selected to be representative of the field and thus to provide the reader with a general broad overview and understanding of the research area. Readers who approach neural rehabilitation engineering for the first time will find the review chapters as an overview of the state of the art, whereas senior researchers or experts within the field may have further interest in the focused chapters that provide a detailed analysis of specific topics with recent solutions. As indicated, the physiology of the neuromuscular system is not presented in this book, which has as its starting point the injuries of the system. Therefore, readers approaching neural engineering for the first time are advised to first consult references on human physiology.

The editors are very grateful to all the contributing authors for enthusiastically accepting the invitation to contribute to this project and to Dr. Antonietta Stango (University Medical Center Göttingen, Germany) for the important contribution of assisting with the editorial tasks.

Dario Farina thanks the European Research Council (ERC), which has awarded him the Advanced Research Grant DEMOVE ("Decoding the Neural Code of Human Movements for a New Generation of Man–Machine Interfaces"; contract #267888). This grant has supported Dario Farina for the time invested in editing this book.

Dario Farina
Winnie Jensen
Metin Akay

REFERENCE

Durand D.M. (2007). What is neural engineering? *J Neural Eng.* 4. http://dx.doi.org/10.1088/1741-2552/4/4/E01.

PART I

INJURIES OF THE NERVOUS SYSTEM

Part I contains three chapters that examine the type of neural injuries that may lead to sensory–motor impairment, as well as aspects of plasticity, as a relatively novel conceptual theme in the field of neural rehabilitation. Damage to the nervous system is typically associated with the loss of motor drive and of afferent input to the central nervous system. The severity of the neural damage depends on the location of the injury, which may lead to adaptation of the movement pattern, paresis, or complete paralysis. Plasticity has been defined as changes in the strength, number, and location of synaptic connections in response to either an environmental stimulus or an alteration in synaptic activity in a network; our fundamental understanding of what underlies neural plasticity is believed to be one of the key elements in devising strategies for rehabilitation or repair of injuries.

Chapter 1, by Popović and Sinkjær, provides a review of the incidence and the pathology of major diseases and injuries within the central nervous system that lead to impairment of the sensory–motor system, such as stroke and spinal cord injury. The chapter also briefly introduces the types of injuries that lead to loss of sensory–motor functions at the peripheral level.

Chapter 2, by Navarro, more specifically examines injuries at the peripheral level that may result in partial or total loss of motor, sensory, and autonomic functions. Functional deficits may be compensated by reinnervation of denervated targets by regenerating the injured axons, by collateral branching of

Introduction to Neural Engineering for Motor Rehabilitation, First Edition. Edited by Dario Farina, Winnie Jensen, and Metin Akay.

undamaged axons, or by remodeling of nervous system circuitries. Plasticity of central connections may compensate functionally for the lack of adequate target reinnervation; however, plasticity has limited effects on disturbed sensory localization or fine motor control after injuries, and it may even result in maladaptive changes, such as neuropathic pain and hyperreflexia.

Obtaining evidence for spinal or cortical plasticity in the human is very difficult without using invasive recording techniques. Chapter 3, by Ivanenko and collaborators, reports on motor primitives to provide a novel perspective on how the neural control system operates under locomotion in healthy subjects and in patients. They find that building blocks with which the central nervous system constructs motor patterns can be preserved in patients with various motor disorders despite the fact that they often modify their muscle activity and adopt *motor equivalent* solutions. Our understanding of these motor primitives may be useful in driving neuroprostheses or entraining locomotor circuits in disabled people in the future.

1

REVIEW CHAPTER

DISEASES AND INJURIES OF THE CENTRAL NERVOUS SYSTEM LEADING TO SENSORY–MOTOR IMPAIRMENT

DEJAN B. POPOVIĆ AND THOMAS SINKJÆR

Center for Sensory–Motor Interaction, Aalborg University, Aalborg, Denmark

SUMMARY

Damage to the central and peripheral nervous systems is associated with a loss of motor drive and a defective afferent input to the central nervous system (CNS). Depending on the location and severity of the neural damage this leads to anything from a complete paralysis to a paresis and a maladaptation of the movement pattern. This chapter starts with a presentation of neuron injury. Such injuries are categorized based on the extent and type of damage to the nerve and the surrounding connective tissue. This chapter addresses sensory–motor deficits that are caused by neuron injury or disease: (a) cerebrovascular accident (CVA), or stroke, which causes impairments due to changes in blood supply to the brain; (b) spinal cord injuries (SCIs), which result in total or partial obstruction of flow of both sensory and motor information between the peripheral and central nervous systems; (c) nontraumatic disorders of the CNS (amyotrophic lateral sclerosis and multiple sclerosis); and (d) cerebral palsy. At the end of the chapter we present the incidence of CNS diseases.

Introduction to Neural Engineering for Motor Rehabilitation, First Edition. Edited by Dario Farina, Winnie Jensen, and Metin Akay.

NEURON INJURY

A **neuron injury** is categorized based on the extent and type of damage to the nerve and the surrounding connective tissue (Fig. 1.1): **neuropraxia**, a nerve injury in which the nerve remains intact but with its signaling ability damaged; **axonotmesis**, in which the nerve remains intact but there with an interruption in conduction of the impulse along the nerve fiber; and **neurotmesis**, which follows a severe contusion, stretch, laceration, or similar damage. In this case both the axon and the encapsulating connective tissue lose their continuity.

In some injuries, the presynaptic neurons that synapse on the damaged cells are also affected. Transneuronal changes of various kinds are important in explaining why a lesion at one site in the central nervous system (CNS) can have effects on sites distant to the lesion, sites that are distributed according to the connections that the lesion interrupts.

The **zone of trauma** is a place where a bundle of axons is cut, either by sectioning of a tract within the CNS or by sectioning a peripheral nerve. The part of the axon still connected to the cell body is the proximal segment, and the part isolated from the rest of the cell is the distal segment.

At a zone of trauma in the CNS, the axon and myelin sheath undergo rapid local degeneration. Because a lesion usually interrupts blood vessels, macrophages from the general circulation can enter the area and phagocytose axonal debris. Astrocytes and microglia proliferate and act as phagocytes. In the CNS, however, the proliferation of fibrous astrocytes leads to the formation of a glial scar around the zone of trauma. Scarring can block the course taken by regenerating axons and establish an effective barrier against the reformation of central connections.

The degeneration spreads in both directions along the axon from the zone of trauma, but only for a short distance in the proximal segment, usually up to the point of origin of the first axon collateral. After few days, a retrograde reaction is seen in the cell body. If the entire cell body dies, then degeneration spreads from the axon hillock down along the remainder of the proximal segment. In the distal segment, outside the zone of trauma, the degeneration

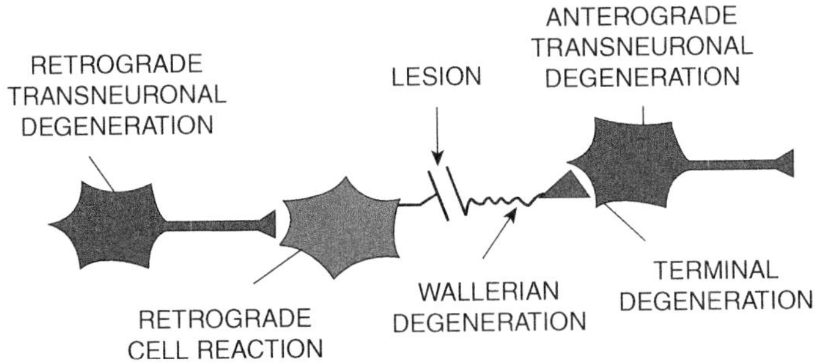

Figure 1.1. Sketch of nerve injury.

first appears in the axon terminal about one day after the occurrence of the lesion. In approximately two weeks, the synapses formed by the distal segment degenerate completely. The process is called terminal degeneration. Degeneration of the distal axon, termed Wallerian degeneration, takes place over a period of about two months. Sometime cells that are prior postsynaptic to the injured neuron may also be affected.

The term **motor unit** refers to the motor neuron in the spinal cord and the population of muscle fibers that it innervates. The motor unit has four functional components: (1) the cell body of the motor neuron, (2) the axon of the motor neuron that runs in the peripheral nerve, (3) the neuromuscular junctions, and (4) the muscle fibers innervated by that neuron.

Most diseases of the motor unit cause weakness and wasting of skeletal muscles. These diseases may differ in other features, however, depending upon which of the four components of the motor unit is primarily affected. A disease can be functionally selective by affecting only the sensory systems or only the motor systems. Motor diseases are regionally selective. They affect only one component of the neuron (e.g., the axon, rather than the cell body).

The clinical consequences of neurogenic disease are most obvious when a peripheral nerve is cut. The muscles innervated by that nerve are immediately paralyzed and then waste progressively. Tendon reflexes are lost immediately, as is the sensation in the area innervated by the nerve because the nerve carries sensory as well as motor fibers. In neurogenic diseases, similar effects of denervation appear more slowly, and the muscles gradually become weak and wasted.

CEREBROVASCULAR ACCIDENT

There is a large population of humans that suffer from impairment caused by changes in blood supply of the brain. Blood flow to the brain is highly protected, yet the brain remains highly susceptible to disturbances of the blood supply, as reflected in the high incidence of symptomatic cerebral vascular disease. The term **stroke**, or **cerebrovascular accident** (CVA), refers to the neurological symptoms and signs, usually focal and acute, which result from diseases involving blood vessels.

Strokes are either occlusive (due to closure of a blood vessel) or hemorraghic (due to bleeding from a vessel), as sketched in Figure 1.2. Insufficiency of blood supply is termed ischemia; if it is temporary, symptoms and signs may clear with little or no pathological evidence of tissue damage. Ischemia reduces blood supply, thereby depriving tissue of oxygen and glucose, and prevents the removal of potentially toxic metabolites such as lactic acid. When ischemia is sufficiently severe and prolonged, neurons and other cellular elements die; this condition is called infarction.

Hemorrhage may occur at the brain surface (extraparenchymal). Alternatively, hemorrhage may be intraparenchymal (e.g., from rupture of vessels

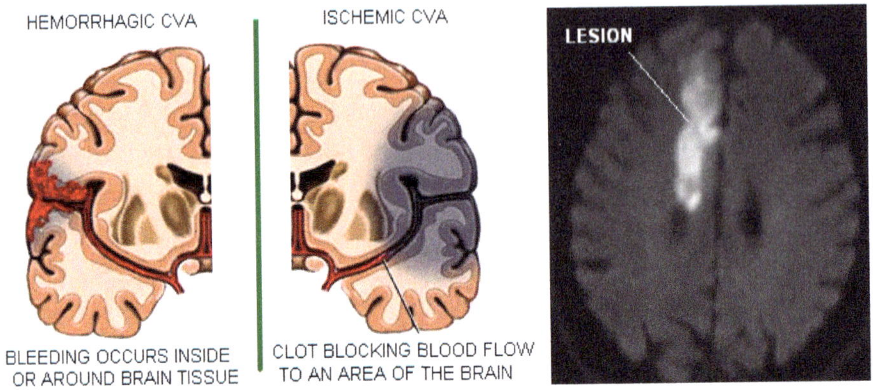

Figure 1.2. Sketch of the hemorrhagic (left) and ischemic (right) cerebrovascular accident. The right panel shows a magnetic resonance imaging (MRI) scan of a brain after an ischemic stroke.

damaged by long-standing hypertension), causing a blood clot or hematoma within the cerebral hemispheres, in the brainstem, or in the cerebellum. Ischemia or infarction may accompany hemorrhage. The mass effect of an intracerebral hematoma may compromise the blood supply of adjacent brain tissue; subarachnoid hemorrhage may, by unclear mechanisms, cause reactive vasospasm of cerebral surface vessels, leading to further ischemic brain damage. Infarcted tissue may also become secondarily hemorrhagic.

Each cerebral hemisphere is supplied by an internal carotid artery, which arises from a common carotid artery beneath the angle of the jaw, enters the cranium through the carotid foramen, traverses the cavernous sinus (giving off the ophthalmic artery), penetrates the dura, and divides into the anterior and middle cerebral arteries (Fig. 1.2 and Fig. 1.3).

Interconnections between blood vessels (anastomoses) protect the brain when part of its vascular supply is compromised. The anterior communicating artery connects the two anterior cerebral arteries; the posterior cerebral arteries are connected to the internal carotid arteries by the posterior communicating arteries.

Middle Cerebral Artery Territory Infarction

Infarction in the territory of the middle cerebral artery (cortex and white matter) causes the most frequently encountered stroke syndrome with contralateral weakness, sensory loss, and visual field cut, and, depending on the hemisphere involved either language disturbance or impaired spatial perception. Weakness and sensory loss affect the face and arm more than the leg because of the somatotopy of the motor and sensory cortex (pre- and postcentral gyri): the face and arm lie on the convexity, whereas the leg resides on

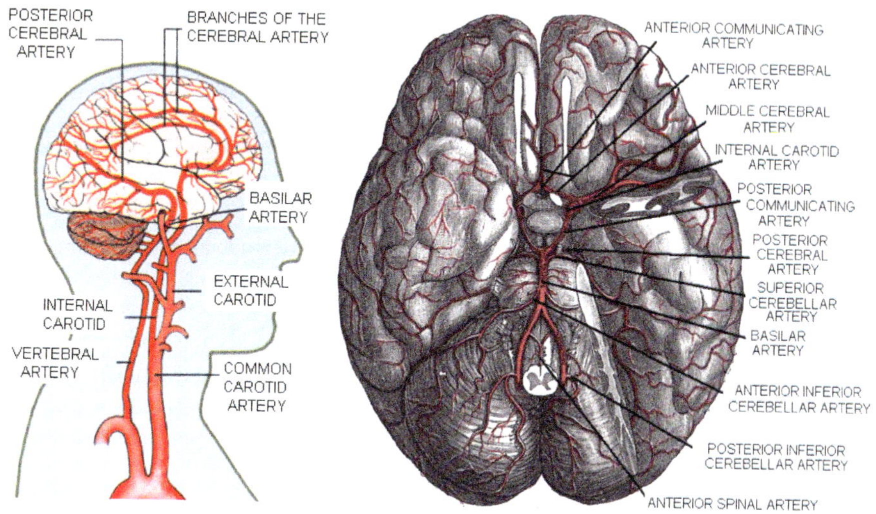

Figure 1.3. Blood supply of the brain. The large surface branches of the anterior cerebral artery supply the cortex and white matter of the inferior frontal lobe, the medial surface of the frontal and parietal lobes, and the anterior corpus callosum. Smaller penetrating branches supply the deeper cerebrum and diencephalon, including limbic structures, the head of the caudate, and the anterior limb of the internal capsule. The large surface branches of the middle cerebral artery supply most of the cortex and white matter of the hemisphere's convexity, including the frontal, parietal, temporal, and occipital lobes, and the insula. Smaller penetrating branches supply the deep white matter and diencephalic structures such as the posterior limb of the internal capsule, the putamen, the outer globus pallidus, and the body of the caudate. After the internal carotid artery emerges from the cavernous sinus, it also gives off the anterior choroidal artery, which supplies the anterior hippocampus and, at a caudal level, the posterior limb of the internal capsule.

the medial surface of the hemisphere. Motor and sensory losses are greatest in the hand, as the more proximal limbs and the trunk tend to have greater representation in both hemispheres. Paraspinal muscles, for example, are hardly ever weak in unilateral cerebral lesions. Similarly, the facial muscles of the forehead and the muscles of the pharynx and jaw are represented bihemispherically and are therefore usually spared. Tongue weakness is variable. If weakness is severe (plegia), the muscle tone is usually decreased initially and is gradually increased over days or weeks to spasticity with hyperactive tendon reflexes. A Babinski sign, reflecting upper motor neuron disturbance, is usually present from the outset. When weakness is mild, or during recovery, there may be clumsiness or slowness of movement out of proportion to loss of strength; such motor disability may resemble Parkinsonian bradykinesia or even cerebellar ataxia.

Acutely, there is often paresis of contralateral conjugate gaze because of damage to the convexity of the cortex anterior to the motor cortex (the frontal eye field). The reason why the gaze palsy persists for only one or two days, although other signs remain severe, is controversial.

Sensory loss tends to involve discriminative and proprioceptive modalities more than affective modalities. Pain and temperature sensation may be impaired or seem altered, but they are usually not lost. Joint position sense, however, may be severely disturbed, causing limb ataxia, and there may be loss of two-point discrimination, astercognosis (inability to recognize a held object by tactual sensation), or failure to appreciate a touch stimulus if another is simultaneously delivered to the normal side of the body (extinction).

Visual field impairment (homonymous hemianopsia) is the result of damage to the optic radiation, the deep fiber tracts connecting the thalamic lateral geniculate nucleus to the visual cortex. Destruction of left opercular cortex in humans causes aphasia, which may take a variety of forms depending on the degree and distribution of the damage. Frontal opercular lesions tend to produce particular difficulty with speech output and writing with relative preservation of language comprehension, whereas infarction of the posterior superior temporal gyrus tends to cause severe difficulty in comprehending spoken speech and reading. When the damage is widespread, there is severe language disturbance of mixed type (global aphasia). Left-hemisphere convexity damage, especially parietal, may also cause motor apraxia, a disturbance of learned motor acts not explained by weakness or incoordination, with the ability to perform the act when the setting is altered.

Right-hemisphere convexity infarction, especially parietal, tends to cause disturbances of spatial perception. There may be difficulty in copying simple pictures or diagrams (constructional apraxia), in interpreting maps or finding one's way about (topographagnosia), or in putting on one's clothes properly (dressing apraxia). Awareness of space and the subject's own body contralateral to the lesion may be particularly affected (hemi-inattention or hemineglect). Subjects may fail to recognize their hemiplegia (anosognosia), left arm (asomatognosia), or any external object to the left of their own midline. Such phenomena may occur independently of visual field defects and in subjects otherwise mentally intact.

Anterior Cerebral Artery Territory Infarction

Infarction in the territory of the anterior cerebral artery causes weakness and sensory loss qualitatively similar to that of convexity lesions, but affects mainly the distal contralateral leg. There may be urinary incontinence, but it is uncertain whether this is due to a lesion of the paracentral lobule (medial hemispheric motor and sensory cortices) or of a more anterior region concerned with the inhibition of bladder emptying. Damage to the supplementary motor cortex may cause speech disturbance, considered aphasic by some and a type of motor inertia by others. Involvement of the anterior corpus callosum may cause apraxia of the left arm (sympathetic apraxia), which is attributed to disconnection of the left (language dominant) hemisphere from the right motor cortex.

Bilateral Anterior Cerebral Artery Territory Infarction

Bilateral infarction in the territory of the anterior cerebral artery (occurring, e.g., when both arteries arise anomalously from a single trunk) may cause a severe behavioral disturbance, with profound apathy, motor inertia, and muteness, attributed variably to destruction of the inferior frontal lobes (orbitofrontal cortex), deeper limbic structures, supplementary motor cortices, or cingulate gyri.

Posterior Cerebral Artery Territory Infarction

Infarction in the territory of the posterior cerebral artery may include, or especially affect, the following structures: the thalamus, causing contralateral hemisensory loss and sometimes spontaneous pain and dysesthesia (thalamic pain syndrome); the subthalamic nucleus, causing contralateral severe proximal chorea (hemiballism); or even the midbrain, with ipsilateral oculomotor palsy and contralateral hemiparesis.

SPINAL CORD INJURIES

Spinal cord injuries (SCIs) or **diseases** are a frequent reason for disability and result in total or partial obstruction of flow of both sensory and motor information instrumental for normal life. Spinal cord injuries are most often caused by trauma, especially following motor vehicle and sports accidents. The resulting syndrome depends on the extent of direct injury of the cord or compression of the cord by displaced vertebrae or blood clots. In extreme cases trauma may lead to complete or partial transection of the spinal cord. Knowledge of the anatomy and physiology of the spinal cord helps in recognizing spinal cord disease and localization of the disease to a particular segment or region of the spinal cord. This allows identification of the nature of the disorder.

Lesions of the spinal cord give rise to motor or sensory symptoms that are often related to a particular sensory or motor segmental level of the spinal cord (Fig. 1.4). Identification of the appropriate level of the motor or sensory loss (called a motor or sensory level) is important for understanding the disability.

When motor roots are involved, or when motor neurons are affected focally, clinical findings may indicate the spinal level of the injury. This clinical evidence would include the typical lower motor neuron signs: weakness, wasting, fasciculation, and loss of tendon reflexes. Because it is clinically difficult to relate the innervation of muscles of the trunk and thorax to specific spinal

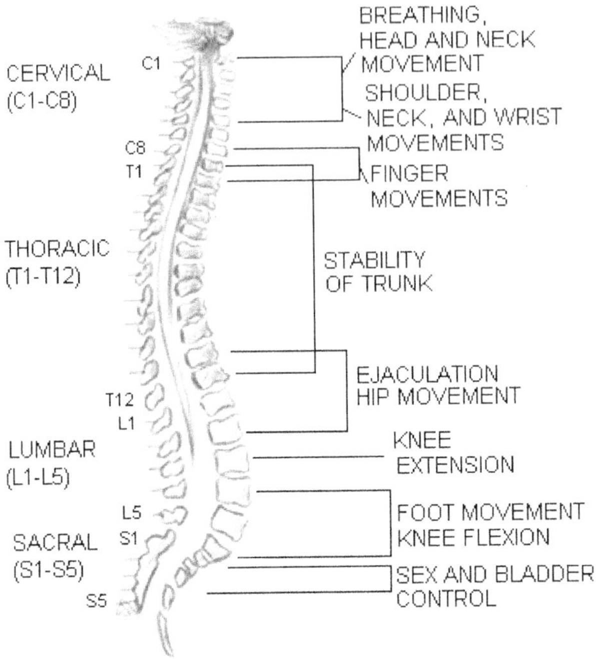

Figure 1.4. Sketch of the spinal cord with the indications of the spinal segments being responsible for specific functions. Numbers point to the spinal nerves.

segments, however, the motor level may not be evident. For instance, a lesion anywhere above the first lumbar segment may cause signs of upper motor neuron disease in the legs. Under these circumstances, sensory abnormalities are more valuable for localizing the lesion.

The characteristic pattern of sensory loss after a transverse spinal cord lesion is loss of cutaneous sensation below the level of the lesion, contralateral to the damaged spinothalamic tract if the lesion is unilateral. The sensory level is often more evident than the motor level. However, sensory loss due to spinal lesions must be differentiated from the pattern of sensory loss caused by lesions of peripheral nerves or isolated nerve roots. In multiple symmetrical peripheral neuropathy (polyneuropathy), there is a glove-and-stocking pattern of impaired perception of pain and temperature. This pattern is attributed to "dying-back" or impaired axonal transport; the parts of the axons most severely affected are those most distant from the sensory neuron cell bodies in the dorsal root ganglia. In injuries of single peripheral nerves, the distribution of sensory loss is more restricted and can be recognized by reference to sensory charts that were originally generated by studies of the long-term effects of traumatic nerve injuries incurred during war.

Nerve root or segmental sensory loss and spinal sensory levels can be identified by the dermatomes typically affected. The spinal cord ends at the base

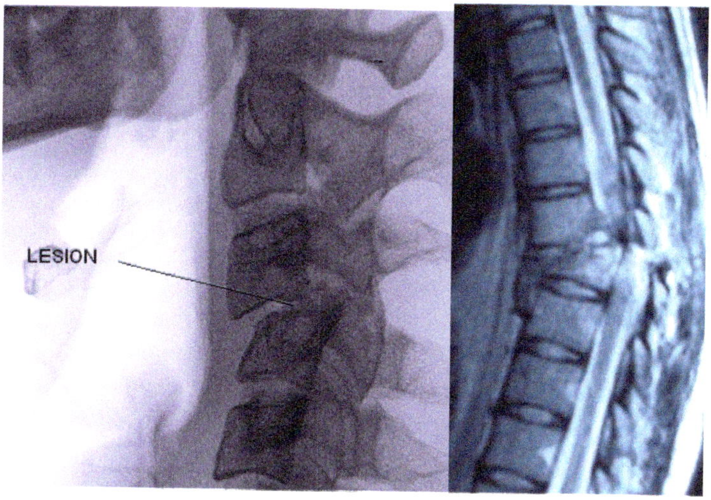

Figure 1.5. Xerograms of two spinal cord injuries.

of the second lumbar (L2) vertebra. Below this level the spinal canal is occupied by the lower nerve roots (the cauda equina).

A spinal cord lesion (Fig. 1.5) arises within the spinal cord (intra-axial or intramedullary) or external to the spinal cord (extra-axial or extramedullary).

Clinical evidence may give some clues that are helpful in making the distinction. For instance, pain is more common in extra-axial lesions because a compressive lesion (such as a tumor) may affect the dura, posterior nerve roots, or blood vessels that are innervated by sensory neurons mediating pain. In contrast, because there are no pain receptors within the spinal cord and the brain, intra-axial lesions may be painless. Intra-axial lesions may be marked by sacral sparing of sensation or may cause a segmental pattern of sensory loss, as in syringomyelia. The bladder function is affected earlier in intra-axial disorders than it is in extra-axial disease.

The following terms are frequently used in communication between people dealing with rehabilitation technology (Maynard et al., 1997):

Tetraplegia. This term refers to impairment or loss of motor and/or sensory function in the cervical segments of the spinal cord (Fig. 1.4 and Fig. 1.6) due to damage of neural elements within the spinal canal. Tetraplegia results in impairment of function in the arms as well as in the trunk, legs, and pelvic organs. It does not include brachial plexus lesions or injury to peripheral nerves outside the neural canal.

Paraplegia. This term refers to impairment or loss of motor and/or sensory function in the thoracic, lumbar, or sacral (but not cervical) segments of the spinal cord, secondary to damage of neural elements within the spinal canal. With paraplegia, arm functioning is spared, but, depending on the level of injury, the trunk, legs, and pelvic organs may be involved. The term is used in referring to cauda equina and conus medullaris injuries, but not to lumbosacral plexus lesions or injury to peripheral nerves outside the neural canal.

Tetraparesis and Paraparesis. These terms are used to describe incomplete lesions, where many functions are preserved.

Dermatome. This term, as illustrated in Figure 1.6, refers to the area of the skin innervated by the sensory axons within each segmental nerve (root).

Myotome. This term refers to the collection of muscle fibers innervated by the motor axons within each segmental nerve (root).

Neurological Level. This term refers to the most caudal segment of the spinal cord with normal sensory and motor function on both sides of the

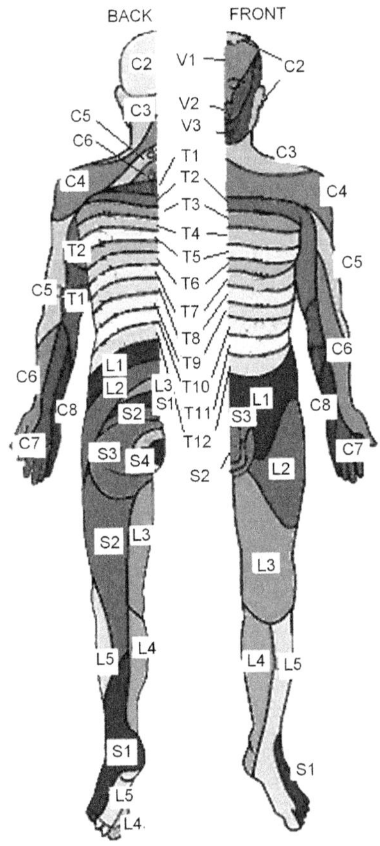

Figure 1.6. Dermatomes on the back and front side of a male subject. The number-and-letter labels correspond to the spinal nerves shown in Figure 1.4.

body. Thus, up to four different segments may be identified in determining the neurological level (i.e., R-sensory, L-sensory, R-motor, L-motor).

Sensory Level. This term refers to the most caudal segment of the spinal cord with normal sensory function on both sides of the body.

Motor Level. The term motor level is defined similarly to sensory level, with respect to motor function.

Skeletal Level. This term refers to the level at which, by radiographic examination, the greatest vertebral damage is found.

Sensory and Motor Scores. These are numerical summary scores that reflect the degree of neurological impairment associated with the SCI.

Incomplete Injury. An incomplete injury is defined in cases where partial preservation of sensory and/or motor functions is found below the neurological level and includes the lowest sacral segment. Sacral sensation includes sensation at the anal mucocutaneous junction as well as deep anal sensation. The test of motor function is the presence of voluntary contraction of the external anal sphincter upon digital examination.

Complete Injury. This is the term used when there is an absence of sensory and motor function in the lowest sacral segment.

Zone of Partial Preservation. The zone of partial preservation (ZPP) refers to those dermatomes and myotomes caudal to the neurological level that remain partially innervated. When some impaired sensory and/or motor function is found below the lowest normal segment, the exact number of segments so affected should be recorded for both sides as the ZPP.

Complete Spinal Cord Transection

The spinal cord may be completely severed acutely in fracture–dislocations of vertebrae or by knife or bullet wounds. Acute transection of the cord may also result from an inflammatory condition called transverse myelitis or from compression due to a tumor, especially metastatic tumors. In myelitis and tumors, symptoms evolve in days or weeks.

Immediately after traumatic section of the cord, however, there is a loss of all sensation and all voluntary movement below the lesion. Bladder and bowel controls are also lost. If the lesion is above C3, breathing may be affected. Although upper motor neuron signs might be expected, tendon reflexes are usually absent—a condition of spinal shock that persists for several weeks. After a while, the reflex activity returns at the levels below the lesion. Hyperactive reflexes, clonus (rapid and repeated contraction and relaxation of passively stretched muscle), and Babinski signs then appear as signs of damage to the corticospinal tract. The legs become spastic; this condition is often preceded by intermittent hypertonia and flexor spasms that occur spontaneously or may be provoked by cutaneous stimuli. Later, flexor and extensor spasms may alternate, and the ultimately fixed posture may be either flexion or extension of the knees and hips. Bladder and bowel function may become automatic, with emptying in response to moderate filling. Automatic bladder emptying may be retarded by severe distention of the bladder or infection in the acute stage, or by damage to lumbar or sacral cord segments.

Partial Transection

In partial transection of the spinal cord, some ascending or descending tracts may be spared. In slowly progressing lesions, as in compression by an extramedullary tumor, the same tracts may be affected, but less severely. Partial function is retained, but specific motor and sensory signs can still be recognized.

Hemisection (Brown–Sequard Syndrome). Because of spinal cord anatomy, hemisection of the right side of the cervical spinal cord (e.g., at C4) has four main clinical consequences:

1. Ipsilateral (right) signs of a lesion in the corticospinal tract result in the weakness of the right arm and leg, with more active tendon reflexes in the right arm and leg. In addition, several abnormal reflexes appear. One is the **Babinski sign**, abnormal extension of the great toe, instead of the normal flexor (downward) plantar reflex in response to a moving stimulus on the lateral border of the sole of the foot. This reflex abnormality reliably indicates a disorder of the corticospinal tract on that side of the spinal cord. Another abnormal reflex is the **Hoffmann sign**, an abnormal flexor reflex of the thumb and other fingers induced by stretching the flexors of the middle finger by flicking the distal phalanx of that finger. Finally, there may be clonus, which is best detected at the ankle when the examiner abruptly moves the subject's foot upward (stretching the gastrocnemius). Sometimes, clonus is so easily evoked that it occurs vigorously in response to a simple tap on the Achilles tendon or when the subject places the foot on the floor. The reaction can be stopped promptly by passively moving the foot down or plantar-flexing the foot, relieving the stretched position of the gastrocnemius.
2. Ipsilateral signs of a posterior column lesion are indicated by a loss of position sense and vibratory sensation.
3. Contralateral loss of pain and temperature perception to the level of C4 follows interruption of the right spinothalamic tract.
4. Loss of autonomic action results in Homer's syndrome (miosis, ptosis) on the same side.

Syringomyelia. Syringomyelia is a condition defined by the formation of cysts within the spinal cord (Fig. 1.7). The cause is unknown, but the lesion affects the central portion of the cord first and then spreads peripherally. Intramedullary tumors may also cause the same clinical syndrome.

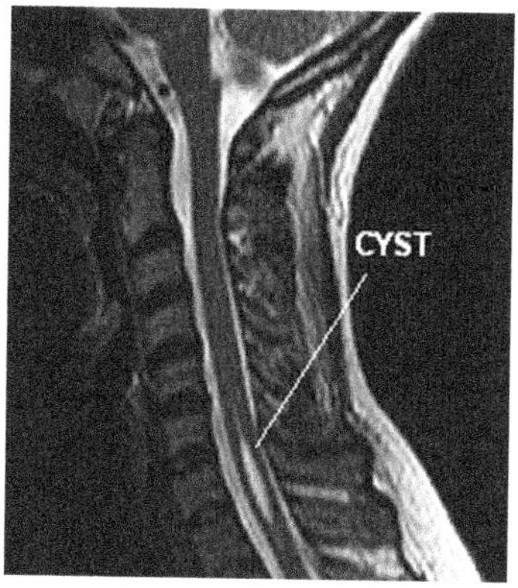

Figure 1.7. MRI scan of spinal cord after syringomyelia.

The clinical picture of syringomyelia is characterized by two unusual patterns of segmental dysfunction (involving cutaneous sensation and motor neurons) as well as interruption of ascending or descending tracts. Because the lesion starts centrally, the first fibers to be affected are those carrying pain and temperature sensations as they cross in the anterior commissure. This usually causes bilateral loss of cutaneous sensation, restricted to the segments involved and resulting in a "shawl" or "cuirass" pattern, affecting a few cervical or thoracic segments and sparing sensation below. Sometimes the segmental sensory loss is unilateral.

The lesion is chronic, and the loss of sensation may lead to painless injuries of the digits or painless burns. Because touch perception is conveyed in posterior columns as well as in spinothalamic tracts, there may be dissociated sensory loss, sparing touch as well as position and vibration sense. If motor neurons in the diseased segment are affected, there are lower motor neuron signs, such as weakness, wasting, and loss of reflexes, in the appropriate area. If the lesion extends laterally, the corticospinal tracts are affected and there may be upper motor neuron signs in the legs.

MULTIPLE SCLEROSIS AND AMYOTROPHIC LATERAL SCLEROSIS

The two most common nontraumatic disorders of the central nervous system are probably amyotrophic lateral sclerosis (ALS) and multiple sclerosis. Upper motor neuron signs and proprioceptive sensory loss are almost always present in advanced cases of multiple sclerosis, although there may be no signs referable to a lesion of the spinal cord. Nonetheless, when subjects who have had these signs come to autopsy, there are usually many small lesions throughout the spinal cord. Some combinations of signs are almost diagnostic of multiple sclerosis; for instance, the combination of proprioceptive sensory loss and signs of upper motor neuron disease together with evidence of either cerebellar dysfunction ataxia, tremor of the arms, disorders of eye movement (nystagmus), difficulty in speaking (dysarthria), or a history or signs of optic neuritis. In addition to signs of disorder elsewhere in the nervous system, there is often a clinical episode of transverse myelitis with corresponding motor and sensory levels.

Amyotrophic lateral sclerosis is a form of motor neuron disease; it predominantly affects the motor neurons, and in the majority of cases the disease does not impair a patient's mind, personality, intelligence, and memory. The defining feature of ALS is the death of both upper and lower motor neurons in the motor cortex of the brain, the brain stem, and the spinal cord. Initially, motor neurons develop proteinaceous inclusions in their cell bodies and axons. These inclusions often contain ubiquitin, and generally incorporate one of the ALS-associated proteins.

The impairment causes muscle weakness and atrophy throughout the body. ALS patients ultimately lose the ability to initiate and control all voluntary movement. In many cases the bladder and bowel sphincters and the muscles responsible for eye movement are spared. The autonomic nervous system (e.g., sweating) are unaffected. Symptoms of upper motor neuron involvement include tight and stiff muscles (spasticity) and exaggerated reflexes (hyperreflexia) including an overactive gag reflex. An abnormal reflex commonly called Babinski's sign (the big toe extends upward and other toes spread out) also indicates upper motor neuron damage. Symptoms of lower motor neuron degeneration include muscle weakness and atrophy, muscle cramps, and fleeting twitches of muscles that can be seen under the skin (fasciculations).

CEREBRAL PARALYSIS

Cerebral palsy (CP) is one of the most common congenital disorders of childhood. It affects by the best estimates about 0.2% of live births. Cerebral palsy refers to permanent disorders leading to compromised motor control. It affects muscle tone, movement, and motor skills (e.g., posture, the ability to move in a coordinated and purposeful way). It could affect other vital functions such as breathing, bladder and bowel control, and eating. One of the common problems is speech disability that follows poor respiratory control, laryngeal and velopharyngeal dysfunction, and oral articulation disorders that are due to restricted movement in the oral–facial muscles. The CP could also lead to impairments of the sensory systems, including vision and hearing, and can affect learning.

The exact causes of most cases of CP are unknown, but many are the result of problems during pregnancy in which the brain is either damaged or does not develop normally. This can be due to infections, maternal health problems, or something else that interferes with normal brain development. Problems during labor and delivery can cause CP in some cases. Premature babies, particularly those who weigh less than 1.5 kg have a higher risk of CP than babies that are carried full-term. Brain damage in infancy or early childhood can also lead to CP. A baby or toddler might suffer this damage because of lead poisoning, bacterial meningitis, malnutrition, being shaken as an infant (shaken baby syndrome), or accident-related brain injury.

Cerebral palsy is usually caused by brain (cerebrum) damage and connections to the cerebellum and other regions of the cortex. The damage occurs before or during a child's birth, or during the first 3 years of a child's life. The three types of CP are as follows:

1. Spastic cerebral palsy causes stiffness and movement difficulties (>70% of all CP cases). Spastic CP can be expressed on one single limb (monoplegia) or both lower extremities (diplegia). Most CP patients with spastic monoplegia and diplegia are fully ambulatory. Flexed knees and hips to varying degrees are common. Hip problems, dislocations, and—in three-quarters of spastic diplegics—strabismus (crossed eyes) can be present as well. In addition, these individuals are often nearsighted. In spastic triplegia three limbs are affected. Finally, in spastic quadriplegia all four limbs are more or less equally affected. People with spastic quadriplegia are the least likely to be able to walk, or if they can, to desire to walk, because their muscles are too tight and it is too much of an effort to do so. Some children with spastic quadriplegia also have hemiparetic tremors, an uncontrollable shaking that affects the limbs on one side of the body and impairs normal movement. In any form of spastic CP, clonus of the affected limb(s) may sometimes result, as well as muscle spasms resulting from pain and/or stress of the tightness experienced.
2. Athetoid cerebral palsy leads to involuntary and uncontrolled movements (10–20% of all CP cases). Athetoid CP patients have difficulties with postural control and often show involuntary motions. Because of their mixed tone and trouble keeping a position they often are not able to hold objects in their hands. The damaged regions in the CNS are in the extrapyramidal motor system, pyramidal tract, and basal ganglia.
3. Ataxic cerebral palsy causes a disturbed sense of balance and depth perception (<10% of all CP cases). Ataxia-type symptoms are caused by damage to the cerebellum. The impairment leads to hypotonia and tremors. Motor skills such as writing, typing, or using fine tools might be affected, but also balance during walking. It is common for individuals to have difficulty with visual and/or auditory processing.

In some cases CP is expressed as hypotonia. Patients with hypotonic CP can move only a little or cannot move at all.

Secondary conditions that accompany CP include seizures, epilepsy, apraxia, dysarthria, communication disorders, and/or behavioral disorders; CP is sometimes accompanied by joint and bone deformities and contractures (permanently fixed, tight muscles and joints).

INCIDENCE OF CNS DISEASE

The data presented in this section come from the most recent data from the statistics published by the U.S. Centers for Disease Control and Prevention (CDC), the Heart Disease and Stroke Statistics of the American Heart Association, the International Cardiovascular Disease Statistics published by the World Health Organization, the National Institute of Neurological Disorders and Strokes (NINDS), and Internet sites.

Stroke is the third leading cause of death. The data suggest that over 140 thousand people die each year from stroke in the United States out of a population of 350 million. Each year, about 800 thousand people get a stroke. About 75% of these are a first CVA, and 25% are recurrent attacks. Nearly 75% of all strokes occur in people over the age of 65. The risk of having a stroke more than doubles each decade after the age of 55. Among adults age 20 and older, the prevalence of stroke in 2005 was 6.5 million (about 40% males and 60% females). Men's stroke incidence rates are greater than women's at younger ages.

According to the World Health Organization, 15 million people suffer stroke worldwide each year. Of these, 5 million die and another 5 million are permanently disabled. High blood pressure contributes to over 12.7 million strokes worldwide. Europe averages approximately 650 thousand stroke deaths each year. In developed countries, the incidence of stroke is declining, largely due to efforts to lower blood pressure and reduce smoking. However, the overall rate of stroke remains high due to the aging of the population.

Among Americans, 250,000 are spinal cord injured. About 52% of spinal cord injured individuals are considered paraplegic and 47% tetraplegic. Approximately 11 thousand new injuries occur each year, 82% in male;. 56% of injuries occur between the ages of 16 and 30. The average age of spinal cord injured persons is 31. Spinal cord injuries are most commonly caused by vehicular accidents (37%), violence (28%), falls (21%), sports accidents (6%), and other (8%). There are an estimated 32 injuries per million population in the United States each year. This is likely significantly under-reported since many injuries are not recorded (cases with little or no remaining neurological deficit, and people who have neurologic problems secondary to trauma but are not classified as SCI). The estimates are that an additional 20 cases per million die before reaching the hospital. The most frequent neurological categories are tetraplegia (complete, 17.5%; incomplete, 31.2%) and paraplegia (complete, 28.2%; incomplete, 23.1%).

Ethnicity is an important factor when analyzing statistics on amyotrophic lateral sclerosis (ALS) data because some studies suggest that ethnic background may modify predisposition to ALS. However, it is widely accepted that the incidence of ALS is uniform across populations of white people in Western Europe and North America (United States and Canada); ALS incidence is about 1 person per 100,000 per year (United States, 1.0; Japan, 0.7/100,000; Denmark, 0.9/100,000; Canada, 1.5/100,000; Europe, 0.7–1.5/100,000). This data is based on the study by Cronin et al. (2007).

It is believed that there are currently about 250 to 350 thousand people in the United States who have been diagnosed with multiple sclerosis. This estimate suggests that approximately 200 new cases are diagnosed each week.

Each year, approximately 50,000 U.S. residents are diagnosed with Parkinson's disease. Statistics indicate that this condition seems to affect men at a slightly higher rate than women, and it appears to affect whites more than African Americans or Asians.

About 10 thousand babies per year in the United States will develop cerebral palsy, meaning that about 2–3 children per 1000 have cerebral palsy. An estimated 800 thousand people have cerebral palsy in United States. Two-thirds of children with cerebral palsy will be mentally delayed.

REFERENCES

Cronin S., Hardiman O., and Traynor B.J. (2007). Ethnic variation in the incidence of ALS—a systematic review. *Neurology* 68:1002–1107.

Maynard F.M. Jr, Bracken M.B., Creasey G., Ditunno J.F. Jr, Donovan W.H., Ducker T.B., Garber S.L., Marino R.J., Stover S.L., Tator C.H., Waters R.L., Wilberger J.E., and Young W. (1997). International standards for neurological and functional classification of spinal cord injury. *American Spinal Injury Association, Spinal Cord* 35(5):266–274.

FURTHER READING

Anonymous (2007). Definition and classification of cerebral palsy. *Dev Med Child Neurol* 49(8):8 (DOI:10.1111/j.1469-8749.2007.tb12610.x).

Compston A. and Coles A. (2002). Multiple sclerosis. *Lancet* 359 (9313):1221–1231.

Compston A. and Coles A. (2008). Multiple sclerosis. *Lancet* 372 (9648):1502–1517.

Davie C.A. (2008). A review of Parkinson's disease. *Br Med Bull* 86:109–127.

Donnan G.A., Fisher M., Macleod M., and Davis S.M. (2008). Stroke. *Lancet* 371 (9624):1612–1623.

Haines D.E. (2004). *Neuroanatomy: An Atlas of Structures, Sections, and Systems*, 7th ed. Philadelphia: Lippincott Williams & Wilkins.

Kandel E.R., Schwartz J.H., and Jessell T.M. (2000). *Principles of Neural Science*, 4th ed. New York: McGraw-Hill.

Kirshblum S., Campagnolo D., and Delisa J. (2001). *Spinal Cord Medicine*. Philadelphia: Lippincott Williams & Wilkins.

Lin V.W.H., Cardenas D.D., Cutter N.C., Frost F.S., and Hammond M.C. (2002). *Spinal Cord Medicine: Principles and Practice*. New York: Demos Medical Publishing.

Lublin F.D. and Reingold S.C. (1996). Defining the clinical course of multiple sclerosis: results of an international survey. National Multiple Sclerosis Society (USA) Advisory Committee on Clinical Trials of New Agents in Multiple Sclerosis. *Neurology* 46(4):907–911.

Mohr J.P., Choi D., Grotta J., and Wolf P. (2004). *Stroke: Pathophysiology, Diagnosis, and Management*. New York: Churchill Livingstone.

National Center on Birth Defects and Developmental Disabilities (2002). *Cerebral Palsy*. Available at http://www.cdc.gov/ncbddd/cp/index.html (accessed April 8, 2013).

NINDS (1999). Stroke: hope through research. National Institute of Neurological Disorders and Stroke. Available at http://www.ninds.nih.gov/disorders/stroke/detail_stroke.htm (accessed April 8, 2013).

NINDS (2012). Brain basics: preventing stroke. National Institute of Neurological Disorders and Stroke. Available at http://www.ninds.nih.gov/disorders/stroke/preventing_stroke.htm (accessed April 8, 2013).

Obeso J.A., Rodriguez-Oroz M.C., Goetz C.G., et al. (2010). Missing pieces in the Parkinson's disease puzzle. *Nat Med* 16(6):653–661.

Popović D.B. and Sinkjær T. (2000). *Control of Movement for the Physically Disabled*. London: Springer.

Samii A., Nutt J.G., and Ransom B.R. (2004). Parkinson's disease. *Lancet* 363(9423):1783–1793.

Warlow C.P., van Gijn J., Dennis M.S., Wardlaw J.M., et al. (2008). *Stroke: Practical Management*, 3rd ed. Hoboken, NJ: Wiley-Blackwell.

World Health Organization (WHO) (1978). *Cerebrovascular Disorders*. Geneva: World Health Organization.

2

FOCUSED CHAPTER

PERIPHERAL AND SPINAL PLASTICITY AFTER NERVE INJURIES

XAVIER NAVARRO[1,2]

[1]*Institute of Neurosciences and Department of Cell Biology, Physiology, and Immunology, Universitat Autònoma de Barcelona, Spain*
[2]*Centro de Investigación en Red sobre Enfermedades Neurodegenerativas (CIBERNED), Spain*

SUMMARY

Injuries to the peripheral nerves result in partial or total loss of motor, sensory, and autonomic functions in the denervated segments of the body, due to the interruption of axons, degeneration of distal nerve fibers, and eventual death of axotomized neurons. Functional deficits caused by nerve injuries can be compensated by reinnervation of denervated targets by regenerating injured axons or by collateral branching of undamaged axons, and remodeling of nervous system circuitry related to the lost functions. Plasticity of central connections may compensate functionally for the lack of adequate target reinnervation; however, plasticity has limited effects on disturbed sensory localization or fine motor control after injuries. It may even result in maladaptive changes, such as neuropathic pain and hyperreflexia.

Introduction to Neural Engineering for Motor Rehabilitation, First Edition. Edited by Dario Farina, Winnie Jensen, and Metin Akay.

INTRODUCTION

The peripheral nervous system consists of neurons, whose cell bodies are located in the spinal cord or in spinal ganglia, and their axons, which extend through peripheral nerves along the body to reach target organs. In the peripheral nerve, axons are accompanied by Schwann cells, which either form a myelin sheath around the axon or just surround the axons, constituting myelinated and unmyelinated nerve fibers (Fig. 2.1).

Peripheral nerves contain several types of nerve fibers, serving different functions. Afferent sensory fibers terminate at the periphery as free endings or in specialized sensory receptors in skin, muscle, and deep tissues. Sensory fibers convey various modalities of sensory inputs, mainly mechanical, thermal, and noxious stimuli. Efferent motor fibers originate from motoneurons in the spinal cord and end in neuromuscular junctions innervating the skeletal muscles. Efferent autonomic nerve fibers in somatic peripheral nerves are unmyelinated sympathetic fibers that innervate smooth muscle and glands. Most of the somatic peripheral nerves are mixed, providing motor, sensory, and autonomic innervation to the corresponding projection territory.

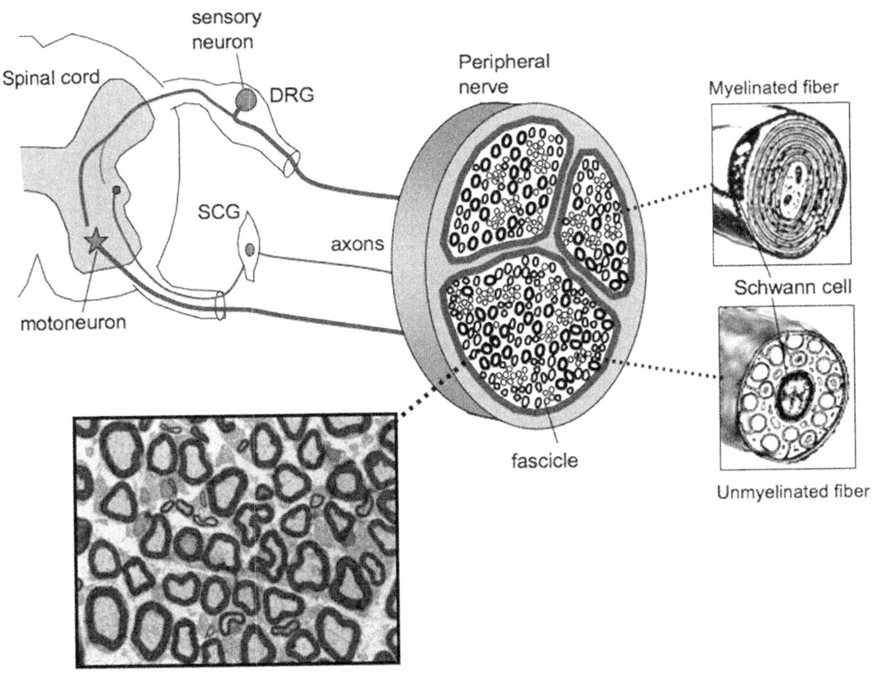

Figure 2.1. Schematic structure of the peripheral nervous system. Motor, sensory, and autonomic neurons send their axons along the peripheral nerves, forming myelinated and unmyelinated nerve fibers. DRG, dorsal root ganglion; SCG, sympathetic chain ganglion. (inset) Microphotograph of a transverse section of a peripheral nerve (×100).

Injuries to the peripheral nerves result in partial or total loss of motor, sensory, and autonomic functions in the denervated segments of the body, due to the interruption and degeneration of nerve fibers distal to the lesion. Injuries to the peripheral nervous system may result in substantial disabilities and decreased quality of life because of permanently impaired sensory and motor functions and secondary problems, such as neuropathic pain.

Functional deficits caused by nerve injuries can be compensated by three neural mechanisms: the reinnervation of denervated organs by regeneration of injured axons; the reinnervation by collateral sprouting of undamaged axons in the vicinity; and the remodeling of central nervous system circuits. However, clinical and experimental evidence demonstrates that such mechanisms do not usually allow for a satisfactory functional recovery after severe nerve injuries (Sunderland, 1991; Lundborg, 2004). Unfortunately, there are no repair techniques that may ensure the recovery of normal sensorimotor functions following transection of a peripheral nerve (Lundborg, 2004).

Despite peripheral nerve fibers being able to regenerate across the injury site and along the distal nerve, reinnervation of target organs does not always lead to adequate recovery of motor and sensory functions. The selectivity of axon–target reconnection plays an important role in the impairment of function after nerve injury and regeneration. In addition, lesioned neurons may require extended periods of time to regenerate axons over long distances, for example in the limbs, to reach the denervated targets, with a consequent decline in neurotrophic support and chronic atrophy of denervated muscles and sensory targets (Gordon et al., 2003). On another hand, collateral reinnervation by intact axons has temporal and spatial constraints, especially for large sensory and motor axons. Thus its contribution for recovery after severe nerve lesions is limited, although it is helpful to recovery of pain sensibility and motor strength in partially denervated muscles. Factors contributing to poor functional recovery after peripheral nerve injuries include the following: (1) damage to the neuronal cell body due to axotomy and retrograde degeneration, excluding the possibility of regeneration; (2) inability of axons to grow, due to the nerve lesion or to underlying diseases; (3) poor specificity of reinnervation by regenerating axons, when target organs become reinnervated by nerve fibers of a different function; (4) changes in the central circuits in which the injured neurons participate, due to plasticity of neural connections, which are particularly evident in the spinal cord (Navarro, 2009) (Fig. 2.2).

The peripheral and central nervous systems are functionally integrated regarding the consequences of a nerve injury: a peripheral nerve lesion not only deprives peripheral organs of neural input, but also results in marked and long-lasting modifications and reorganization at central levels (for reviews see Lundborg, 2000; Wall et al., 2002; Kaas and Collins, 2003; Navarro et al., 2007). Plastic changes and reorganization of spinal and brain circuits linked with the lesioned neurons may result in beneficial adaptive functional recovery, but also in maladaptive changes that lead to positive symptoms, such as pain, abnormal sensations (dysesthesia), and increased reflex responses (hyperreflexia).

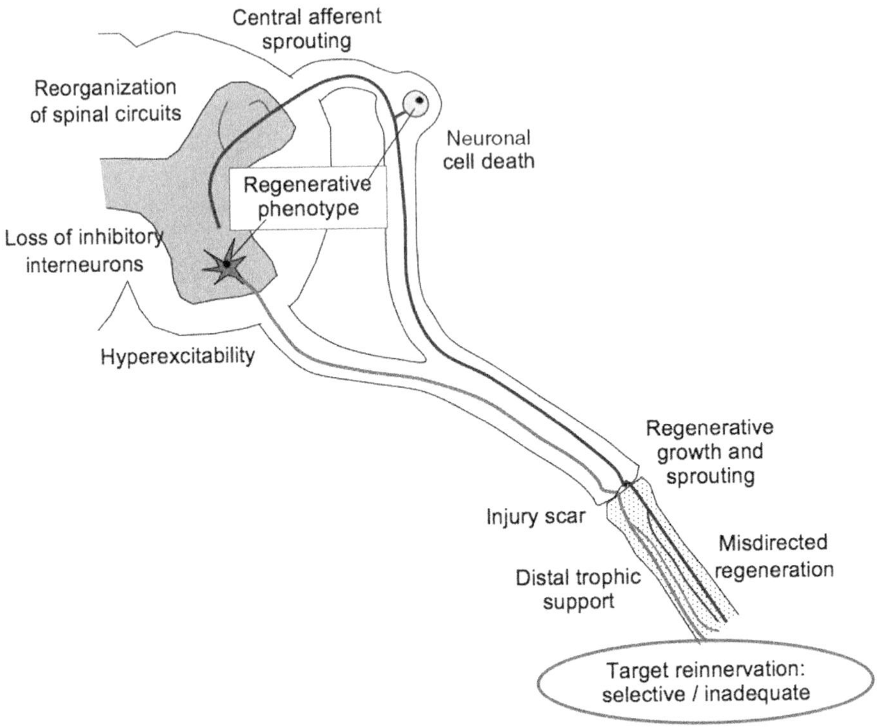

Figure 2.2. Schematic summary of the main plastic changes that develop after nerve injury at the peripheral nerve and spinal cord levels. See text for details.

NERVE DEGENERATION AND REGENERATION

Peripheral neurons are some of the largest and most complex cells in the body. A focal axonal injury, which may occur centimeters from the neuron body, induces a response that involves the entire cell and its associated glial cells. The functional significance of regeneration is to replace the distal nerve segment lost during degeneration, allowing reinnervation of target organs and restitution of their functions (for reviews see Fu and Gordon, 1997; Verdú and Navarro, 1998; Hall, 2001; Makwana and Raivich, 2005).

The success of nerve regeneration depends at a first instance on the capacity of axotomized neurons to survive and shift toward a regenerative phenotype. The availability of neurotrophic factors influences the response of axotomized neurons (Terenghi, 1999; Boyd and Gordon, 2003). The proportion of neuronal death among sensory neurons of the dorsal root ganglia (DRG) has been reported to be between 10% and 30%, affecting more small than large neurons (Ygge, 1989; Groves et al., 1997), whereas motoneurons are more resistant (Vanden-Noven, et al. 1993; Valero-Cabré et al., 2001). Adult neurons are less susceptible to death than immature neurons, and lesions near the cell body cause a higher proportion of neuronal death than distal lesions.

To regenerate, adult neurons should have the intrinsic capacity to survive the traumatic lesion, and to activate the cell regrowing program. This program is influenced by interactions between neurons and neighboring Schwann cells, mediated by cell- and substratum-adhesion molecules and their receptors, and by several trophic factors secreted into the extracellular space. Regenerating neurons change the expression of hundreds of genes to compensate for the loss of target-derived neurotrophic factors, and in order to regrow their axons back into the periphery. Axotomized neurons undergo profound changes in their body, known as neuronal reaction and *chromatolysis* (Kreutzberg, 1995). The intensity and duration of the neuronal response are influenced by the severity of the injury, the distance between lesion and cell body, type of neuron, and age.

Axons and myelin sheaths distal to the lesion are degraded by a process named **Wallerian degeneration**, which starts within hours after injury and continues for about 1–2 weeks (Fig. 2.3). The degenerative end products are eliminated by the cooperative action of Schwann cells and infiltrating macrophages (Stoll and Müller, 1999). Denervated Schwann cells detach their myelin sheaths and are able to phagocytize myelin to some extent; however, recruitment of macrophages is the main pathway for eliminating myelin and axonal debris (Tanaka et al., 1992). From 2–3 days after injury there is an important infiltration of macrophages into the degenerating nerve, attracted by several cytokines secreted by the reactive Schwann cells (Tofaris et al., 2002). Schwann cells in the distal nerve are stimulated by the loss of axonal contact, and later by cytokines secreted by macrophages, to proliferate after injury, and thus increase in number (Salonen et al., 1988). De-differentiated Schwann cells line up within the endoneurial tubules to form **bands of Büngner**, which will provide support for regenerating axons.

Proximal to the lesion, the sectioned axons emit several sprouts, known as growth cones, that elongate if they find favorable terrain. In the absence of a guide, such as the distal nerve, regenerating axons make a tortuous course and form a neuroma, composed of immature axonal sprouts and connective tissue. In contrast, if regenerating axons gain the distal nerve, they elongate within the endoneurial tubules in association with Schwann cells, constituting regenerative units (Fig. 2.3). The rate of axonal regeneration is initially slow across the injury site; it later increases to a value of about 1–3 mm/day from 3 to 4 days after injury. The factors that stimulate and control axonal regeneration originate from multiple sources, but the most important influences derive from the local environment of the lesion. Axonal regeneration requires an adequate substrate of trophic factors, provided by reactive Schwann cells and the extracellular matrix within the degenerated nerve stump. Several sprouts emerge initially from each parent axon (Witzel et al., 2005) and may advance in the distal nerve segment. Thus, the total number of axons in the distal segment may exceed the number of parent axons in the proximal nerve even for long

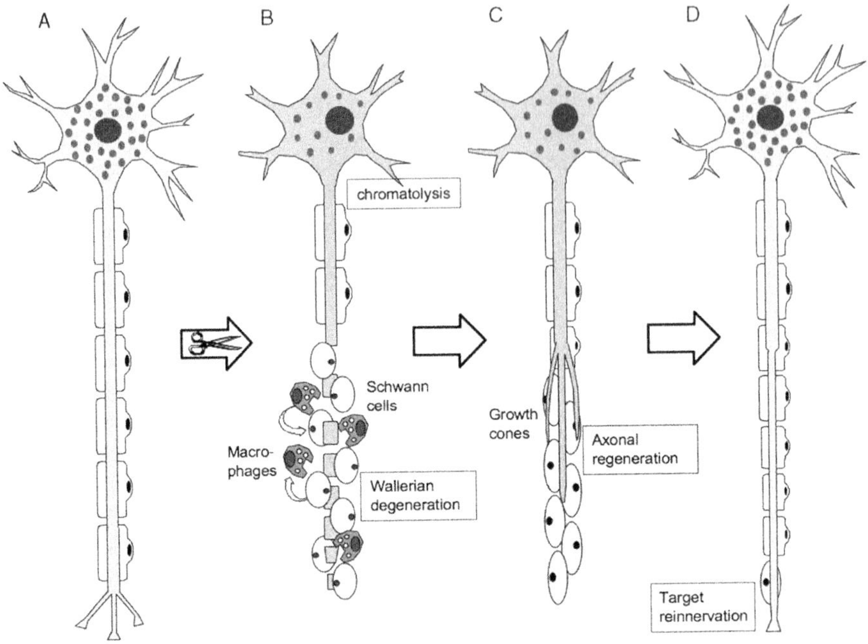

Figure 2.3. Degeneration and regeneration after peripheral nerve injury. (A) Normal neuron and nerve fiber. (B) Transection of the axon induces neuron body reaction and chromatolysis. The distal axon and myelin sheaths undergo degeneration. Schwann cells proliferate, and macrophages invade the distal segment. (C) Growth sprouts emerge from the proximal axonal end and elongate in association with Schwann cells, lined up in bands of Büngner, in the distal nerve. (D) Axonal reconnection with target organs, and maturation and myelination of the regenerated nerve fiber.

times (Jenq and Coggeshall, 1985). When axons reconnect with peripheral targets, supernumerary axonal sprouts tend to withdraw.

The regenerative process usually does not reconstitute a normal nerve structure nor allows for normal function, especially when the lesion is severe. After nerve injury and repair, the diameter of regenerated axons, their conduction velocity, and their excitability remain below normal levels for a long time (Fields and Ellisman, 1986a,b); consequently, recovery of reinnervated organs is incomplete. The limitation of nerve regeneration is more marked when the lesion creates loss of continuity in the nerve, and the amount of regeneration is lower with increasing length of the gap in the nerve after surgical repair (Butí et al., 1996; Krarup et al., 2002).

Plastic Changes in Peripheral Targets

Denervation leads to progressive atrophy of skeletal muscle, which is characterized by a marked loss of muscle mass, decreased area of muscle fibers, and changes in the mechanical and biochemical muscle characteristics. These changes are most evident in muscles that are normally more active—such as the slow-twitch ones—and, among the fast-twitch muscles, in the extensors versus flexors of the limbs (Roy et al., 1999). When the muscle is reinnervated, the normal characteristics are partially restored, indicating the importance of the neural induced activity in changing the muscle fiber function (Wang et al., 2002). Artificially maintaining muscle activity during periods of denervation–paralysis may improve neuromuscular rehabilitation. Electrostimulation of denervated muscles, passive exercise, and locomotion training have been shown as effective strategies to retard muscle atrophy and to improve recovery after reinnervation (Udina et al., 2011). Muscle activity induced by stimulation could induce the release of muscle trophic factors, thus avoiding the disuse effect following nerve or central lesions.

Regarding sensory receptors, histological studies have provided knowledge about their changes following denervation and reinnervation. Corpuscular terminals in the skin disappear after denervation and they are restored when regenerating axons reach the target, although their number is reduced compared with normal skin (Nurse et al., 1984; Zelená, 1984; Verdú and Navarro, 1997). Thus, some receptors remain vacant, whereas reinnervated ones often show an abnormal configuration, which may affect sensory stimuli transduction in complex receptors, such as Meissner touch receptors and muscle spindle proprioceptive receptors (DeSantis and Norman, 1993; Verdú and Navarro, 1997).

Specificity of Nerve Regeneration

In spite of the fact that peripheral axons can regenerate through the injury site toward distal territories, reinnervation of targets does not always lead to adequate recovery of motor and sensory functions. When motor axons reinnervate incorrect muscles, it results in uncoordinated movements and synkinesia (involuntary contractions of muscles accompanying a wished action), whereas misrouted regenerating sensory axons cannot correctly transduce sensory stimuli and may cause dysesthesia and mislocalization. When mixed nerves are injured and repaired, motor and sensory axons regenerate at random; thus regenerating motor axons are often misrouted to sensory end-organs, and sensory axons formerly innervating skin are often misrouted to muscle. The selectivity of axon–target reconnection thus plays an important role in the impairment of functions after nerve injury and regeneration. Fascicular specificity, the preferential regeneration through the original nerve fascicle to which axons belonged, was suggested in some early studies but not confirmed by most investigations. Target-organ specificity, or adequate reinnervation of each type of end organ (muscle, sensory receptor) by their original axons, is far from perfect after nerve transection even if refined surgical repair is applied (Molander and Aldskogius, 1992; Bodine-Fowler et al., 1997; Valero-Cabré et al., 2004; Ruiter et al., 2008). Nonetheless, it has been reported that motor axons have the ability to reinnervate motor pathways more accurately than sensory axons to cutaneous pathways, a phenomenon called **preferential motor reinnervation** (Brushart, 1993). Other studies propose that it is the relative level of trophic support provided by each nerve branch, dependent on target contact and relative number of Schwann cells, that determines whether motor axons will remain in that particular branch (Madison et al., 2007).

There are as yet no effective neurobiological strategies to allow for a significant degree of specificity of regeneration and target reinnervation by motor and sensory axons after transection of peripheral nerves in adult mammals. In experimental studies, the recovery of walking pattern was inversely correlated with the percentage of misdirected reinnervation, indicating that reinnervation of muscles by the wrong motoneurons might be a significant factor in the deficient recovery of function after section and repair of the nerve (Valero-Cabré and Navarro, 2002a). Even a long time after nerve injury and reinnervation, abnormal coactivation of limb antagonistic muscles occurs during locomotion and it has been attributed to misdirected muscle reinnervation (Gramsbergen et al., 2000). In addition, misdirection of reinnervated proprioceptive sensory axons and the reorganization of spinal reflexes can play a role in impairing locomotor recovery by altering the ordered recruitment of flexor and extensor motor units (Wasserschaff, 1990; Valero-Cabré and Navarro, 2001).

Collateral Spouting and Reinnervation

Axonal sprouting, another remarkable neural regenerative phenomenon, is a natural compensatory mechanism that comes into play following partial denervation due to nerve trauma or degenerative neuronal loss (Fig. 2.4).

This growth response has different characteristics from regenerative sprouting of transected axons. Collateral sprouting is a characteristic of undamaged nerve fibers in response to elimination of other axons within the same target tissues (Diamond and Foerster, 1992) and is dependent on local secretion of neurotrophic factors by the denervated cells within the tissue. For example, administration of nerve growth factor (NGF) and other neurotrophins *in vivo* leads to potentiation of collateral sprouting in denervated skin (Diamond et al., 1992) and in the spinal cord (Schnell et al., 1994). In contrast, application of NGF does not affect the rate of axonal elongation after axotomy but may even delay the onset of regeneration and reduce the neuronal body response to injury (Mohiuddin et al., 1999).

Collateral reinnervation has been widely studied in muscle. Intact motoneurons are able to expand the size of their motor units by axonal sprouting to reinnervate denervated muscle fibers so as to compensate for reduced strength, although with less precision control. However, electrophysiological analyses revealed that motor unit enlargement by axonal sprouting is restricted to a limit of 3- to 5-fold, such that axonal sprouting compensates for loss of as

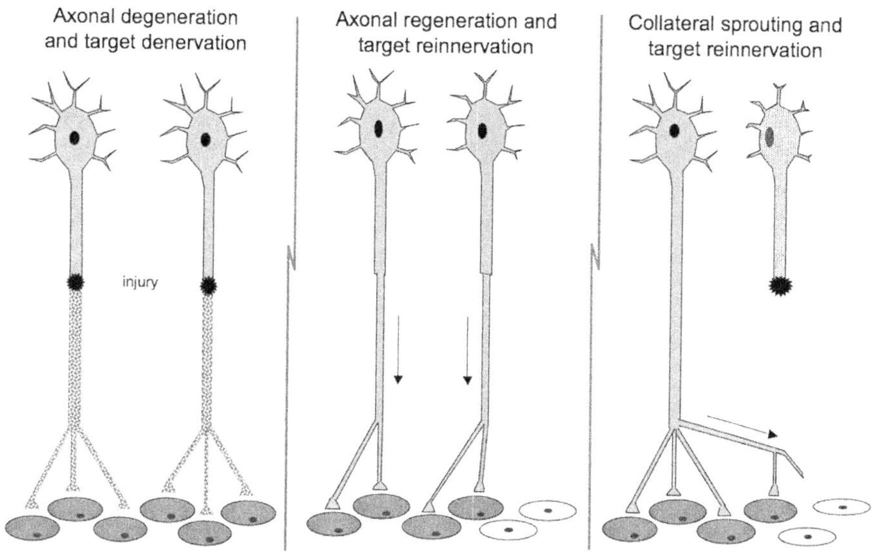

Figure 2.4. Peripheral nerve injury leads to degeneration of the distal axons with consequent denervation of target cells (muscle cells, sensory receptors, etc.). Reinnervation can be achieved by regeneration of the injured axons, although it is usually incomplete and inaccurate (mismatch of neuron-targets). Intact axons within a partially denervated tissue (muscle or skin) are able to emit collateral sprouts and reinnervate nearby denervated cells, thus increasing the number of targets per neuron.

many as 85% of the normal number of motor units (Brown and Ironton, 1978; Tam et al., 2001). The new axonal sprouts emerge from the intact axons either at the motor endplates to give rise to terminal sprouts or at the nodes of Ranvier to give rise to nodal sprouts. Perisynaptic Schwann cells are induced to produce processes at denervated endplates and play a fundamental role in directing axonal sprouts by bridging between denervated and innervated endplates (Son and Thompson, 1995; Tam and Gordon, 2003)

There are important differences in sprouting capabilities of different nerve fibers; small axons are able to sprout and reinnervate larger territories than large myelinated axons. Functional studies following partial denervation of the hind paw showed that thin nociceptors, but not large low-threshold mechanoreceptors, produce functionally effective collateral reinnervation of adjacent denervated skin (Jackson and Diamond, 1984; Wiesenfeld-Hallin et al., 1989), indicating that large sensory fibers have more constraints on reinnervation than small ones. Similarly, unmyelinated sympathetic axons are able to collaterally reinnervate target organs denervated in a surrounding skin area, doubling their corresponding territory (Kennedy et al., 1988; Udina et al., 2003), whereas sprouts from motor axons do not reinnervate adjacent denervated muscles and branch only within restricted areas in their parent muscle (Brown et al., 1980). These differences in reinnervation abilities seem closely related to the functions conveyed by the axon types. The constraints on collateral and terminal sprouting of large nerve fibers to mechanoreceptors and to muscle exist to preserve the precise spatial organization of organ innervation necessary to retain correct discrimination and coordination of motion. On the other hand, the warning aspects of nociceptive and thermal sensitivity and the thermoregulatory responses can be well accomplished in a more diffuse manner, without a precise topographical innervation. This favors the assumption that the number of regenerating fibers is more essential for thin nerve fibers than reconnection with specific end organs, as is the case for myelinated fibers, for an appropriate functional restitution.

PLASTICITY AND REMODELING AT THE SPINAL CORD

Nerve injury may sensitize and induce remodeling of central neural structures. Experimental evidence of these changes includes the development of windup, classic central sensitization, long-term potentiation, and distortion of receptive fields of CNS neurons, as well as the enhancement of spinal reflexes and the persistence of pain or hyperalgesia (excessive pain sensation) (see Fig. 2.2).

Plastic Changes of the Injured Neurons

One of the most relevant changes in neurons after axotomy is the decrease of neurotransmitters and transmitter-related proteins. Axotomy also triggers plastic changes in the expression of ion channels and receptors in the injured neurons, which are related to modifications of neuronal function, including hyperexcitability.

Following nerve injury, the expression of tetrodotoxin (TTX)–sensitive sodium channels $Na_v1.3$, normally expressed only during development, is upregulated in sensory neurons, whereas other sodium channels are downregulated (Waxman et al., 2000; Wood et al., 2004). Increase in excitability in sensory neurons occurs within weeks of axotomy, with significant reduction of excitation threshold and increase of spontaneous activity (Zhang et al., 1997; Abdulla and Smith, 2001), abnormalities that have been related to signs of pain. Axotomized motoneurons also exhibit changes in membrane electrical properties (Titmus and Faber, 1990). When muscle reinnervation is not allowed, these changes are maintained chronically, whereas following reinnervation normal electrical properties recover.

In parallel with the neuronal body reaction, there is retraction of the central dendrites and reduction of the synapses received by axotomized neurons, morphological changes that account for a functional isolation of the injured neurons from the remaining neural circuits. Following reinnervation of target muscles, the dendritic volume of the motoneurons is normalized, although the number of dendritic branches remains reduced (Brannstrom et al., 1992). In parallel with dendrite retraction, the number and area of synaptic terminals on motoneurons is reduced, due to detachment of synapses (Aldskogius et al., 1999). Synaptic restoration also takes place in parallel with peripheral target reinnervation. In contrast, after permanent axotomy synaptic contacts remain reduced.

Changes in Spinal Reflexes after Nerve Lesions

Marked plastic changes in the connections and function of spinal reflexes occur after nerve injuries, in parallel with axonal regeneration and reinnervation. Such changes may have important effects on movement control and sensory processing. The monosynaptic spinal stretch reflex shows complete recovery after nerve crush lesion but, in contrast, is permanently abolished after nerve section (Cope et al., 1994), probably due to either stripping of central synapses or poor muscle spindle reinnervation (Verdú and Navarro, 1997).

Recovery of the H-wave reflex, the electrical counterpart of the monosynaptic stretch reflex, induced by electrical stimulation of afferent fibers proximal to the injury indicates that spinal reflex circuits are functional as soon as the muscle is reinnervated. The H-reflex is highly facilitated at early stages of reinnervation, resulting in an increase of the H/M amplitude ratio, thus indicating a facilitated response of the sensory axon Ia–motoneuron synapse. Muscle reinnervation tends to revert this facilitatory effect, but the H/M ratio remains higher than control during months following nerve section and repair (Valero-Cabré and Navarro, 2001, 2002a). Interestingly, the reflex facilitation affects also adjacent spinal cord segments, where intact motoneurons undergo compensatory synaptic rearrangements, receiving more excitatory boutons (Holmberg and Kellerth, 2000; Havton and Kellerth, 2004). After sciatic nerve injury, crossed extensor reflex responses conveyed by myelinated afferent fibers also show a marked increase in amplitude during the first stages of regeneration, and they tend to decline with reinnervation. In contrast, the slow reflex component mediated by unmyelinated afferents is normal or absent (Valero-Cabré and Navarro, 2002b). Therefore, after nerve injuries reflexes mediated by myelinated afferents exhibit a higher degree of facilitation than those mediated by unmyelinated fibers. Different mechanisms may account for the facilitation of reflex responses following nerve injuries, including increase of neuronal and synaptic excitability, decrease of the effectiveness of inhibitory systems, and remodeling of sensory afferent projections within the spinal cord.

Remodeling of Spinal Cord Projections and Circuits

Peripheral regeneration following nerve section results in the mismatch of the connections between sensory receptors and afferent fibers with second-order neurons in the dorsal horn of the spinal cord, and it alters the somatotopy of the body representation at the spinal cord level. These alterations result in loss or decrease of tactile acuity and discrimination, and underlie in part positive symptoms such as dysesthesia and neuropathic pain. Due to the lack of specificity in peripheral reinnervation after complete nerve section, regenerating sensory axons can reinnervate foreign target territories and inappropriate receptors, including muscle receptors by cutaneous axons and vice versa, or tactile receptors by nociceptive axons.

Peripheral nerve lesions produce changes in the efficacy of central connections and the projections of afferent fibers in the spinal cord, so that the properties of the central projections of regenerated sensory neurons are not well matched with the peripheral receptor properties. After nerve section and repair, the receptive fields of dorsal horn cells are larger than normal and distorted, in agreement with the random and disperse nature of axonal regeneration. At long intervals the receptive fields shrink, approaching normal sizes, but the somatotopic organization remains scrambled. This plastic process implies changes in synaptic efficacy, with some synaptic inputs being strengthened while others weaken or disappear, and new synaptic inputs may be established in order to make the receptive fields more continuous (Koerber et al., 2006).

There has been extensive controversy regarding the rearrangement of central projections of injured and intact sensory afferents in the spinal cord following nerve lesions, and its potential contribution to neuropathic pain and mechanical hypersensitivity. After peripheral nerve lesions, the central axons of injured myelinated Aβ fibers were found to sprout from their normal termination territory within deep laminae III–V of the dorsal horn to partially invade laminae II and I, which normally receive inputs from C and Aδ sensory fibers (Shortland and Woolf, 1993; Woolf et al., 1995). The death or atrophy of a number of small sensory DRG cells and degeneration of their terminations within the dorsal horn would induce sprouting of Aβ and Aδ afferents into the empty laminae I and II. Nevertheless, more recent studies found only a very limited number of large A fiber sprouts into inner lamina II, normally innervated by C-fibers, but not into outer lamina II or lamina I, and no obvious changes in the laminar locations of axonal collaterals (Hughes et al., 2003; Shehab et al., 2003; Koerber et al., 2006). Sprouting of the central projections of undamaged small C and Aδ fibers into the region occupied by central projection of fibers whose peripheral projections had been interrupted has been demonstrated (Hu et al., 2004). The somatotopic reorganization following nerve lesions in adult animals might be better explained by an increased synaptic efficacy and formation of new synaptic boutons of existing silent projections, which can activate second-order neurons that were originally outside the corresponding receptive field, without notable structural changes or collateral sprouting of central fibers (Koerber et al., 2006). This functional reorganization of the sensory circuitry may constitute an underlying mechanism for sensory abnormalities following peripheral nerve injuries.

In addition, peripheral nerve injuries also alter spinal pathways. An early and progressive decrease of inhibitory interneurons in the spinal cord has been reported after nerve lesions (Castro-Lopes et al., 1993; Sanna et al., 1993). The reduction of activity of inhibitory interneurons in the spinal cord may therefore contribute to the increased excitation of ascending sensory pathways and motor reflex responses.

CONCLUSION

Peripheral nerve injuries induce a cascade of events, at the molecular, cellular, and system levels, that support axonal regeneration and target reinnervation in the periphery, but also imply plastic changes at the spinal cord and the brain. Mechanisms involved in these changes include alterations of excitatory and inhibitory synaptic connections, sprouting of new connections, and reorganization of sensory and motor maps in the nervous system. All these changes have to be considered when attempting to interact with the injured nervous system for rehabilitation strategies.

REFERENCES

Abdulla F.A. and Smith P.A. (2001). Axotomy- and autotomy-induced changes in the excitability of rat dorsal root ganglion neurons. *J Neurophysiol* 85:630–643.

Aldskogius H., Liu L., and Svensson M. (1999). Glial responses to synaptic damage and plasticity. *J Neurosci Res* 58:33–41.

Bodine-Fowler S.C., Meyer R.S., Moscovitz A., Abrams R., and Botte J. (1997). Inaccurate projection of rat soleus motoneurons: a comparison of nerve repair techniques. *Muscle Nerve* 20:29–37.

Boyd J.G. and Gordon T. (2003). Neurotrophic factors and their receptors in axonal regeneration and functional recovery after peripheral nerve injury. *Mol Neurobiol* 27:277–324.

Brannstrom T., Havton L., and Kellerth J.O. (1992). Restorative effects of reinnervation on the size and dendritic arborization patterns of axotomized cat spinal alpha-motoneurons. *J Comp Neurol* 318:452–461.

Brown M.C. and Ironton R. (1978) Sprouting and regression of neuromuscular synapses in partially denervated mammalian muscles. *J Physiol* 278:325–348.

Brown M.C., Holland R.L., and Ironton R. (1980). Nodal and terminal sprouting from motor nerves in fast and slow muscles in the mouse. *J Physiol* 306:493–510.

Brushart T.M. (1993). Motor axons preferentially reinnervate motor pathways. *J Neurosci* 13:2730–2738.

Butí M., Verdú E., Labrador R.O., Vilches J.J., Forés J., and Navarro X. (1996). Influence of physical parameters of nerve chambers on peripheral nerve regeneration and reinnervation. *Exp Neurol* 137:26–33.

Castro-Lopes J.M., Tavara I., and Coimbra A. (1993). GABA decreases in the spinal cord dorsal horn after peripheral neurectomy. *Brain Res* 620:287–291.

Cope T.C., Bonasera J., and Nichols T.R. (1994). Reinnervated muscles fail to produce stretch reflexes. *J Neurophysiol* 71:817–820.

DeSantis M. and Norman W.P. (1993). Location and completeness of reinnervation by two types of neurons at a single target: the feline muscle spindle. *J Comp Neurol* 336:66–76.

Diamond J. and Foerster A. (1992). Recovery of sensory function in skin deprived of its innervation by lesion of the peripheral nerve. *Exp Neurol* 115:100–103.

Diamond J., Holmes M., and Coughlin M. (1992). Endogenous NGF and nerve impulses regulate the collateral sprouting of sensory axons in the skin of the adult rat. *J Neurosci* 12:1454–1466.

Fields R.D. and Ellisman M.H. (1986a). Axons regenerated through silicone tube splices. I. Conduction properties. *Exp Neurol* 92:48–60.

Fields R.D. and Ellisman M.H. (1986b). Axons regenerated through silicone tube splices. II. Functional morphology. *Exp Neurol* 92:61–74.

Fu S.Y. and Gordon T. (1997). The cellular and molecular basis of peripheral nerve regeneration. *Mol Neurobiol* 14:67–116.

Gordon T., Sulaiman O., and Boyd J.G. (2003). Experimental strategies to promote functional recovery after peripheral nerve injuries. *J Peripher Nerv System* 8: 236–250.

Gramsbergen A., Ikjmena-Paassen J., and Meek M.F. (2000). Sciatic nerve transection in the adult rat: abnormal EMG patterns during locomotion and aberrant innervation of hindleg muscles. *Exp Neurol* 161:183–193.

Groves M.J., Christopherson T., Giometto B., and Scaravilli F. (1997). Axotomy-induced apoptosis in adult rat primary sensory neurons. *J Neurocytol* 26:615–624.

Hall S. (2001). Nerve repair: a neurobiologist's view. *J Hand Surg* 26B:129–136.

Havton L.A. and Kellerth J.O. (2004). Plasticity of lumbosacral monosynaptic reflexes after a ventral root transection injury in the adult cat. *Exp Brain Res* 155:111–114.

Holmberg P. and Kellerth J.O. (2000). Do synaptic rearrangements underlie compensatory reflex enhancement in spinal motoneurons after partial cell loss? *Synapse* 38:384–391.

Hu J., Mata M., Hao S., Zhang G., and Fink D.J. (2004). Central sprouting of uninjured small fiber afferents in the adult rat spinal cord following spinal nerve ligation. *Eur J Neurosci* 20:1705–1712.

Hughes D.I., Scott D.T., Todd A.J., and Riddell J.S. (2003). Lack of evidence for sprouting of Abeta afferents into the superficial laminas of the spinal cord dorsal horn after nerve section. *J Neurosci* 23:9491–9499.

Jackson P.C. and Diamond J. (1984). Temporal and spatial constraints on the collateral sprouting of low-threshold mechanosensory nerves in the skin of rats. *J Comp Neurol* 226:336–345.

Jenq C.-B. and Coggeshall R.E. (1985). Numbers of regenerating axons in parent and tributary peripheral nerves in the rat. *Brain Res* 326:27–40.

Kaas J.H. and Collins C.E. (2003). Anatomic and functional reorganization of somatosensory cortex in mature primates after peripheral nerve and spinal cord injury. *Adv Neurol* 93:87–95.

Kennedy W.R., Navarro X., and Kamei H. (1988). Reinnervation of sweat glands in the mouse: axonal regeneration versus collateral sprouting. *Muscle Nerve* 11: 603–609.

Koerber H.R., Mirnics K., and Lawson J.J. (2006). Synaptic plasticity in the adult spinal dorsal horn: the appearance of new functional connections following peripheral nerve regeneration. *Exp Neurol* 200:468–479.

Krarup C., Archibald S.J., and Madison R.D. (2002). Factors that influence peripheral nerve regeneration: an electrophysiological study of the monkey median nerve. *Ann Neurol* 51:69–81.

Kreutzberg G.W. (1995). Reaction of the neuronal cell body to axonal damage. In: *The Axon: Structure, Function and Pathophysiology* (Waxman S.G., Kocsis J.D., and Stys P.K., Eds.), pp. 355–374. Oxford, U.K.: Oxford University Press.

Lundborg G. (2000). Brain plasticity and hand surgery: an overview. *J Hand Surg* 25B:2422–2252.

Lundborg G. (2004). *Nerve Injury and Repair: Regeneration, Reconstruction, and Cortical Remodeling*, 2nd ed. Philadelphia: Elsevier.

Madison R., Robinson G.A., and Chadaram S.R. (2007). The specificity of motor neurone regeneration (preferential reinervation). *Acta Physiol* 189:201–206.

Makwana M. and Raivich G 2005. Molecular mechanisms in successful peripheral regeneration. *FEBS J* 272:2628–2638.

Mohiuddin L., Delcroix J.D., Fernyhough P., and Tomlinson D.R. (1999). Focally administered nerve growth factor suppresses molecular regenerative responses of axotomized peripheral afferents in rats. *Neuroscience* 91:265–271.

Molander C. and Aldskogius H. (1992). Directional specificity of regenerating primary sensory neurons after peripheral nerve crush or transection and epineurial suture. A sequential double-labelling study in the rat. *Restor Neurol Neurosci* 4:339–344.

Navarro X. (2009). Neural plasticity after nerve injury and regeneration. *Int Rev Neurobiol* 87:483–505.

Navarro X., Vivó M., and Valero-Cabré A. (2007). Neural plasticity after peripheral nerve injury and regeneration. *Prog Neurobiol* 82:163–201.

Nurse C.A., Macintyre L., and Diamond J. (1984) Reinnervation of the rat touch dome restores the Merkel cell population reduced after denervation. *Neuroscience* 13:563–571.

Roy R.R., Talmadge R.J., Hodgson J.A., Oishi Y., Baldwin K.M., and Edgerton V.R. (1999). Differential response of fast hindlimb extensor and flexor muscles to exercise in adult spinalized cats. *Muscle Nerve* 22:230–241.

Ruiter G.C.W., Malessy M.J.A., Alaid A.O., Spinner R.J., Engelstad J.K., Sorenson E.J., Kaufman K.R., Dyck P.J., and Windebank A.J. (2008). Misdirection of regenerating motor axons after nerve injury and repair in the rat sciatic nerve model. *Exp Neurol* 211:339–350.

Salonen V., Aho H., Röyttä M., and Peltonen J. (1988). Quantitation of Schwann cells and endoneurial fibroblast-like cells after experimental nerve trauma. *Acta Neuropathol (Berlin)* 75:331–336.

Sanna P.P., Celio M., Bloom F., and Rende M. (1993). Presumptive Renshaw cells contain decreased calbindin during recovery from sciatic nerve lesions. *Proc Natl Acad Sci U S A* 90:3048–3052.

Schnell L., Schneider R., Kolbeck R., Barde Y., and Schwab M.E. (1994). Neurotrophin-3 enhances sprouting of corticospinal tract during development and after adult spinal cord lesion. *Nature* 367:170–173.

Shehab S.A., Spike R.C., and Todd A.J. (2003). Evidence against cholera toxin B subunit as a reliable tracer for sprouting of primary afferents following peripheral nerve injury. *Brain Res* 964:218–227.

Shortland P. and Woolf C.J. (1993). Chronic peripheral nerve section results in a rearrangement of the central axonal arborizations of axotomized A-beta primary afferent neurons in the rat spinal cord. *J Comp Neurol* 330:65–82.

Son Y.J. and Thompson W.J. (1995). Nerve sprouting in muscle is induced and guided by processes extended by Schwann cells. *Neuron* 14:133–141.

Stoll G. and Müller H.W. 1999. Nerve injury, axonal degeneration and neural regeneration: basic insights. *Brain Pathol* 9:313–325.

Sunderland S. (1991). *Nerve Injuries and Their Repair: A Critical Appraisal.* Edinburgh: Churchill Livingstone.

Tam S.L. and Gordon T. (2003). Mechanisms controlling axonal sprouting at the neuromuscular junction. *J Neurocytol* 32:961–974.

Tam S.L., Archibald V., Tyreman N., Jassar B., and Gordon T. (2001). Increased neuromuscular activity reduces sprouting in partially denervated muscles. *J Neurosci* 21:654–667.

Tanaka K., Zhang Q.L., and Webster H.F. (1992). Myelinated fiber regeneration after sciatic nerve crush: morphometric observations in young adult and aging mice and effects of macrophage suppression and conditioning lesions. *Exp Neurol* 118:53–61.

Terenghi G. (1999). Peripheral nerve regeneration and neurotrophic factors. *J Anat* 194:1–14.

Titmus M.J. and Faber D.S. (1990). Axotomy-induced alterations in the electrophysiological characteristics of neurons. *Prog Neurobiol* 35:1–51.

Tofaris G.K., Patterson P.H., Jessen K.R., and Mirsky R. (2002). Denervated Schwann cells attract macrophages by secretion of leukemia inhibitory factor (LIF) and monocyte chemoattractant protein-1 in a process regulated by interleukin-6 and LIF. *J Neurosci* 22:6696–6703.

Udina E., Ceballos D., Gold B.G., and Navarro X. (2003). FK506 enhances reinnervation by regeneration and by collateral sprouting of nerve fibers. *Exp Neurol* 183:220–231.

Udina E., Cobianchi S., Allodi I., and Navarro X. (2011). Effects of activity-dependent strategies on regeneration and plasticity after peripheral nerve injuries. *Ann Anat* 193:347–353.

Valero-Cabré A. and Navarro X. (2001). H reflex restitution and facilitation after different types of peripheral nerve injury and repair. *Brain Res* 919:302–312.

Valero-Cabré A. and Navarro X. (2002a). Functional impact of axonal misdirection after peripheral nerve injuries followed by graft or tube repair. *J Neurotrauma* 19:1475–1485

Valero-Cabré A. and Navarro X. (2002b). Changes in crossed spinal reflexes after peripheral nerve injury and repair. *J Neurophysiol* 87:1763–1771.

Valero-Cabré A., Tsironis K., Skouras E., Perego G., Navarro X., and Neiss W.F. (2001). Superior muscle reinnervation after autologous nerve graft or poly-l-lactide-ε-caprolactone (PLC) tube implantation in comparison to silicone tube repair. *J Neurosci Res* 63:214–223.

Valero-Cabré A., Tsironis K., Skouras E., Navarro X., and Neiss W.F. (2004). Peripheral and spinal motor reorganization after nerve injury and repair. *J Neurotrauma* 21:95–108.

Vanden-Noven S., Wallace N., Muccio D., Turtz A., and Pinter M.J. (1993). Adult spinal motoneurons remain viable despite prolonged absence of functional synaptic contact with muscle. *Exp Neurol* 123:147–156.

Verdú E. and Navarro X. (1997). Comparison of immunohistochemical and functional reinervation of skin and muscle after peripheral nerve injury. *Exp Neurol* 146:187–198.

Verdú E. and Navarro X. (1998). The role of Schwann cell in nerve regeneration. In: *Understanding Glial Cells* (Castellano B., González B., and Nieto-Sampedro M., Eds.), pp. 319–359. Boston: Kluwer Academic.

Wall J.T., Xu J., and Wang X. (2002). Human brain plasticity: an emerging view of the multiple substrates and mechanisms that cause cortical changes and related sensory dysfunctions after injuries of sensory inputs from the body. *Brain Res Rev* 39:181–215.

Wang L., Copray S., Brouwer N., Meek M.F., and Kernell D. (2002). Regional distribution of slow-twitch muscle fibers after reinnervation in adult rat hindlimb muscles. *Muscle Nerve* 25:805–815.

Wasserschaff M. (1990). Coordination of reinnervated muscle and reorganization of spinal cord motoneurons after nerve transection in mice. *Brain Res* 515:241–246.

Waxman S.G., Cummins T.R., Dib-Hajj S.D., and Black J.A. (2000). Voltage-gated sodium channels and the molecular pathogenesis of pain: a review. *J Rehabil Res Dev* 37:517–528.

Wiesenfeld-Hallin Z., Kinnman E., and Aldskogius H. (1989). Expansion of innervation territory by afferents involved in plasma extravasation after nerve regeneration in adult and neonatal rats. *Exp Brain Res* 76:88–96.

Witzel C., Rohde C., and Brushart T.M. (2005). Pathway sampling by regenerating peripheral axons. *J Comp Neurol* 485:183–190.

Wood J.N., Boorman J.P., Okuse K., and Baker M.D. (2004). Voltage-gated sodium channels and pain pathways. *J Neurobiol* 61:55–71.

Woolf C.J., Shortland P., Reynolds M., Ridings J., Doubell T., and Coggeshall R.E. (1995). Reorganization of central terminals of myelinated primary afferents in the rat dorsal horn following peripheral axotomy. *J Comp Neurol* 360:121–131.

Ygge J. (1989). Neuronal loss in lumbar dorsal root ganglia after proximal compared to distal sciatic nerve resection: a quantitative study in the rat. *Brain Res* 478: 193–195.

Zelená J. (1984). Multiple axon terminals in reinnervated Pacinian corpuscles of adult rat. *J Neurocytol* 13:665–684.

Zhang J.M., Donnelly D.F., Song X.J,. and Lamotte R.H. (1997). Axotomy increases the excitability of dorsal root ganglion cells with unmyelinated axons. *J Neurophysiol* 78:2790–2794.

3

FOCUSED CHAPTER

MOTOR CONTROL MODULES OF HUMAN MOVEMENT IN HEALTH AND DISEASE

YURI P. IVANENKO,[1] GERMANA CAPPELLINI,[1] MARCO MOLINARI,[2] AND FRANCESCO LACQUANITI[1,3]

[1] *Laboratory of Neuromotor Physiology, Santa Lucia Foundation, Rome, Italy*
[2] *Spinal Cord Unit, Santa Lucia Foundation, via Ardeatina, Rome, Italy*
[3] *Department of Neuroscience and Centre of Space BioMedicine, University of Rome Tor Vergata, Rome, Italy*

SUMMARY

Human movements exhibit considerable variability and are highly complex in terms of both neural activation and biomechanical output. However, much evidence now indicates that a relatively low-dimensional organizational structure may underlie the rich complexity of neuromechanical output by using a few motor modules or motor primitives that may be combined in various ways to produce movements. Consistent with some of the ideas of motor primitives, many authors have shown that muscle activity during human locomotion is driven by a few temporal activation components in several conditions. Spatiotemporal maps of spinal cord motoneuron activation also show discrete periods of activity. This intermittent control of locomotor movements may stem from a general property of rhythm generation circuits consisting of a few half-center oscillators synchronized bilaterally. As a consequence, the selection of muscle synergies

Introduction to Neural Engineering for Motor Rehabilitation, First Edition. Edited by Dario Farina, Winnie Jensen, and Metin Akay.

appears to be downstream from the processes that generate activation timings. A similar concept of a limited basic set of motor modules was introduced for balance control and for reaching movements. The building blocks with which the central nervous system constructs motor patterns can be preserved in patients with various motor disorders despite the fact that they often modify their muscle activity and adopt motor equivalent solutions. Motor primitives provide a novel perspective on the neural control system for locomotion, and in the future they could be useful to drive neuroprostheses or entrain locomotor circuits in disabled people.

INTRODUCTION

Modularity can be defined as the use of designs employing standardized components or units, allowing easy assembly, repair, and flexible arrangements of the components (Giszter et al., 2010). Recent evidence indicates that movement may not be controlled muscle-by-muscle, but the activity of many muscles is coordinated to produce global goals. One hypothesis is that movements are based on the flexible combination of a small number of functional units, although the exact definition of such units remains controversial (Bizzi et al., 2000; Grillner, 1981; Hultborn, 2001; Lacquaniti et al., 1999; Orlovsky et al., 1999; Stein et al., 2002; Latash and Anson, 2006).

In a general sense, the fundamental units of the behavior of complex living things are structural–functional organizations that can be observed at multiple levels (cellular, neuronal, bodily, cognitive, emotional, social, etc.) (Kelso, 2009). In the context of motor control, synergies arose because of the need to coordinate and exploit large numbers of degrees of freedom in complex systems. Evidence supporting the notion of modules for movement construction has come from several studies examining muscle patterns across a range of tasks, in many contexts and in most species (Cheung et al., 2009).

In this chapter, we focus on locomotion as one of the basic activities of everyday life. While biomechanical and neural aspects of human locomotion have been documented in many papers, in both normal and pathological gait, the architecture of neuronal circuits and the nature of descending neural signals that are involved in locomotor control remain elusive in humans. To date, little work has been completed on characterizing the neural substrates for modularity in both healthy individuals and patients with different motor disorders. A number of recent studies explored the bases of central motor programming by decomposing muscle activation patterns as a means to look backward from the periphery to the central nervous system (Davis and Vaughan, 1993; d'Avella and Bizzi, 2005; Giszter et al., 2007; Ivanenko et al., 2004).

Different statistical methods have been developed to assess a linear decomposition of the original set of data based on different assumptions: for example, factor analysis (FA), independent component analysis (ICA), and nonnegative matrix factorization (NMF) (Ivanenko et al., 2005; Tresch et al., 2006). Thus, FA with varimax rotation constrains the analysis to orthogonal (uncorrelated) factors. Unlike FA, ICA aims at extracting unknown hidden components from multivariate data using only the assumption that the unknown components are mutually independent. The NMF analysis constrains the temporal components and weighting coefficients to be nonnegative. The rationale is that the data consists of nonnegative values (rectified electromyographic activity) and the NMF allows only additive and not subtractive combinations. In contrast to NMF, FA and ICA allow negative values for both the basic temporal components and the weights, under the assumption that the neuronal circuits include subtractive operations (e.g., inhibitory synapses). While different studies use different decomposition techniques, the common message is the emphasis on modular organization of the motor output. Furthermore, these techniques often converge to a similar solution (Ivanenko et al., 2005; Tresch et al., 2006). The data and concepts discussed in this chapter offer a new approach, based on the modular architecture of neuronal networks, to characterizing the mechanisms underlying control of human movements that may potentially benefit the study of pathological gait and the ability of current therapeutic exercises to change or optimize patient outcomes (Ivanenko et al., 2006b; Giszter et al., 2010).

STATE OF THE ART

Central Pattern Generators for the Control of Locomotor Movements

In vertebrates, rhythmic locomotor movements are thought to be produced by spinal central pattern generators (CPG) that are driven or modulated by descending command signals and sensory feedback (Grillner, 1981). The capacity of neural circuits to generate rhythmic activity in the absence of peripheral and supraspinal inputs is a well-established notion in animal models (Feldman and Orlovsky, 1975; Grillner and Zangger, 1979; Pearson and Rossignol, 1991) and there is indirect evidence that CPGs may also be a feature of the human spinal cord (Dietz, 2003). It may, in fact, represent the common core for various rhythmic movements in humans (Zehr, 2005), even though the natural locomotor output is highly adaptable to specific conditions. Afferent information influences the central pattern and, conversely, CPGs select the appropriate afferent information according to external requirements (Dietz, 2003). The convergence of spinal reflex pathways and descending pathways on common spinal interneurons seems to play an integrative role.

Spinal pattern generators for locomotion have now been studied in several mammals (Orlovsky et al., 1999). However, the details of such circuitry in the human spinal cord are still largely unknown (Winter, 1989; Duysens and Van de Crommert, 1998; Edgerton et al., 2008; Dietz and Colombo, 2004; Grasso

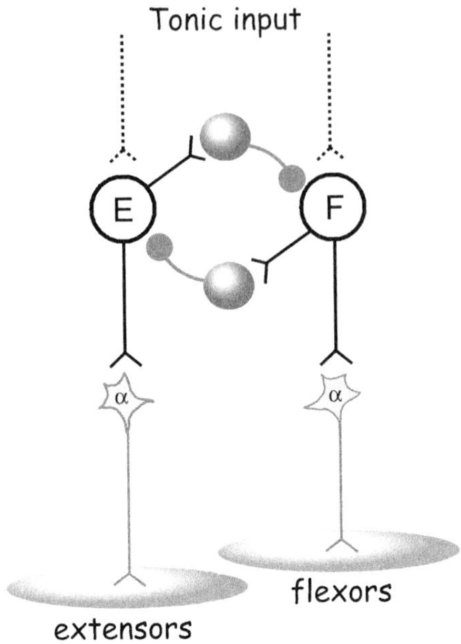

Figure 3.1. Schematic representation of a half-center CPG model that transforms a tonic input into a rhythmic output of agonist and antagonist muscles. This represents a simplistic rendering of Brown's (1911) half-center model. Excitatory and inhibitory connections are shown by open endings and small filled circles, respectively. E, extensor half-center; F, flexor half-center; α, alpha motoneurons. Interconnected half-centers may form a chain of segmental oscillatory networks (Grillner, 1981; Zehr, 2005; McCrea and Rybak, 2008).

et al., 2004a,b). The current understanding of these spinal circuits is based largely on studies in the cat. Pattern generating oscillators are believed to be primarily localized in the cervical and lumbosacral segments of the cord, where they control the locomotion movements of the upper and lower limbs, respectively. Early studies by Brown (1911) suggested there are one or more half-center CPGs (Fig. 3.1) for each limb (see also Jankowska et al., 1967; Grillner and Zangger 1979). More recent data suggest a major site for CPG activity in the upper lumbar segments that may serve a pacemaker role, supplemented by separate generator sites in more caudal segments (Lev-Tov et al., 2000; Cazalets and Bertrand, 2000). While the CPGs organize the sequencing of motor activity in locomotion, they are also controlled in turn by both central commands and sensory feedback (Drew et al., 2004; Pearson, 2004). For example, electrical stimulation applied to the mesencephalic locomotor region in the cat (Shik, 1983) can elicit locomotion at various speeds, depending on the intensity of the stimulation. Moreover, the transition period from rest to locomotion, which may be characterized by a specific tuning of the spinal cord circuitry, is also strongly influenced by descending activity (Shik, 1983; Mori et al., 1982). In addition to the segmental feedback from sensory receptors, it has also been suggested that motoneurons (MNs) themselves may be integral elements of CPGs (Marder, 1991; O'Donovan et al., 1998). However, opinions diverge as to whether the mammalian CPGs are localized or distributed (Kiehn et al., 1998; Orlovsky et al., 1999).

Spatiotemporal Organization of the Motor Output

There is relatively little data relevant to understanding the organization of the CPG circuitry in humans. However, it is possible to visualize α-motoneuron activation during locomotion by mapping the activity patterns from a large number of simultaneously recorded muscles onto the approximate rostrocaudal location of the MN pools in the human spinal cord (Fig. 3.2; Ivanenko et al., 2006a). This technique, similar to one previously reported for the cat (Yakovenko et al., 2002), does not show the organization of the CPG directly, but it does show how the CPG output is directed to the MN pools within each spinal segment. We applied this technique by recording electromyographic (EMG) activity from 32 ipsilateral leg and trunk muscles during locomotion, and adding up the contribution of each muscle to the total activity in each spinal segment according to the published myotomal charts of segmental innervation in humans (Sharrard, 1964; Kendall et al., 1993). The implicit assumption is that the rectified EMG provides an indirect measure of the net firing of spinal MNs innervating that muscle. Although it is a simplification (Yakovenko et al., 2002), this is a reasonable hypothesis because mean EMG seems to increase fairly linearly with the net motor unit firing rate (Day and Hulliger, 2001; Hoffer et al., 1987).

This approach also allows assessment of the relative intensities of MN activation in various phases of the gait cycle. Motoneurons of different muscles in the lumbosacral enlargement are arranged in columns whose disposition is determined at any given segmental level (Sharrard, 1964). In general, each muscle is innervated by several spinal segments, and each segment supplies several muscles. Schematic representation of the location of the motor cell columns of all muscles in the lumbosacral segments is illustrated in Figure 3.3. We focused on the lumbosacral enlargement (Fig. 3.3), because it contains the MNs of all the leg muscles. Although the classification of muscles as being active in stance phase versus swing phase is clear in most cases, some of the stance-phase muscles actually start activity during the late swing phase, whereas some of the swing-phase muscles start or increase their activity during stance. Nevertheless, for illustrative purposes, the motor cell columns that display their major activity during the stance and swing phases are shown schematically in Figure 3.3A. From this schematic picture, it is evident that each lumbosacral segment exhibits activity during both stance and swing (Ivanenko et al., 2008). Interestingly, the distribution of activity between swing-phase and stance-phase muscles is relatively independent of speed, and

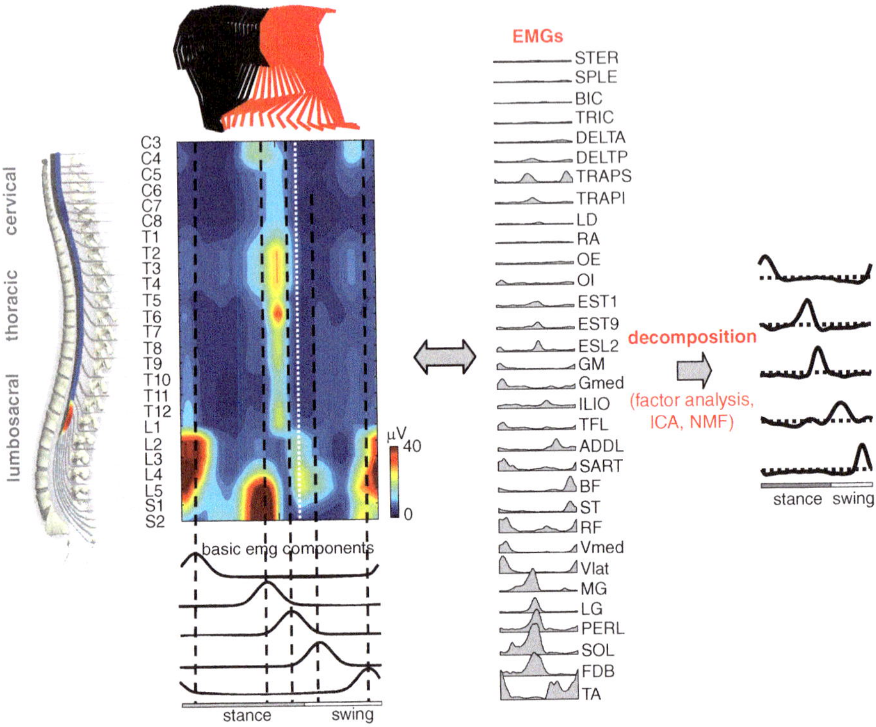

Figure 3.2. Spatiotemporal patterns of ipsilateral α-motoneuron (MN) activity along the rostrocaudal axis of the spinal cord during walking on a treadmill at 5 km/h. Pattern is plotted versus normalized gait cycle. Output pattern was constructed by mapping the recorded electromyographic (EMG) waveform of 32 ipsilateral limb and trunk and shoulder muscles (nonnormalized method, adapted from Ivanenko et al., 2006a, Am Physiol Soc, used with permission) onto the known charts of segmental localization. These EMG waveforms can also be decomposed into a smaller set of five basic temporal components using different statistical methods (factor analysis, independent component analysis, and nonnegative matrix factorization; see Ivanenko et al., 2005), each component being characterized by a relatively narrow peak of activation (Gaussian-like) at a particular phase of the cycle. Components are designated in chronological order. Five Gaussian activation components that correspond to five major discrete periods of activity and account for ~90% of total variance are shown below the spinal map.

swing-related intensity of MN activation constitutes as much as 40% of the total activity (Fig. 3.3B), which is compatible with energy consumption estimation in guinea fowl (Marsh et al., 2004). Thus, lumbosacral segments exhibit prominent activity during both stance and swing.

Maps constructed from data recorded during treadmill locomotion at several different speeds show a number of common features (Ivanenko et al., 2006a). One is that MN activity tends to occur in bursts that are temporally

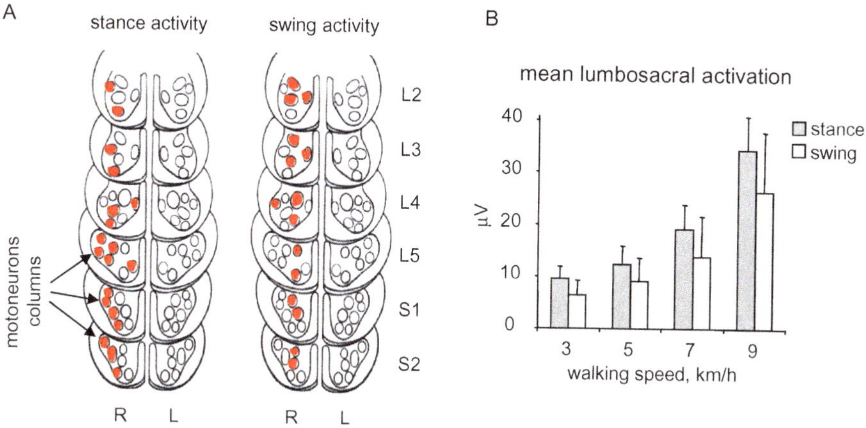

Figure 3.3. Lumbosacral activation during the stance and swing phases of walking. (A) schematic representation of the motor cell columns in the lumbosacral enlargement. Columns that display their major activity during the stance (A, left panel) and swing (A, right panel) phases of the right leg are marked in red. (B) Mean levels (±SD) of lumbosacral MN activation (nonnormalized method, in microvolts; Ivanenko et al., 2008) during the stance and swing phases of walking at different constant speeds. Note that the swing-related activity constitutes ~40% of the total MN activity in the gait cycle at all walking speeds.

aligned across several spinal segments (Fig. 3.2). These maps are relatively invariant across subjects, especially at high walking speeds (Ivanenko et al., 2006a), in spite of the fact that the root innervation of many muscles does show interindividual variations (Wilbourn and Aminoff, 1998). A simple inspection of the maps suggests that almost all of the EMG activity during locomotion can be represented by five separate periods of MN activation. The timing of these activity peaks tends to be associated with major kinematic and kinetic events in the gait cycle: weight acceptance; loading and propulsion; trunk stabilization during double support phase; liftoff; and heel strike (see Fig. 3.2, bottom left). Thus it appears that the activation patterns may represent the drive provided by spinal pattern generators and/or sensory feedback.

Locomotion Motor Program as a Characteristic Timing of Muscle Activations

Another line of evidence relating to the nature of the locomotion motor program has come from statistical analyses of multimuscle EMG activity. A few studies in the 1980s and 1990s showed that the activity patterns of leg muscles during locomotion could be accounted for by specific bursts of activation (Patla et al., 1985; Davis and Vaughan, 1993). That is, the activity patterns resulting from the modulation of leg muscle activity over a step cycle could be reconstructed for each muscle from a weighted sum of a few basic temporal components (Fig. 3.2, right panel):

$$m_i(t) = \sum_{j=1}^{k} w_{ij} \cdot c_j(t) + r_i(t)$$

where $m_i(t)$ is the ith element of the n-dimensional vector of the original EMG data (n = number of recorded muscles), t = time samples over the gait cycle, $c_j(t)$ is the jth element of the k-dimensional vector of the basic temporal components ($k < n$ is the number of basic temporal components), w_{ij} is the ijth element of the matrix of weights ($n \times k$ matrix), and $r_i(t)$ is the ith element of the n-dimensional error vector, typically related to noise. We showed that this analysis (e.g, factor analysis, Ivanenko et al., 2004, 2005) gives the same basic result when trunk and arm muscles are included in the analysis and independently of walking speed or body weight support.

Furthermore, since the early 1990s a number of statistical analysis methods have been developed and applied to decomposing multiple EMG records, each based on different assumptions (Davis and Vaughan, 1993; Bell and Sejnowski, 1995; Lee and Seung, 1999; Tresch et al., 2006). We have compared three of these methods with the EMGs recorded during locomotion: factor analysis (FA), independent component analysis (ICA), and nonnegative matrix factorization (NMF). Although each of these statistical approaches places different restrictions on the outcomes (e.g., nonnegative values in NMF), they all converge to a similar solution about the temporal structure of the EMG activity pattern during human locomotion (Ivanenko et al., 2005). The same five basic components accounted for about 90% of the variance in the original data set regardless of the statistical analysis used. Although the number of modules or components may vary slightly from study to study due in part to the somewhat different sets of muscle EMGs recorded, it is important to note that the four main components are present throughout the literature (Olree and Vaughan, 1995; Ivanenko et al., 2004; McGowan et al., 2010; Monaco et al., 2010; Gizzi et al., 2011).

Another reason to suppose that the result is not a statistical artifact is that the timing of muscle activation revealed by the statistical analysis aligns with the activation loci observed in the spinal maps (Fig. 3.2, bottom). The maps also show that each activation period is associated with MNs in particular spinal segments and not in others.

The temporal components are each characterized by the same relatively narrow peak of activation (Gaussian-like) occurring at a specific point in the normalized gait cycle (Fig. 3.4A). The implication of this is that the activity drive represented by each component is scaled by the cycle duration so that muscle activation has a short duration at higher speeds (about 90 ms for walking at 9 km/h) and a much longer duration at low speeds (about 250 ms at 1 km/h). If these periods of muscle activation represent the activity of a CPG, then they imply that the fraction of the cycle over which the CPG is active is considerably less than 50%. This is unlike the classic half-center oscillator that is postulated to be active for one-half the cycle and inhibited for the other half. Once again, however, this behavior might have its origin in the sensory input, which would naturally be scaled to the biomechanical events.

Invariance of Activation Timing

Even though CPGs can provide the basis for rhythmic and reciprocal muscle activation, the resultant motor output is complex. It engages tens of muscles in each limb and is speed- and gait-dependent. Does the variance in muscle activation across subjects or locomotion context imply a variance in the motor program also? To study this issue, we examined motor patterns at different locomotion speeds and under weight-supported conditions. For a speed range of 1–9 km/h we found that the same 5 activation components accounted for the EMG activation of up to 32 ipsilateral muscles (Ivanenko et al., 2004, 2008). There was, however, a consistent speed-dependent phase shift of the peak of each component that corresponded to the shift in relative stance duration with walking speed. This again might imply some modulation by sensory input.

By supporting various levels of body weight, we also placed different biomechanical demands on the limbs during locomotion. While this might be expected to alter the activation patterns of specific muscles (and it did), we again found that the overall muscle activation could be accounted for by roughly the same five activation components (Fig. 3.4A). Once again, each activation component described a short period of synchronous activation or relaxation of a particular set of muscles. It should be noted, however, that at low speeds (1 km/h) and high levels of body unloading (>75%) there tended to be some distortion in the components, possibly due to very low activity in some muscles under these conditions. Nonetheless, the results supported the idea that the same sequence of activation components (Fig. 3.4B) provides a stable motor program that is basically unaltered by changes in locomotion speed or by body weight support during locomotion (Ivanenko et al., 2004; McGowan et al., 2010).

A further test of this concept of a stable, temporally defined motor program for locomotion was the extent to which it might be preserved when the basic locomotion pattern is perturbed by voluntary movements. To gain some insight into how this interaction might take place, we examined several voluntary tasks added to overground walking in which subjects kicked a ball, stepped over an obstacle, or reached down and grasped an object on the floor (with weight support on either the right or the left foot). Our basic finding was that the five basic activation components underlying locomotion were invariantly present when voluntary tasks were performed during these tasks (Ivanenko et al., 2005). The voluntary tasks were accompanied by an additional sixth component timed to the voluntary task and by different loadings on the five basic temporal components.

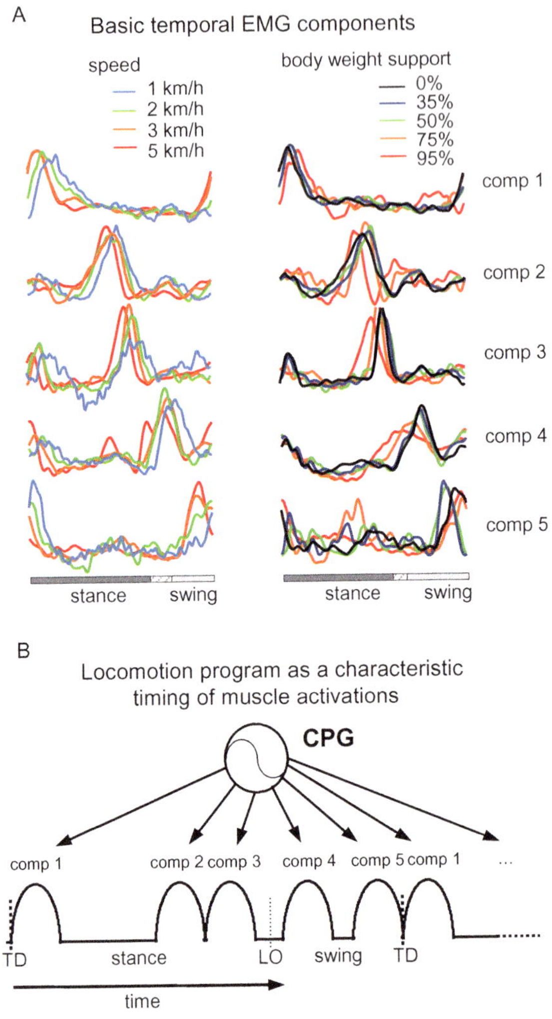

Figure 3.4. (A) Effect of walking speed (left panel) and body weight support (right panel) on basic EMG components obtained using factor analysis of averaged activity patterns of 16 ipsilateral leg and trunk muscles during treadmill locomotion (adapted from Ivanenko et al., 2004). Five common components across conditions are designated in "chronological" order of their main peak in the cycle beginning with touchdown. (B) Proposed locomotion motor program as a sequence of activation pulses. TD, touchdown; LO, liftoff; CPG, central pattern generator.

The idea that motor tasks employ synergies specific to the task is long-standing in the motor control literature. A **muscle synergy** is usually considered as the set of muscles that are synchronously activated in performing a task. For instance, a mixed muscle synergy (hip and knee extensors plus ankle flexors) becomes active in locomotion at the moment of heel strike, while a synchronous activation of the ankle extensors comes into action by the end of the stance phase during normal locomotion. Another set of examples that has been well documented in the literature is the set of muscle synergies participating in automatic postural responses to stance perturbations (e.g., the so-called hip or ankle strategies, Horak and Nashner, 1986; see also Krishnamoorthy et al., 2003; Ting, 2007). In addition, specific correcting synergies seem to be incorporated into the basic locomotor program (Nashner, 1980; Nashner and Forssberg, 1986). However, even though fixed muscle synergies for locomotion may exist, they should be temporally organized to produce the full locomotor pattern. It appears then that the question of muscle activation during locomotion can be divided into two parts, the timing of activation and the specification of which muscles to activate. In fact, the necessity for such a dual motor pattern representation has given rise to a nontraditional definition of muscle synergy—a time-varying muscle synergy (d'Avella and Bizzi, 2005).

The sequence of activation components in locomotion bears a resemblance to the intermittent control of biological movements (Gawthrop et al., 2011) and to the sequential activation of "premotor drives" or muscle synergies activated with pulses in decerebrate and spinal frogs in a variety of different motor behaviors (Hart and Giszter, 2010). Although different speeds of movement were not explored in this system, the result seems to be different from the locomotion result where the duration of the "activation pulses" was scaled to the duration of the gait cycle. Another result of the frog study was the finding that each premotor drive was associated with a small core group of muscles or a muscle synergy. The linkage between these "pulse" events and the muscle activation during locomotion is not so clear, however. For example, a temporal component might be loaded on proximal leg muscles under one condition, while the same component is loaded on more distal leg muscles under another condition (Ivanenko et al., 2004, 2005, 2008).

A straightforward interpretation of our results is that locomotion timing generation networks activate motoneurons via a distribution network that sets weightings according to various feedback and feedforward parameters (Prentice et al., 1998; Poppele and Bosco, 2003; Ivanenko et al., 2003; 2004). A major part of the proprioceptive feedback, along with the cyclic excitability of spinal interneurons and ascending and descending pathways (Orlovsky et al., 1999), may interact to shape the motor output in order to provide an appropriate kinematic pattern (Lacquaniti et al., 1999). Our findings suggest that the selection of muscle synergies may be downstream from the processes that generate activation timings.

Reorganization of Motor Output in Patients after Stroke and Spinal Cord Injury

We have just described the general design of locomotor output in healthy subjects. Muscle activity during normal gait has both invariant and variant features. If someone aims to reactivate (in patients) the "normal" motor patterns

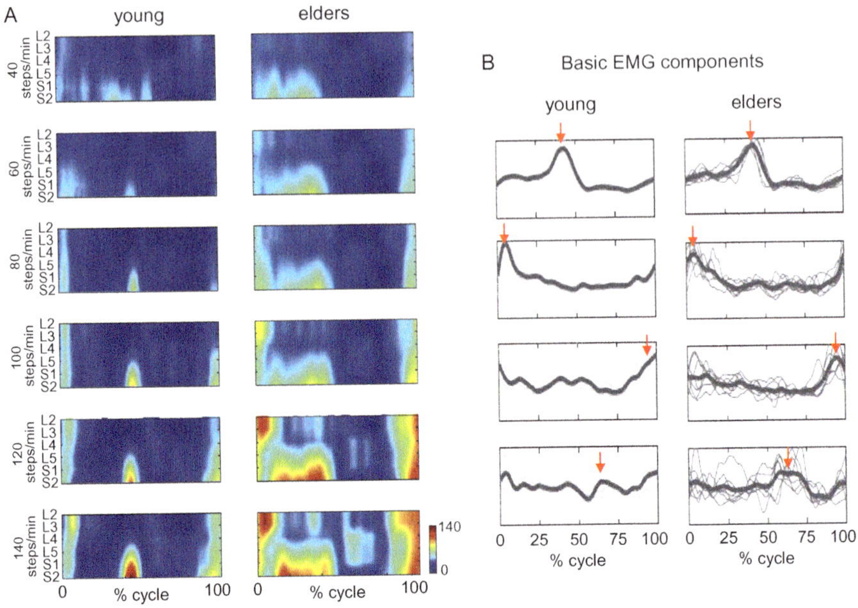

Figure 3.5. Motor patterns in young and elderly people during walking. (A) Motoneuron activity of two representative subjects while walking at different cadences. (B) Four most significant EMG components of young (light gray) and elderly (dark gray) subjects walking at 10 steps per minute. For both groups, the averaged basic temporal components are represented in black. Note similar timing of basic components in both groups (marked by the orange arrow). Adapted from Monaco et al. (*J Neurophysiol*, 2010). Am Physiol Soc, used with permission.

and extrapolate them to any walking condition, he or she might be very surprised in some cases since muscle activity in healthy subjects may show very nonlinear changes in both amplitude and temporal envelope (e.g., Ivanenko et al., 2009), even while kinematic patterns remain similar. The most variable patterns are observed in the proximal and bi-articular muscles. Nevertheless, a few basic temporal components may account for about 90% of variance across different locomotor conditions (Fig. 3.4). In the elderly, the pattern seems to be more variable, or "noisy"; however, the three most significant temporal components are still preserved (Fig. 3.5; Monaco et al., 2010). The high-order components are less stable, possibly due to a limited set of recorded muscles in that study. The question arises as to whether the rhythmic patterning elements are invariant in conditions where muscle activation patterns are compromised by spinal cord lesions, brain damage, and other motor disturbances. A few recent studies provide some clues on this topic.

Impaired locomotor coordination in post-stroke patients may be accompanied by fewer modules (Clark et al., 2010), though in a more recent study Gizzi et al. (2011) have argued that impulses of activation rather than motor modules (or muscle synergies) are preserved in the locomotion of subacute stroke patients. The authors concluded that identification of motor modules may lead to new insight into how nervous system injury may alter the organization of motor modules and their biomechanical outputs. Furthermore, entraining appropriate motor modules can be of major importance for neurorehabilitation of gait in these patients since many of them develop an abnormal stereotype of movement during walking, which is difficult to correct. A similar conclusion has been reached in a study on upper-limb control (Cheung et al., 2009). The robustness of muscle synergies observed in that study supports the notion that descending cortical signals represent neuronal drives that select, activate, and flexibly combine muscle synergies specified by networks in the spinal cord and/or brain stem and suggest an approach to stroke rehabilitation by focusing on those synergies with altered activations after stroke. Nevertheless, all these studies suggest a common modular organization of muscle coordination underlying motor control in both healthy and post-stroke subjects.

We studied patients with spinal cord injury (SCI) trained to step on a treadmill with body weight support (BWS; Ivanenko et al., 2003; Grasso et al., 2004b). Possible beneficial effects of locomotor training in SCI patients have been shown in many clinical trials. At the beginning of training or in case of severe SCI disorders, the patients have to be suspended by a harness with their body partially unloaded. In addition, their leg movements need to be assisted if necessary. These patients frequently show EMG patterns different from those of healthy individuals. Therefore, while the human spinal cord can interpret loading- or velocity-dependent sensory input during stepping (Beres-Jones and Harkema, 2004), this interpretation may be at variance with that of healthy subjects. The corresponding spatiotemporal maps of MN activity are also generally different from those of healthy subjects (Grasso et al., 2004a). Even in the patients with motor-incomplete paraplegia who recovered independent control of their limb at full loading (BWS = 0), the lumbosacral enlargement could often show "abnormal" loci of activation (Fig. 3.6B), for instance, due to an "abnormal" activation of the vastus lateralis muscle during midstance (Maegele et al., 2002). Patients can be trained to step with BWS, unassisted under laboratory conditions, but they often use new coordinative strategies. With training, patients learned to produce foot kinematics similar to that of healthy subjects but with activity patterns in individual muscles that were generally different from the control group (Pepin et al., 2003; Grasso et al., 2004a). Most patients with incomplete paraplegia recovered independent control of leg muscles sufficient to propel the limbs in swing and to support body weight in stance. Those with complete paraplegia, instead, did not recover independent control and they typically used their arms and body to assist the leg movements. SCI patients largely relied on proximal and axial muscles to lift the foot and to project the limb forward (Fig. 3.6A).

Taken together, the data support the idea of plasticity and distributed networks for controlling human locomotion (Scivoletto et al., 2007; Ivanenko

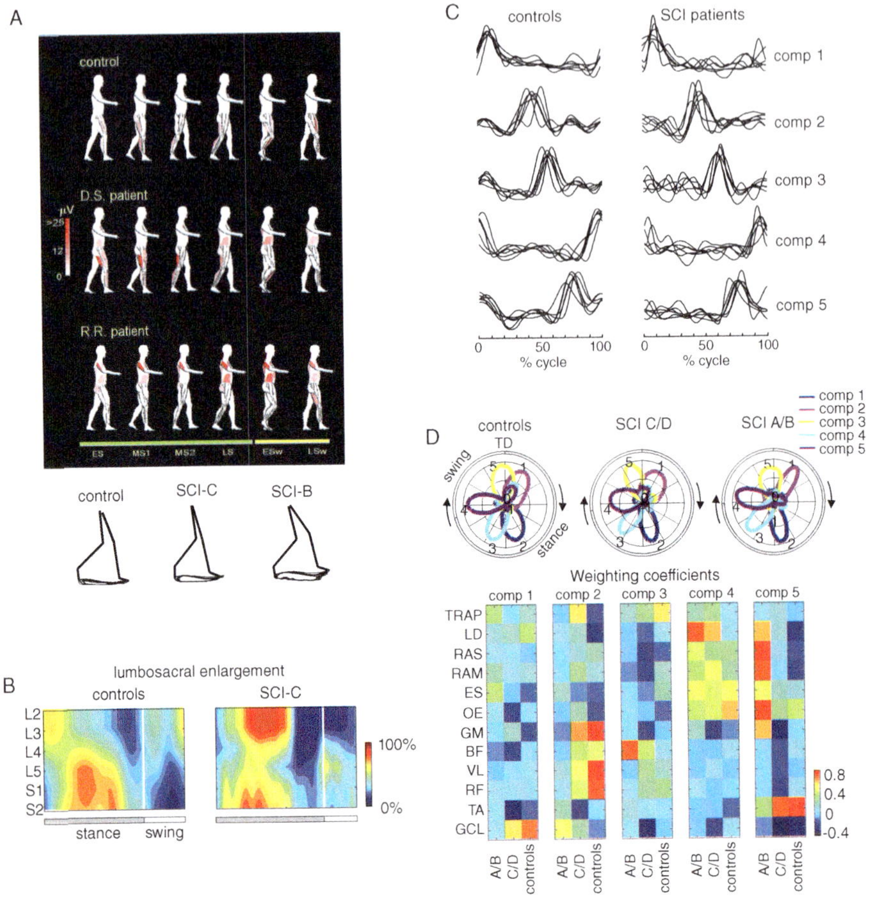

Figure 3.6. Motor patterns in SCI patients. (A) Distribution of activity across different limb and body muscles as a function of gait phase (ES, early stance; MS1, midstance first half; MS2, midstance second half; LS, late stance; ESw, early swing; LSw, late swing). Rectified EMG was integrated over each epoch and plotted in color scale. Data from a typical control, one American Spinal Injury Association (ASIA)–C patient, and one ASIA-A patient are plotted in the top, middle, and bottom panels, respectively. (B) Examples of spatiotemporal patterns of α-motoneuron (MN) in the lumbosacral enlargement in controls and SCI-C patient during walking on a treadmill at 1 km/h. Output pattern for each segment was reconstructed by mapping the recorded EMG waveforms (normalized method; see Ivanenko et al., 2006a) onto the known charts of segmental localization. Pattern is plotted vs. normalized gait cycle in a color scale (right calibration bar) using a filled contour plot. White vertical lines denote stance-to-swing transition time. (C) Time course of the temporal components in controls and patients for stepping at 2 km/h, 0–75% BWS. The components were extracted by factor analysis from individual subjects. Note similar basic EMG components in controls and patients as opposed to quite different EMG patterns. (D) Polar plots of the components obtained from the normal ensemble, grouped data from all ASIA-C/D patients, and grouped data from all ASIA-A/B patients stepping at 2 km/h, 75%-BWS (same format as Fig. 3.5A). Lower panel illustrates weighting coefficients of the temporal components in individual activity patterns of 12 muscles for all groups of subjects in a color-coded scale. Panels A, C, and D are adapted from Ivanenko et al. (*J Neurophysiol*, 2003), Am Physiol Soc, used with permission.

et al., 2009; Molinari, 2009). Spinal lesions probably trigger multiple forms of plasticity, including modified synaptic strengths (synaptic plasticity), sprouting and anatomical development of new circuits (anatomical plasticity), as well as plasticity of unlesioned descending pathways, including both subcortical and cortical motor areas. In particular, stepping may depend more heavily on cortical (and voluntary) control after severe spinal lesions than it does in healthy subjects, where locomotion may be more automatic. Yet, we found a similar set of five temporal components in the EMG activity of the patients, implying that a flexible combination of those components could account for the wide range of muscle patterns we observed (Fig. 3.6C,D). Moreover, muscles both rostral and caudal to the lesions could be strongly weighted on a given component. It seems that the invariance of activation timing in these patients reflects in some way the global kinematic goal (a motor equivalent solution; Grasso et al., 2004a).

Thus, growing consensus indicates that motor control is achieved by interaction among central controllers, a few motor primitives located downstream, and an intermediate control process influenced by the periphery. This theoretical framework imposes a reconsideration of neurorehabilitation treatments providing solid ground for the development of experimental evidence. Both top-down and bottom-up approaches find their ground, and lesions at different levels of the neuraxis would differently affect motor control. Rehabilitation of gait after cortical, subcortical, or spinal cord damage might represent the experimental field in which one might test the therapeutic relevance of motor-primitives theory. After lesion of the cortical motor areas, gait is impaired in a characteristic way with gait asymmetries and prevalence of extensor over flexor muscles in the affected limb. In this context, a lesion of the central command modules with preservation of the controllers and of the motor-primitive modules can be hypothesized. Conversely, a lesion of the spinal cord would imply preservation of the central command modules and of the controllers with loss of the motor primitives. A third condition can be envisaged by considering damage to subcortical nonspinally located motor centers. An example of such damage can be hypothesized after cerebellar focal lesions. In this case one would expect to see a preservation of central command and motor primitive modules with impairment of the controllers. All in all, the comparative study of motor learning under these conditions may provide experimental data to evaluate reciprocal influences of the three motor control levels. In fact, by specific training protocols the capacities of the three control levels in sustaining recovery and in controlling postlesion gait could be evaluated.

CURRENT CHALLENGES AND FUTURE DIRECTIONS

A modular motor organization may be needed to solve the degrees-of-freedom problem in biological motor control (Giszter et al., 2007). Nevertheless there are still many open questions related to the choice of appropriate modules, their task dependence, and their adaptation to the malfunctioning of neuronal networks in the case of different pathologies.

Although many studies have successfully decomposed motor patterns into a few "motor modules," nevertheless, the way in which the central nervous system combines them together to perform a movement and how and where the weighting coefficients are encoded are not understood. Moreover, each individual interneuron in the spinal cord, as a rule, receives a broad convergence from a large variety of sensory inputs, as well as inputs from one or more descending tracts. At the same time, each motoneuron participates in a vast repertoire of motor activities. Nevertheless, there is now much evidence in favor of the view that the many classes of spinal interneurons may be seen as "functional units" representing different levels of muscle synergies, parts of movements, or even more integrated motor behavior (Hultborn, 2001). In the case of locomotion, CPGs and rhythm generation networks (Fig. 3.4) can be considered as important functional units or modules that determine a variety of locomotor movements in health and disease.

Most CPGs are quiescent under resting condition and become recruited only by neurons with command function (Grillner, 2006). Yet, the nature of this central command is unclear (Jordan et al., 2008). In particular, for a command to trigger an action, the physiological state of the spinal network needs to be properly prepared (Edgerton et al., 2008) since the same interneurons may contribute to a vast repertoire of motor actions. Thus, the selection of appropriate functional units may involve an appropriate preparedness of the low-level structures. Given that the spinal cord is not simply a relay structure for information transmitted to and from supraspinal systems (Collins et al., 2001; Hultborn, 2001; Pearson, 2004; Poppele and Bosco, 2003), one may even argue that motor control systems make use of several types of command signals (Jordan et al., 2008): for example, one for generating tonic readiness and the other for performing the movement. The importance of investigating the tonic "state" of the spinal circuits has been stressed in several studies (Edgerton et al., 2008; Selionov et al. 2009).

Uncovering a common underlying neural framework for the modular control of human locomotion and its development represent an interesting avenue for the future work. Motor primitives may reflect in some way how the nervous system develops, by building up or modifying modules as it matures. Some functional units are likely inborn; others may develop later or be dependent on individual body size or experience.

Finally, such investigations may have important implications related to the construction of biologically inspired robots (Pfeifer et al., 2007) and clinical studies (Ivanenko et al., 2003; Cheung et al., 2012; Clark et al., 2010; Kuo and Donelan, 2010). The problem of motor neurorehabilitation after stroke or spinal cord injury is significant and complex. Numerous studies have shown that motor activity after brain damage plays a critical role in anatomophysiological reorganization (Solopova et al., 2011), which may occur in the areas adjacent to the damage. The lack of adequate rehabilitation may lead to the appearance of irreversible anatomical and functional changes, as well as progressive weakness in the affected limbs. Many patients develop an abnormal stereotype of movement during walking, which is difficult to correct. A new generation of robotic devices should take advantage of this feature by providing the opportunity for patients to generate and correct limb movements using assist-as-needed technology or biofeedback. We have also to learn the rules of human–machine interaction applied to a rehabilitation setting. Patients with spinal cord injury can learn to produce foot movements similar to those of healthy subjects but with activity patterns of individual muscles generally different from the healthy subjects. Nevertheless, because the activation components are preserved, in the future they could be useful to drive neuroprostheses for people with spinal cord injury. Thus appropriately weighted components could be distributed by neural network algorithms running on computers to a set of stimulators imbedded in the patient's paralyzed muscles. In summary, training based on reactivating motor modules may provide a more successful rehabilitation than training based on just adapting muscle activation to the fixed kinematic template imposed by a gait orthosis.

CONCLUSIONS

Muscle activity during human locomotion seems to be driven by a fixed set of temporal activation components distributed to various muscles in a variable manner, depending on the kinetic and kinematic demands of the specific task and conditions. In addition to a "spatial" architecture of neuronal networks that selects the appropriate muscle patterns for compound movements, there may also be a "temporal" architecture that utilizes discrete timing intervals to coordinate the elements of the movements. An intermittent type of control of locomotor movements in health and disease may stem from a general basic property of rhythm generation circuits, consisting of half-center oscillators synchronized on both sides of the body (Fig. 3.4). It also opens up new possibilities for the design of robots and prostheses. Motor primitives provide a novel perspective on the neural control system for locomotion, by revealing fundamental motor activation patterns that can account for the varied and complex activations of specific muscles. In the future they could be useful to drive neuroprostheses or entrain locomotor circuits in disabled people.

REFERENCES

Bell A.J. and Sejnowski T.J. (1995). An information-maximization approach to blind separation and blind deconvolution. *Neural Comput* 7(6):1129–1159.

Beres-Jones J.A. and Harkema S.J. (2004). The human spinal cord interprets velocity dependent afferent input during stepping. *Brain* 127:2232–2246.

Bizzi E., Tresch M.C., Saltiel P., and d'Avella A. (2000). New perspectives on spinal motor systems. *Nat Rev Neurosci* 1(2):101–108.

Brown T.G. (1911). The intrinsic factors in the act of progression in the mammal. *Proc Roy Soc London B* 84:308–319.

Cazalets J.R. and Bertrand S. (2000). Ubiquity of motor networks in the spinal cord of vertebrates. *Brain Res Bull* 53(5):627–634.

Cheung V.C., Piron L., Agostini M., Silvoni S., Turolla A., and Bizzi E. (2009). Stability of muscle synergies for voluntary actions after cortical stroke in humans. *Proc Natl Acad Sci U S A* 106(46):19563–19568.

Cheung V.C., Turolla A., Agostini M., Silvoni S., Bennis C., Kasi P., Paganoni S., Bonato P., and Bizzi E. (2012). Muscle synergy patterns as physiological markers of motor cortical damage. *Proc Natl Acad Sci U S A* 109(36):14652–14656.

Clark D.J., Ting L.H., Zajac F.E., Neptune R.R., and Kautz S.A. (2010). Merging of healthy motor modules predicts reduced locomotor performance and muscle coordination complexity post-stroke. *J Neurophysiol* 103(2):844–857.

Collins D.F., Burke D., and Gandevia S.C. (2001). Large involuntary forces consistent with plateau-like behavior of human motoneurons. *J Neurosci* 21:4059–4065.

d'Avella A. and Bizzi E. (2005). Shared and specific muscle synergies in natural motor behaviors. *Proc Natl Acad Sci U S A* 102(8):3076–3081.

Davis B.L. and Vaughan C.L. (1993). Phasic behavior of EMG signals during gait: use of multivariate statistics. *J EMG Kinesiol* 3:51–60.

Day S.J. and Hulliger M. (2001). Experimental simulation of cat electromyogram: evidence for algebraic summation of motor-unit action-potential trains. *J Neurophysiol* 86:2144–2158.

Dietz V. (2003). Spinal cord pattern generators for locomotion. *Clin Neurophysiol* 114:1379–1389.

Dietz V. and Colombo G. (2004). Recovery from spinal cord injury—underlying mechanisms and efficacy of rehabilitation. *Acta Neurochir Suppl* 89:95–100.

Drew T., Prentice S., and Schepens B. (2004). Cortical and brainstem control of locomotion. *Prog Brain Res* 143:251–261.

Duysens J. and Van de Crommert H.W. (1998). Neural control of locomotion: the central pattern generator from cats to humans. *Gait Posture* 7(2):131–141.

Edgerton V.R., Courtine G., Gerasimenko Y.P., Lavrov I., Ichiyama R.M., Fong A.J., Cai L.L., Otoshi C.K., Tillakaratne N.J., Burdick J.W., and Roy R.R. (2008). Training locomotor networks. *Brain Res Rev* 57:241–254.

Feldman A.G. and Orlovsky G.N. (1975). Activity of interneurons mediating reciprocal 1a inhibition during locomotion. *Brain Res* 84:181–194.

Gawthrop P., Loram I., Lakie M., and Gollee H. (2011). Intermittent control: a computational theory of human control. *Biol Cybern* 104(1–2):31–51.

Giszter S., Patil V., and Hart C. (2007). Primitives, premotor drives, and pattern generation: a combined computational and neuroethological perspective. *Prog Brain Res* 165:323–346.

Giszter S.F., Hart C.B., and Silfies S.P. (2010). Spinal cord modularity: evolution, development, and optimization and the possible relevance to low back pain in man. *Exp Brain Res* 200(3–4):283–306.

Gizzi L., Nielsen J.F., Felici F., Ivanenko Y.P., and Farina D. (2011). Impulses of activation but not motor modules are preserved in the locomotion of subacute stroke patients. *J Neurophysiol* 106(1):202–210.

Grasso R., Ivanenko Y.P., Zago M., Molinari M., Scivoletto G., Castellano V., Macellari V., and Lacquaniti F. (2004a). Distributed plasticity of locomotor pattern generators in spinal cord injured patients. *Brain* 127(Pt 5):1019–1034.

Grasso R., Ivanenko Y.P., Zago M., Molinari M., Scivoletto G., and Lacquaniti F. (2004b). Recovery of forward stepping in spinal cord injured patients does not transfer to untrained backward stepping. *Exp Brain Res* 157(3):377–382.

Grillner S. (1981). Control of locomotion in bipeds, tetrapods, and fish. In: *Handbook of Physiology. The Nervous System. Motor Control* (Brookhart J.M. and Mountcastle V.B., Eds.), Sect. 1, Vol. 2, Part 1, pp. 1179–1236. Bethesda, MD: American Physiological Society.

Grillner S. and Zangger P. (1979). On the central generation of locomotion in the low spinal cord. *Exp Brain Res* 34:241–261.

Hart C.B. and Giszter S.F. (2010). A neural basis for motor primitives in the spinal cord. *J Neurosci* 30(4):1322–1336.

Hoffer J.A., Sugano N., Loeb G.E., Marks W.B., O'Donovan M.J., and Pratt C.A. (1987). Cat hindlimb motoneurons during locomotion. II. Normal activity patterns. *J Neurophysiol* 57:530–553.

Horak F.B., and Nashner L.M. (1986). Central programming of postural movements: adaptation to altered support-surface configurations. *J Neurophysiol* 55(6): 1369–1381.

Hultborn H. (2001). State-dependent modulation of sensory feedback. *J Physiol* 533:5–13.

Ivanenko Y.P., Grasso R., Zago M., Molinari M., Scivoletto G., Castellano V., Macellari V., and Lacquaniti F. (2003). Temporal components of the motor patterns expressed by the human spinal cord reflect foot kinematics. *J Neurophysiol* 90(5):3555–3565.

Ivanenko Y.P., Poppele R.E., and Lacquaniti F. (2004). Five basic muscle activation patterns account for muscle activity during human locomotion. *J Physiol* 556(Pt 1):267–282.

Ivanenko Y.P., Cappellini G., Dominici N., Poppele R.E., and Lacquaniti F. (2005). Coordination of locomotion with voluntary movements in humans. *J Neurosci* 25(31):7238–7253.

Ivanenko Y.P., Poppele R.E., and Lacquaniti F. (2006a). Spinal cord maps of spatiotemporal alpha-motoneuron activation in humans walking at different speeds. *J Neurophysiol* 95(2):602–618.

Ivanenko Y.P., Poppele R.E., and Lacquaniti F. (2006b). Motor control programs and walking. *Neuroscientist* 12(4):339–348.

Ivanenko Y.P., Cappellini G., Poppele R.E., and Lacquaniti F. (2008). Spatiotemporal organization of alpha-motoneuron activity in the human spinal cord during different gaits and gait transitions. *Eur J Neurosci* 27(12):3351–3368.

Ivanenko Y.P., Poppele R.E., and Lacquaniti F. (2009). Distributed neural networks for controlling human locomotion: lessons from normal and SCI subjects. *Brain Res Bull* 78(1):13–21.

Jankowska E., Jukes M.G., Lund S., and Lundberg A. (1967). The effect of DOPA on the spinal cord. 6. Half-centre organization of interneurones transmitting effects from the flexor reflex afferents. *Acta Physiol Scand* 70(3):389–402.

Jordan L.M., Liu J., Hedlund P.B., Akay T., and Pearson K.G. (2008). Descending command systems for the initiation of locomotion in mammals. *Brain Res Rev* 57:183–191.

Kelso J.A. (2009). Synergies: atoms of brain and behavior. *Adv Exp Med Biol* 629:83–91.

Kendall F.P., McCreary E.K., and Provance P.G. (1993). *Muscles: Testing and Function*. Baltimore, MD: Williams and Wilkins.

Kiehn O., Harris-Warrick R.M., Jordan L.M., Hultborn H., and Kudo N., Eds. (1998). Neuronal mechanisms for generating locomotor activity. *Ann N Y Acad Sci* 860:1–573.

Krishnamoorthy V., Latash M.L., Scholz J.P., and Zatsiorsky V.M. (2003). Muscle synergies during shifts of the center of pressure by standing persons. *Exp Brain Res* 152(3):281–292.

Kuo A.D. and Donelan J.M. (2010). Dynamic principles of gait and their clinical implications. *Phys Ther* 90:157–114.

Lacquaniti F., Grasso R., and Zago M. (1999). Motor patterns in walking. *News Physiol Sci* 14:168–174.

Latash M.L. and Anson J.G. (2006). Synergies in health and disease: relations to adaptive changes in motor coordination. *Phys Ther* 86(8):1151–1160.

Lee D.D. and Seung H.S. (1999). Learning the parts of objects by non-negative matrix factorization. *Nature* 401(6755):788–791.

Lev-Tov A., Del Volve I., and Kremer E. (2000). Sacrocaudal afferents induce rhythmic efferent bursting in isolated spinal cords of neonatal rats. *J Neurophysiol* 83(2): 888–894.

Maegele M., Muller S., Wernig A., Edgerton V.R., and Harkema S.J. (2002). Recruitment of spinal motor pools during voluntary movements versus stepping after human spinal cord injury. *J. Neurotrauma* 19:1217–1229.

Marder E. (1991). Modifiability of pattern generator. *Curr Opin Neurobiol* 1: 571–576.

Marsh R.L., Ellerby D.J., Carr J.A., Henry H.T., and Buchanan C.I. (2004). Partitioning the energetics of walking and running: swinging the limbs is expensive. *Science* 303:80–83.

McCrea D.A. and Rybak I.A. (2008). Organization of mammalian locomotor rhythm and pattern generation. *Brain Res Rev* 57(1):134–146.

McGowan C.P., Neptune R.R., Clark D.J., and Kautz S.A. (2010). Modular control of human walking: adaptations to altered mechanical demands. *J Biomech* 43(3): 412–419.

Molinari M. (2009). Plasticity properties of CPG circuits in humans: impact on gait recovery. *Brain Res Bull* 78(1):22–25.

Monaco V., Ghionzoli A., and Micera S. (2010). Age-related modifications of muscle synergies and spinal cord activity during locomotion. *J Neurophysiol* 104(4): 2092–2102.

Mori S., Kawahara K., Sakamoto T., Aoki M., and Tomiyama, T. (1982). Setting and resetting of level of postural muscle tone in decerebrate cat by stimulation of brain stem. *J Neurophysiol* 48:737–748.

Nashner L.M. (1980). Balance adjustments of humans perturbed while walking. *J Neurophysiol* 44(4):650–664.

Nashner L.M. and Forssberg H. (1986). Phase-dependent organization of postural adjustments associated with arm movements while walking. *J Neurophysiol* 55 (6):1382–1394.

O'Donovan M.J., Wenner P., Chub N., Tabak J., and Rinzel J. (1998). Mechanisms of spontaneous activity in the developing spinal cord and their relevance to locomotion. *Ann N Y Acad Sci* 860:130–141.

Olree K.S. and Vaughan C.L. (1995). Fundamental patterns of bilateral muscle activity in human locomotion. *Biol Cybern* 73:409–414.

Orlovsky G.N., Deliagina T.G., and Grillner S. (1999). *Neural Control of Locomotion. From Mollusc to Man*. Oxford, U.K.: Oxford University Press.

Patla A.E., Calvert T.W., and Stein R.B. (1985). Model of a pattern generator for locomotion in mammals. *Am J Physiol* 248(4 Pt 2):R484–494.

Pearson K.G. (2004). Generating the walking gait: role of sensory feedback. *Prog Brain Res* 143:123–129.

Pearson K.G. and Rossignol S. (1991). Fictive motor patterns in chronic spinal cats. *J Neurophysiol* 66:1874–1887.

Pepin A., Norman K.E., and Barbeau H. (2003). Treadmill walking in incomplete spinal cord–injured subjects. 1. Adaptation to changes in speed. *Spinal Cord* 41:257–270.

Pfeifer R., Lungarella M., and Iida F. (2007). Self-organization, embodiment, and biologically inspired robotics. *Science* 318:1088–1093.

Poppele R. and Bosco G. (2003). Sophisticated spinal contributions to motor control. *Trends Neurosci* 26(5):269–276.

Prentice S.D., Patla A.E., and Stacey D.A. (1998). Simple artificial neural network models can generate basic muscle activity patterns for human locomotion at different speeds. *Exp Brain Res* 123(4):474–480.

Scivoletto G., Ivanenko Y., Morganti B., Grasso R., Zago M., Lacquaniti F., Ditunno J., and Molinari M. (2007). Plasticity of spinal centers in spinal cord injury patients: new concepts for gait evaluation and training. *Neurorehabil Neural Repair* 21(4):358–365.

Selionov V.A., Ivanenko Y.P., Solopova I.A., and Gurfinkel V.S. (2009). Tonic central and sensory stimuli facilitate involuntary air-stepping in humans. *J Neurophysiol* 101(6):2847–2858.

Sharrard W.J.W. (1964). The segmental innervation of the lower limb muscles in man. *Ann Roy Coll Surg Engl* 35:106–122.

Shik M.L. (1983). Action of the brainstem locomotor region on spinal stepping generators via propriospinal pathways. In: *Spinal Cord Reconstruction* (Kao C.C., Bunge R.P., and Reier P.J., Eds.), pp. 421–434. New York: Raven Press.

Solopova I.A., Tihonova D.Y., Grishin A.A., and Ivanenko Y.P. (2011). Assisted leg displacements and progressive loading by a tilt table combined with FES promote gait recovery in acute stroke. *NeuroRehabilitation* 29(1):67–77.

Stein R.B., Aoyagi Y., Mushahwar V.K., and Prochazka A. (2002). Limb movements generated by stimulating muscle, nerve and spinal cord. *Arch Ital Biol* 140: 273–281.

Ting L.H. (2007). Dimensional reduction in sensorimotor systems: a framework for understanding muscle coordination of posture. *Prog Brain Res* 165:299–321.

Tresch M.C., Cheung V.C.K., and d'Avella A. (2006). Matrix factorization algorithms for the identification of muscle synergies: evaluation on simulated and experimental data sets. *J Neurophysiol* 95(4):2199–2212.

Wilbourn A.J. and Aminoff M.J. (1998). AAEM minimonograph 32: the electrodiagnostic examination in patients with radiculopathies. American Association of Electrodiagnostic Medicine. *Muscle Nerve* 21(12):1612–1631.

Winter D.A. (1989). Biomechanics of normal and pathological gait: implications for understanding human locomotor control. *J Mot Behav* 21:337–355.

Yakovenko S., Mushahwar V., VanderHorst V., Holstege G., and Prochazka A. (2002). Spatiotemporal activation of lumbosacral motoneurons in the locomotor step cycle. *J Neurophysiol* 87:1542–1553.

Zehr E.P. (2005). Neural control of rhythmic human movement: the common core hypothesis. *Exerc Sport Sci Rev* 33:54–60.

PART II

SIGNAL DETECTION AND CONDITIONING

Neural and muscular signals are used for interfacing humans with external devices in most neurorehabilitation technologies. Part II introduces methods for biosignal detection and preprocessing for interfacing the neuromuscular system. The methods described here are used for the techniques introduced in the rest of the book.

The first four chapters of Part II focus on electrode technology for interfacing nerves, the brain, and muscles. In Chapter 4, Qiao and collaborators review the technologies for nerve interfacing and present the open issues in neural recordings and the potential applications for controlling prosthetic devices. Hofmann and collaborators, in Chapter 5, present electrodes for brain recordings. Specifically, they introduce probes that include both electrodes for neural electrical activity and glass fibers that can provide additional information from optical parameters. The remaining two chapters on interfacing present noninvasive methods. In Chapter 6, Botter and collaborators describe methods for recording and conditioning surface electromyogram (EMG) signals from muscles. They provide a broad overview that covers both classic recording systems (i.e., bipolar EMG) and recent advances, such as high-density surface EMG grids. Guger and Edlinger, in Chapter 7, present noninvasive methods for electroencephalographic (EEG) recording, providing an overview of both electrode types and amplifiers for multichannel EEGs.

Introduction to Neural Engineering for Motor Rehabilitation, First Edition. Edited by Dario Farina, Winnie Jensen, and Metin Akay.

The last four chapters of Part II focus on signal processing methods for conditioning neural and muscular recordings. In Chapter 8, Ge and Farina describe the decomposition of multiunit signals into the contributions of the individual sources, a process referred to as spike sorting. This processing step can be applied to muscle, nerve, and brain recordings for detecting the activities of muscle fibers or neural cells. In Chapter 9, Citi and Micera present a more specific example of signal conditioning for denoising neural signals. The methods presented are based on the discrete wavelet transform. In Chapter 10, Paiva and collaborators address the important issue of quantifying the degree of correlation between spike trains. Methods for quantifying the common input received by neural cells are usually based on time averaging whereas the authors propose the alternative use of ensemble averages. Chapter 11, by Biessmann and collaborators, finishes Part II with a review of unsupervised learning techniques used for neural data analysis, focusing on multimodal signals.

4

REVIEW CHAPTER

PROGRESS IN PERIPHERAL NEURAL INTERFACES

SHAOYU QIAO,[1] KEVIN A. MAUSER,[2] AND KEN YOSHIDA[2,1,3]

[1]*Weldon School of Biomedical Engineering, Purdue University, West Lafayette, Indiana, USA*

[2]*Biomedical Engineering Department, Indiana University–Purdue University Indianapolis, Indianapolis, Indiana, USA*

[3]*Center for Sensory–Motor Interaction, Department of Health Science and Technology, Aalborg University, Aalborg, Denmark*

SUMMARY

Current developments in peripheral neural interfaces are facilitating their translation to clinical applications such as devices for use in human–machine interfaces (HMI) for advanced active prostheses. Despite significant advances, several challenges, such as (i) interface selectivity, (ii) device reach, and (iii) implanted device long-term stability and biocompatibility, still need to be resolved in their translation from experimental to clinical devices. Since the early 1980s, various devices have been developed and implanted in and around the peripheral nerve in both animal models and human subjects. Current devices can be divided into three broad categories using three different design concepts: the extrafascicular approach, the intrafascicular approach, and the nerve transection approach. The first two categories represent two different strategies to meet the three challenges of peripheral neural interface devices without disrupting the nerve; the third category involves transecting and transferring the nerve. In this chapter, we

Introduction to Neural Engineering for Motor Rehabilitation, First Edition. Edited by Dario Farina, Winnie Jensen, and Metin Akay.

will briefly introduce as background the anatomy of the nervous system and then focus mainly on the progress and methods used to address the three challenges just outlined for the extrafascicular and intrafascicular approaches. The state of the art of devices for each approach is briefly discussed. Finally, key advances such as the application of microfabrication techniques and the translation from animal to human are highlighted. Advances in device modeling and cursory investigations of the use of nanotechnologies and the relevance of each in overcoming current issues of the state of the art are also discussed.

INTRODUCTION

Neural interfaces are portals to the nervous system through which we are able to sample, monitor, access, and even interact with a small part of the information coursing through the nervous system. This is accomplished through the placement of electrochemical interfaces/transducers—electrodes—around, between, or within neural tissues. The electrodes themselves are usually deployed on structures or substrates, which secure their location in the tissue and act as the means to make electrical contact to the electrode sites. One or more electrodes can be on each neural interface structure, which are commonly called electrodes. In this chapter, however, we will reserve the term **electrode** to indicate the electrochemical site(s), while referring to the substrate or structure as the **neural interface structure** and to the entire device as the **neural interface device** or **neural prosthetic device**.

In a practical expression of Helmholtz's reciprocity theorem (Helmholtz, 1853), electrodes can be used to detect the bioelectrical activity from active neurons (recording) or can be used as a means to locally inject current to effect changes in the electric potential in the extracellular space to induce activity in neurons (stimulation). Although this chapter focuses on progress in neural prosthetic devices implanted in or around the peripheral nerve, neural prosthetic devices in other locations include sensors for the special senses (e.g., cochlea, retina) and tissues of the central nervous system (CNS) such as spinal cord and brain. Additional reviews and book chapters on the development of devices can be found in Yoshida and Riso (2004), Navarro et al. (2005), and Yoshida et al. (2010).

Background Anatomy

The peripheral nervous system (PNS) is typically defined as the part of the nervous system caudal to the spinal cord and brain. The peripheral nerve is a cable-like structure consisting of nerve fibers (axons), support cells, vessels, and connective tissues. Classically, it is described as a cable held together by a loose layer of connective tissue called the epineurium and consisting of bundles called the nerve fascicles delineated by a tough sheet of connective tissue called the perineurium. The nerve fibers themselves and their support cells are contained within the nerve fascicle (Fig. 4.1).

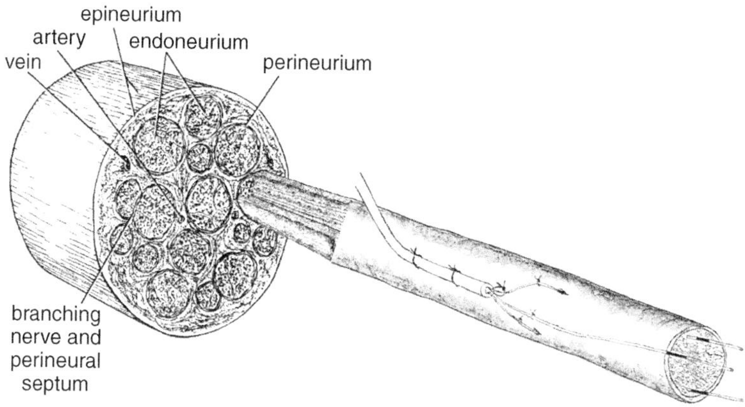

Figure 4.1. Structure of the peripheral nerve and how the nerve fibers are bundled into perineurium ensheathed nerve fascicles. Longitudinal intrafascicular electrodes (LIFEs) are implanted parallel with the nerve fiber and fascicle. (Reproduced with permission from Yoshida et al., 2010.)

Fibers in the peripheral nerve originate centrally in the brain stem, spinal cord, or paravertebral column and project peripherally to the end organ. Information flow peripherally, such as motor commands, is mediated by efferent fibers; information flow centrally, such as sensory information, is mediated by afferent fibers. A single neuron consisting of a cell body (soma), presynaptic branches (axons), and postsynaptic branches (dendrites) contains information to and from a single effector or sensor. It is the smallest unit or channel of information in the nervous system and is called a **single unit**.

In the somatic nervous system where sensory–motor rehabilitation neuroprostheses target efferent fibers (axons), fibers emerge from the ventral horn of the spinal cord where their somas aggregate and cluster to form ventral roots from each spinal cord segment. The cell bodies of the sensory units reside in the dorsal root ganglion (DRG). These cells are pseudounipolar and have a modified axon, which bifurcates at the DRG with one branch projecting centrally to the dorsal horn of the spinal cord through the dorsal roots, and the other branch projecting peripherally to the sensory end organ. The dorsal root and ventral root bundles aggregate to form the spinal nerves as they emerge from the spinal column. Once the spinal nerves emerge from the spinal cord, they initially aggregate but branch as they course peripherally to their end organs.

The spinal cord is roughly topologically organized (Prochazka et al., 2002) such that units and neural circuitry related to similar parts of the body are clustered together. Thus, sensory and motor information from and to similar parts of the body cluster as they emerge from the dorsal and ventral cord, respectively, and from each spinal cord segment. This topological organization appears to be preserved in the peripheral nerve such that units projecting to similar locations in the body are bundled together within the nerve trunk and nerve fascicles (Sunderland, 1978; Gustafson et al., 2006).

Most of the processing power of the nervous system lies in the massive interconnection of the neural-networks circuitry and their synaptic modulation. The PNS, generally, only consists of bundles of axons and the end organs and neuromuscular synapses to which they project. The axons do not influence each other, except in pathological cases, so the information carried in each can be considered isolated from information in the others. The peripheral nerves can be considered conduits of information—information containing the inputs to and outputs from the CNS—and, therefore, almost all interactions to and from the environment are conveyed to us through the PNS. Thus a means to tap or modify that information would give us the ability to directly interact with the sensations and motor commands, as well as study how the CNS uses and processes information. Peripheral neural prosthetic devices are the means by which we are able to interact with the body.

Advantages and Challenges

There are several advantages to implanting devices in or around the peripheral nerves. Peripheral nerves are not encased in bone and have relatively easy surgical access. Injury to the nerve only injures the axon and does not injure the cell body, leaving the processing power of the CNS untouched and intact. Moreover, the axons have regenerative capacity and can regenerate if injured during implantation or by the implant. With a peripheral device infections introduced during implantation are also peripheral, meaning that the devices can be easily removed and allowing for the infection to be resolved in time. In contrast, central neural prosthetic devices cross the meningeal barrier and any infections could quickly become life threatening. Thus, peripheral devices are clearly less invasive and are a less radical intervention than devices for the CNS.

It is important to indicate that, anatomically, these nerve bundles typically contain many hundreds of units and are confined within one or more nerve fascicles. The units a nerve bundle contains might project to vastly different parts of the body or limb and not only contain a mixture of afferent and efferent fibers, but also project to end organs that have vastly different functions. Ultimately, the overall information that the nerve contains is mixed. For example, motor nerves, whose nerve bundles lead to muscles, are mixed and contain both sensory and motor fibers; paradoxically, they contain more afferent units than efferent units (Eccles and Sherrington, 1930). Interestingly, this mixture of information generally becomes more and more specific as one moves down to smaller and smaller nerve bundles, and closer to the target regions. For example, the information of nerve bundles of the sciatic nerve at the level of the hip contains information projecting to and from the entire leg, while smaller branches of the sciatic nerve near specific muscles contain specific information to and from that muscle. Thus, placing neural interface devices in the peripheral nerve has its own set of challenges.

A major challenge for neural prosthetic devices is to be sufficiently *selective* and thus be able to interface to meaningful clusters of the entire nerve bundle population so that the information drawn or added is sufficiently specific, coherent, and useful. Balancing the need for higher selectivity of the device (i.e., influencing smaller populations of fibers) is the need to be able to *reach* and span a sufficiently large portion of the entire information space contained in the nerve (i.e., influence as large a population of fibers as necessary) to make a difference.

Residing outside of the bony protection encasing the CNS, peripheral nerve bundles are generally unprotected by hard tissues, but rather are surrounded by muscles and other bioelectrically active tissues. This has specific consequences in terms of recording activity and biocompatibility in the PNS. Electrodes must contend with biological noise from other bioelectrically active tissues such as the heart and skeletal muscles since the interaction between electrodes and bioelectrical activity is not restricted to peripheral nerves. This becomes a particular challenge in the case of recording where signals from the nerves are in the tens of microvolts, whereas signals picked up from nearby active muscles and the heart can be one or more orders of magnitude larger (Popovic et al., 1993). Thus recording from the PNS leads to a major consequence in the signal-to-noise ratio (SNR) of the neural interface.

Finally, the PNS responds far more aggressively to foreign bodies compared with the CNS. This is a consequence of the need of the PNS to repair itself and retain connectivity in response to the wear and tear that it sustains from its environment due to suboptimal biocompatibility. Neural prosthetic devices are aggressively encapsulated in ensheathing tissue, leading to decreased chronic performance in selectivity. The need for robustness extends to peripheral neural prosthetic devices and their support structures. Thus devices must be robust yet modified to increase their *biocompatibility*.

The remainder of the chapter is devoted to a review of advances and progress in peripheral nerve prosthetic devices and techniques that address the three challenges just described: interface selectivity, device reach, and biocompatibility.

STATE OF THE ART

Peripheral neural prosthetic devices are typically deployed on nerve trunks or branches, placing them midaxonally relative to the neuronal unit. Since the early 1980s, various devices have been developed for use as neuroprostheses and have been implanted in and around the peripheral nerve of animal models or human subjects either acutely or chronically. Current devices can be divided into three broad categories:

1. *Extrafascicular Approach.* Extraneural recordings are made using extrafascicular electrodes such as the cuff electrode (Stein et al., 1975; Hoffer and Loeb, 1980; Naples et al., 1988; Veraart et al.,1993; Grill and Mortimer, 1996; Fisher et al., 2008, 2009; Polasek et al., 2009a,b) and the flat interface nerve electrode (FINE; Tyler and Durand, 2002; Leventhal and Durand, 2003; Tyler and Durand, 2003; Yoo et al., 2004; Yoo and Durand, 2005; Leventhal et al., 2006; Schiefer et al., 2006; Lertmanorat et al., 2009; Schiefer et al., 2009, 2010) that do not penetrate but wrap around the circumference of the nerve bundle.
2. *Intrafascicular Approach.* Intrafascicular recordings are made using intrafascicular electrode arrays such as the longitudinal intrafascicular electrode (LIFE; Malagodi et al., 1989; Goodall et al., 1991; Lefurge et al., 1991; Yoshida and Horch, 1993a,b; McNaughton and Horch, 1996; Yoshida and Horch, 1996; Malmstrom et al., 1998; Yoshida and Stein, 1999; Yoshida et al., 2000a,b; Lawrence et al., 2002; Dhillon et al., 2004; Lawrence et al., 2004; Dhillon and Horch, 2005; Dhillon et al., 2005; Yoshida et al., 2006; Bossi et al., 2007; Lago et al., 2007; Yoshida et al., 2007; Farina et al., 2008; Micera et al., 2008; Bossi et al., 2009; Benvenuto et al., 2010), the transverse intrafascicular multichannel electrode (TIME; Boretius et al., 2010; Jensen et al., 2010; Kundu et al., 2010; Badia et al., 2011a,b), and the Utah electrode array (UEA) and the Utah slanted electrode array (USEA) (Branner and Normann, 2000; Branner et al., 2001; Aoyagi et al., 2003; Branner et al., 2004; McDonnall et al., 2004; Stein et al., 2004a,b; Weber et al., 2006, 2007; Dowden et al., 2009) that, in contrast, penetrate the nerve fascicle but not the nerve fiber.
3. *Nerve Transection Approach.* This describes those techniques that require transectioning of the nerve and taking advantage of the regenerative capacity of the nerve (Kuiken et al., 2004; Hijjawi et al., 2006; Kuiken et al., 2007a,b; Miller et al., 2008a,b; O'Shaughnessy et al., 2008; Kuiken et al., 2009; Marasco et al., 2009; Schultz et al., 2009). This approach is discussed later in this volume in Chapter 15 ("Myoelectric Prostheses and Targeted Reinnervation").

The extrafascicular and intrafascicular approaches, the first two categories of neural prosthetic devices just defined, represent two different strategies to meet the three challenges for PNS neural interface devices without disrupting the nerve. To provide a broader overview of the usage of extrafascicular and intrafascicular devices, the characteristics and applications of these devices are summarized and compared in Table 4.1.

We review the extrafascicular and intrafascicular approaches by providing a brief synopsis of the state of the art of peripheral neural prosthetic devices found within each approach while also highlighting key advances such as the application of microfabrication techniques and the translation from animal to human. Advances in device modeling and cursory investigations of the use of

TABLE 4.1. Overview of the Characteristics and Applications of Extrafascicular and Intrafascicular Neural Prosthetic Devices

Category	Electrodes	Electrode Base Materials	Design Features	Target Sites	Experiment Subjects
Extrafascicular electrodes	Cuff	Silastic tube	Teflon-coated Ag and Pt–Ir wire No. of channels, 3 or more	Hind limb, sciatic, LG, tibial, and sural nerves	Chronic cat (Stein et al., 1977)
Low selectivity and invasiveness	Spiral Cuff	Silicone sheeting Ovoid sites of platinum foil	Inner diameter, 6–10 mm No. of channels, 4 Sites evenly spaced	Femoral nerve	Human with femoral–popliteal bypass surgery (Polasek et al., 2009b)
				Radial and musculocutaneous nerves	Chronic human upper extremity (Polasek et al., 2009a)
	Multi-contact cuff	Silicone rubber; platinum sites	Ratio of cuff diameter to nerve diameter, 0.72–0.8 No. of channels, 12	Sciatic nerve	Acute cat (Grill and Mortimer, 1996)
	FINE	Silicone elastomer; platinum foil sites	Height of central opening, 1–1.5 mm Width of central opening, 10 mm Wall thickness, 2 mm Size of sites, 0.5 × 0.5 × 0.05 mm No. of channels, 8–12, 4, or 6 evenly distributed on the top and bottom	Sciatic nerve	Chronic rat (Tyler and Durand, 2003) Chronic cat (Leventhal et al., 2006) Acute cat (Tyler and Durand, 2002)
				Hypoglossal nerve	Acute beagle (Yoo et al., 2004)
				Femoral nerve	Human (Schiefer et al., 2010)

(Continued)

TABLE 4.1. (*Continued*)

Category	Electrodes	Electrode Base Materials	Design Features	Target Sites	Experiment Subjects
Intrafascicular electrodes High selectivity and invasiveness	LIFE	Teflon-insulated 90%Pt–10%Ir wire	Electrode diameter, 25 μm No. of channels, 1 Handmade	Sciatic, ulnar, and radial nerves	Acute cat (Malagodi et al., 1989) Chronic cat (Lefurge et al., 1991)
		Teflon-insulated 90%Pt–10%Ir wire	Electrode diameter, 25 μm Length of site, 0.5–1 mm	Sciatic nerve	Acute cat (Yoshida and Horch, 1993a,b)
			Length of tungsten needle, 25 mm	Radial nerve	Acute cat (McNaughton and Horch, 1996)
			Diameter of needle, 50 μm No. of channels, 2 Two wires glued to one needle for implant Handmade	LG/S nerve and/or MG nerve	Acute cat (Yoshida and Stein, 1999)
	PolyLIFE	Kevlar filament; Pt–PtCl sites	12-μm-diameter Kevlar fiber metalized with three layers (Ti, Au, and Pt)	Tibial and medial gastrocnemius branches of sciatic nerve	Acute rabbit (Lawrence et al., 2004)
			Length of site, 0.5–1 mm No. of channels, 1 Handmade	Sciatic nerve	Acute rat (Lawrence et al., 2004) Chronic rabbit (Lawrence et al., 2002)
			First human implants of LIFEs	Detached nerves 3–8 cm from the distal end of the nerve stump	Human amputee (Dhillon et al., 2004; Dhillon and Horch, 2005; Dhillon et al., 2005)

tfLIFE	Polyimide; gold sites	Size of site, 50 μm × 50 μm Intersite spacing, 2 mm No. of channels, 4 Hand-assembly to leadout wire; first microfabricated tfLIFE	MG nerve	Acute rabbit (Yoshida et al., 2000b)
	Polyimide; Pt–PtCl sites	Site diameter, 40 μm No. of channels, 8 Large ceramic interconnect	Sciatic nerve	Chronic rat (Lago et al., 2007)
	Polyimide; Pt–PtCl sites	Site diameter 40 μm No. of channels, 8 Decreased interconnect size	Sciatic nerve Median and ulnar nerves	Acute rabbit (Yoshida et al., 2007) Human amputee (Rossini et al., 2010)
Muscle LIFE	Polyimide; Pt–PtCl sites	Width, 200 μm; length, 1.5 cm Thickness, 10 μm Intersite spacing, 1500 μm Site diameter, 40 μm No. of channels, 8–10	Gastrocnemius muscle	Acute rabbit (Farina et al., 2008)
TIME	Polyimide; Pt or Pt-black sites	Intersite spacing, 230 μm Site diameter, 60 μm Ground site length, 400 μm No. of channels, 16 Microfabricated corrugated structure	Sciatic nerve Median nerve ~2–3 cm above the elbow	Acute rat (Boretius et al., 2010; Badia et al., 2011a) Chronic rat (Badia et al., 2011b) Acute pig (Kundu et al., 2010)

(*Continued*)

TABLE 4.1. (*Continued*)

Category | Electrodes	Electrode Base Materials	Design Features	Target Sites	Experiment Subjects
UEA/ USEA	Silicon substrate; platinum-plated tips	Electrode array, 5 × 10 to 10 × 10 Length of electrodes, 0.5–1.5 mm, 0.1 mm difference in length between rows of adjacent electrodes Interelectrode spacing, 400 μm No. of channels, 50–100	Sciatic nerve	Acute cat (Branner et al., 2001; Aoyagi et al., 2003; McDonnall et al., 2004; Dowden et al., 2009; Frankel et al., 2011) Chronic cat (Branner and Normann, 2000; Branner et al., 2004)
			DRG	Acute cat (Stein et al., 2004b; Weber et al., 2007; Wagenaar et al., 2011; Bruns et al., 2011) Chronic cat (Weber et al., 2006; Rigosa et al., 2011)

LG, lateral gastrocnemius nerve; LG/S, lateral gastrocnemius soleus nerve; MG, medial gastrocnemius nerve.

nanotechnologies and the relevance of each in overcoming current issues of the state of the art are also discussed.

Extrafascicular Approach

The strength of the extrafascicular approach is its relative simplicity, robustness, and ability to reach a large part of the nerve. The archetype of this approach is the cuff electrode: the electrodes are placed on the inside of an insulating tube, which is placed around the nerve trunk. This places the electrodes in contact with the nerve epineurium. If sized properly (Naples et al., 1988), the nerve cuff has excellent biocompatibility and stability. It is considered a mature technique and has had extensive use in human applications.

Extrafascicular: Translation from Animal to Human. Since the late 1990s, the nerve cuff electrode has been translated to human studies in applications of vision (Veraart et al., 1998; Delbeke et al., 2003; Veraart et al., 2003; Brelen et al., 2005), lower-extremity (Fisher et al., 2008, 2009; Polasek et al., 2009b), and upper-extremity (Polasek et al., 2009a) neuroprostheses. A 16-channel implanted stimulator–telemeter (IST-16) system integrated with the Case Western Reserve University (CWRU) spiral nerve cuff electrodes, for example, has shown its ability to generate sufficient knee extension moment for prolonged standing by stimulating the femoral nerves of a patient with spinal cord injury (SCI) at T6 level (Fisher et al., 2008). A subsequent 63-week human study (Fisher et al., 2009) and an intraoperative evaluation on the human femoral nerve trunk (Polasek et al., 2009b) demonstrated that the CWRU spiral nerve cuff electrode can be used as a chronically stable peripheral neural interface for selectively activating specific muscle groups for lower-extremity neuroprostheses. The nerve cuff electrodes were also implanted in upper-extremity human peripheral nerves to restore or enhance shoulder, elbow, and hand functions (Polasek et al., 2009a). This study showed that the implanted nerve cuff electrodes do not cause adverse physiological changes in the nerve or adverse sensory effects up to three years, and the selectivity was increased using field steering techniques.

Despite the ability of the cuff electrode to provide a relatively stable and selective recording and stimulation, the major challenges of recording-and-stimulation selectivity and recording signal amplitude limit its widespread utility with multichannel human–machine interfaces (HMI). The FINE (Fig. 4.2) is a novel structure derived from the cuff electrode, designed to overcome many of the limitations of the classic cuff electrode. Following implantation, it reshapes the peripheral nerves into a flat geometry, increasing the surface area and redistributing axonal populations in the center of the nerve to its surface (Tyler and Durand, 2002). The reshaping increases the selectivity of the interface. Animal studies have demonstrated selective stimulation (Tyler and Durand, 2002) and recording (Yoo and Durand, 2005) with FINE and also showed that reshaping the nerve with a small chronically applied force does

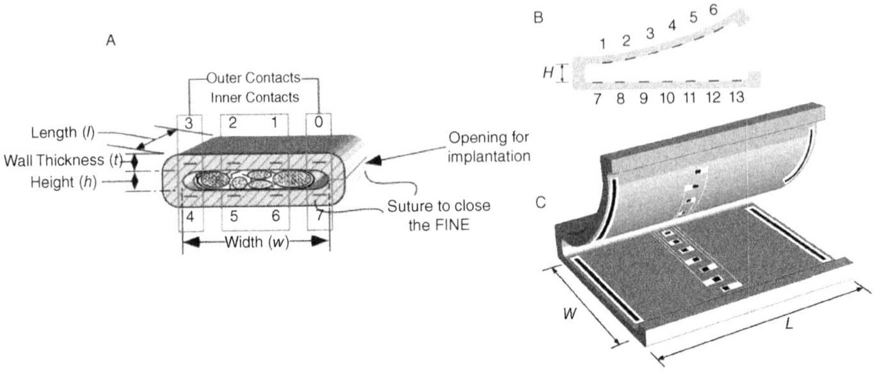

Figure 4.2. FINE design concept. (A) Schematic cross section of a FINE on a nerve (reproduced with permission from Tyler and Durand, 2003). (B) Cross section of the FINE with all 13 recording positions (i.e., cathodes) and height of electrode opening labeled 1–13 and *H*, respectively (reproduced with permission from Yoo and Durand, 2005). (C) Rotated drawing of the FINE showing the spot-welded platinum pieces (white) and the exposed cathodic (0.5 mm × 0.5 mm) and anodic (black; 0.5 mm × 6.5 mm) contacts. Tripolar recording is achieved by measuring the differential voltage between the center cathode and the electrically shorted anodes. The width (*w*) and length (*l*) of the space inside the FINE are labeled accordingly (reproduced with permission from Yoo and Durand, 2005).

not cause nerve damage (Tyler and Durand, 2003; Leventhal et al., 2006). A human study by Schiefer and co-workers (Schiefer et al., 2010) showed that a single FINE with eight contacts placed around the femoral nerve trunk is able to selectively recruit specific muscles innervated by the femoral nerve, and at least one hip flexor and two knee extensors can be selectively activated in each human subject, indicating that a multichannel FINE implanted on the femoral nerve could provide a promising solution to lower extremity neuroprostheses.

Intrafascicular Approach

The intrafascicular approach places the electrode contact within the nerve fascicle in intimate contact with the endoneural space and nerve fibers. The advantage of the intrafascicular approach is its relatively high selectivity and signal amplitude. This is achieved at the cost of greater invasiveness, sensitivity to electrode movement, and surgical skill required for implanting the device (Navarro et al., 2005; Yoshida et al., 2010).

The LIFE is an implantable intrafascicular microelectrode (Fig. 4.1) that has been under development as a PNS neural interface since the late 1980s (Malagodi et al., 1989; Goodall et al., 1991; Yoshida and Horch, 1993b). The electrodes were fabricated from Teflon-insulated 25-μm-diameter, 90% Pt–10% Ir microwires, where the active stimulation/recording site was produced by manually deinsulating a 0.5- to 1-mm-long Pt–Ir section and electrodepositing platinum black. Since the original devices were handmade, interelectrode variations contributed to differences in electrode selectivity and contact impedance mismatch, resulting in degradation of the recording SNR. Moreover, since electrode site selectivity is inversely proportional to its range of influence, the increased selectivity of the electrode comes at the cost of decreased reach of the device. The effect of mechanical flexibility to reduce the mechanical mismatch between the electrode and endoneurium was explored following the implementation of a polymer-based electrode, the polyLIFE (Malmstrom et al., 1998; Lawrence et al., 2002, 2004). The results from these studies showed that these single-electrode intrafascicular devices can record and resolve single units (Malagodi et al., 1989; Goodall et al., 1991; Goodall and Horch, 1992; McNaughton and Horch, 1994) and selectively stimulate subsections of the nerve (Yoshida and Horch, 1993b). However, to reach and span the extent of the fibers in the nerve trunk would require multiple distributed implants of unitary sited devices.

Microfabrication. To address some of the shortcomings of the initial intrafascicular devices, techniques of microfabrication were implemented, which led to the development of electrode array structures such as the USEA, thin-film LIFE (tfLIFE), and TIME. Groups such as Anderson's and Kipke's groups at the University of Michigan (BeMent et al., 1986; Drake et al., 1988; Anderson et al., 1989; Hoogerwerf and Wise, 1994; Bai et al., 2000; Rousche et al., 2001; Kipke et al., 2003; Ludwig et al., 2006; Seymour and Kipke, 2007) and Normann's laboratory at the University of Utah (Campbell et al., 1989, 1991; Nordhausen et al., 1996; Rousche and Normann, 1998) first applied the techniques of microfabrication to cortical neural prosthetics. Benefits gained by microfabricating devices include the ability to implement devices with precisely defined microstructures, the ability to achieve cleanliness, uniformity, and reproducibility of structures, the ability to add processes in the clean room, and the potential to add active circuitry directly on devices.

Intracortical devices, like intrafascicular devices, rely on high-selectivity electrodes to attempt to interface with small populations of single units. Like intrafascicular devices they extract and input high-resolution information from and to these single units in order to give detailed information to and from a very small part of the cortical representation. Thus, a way to distribute many high-resolution sites over as much of the cortical representation as possible was required. Microfabrication techniques married the ability to reproduce multiple microscaled details with the ability to regularly distribute these details across a relatively large area. The first applications of microfabricated devices to the peripheral nerve appeared nearly a decade later with the development of the USEA and tfLIFE, but only after many of the initial technical difficulties with the cortical devices were overcome.

UEA and USEA. The UEA and its variation, the USEA, came directly from the Utah intracortical array program (Campbell et al., 1991) as an application for the peripheral nerve (Branner and Normann, 2000; Branner et al., 2001). Constructed by micromachining silicon, a three-dimensional (3D) array of microneedles is implemented. Such structures were developed for use in peripheral nerves (Branner and Normann, 2000) and DRG (Aoyagi et al., 2003). Animal studies have shown that higher stimulation selectivity with much lower currents were achievable by UEA or USEA in peripheral nerves compared with cuff electrodes (Branner and Normann, 2000; Branner et al., 2001) and that more graded recruitment of individual muscles could be realized using USEA with microelectrodes of varying length (Branner et al., 2001). Additionally, a study showed that USEAs could also be used to block selected muscles independently of activating other muscles, using high-frequency alternating current (HFAC) (Dowden et al., 2010). USEA chronically implanted in cat sciatic nerve provided relatively stable stimulation but unstable sensory recordings yet did not cause nerve damage around the implant, based on histological examination (Branner et al., 2004). Unstable sensory recordings were due to only 10–20% of the electrodes on the array actually recording the sensory signals (Branner and Normann, 2000; Branner et al., 2001). To resolve this drawback, researchers found that DRG is an attractive location (Aoyagi et al., 2003) due to larger electrical activities generated by the somas of sensory nerve fibers and better protection from signal contamination. Microstimulation applied through the UEA in the DRG, which could provide selective recruitment of afferent fibers from a range of sensory modalities (Gaunt et al., 2009), has been demonstrated as one approach to convey natural tactile and kinesthetic sensations by encoding and decoding the information of sensory feedback (Stein et al., 2004a,b; Weber et al., 2006). Several models using DRG recordings have been proposed to predict the kinematic state of the hind limb by decoding the afferent neuron firing rate (Stein et al., 2004b; Weber et al., 2006, 2007; Wagenaar et al., 2011). The integrated algorithm based on DRG recordings for predicting the state feedback information would play an important role in providing stable and adaptive movements of extremities.

In human studies, the UEA was implanted in the median nerve fibers of a healthy human subject for 3 months (Warwick et al., 2003). It is reported that the blindfolded subject was able to receive feedback information from force and slip sensor on the prosthetic hand and subsequently control the hand using the implanted device to grasp an unseen object. The subject was also able to control an electric wheelchair using decoded signals from the implanted UEA to determine the intended direction. Intention was signaled by opening and closing his hand. Although the implant was removed eventually due to gradual degradation and mechanical fatigue of the electrode wire bundle, this study provided proof that the UEA has a potential to be used as a bidirectional PNS neural interface.

tfLIFE and TIME. The tfLIFE is a micropatterned thin film on polymer film structure (Fig. 4.3) developed to overcome the shortcomings of LIFE devices.

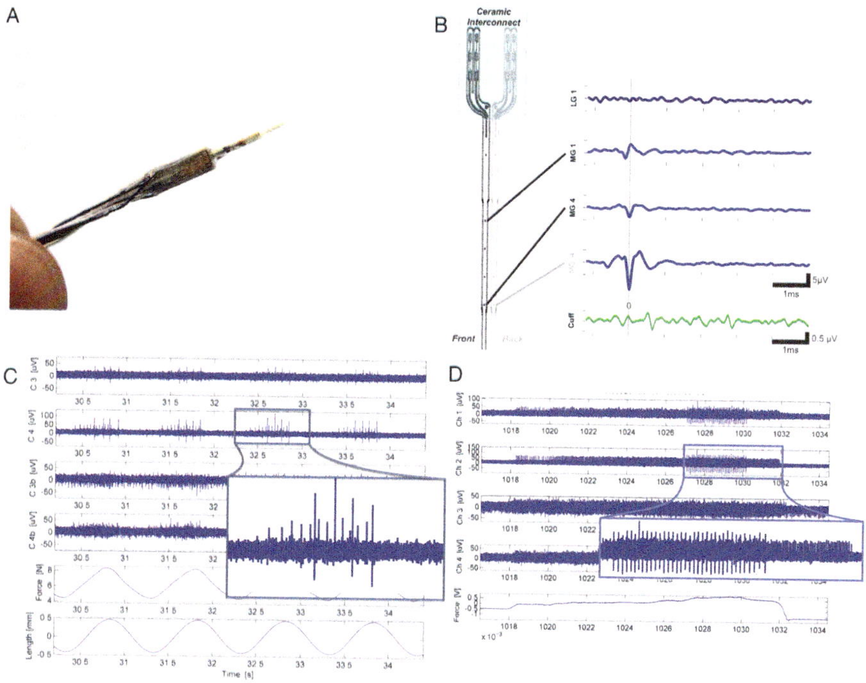

Figure 4.3. Thin-film LIFE (tfLIFE). (A) The tfLIFE is a microfabricated, two-sided structure containing eight intrafascicular active sites. The third-generation tfLIFE is shown here. (B) Effect of position and side on the amplitude of the simultaneously recorded unit. The back side contact has the largest amplitude. The spike recorded at MG 1 leads those recorded at MG 4 and MG –4 indicating propagation of the unit from distal to proximal. (C) Example showing four tfLIFE channels recording proprioceptor activity from the MG nerve in response to stretching the MG muscle with a sinusoidal stretch. (D) Response of two different units responding to a slow ramp stretch. Panels (C) and (D) appear courtesy of M. Kurstjens (Aalborg University). (Reproduced with permission from Yoshida et al., 2010.)

It came with the discovery that flexible planar thin films could be implanted in the nerve fascicles (Yoshida et al., 2000b) without tearing the substrate film or injuring the nerve. The technique is described in detail in Lago et al. (2007). This demonstration opened the way to the possibility that planar flexible thin-film micropatterned structures could be developed and implanted without damage to the nerve or structure. The result was the micropatterned and fabricated tfLIFE (Bossi et al., 2007; Lago et al., 2007; Bossi et al., 2009) and TIME (Boretius et al., 2010) electrode arrays. Microfabrication through microelectromechanical systems (MEMS) allows implementation of arrays of microelectrode sites distributed along a flexible polymer microribbon cable. Thus, a single implant could place multiple sites distributed within the nerve trunk to achieve multiple high-selectivity access points spanning the range of the nerve. Developments of the tfLIFE and TIME arrays are described further in Bossi et al. (2007), Lago et al. (2007), Bossi et al. (2009), Boretius et al. (2010), Jensen et al. (2010), and Badia et al. (2011a,b). Some of the challenges include the increased cost to implement prototypes, uncertainties about the reliability of devices, and robustness of thin-film conductors.

Intrafascicular: Translation from Animal to Human. The family of LIFEs is a candidate neural interface for use in direct neural-controlled cybernetic prosthetic devices and several studies have been undertaken using LIFEs in human subjects (Dhillon et al., 2004; Dhillon and Horch, 2005; Dhillon et al., 2005; Rossini et al., 2010). Dhillon and Horch (Dhillon et al., 2004; Dhillon and Horch, 2005; Dhillon et al., 2005) reported that LIFEs implanted in long-term human amputee subjects ranging from 0.83 to 30 years (average duration ~7.3 yr) during short-term (2 wk) implantation were able to induce different sensations by stimulating afferent nerves, and tactile and proprioceptive sensations could be discriminated, whereas efferent signals were used to control a one-degree-of-freedom (DOF) prosthesis. A four-week human study by Rossini et al. showed evidence that tfLIFEs implanted in the human amputee were able to discriminate patterns of recorded nerve signals for controlling independent types of hand grip as well as elicit sensations (Rossini et al., 2010). Thus, several DOFs of a prosthetic limb could be controlled using the nerve signals recorded from current LIFE and tfLIFE designs.

In addition to the application to neural recordings, the concept of LIFEs used for intramuscular electromyogram (iEMG) recordings has been demonstrated (Farina et al., 2008). A prototype of thin-film wire electrode array for multichannel iEMG recordings (called muscle LIFE) was developed in order to achieve concurrent EMG recordings from multiple locations inside the muscle using a single wire electrode array. The prototype thin-film wire electrode array for multichannel iEMG recordings adopts the design concept of the thin-film system developed for nerve recordings (i.e., tfLIFE; Farina et al., 2008). This prototype was tested in six acute experiments on the medial head of the gastrocnemius muscles of rabbits, showing that asynchronous motor unit activities can be detected by the muscle LIFE during cyclical crushes of the sciatic nerve and many motor units can be identified from these highly selective recordings (Farina et al., 2008). Although the noise level does not significantly vary over 40-min intervals in this study, the chronic characterization of this prototype system in terms of selectivity still needs to be further investigated. Additionally, a study by Kamavuako and co-workers (Kamavuako et al., 2009) showed that the grasping force can be estimated by the features of the single-channel iEMG signals and the iEMG signals can be applied for proportional control of prosthetic devices. Compared with the widely used surface EMG (sEMG) signals in the myoelectric prostheses (DARPA; Otto-Bock), which is limited to fewer DOFs, the iEMG signals recorded by the multichannel LIFE structure can theoretically provide independent and multiple-DOFs control for advanced active prosthetic limbs.

CURRENT CHALLENGES AND FUTURE DIRECTIONS

The stability and biointegration of implanted devices, particularly of microdevices, remains a challenge for the future. With developments in biomaterials, optimization and iteration of the design to improve device stability and improved biocompatibility through the integration of newer and more-biocompatible materials is a new horizon facing peripheral neural prosthetics to help with current challenges. Some of the current challenges are being met by research into more realistic computational models and by the development and integration of nanomaterials and bioactive materials.

Integration of Computational Modeling

The translation of these technologies toward use in humans and the use of microfabrication techniques are associated with increased costs. In the case of microfabrication, every redesign has associated with it significant costs of making new masks and adapting processes. In the case of human work, increased regulatory load and safety concerns limit the number of changes that can be made to the device once a protocol is approved. Even in work with animal models, the ethical responsibility and pressure to reduce the number of animals used limits the number of designs that can be tested. There is less room to modify devices and search for more optimal designs and implementations.

The continuous increase in the performance of the average desktop computer and the availability of relatively inexpensive but powerful modeling packages have brought the possibility of creating realistic computational models to the field. Design guidelines can be developed using these tools to minimize the number of iterations implemented and increase the successful design of newer interfaces. Chief among these methods is finite element method (FEM) modeling.

The FEM has been widely used in modeling the selective neural interfaces in volume conductors. FEM models that predict the bioelectrical phenomena of the electrode in given environments will improve the understanding of the underlying biophysical principles for the electrode and serve as a design tool for future generations of selective neural interfaces. Model validation will eventually reduce the number of animal trials needed to arrive at an optimized electrode design to be used in preclinical and clinical trials. As models become more and more refined, they will begin to play a larger role in electrode design and optimization. Several integrated computational models of extrafascicular electrodes (cuff electrode, FINE) and intrafascicular electrodes (LIFE, tfLIFE, TIME) are summarized in Table 4.2.

TABLE 4.2. Summary of the Representative Integrated Computational Models of Extrafascicular and Intrafascicular Neural Prosthetic Devices

Electrode	Study	Aim	Model/Major Electrical Properties[a]	Findings/Limitations
Cuff	(Choi et al., 2001) FEM model for selectivity of multicontact cuff electrodes	Compare the selectivity of the round and flat cuff electrodes	3D volume conductor models of the nerve bundle wrapped with round and flat cuff structures Perineurium, $\rho = 47.8$ Epineurium, $\rho = 1.211$ Saline, $\rho = 0.05$ Silicon rubber, $\rho = 1 \times 10^9$ Endoneurium: $\rho = 1.211$ (transverse); $\rho = 0.175$ (longitudinal) Axon diameter, fascicle diameter, and position of the nodes of Ranvier are randomly distributed.	*Findings:* (a) Selectivity index (SI) was introduced; (b) reshaping the nerve into a flat configuration is possible to increase the functional selectivity of a nerve cuff; (c) the selectivity using the flat shape could be significantly improved by increasing the contact density around the nerve cuff. *Limitations:* (a) Some confounding variables such as fascicle shape are ideally eliminated in the model; (b) a realistic model could change the predicted results.
FINE	(Yoo and Durand, 2005) FEM model for recording selectivity of canine hypoglossal nerve using FINE	Investigate the capability of FINE to achieve selective recordings at various physiological SNRs	A 3D model of a beagle hypoglossal nerve with an implanted FINE Endoneurium: $\sigma = 0.0826$ (transverse); $\sigma = 0.571$ (longitudinal); Perineurium, $\sigma = 0.0021$ Epineurium, $\sigma = 0.00826$ Silicon, $\sigma = 1 \times 10^{-7}$ Saline, $\sigma = 2$	*Findings:* (a) Recording selectivity of FINE depends on the distance between the active sources; (b) active contact density influences the recording SI; (c) the minimum interfiber distance selectively recorded with the FINE can be estimated. *Limitations:* naturally generated neural activity (SNR < 3 dB) and signal extraction algorithm should be further investigated.

FINE	(Schiefer et al., 2008) FEM model for selective activation of the femoral nerve with FINE for lower extremity neuroprosthesis	Investigate the performance of a FINE structure on the proximal femoral nerve for standing and stepping applications	3D realistic models of proximal human femoral nerves (containing 47, 25, and 22 fascicles, respectively) developed from digitized cross sections from cadaver femoral nerve specimens. FINE opening heights were 3.8, 2.3, and 1.4 mm. Four general FINE configurations in terms of the number of active contacts: 22 contacts (11 on top and 11 on bottom); 11 contacts (only on either the upper or the lower surface of FINE); 8 contacts (4 on the upper and 4 on the lower surfaces); 6 contacts (3 on the upper and 3 on the lower surfaces). Endoneurium: $\sigma = 0.0826$ (transverse); $\sigma = 0.571$ (longitudinal) Perineurium, $\sigma = 0.002$ Epineurium, $\sigma = 0.0083$; saline, $\sigma = 2$	*Findings:* (a) Possible to selectively stimulate the six muscles innervated by the femoral nerve using a FINE; (b) 8-contact FINE (contacts on the upper surface are staggered relative to those on the lower surface) could achieve sufficient selectivity with fewer contacts for standing and gaiting prosthesis (stimulate 60% muscles above the threshold and produce 60% of hip flexion moment). *Limitations:* (a) Equal axonal fraction to percentage muscle activation was assumed; (b) the costs of each muscle recruited should be weighted by its contribution to the functional output; (c) encapsulation effect was ignored.

(*Continued*)

TABLE 4.2. (***Continued***)

Electrode	Study	Aim	Model/Major Electrical Properties[a]	Findings/Limitations
LIFE, tfLIFE	(Chemineau et al., 2004) FEM model for selectivity of LIFE and tfLIFE	Investigate and compare the spatial selectivity of LIFE and tfLIFE	3D volume conductor models of LIFE and tfLIFE structures residing in the nerve fascicle Fascicle: $\sigma = 0.083$ (transverse); $\sigma = 0.06$ (longitudinal) Perineurium, $\sigma = 0.0034$ Medium, $\sigma = 2$	*Findings:* A significant improvement in spatial nerve recording selectivity was seen with the tfLIFE compared with the traditional LIFE. *Limitations:* (a) Nerve fibers are assumed to be parallel to the electrode; (b) electrode structures and the active site were placed in the center of the model.
TIME	(Raspopovic et al. 2011) FEM model for performance of TIME implanted in rat sciatic nerve	Investigate the influence of electrode geometry and positioning for new intraneural TIMEs	3D realistic FEM model of TIME with 10 active sites implanted in rat sciatic nerve integrated with field-neuron model. Endoneurium: $\sigma = 0.0826$ (transverse); $\sigma = 0.57$ (longitudinal) Perineurium, $\sigma = 0.0021$ Epineurium, $\sigma = 0.0083$ Polyimide, $\sigma = 6.67 \times 10^{-14}$ Saline, $\sigma = 2$	*Findings:* (a) TIME can provide highly selective stimulation; (b) TIME can selectively activate a different axonal population with different functions; (c) charge required for activation is significantly lower than for extraneural electrodes. *Limitations:* (a) Encapsulation of the electrode was ignored; (b) depolarizing afterpotential (DAP) was not considered in the model.

[a] σ = electrical conductivity (S/m); ρ = resistivity (kΩcm).

Emphasis on Nanoscale and Bioactivity

As the complexity of devices increases, greater demand is being placed upon the materials supporting the techniques. In particular, greater attention must be given to the charge density requirements for advanced neural prosthetics. Glial scarring and the foreign-body response still play a major role in the success of the implant. Greater insights from the area of nanoscience, on the chemical and signaling basis of the inflammatory response and glial scarring, can help to develop directions to influence the foreign-body response. Furthermore, the development of new materials and engineering of surfaces at the nanoscale could lead to increases in the performance of electrodes and greater biocompatibility and integration.

Neural-tissue interfaces are developed with biocompatibility as well as optimal electrode characteristics in mind. The recent state-of-the-art materials of neural interfaces have been iridium oxide (IrOx) or poly(3,4-ethylenedioxythiophene) (PEDOT; Fig. 4.4) (Weiland and Anderson, 2000; Weiland et al., 2002; Cogan et al., 2004; Cui and Zhou, 2007; Gawad et al., 2009; Wilks et al., 2009; Göbbels et al., 2010; Negi et al., 2010; Ludwig et al., 2011). Specifically, sputtered IrOx coatings on electrodes have been shown to lower the impedance 10–20 times that of platinum electrodes, to increase the effective surface area of individual electrodes, and to have a higher charge-injection

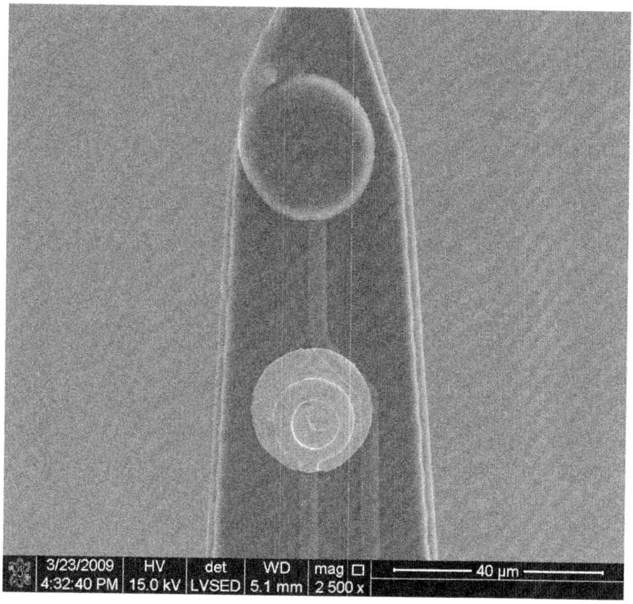

Figure 4.4. Scanning electron microscopy (SEM) images of intracortical microelectrode surfaces with coatings of PEDOT (top) versus IrOx (bottom). (By courtesy of S.J. Wilks and K.J. Otto from Purdue University.)

capacity than platinum-coated or titanium nitride electrodes (Weiland et al., 2002; Cogan et al., 2004; Gawad et al., 2009; Negi et al., 2010). Since IrOx has been demonstrated as biocompatible in several cases (Weiland and Anderson, 2000; Thanawala et al., 2007; Göbbels et al., 2010), iridium-coated electrodes are viable candidates for neural stimulation and recording.

There have also been many reports since the late 1990s on the use of conductive polymers such as polypyrrole (Ppy), PEDOT, and derivatives of PEDOT (Schmidt et al., 1997; Cui and Zhou, 2007; Abidian and Martin, 2008; Wilks et al., 2009). It was determined that rat PC12 cell adhesion and neurite growth was promoted on thick oxidized polypropylene films both *in vitro* and *in vivo* (Schmidt et al., 1997). It was also shown that PEDOT coatings lower the impedance of platinum electrodes by several orders of magnitude and are comparable to IrOx coatings with regard to charge-injection capacity (Cui and Zhou, 2007). Additionally, PEDOT coatings create superior microneural electrodes compared with IrOx-coated electrodes due to their lower impedance and thus higher SNR (Wilks et al., 2009; Abidian and Martin, 2008; Ludwig et al., 2011). There are also reports that IrOx coatings age or degrade even over short-term use according to several studies (Weiland and Anderson, 2000; Cui and Zhou, 2007; Wilks et al., 2009) but PEDOT coatings have been shown to be more stable (Wilks et al., 2009), making them more likely candidates for use in neural-tissue interface.

Since the late 1990s, the continued pursuit of improved biocompatibility and optimal electrode characteristics has pushed the world of neural interfacing to new forms of nanotechnology and nanomaterials. Carbon nanotube (CNT) electrodes and CNT coatings on electrodes can be used to outperform the IrOx or conductive polymer electrodes in some ways. More recent studies (Keefer et al., 2008; Jan et al., 2009) have shown that the combination of CNT and conductive polymers resulted in reduced impedance and increased charge-injection capacity compared with standard PEDOT or IrOx electrodes. The impedance at a CNT layer on a platinum–iridium (PtIr) electrode with a thickness of 100 nm was 277 Ω, which was reduced from 17.4 kΩ originally measured from the PtIr electrode (Jan et al., 2009). Functionalized CNT coatings have also been shown to provide higher charge-injection capacities than metal electrodes (Wang et al., 2006). These result in a higher SNR than those provided by IrOx or PEDOT electrodes (Lin et al., 2009). A promising reported observation of CNT coatings on electrodes is the ability to stimulate neural growth and adhesion (Lovat et al., 2005; Mazzatenta et al., 2007). CNT fibers are also reported as being biocompatible with mammalian cells and neurons (Dubin et al., 2008). Despite these observations, the long-term biocompatibility of CNT-coated electrodes has yet to be established. Other materials such as sol–gels that provide improved biocompatibility and even local drug delivery methods could be used as electrode coatings (Avnir et al., 2006; Pierce et al., 2009). These sol–gel coatings have also been reported to provide slight improvements in impedance and charge-injection capacity (Pierce et al., 2009). Work is also being done to understand the effect of microscale and nanoscale topology and morphology in order to improve the neural response to electrode surfaces (Pennisi et al., 2009; Tonazzini et al., 2010).

CONCLUSIONS

With rapid advances in peripheral neural interface technologies, we see the nearing possibility of the realization of peripheral neural prosthetic devices in use as advanced HMIs. The promise of reconnecting the sensory or motor pathways and restoring functions by a functional electrical stimulation (FES) system can to some degree be realized. However, chronic performance of these devices, the utility of the limited selectivity, channel density, and reach they achieve, the relationship of these parameters to practical use, and utility of these devices have yet to be resolved. The use of advanced microfabrication and nanoscaled surface techniques along with the parallel development of more-sophisticated realistic computational models will continue to play an important role in improving the performance of future devices.

REFERENCES

Abidian M.R. and Martin D.C. (2008). Experimental and theoretical characterization of implantable neural microelectrodes modified with conducting polymer nanotubes. *Biomaterials* 29:1273–1283.

Anderson D.J., Najafi K., Tanghe S.J., Evans D.A., Levy K.L., Hetke J.F., Xue X.L., Zappia J.J., and Wise K.D. (1989). Batch-fabricated thin-film electrodes for stimulation of the central auditory system. *IEEE Trans Biomed Eng* 36:693–704.

Aoyagi Y., Stein R.B., Branner A., Pearson K.G., and Normann R.A. (2003). Capabilities of a penetrating microelectrode array for recording single units in dorsal root ganglia of the cat. *J Neurosci Methods* 128:9–20.

Avnir D., Coradin T., Lev O., and Livage J. (2006). Recent bio-applications of sol-gel materials. *J Mater Chem* 16:1013–1030.

Badia J., Boretius T., Andreu D., Azevedo-Coste C., Stieglitz T., and Navarro X. (2011a). Comparative analysis of transverse intrafascicular multichannel, longitudinal intrafascicular and multipolar cuff electrodes for the selective stimulation of nerve fascicles. *J Neural Eng* 8:036023.

Badia, J., Boretius T., Udina E., Stieglitz T., and Navarro X. (2011b). Biocompatibility of chronically implanted transverse intrafascicular multichannel electrode (TIME) in the rat sciatic nerve. *IEEE Trans Biomed Eng* 58:2324–2332.

Bai Q., Wise K.D., and Anderson D.J. (2000). A high-yield microassembly structure for three-dimensional microelectrode arrays. *IEEE Trans Biomed Eng* 47:281–289.

BeMent S.L., Wise K.D., Anderson D.J., Najafi K., and Drake K.L. (1986). Solid-state electrodes for multichannel multiplexed intracortical neuronal recording. *IEEE Trans Biomed Eng* 33:230–241.

Benvenuto A., Raspopovic S., Hoffmann K.P., Carpaneto J., Cavallo G., Di Pino G., Guglielmelli E., Rossini L., Rossini P.M., Tombini M., and Micera S. (2010).

Intrafascicular thin-film multichannel electrodes for sensory feedback: evidences on a human amputee. *Conf Proc IEEE Eng Med Biol Soc* 1:1800–1803.

Boretius T., Badia J., Pascual-Font A., Schuettler M., Navarro X., Yoshida K., and Stieglitz T. (2010). A transverse intrafascicular multichannel electrode (TIME) to interface with the peripheral nerve. *Biosens Bioelectron* 26:62–69.

Bossi S., Menciassi A., Koch K.P., Hoffmann K.P., Yoshida K., Dario P., and Micera S. (2007). Shape memory alloy microactuation of tf-LIFEs: preliminary results. *IEEE Trans Biomed Eng* 54:1115–1120.

Bossi S., Kammer S., Dorge T., Menciassi A., Hoffmann K.P., and Micera S. (2009). An implantable microactuated intrafascicular electrode for peripheral nerves. *IEEE Trans Biomed Eng* 56:2701–2706.

Branner A. and Normann R.A. (2000). A multielectrode array for intrafascicular recording and stimulation in sciatic nerve of cats. *Brain Res Bull* 51:293–306.

Branner A., Stein R.B., and Normann R.A. (2001). Selective stimulation of cat sciatic nerve using an array of varying-length microelectrodes. *J Neurophysiol* 85: 1585–1594.

Branner A., Stein R.B., Fernandez E., Aoyagi Y., and Normann R.A. (2004). Long-term stimulation and recording with a penetrating microelectrode array in cat sciatic nerve. *IEEE Trans Biomed Eng* 51:146–157.

Brelen M.E., Duret F., Gerard B., Delbeke J., and Veraart C. (2005). Creating a meaningful visual perception in blind volunteers by optic nerve stimulation. *J Neural Eng* 2:S22–S28.

Bruns T.M., Gaunt R.A., and Weber D.J. (2011). Multielectrode array recordings of bladder and perineal primary afferent activity from the sacral dorsal root ganglia. *J Neural Eng* 8:056010.

Campbell P.K., Normann R.A., Horch K.W., and Stensaas S.S. (1989). A chronic intracortical electrode array: preliminary results. *J Biomed Mater Res* 23:245–259.

Campbell P.K., Jones K.E., Huber R.J., Horch K.W., and Normann R.A. (1991). A silicon-based, three-dimensional neural interface: manufacturing processes for an intracortical electrode array. *IEEE Trans Biomed Eng* 38:758–768.

Chemineau E.T., Schnabel V., and Yoshida K. (2004). A modeling study of the recording selectivity of longitudinal intrafascicular electrodes. 9th Annual Conference of the International FES Society, Bournemouth, UK.

Choi A.Q., Cavanaugh J.K., and Durand D.M. (2001). Selectivity of multiple-contact nerve cuff electrodes: a simulation analysis. *IEEE Trans Biomed Eng* 48:165–172.

Cogan S.F., Plante T.D., and Ehrlich J. (2004). Sputtered iridium oxide films (SIROFs) for low-impedance neural stimulation and recording electrodes. *Conf Proc IEEE Eng Med Biol Soc* 6:4153–4156.

Cui X.T. and Zhou D.D. (2007). Poly (3,4-ethylenedioxythiophene) for chronic neural stimulation. *IEEE Trans Neural Syst Rehabil Eng* 15:502–508.

DARPA. Revolutionizing prosthetics. Available at http://www.jhuapl.edu/prosthetics/default.asp (accessed April 8, 2013).

Delbeke J., Oozeer M., and Veraart C. (2003). Position, size and luminosity of phosphenes generated by direct optic nerve stimulation. *Vision Res* 43:1091–1102.

Dhillon G.S. and Horch K.W. (2005). Direct neural sensory feedback and control of a prosthetic arm. *IEEE Trans Neural Syst Rehabil Eng* 13:468–472.

Dhillon G.S., Lawrence S.M., Hutchinson D.T., and Horch K.W. (2004). Residual function in peripheral nerve stumps of amputees: implications for neural control of artificial limbs. *J Hand Surg Am* 29:605–615; discussion 29:616–618.

Dhillon G.S., Kruger T.B., Sandhu J.S., and Horch K.W. (2005). Effects of short-term training on sensory and motor function in severed nerves of long-term human amputees. *J Neurophysiol* 93:2625–2633.

Dowden B.R., Wilder A.M., Hiatt S.D., Normann R.A., Brown N.A., and Clark G.A. (2009). Selective and graded recruitment of cat hamstring muscles with intrafascicular stimulation. *IEEE Trans Neural Syst Rehabil Eng* 17:545–552.

Dowden B.R., Wark H.A., and Normann R.A. (2010). Muscle-selective block using intrafascicular high-frequency alternating current. *Muscle Nerve* 42:339–347.

Drake K.L., Wise K.D., Farraye J., Anderson D.J., and BeMent S.L. (1988). Performance of planar multisite microprobes in recording extracellular single-unit intracortical activity. *IEEE Trans Biomed Eng* 35:719–732.

Dubin R.A., Callegari G., Kohn J., and Neimark A. (2008). Carbon nanotube fibers are compatible with mammalian cells and neurons. *IEEE Trans Nanobiosci* 7:11–14.

Eccles J.C. and Sherrington C.S. (1930). Numbers and contraction values of individual motor-units examined in some muscles of the limb. *Proc Roy Soc* 106:326–357.

Farina D., Yoshida K., Stieglitz T., and Koch K.P. (2008). Multichannel thin-film electrode for intramuscular electromyographic recordings. *J Appl Physiol* 104:821–827.

Fisher L.E., Miller M.E., Bailey S.N., Davis J.A., Anderson J.S., Rhode L., Tyler D.J., and Triolo R.J. (2008). Standing after spinal cord injury with four-contact nerve-cuff electrodes for quadriceps stimulation. *IEEE Trans Neural Syst Rehabil Eng* 16:473–478.

Fisher L.E., Tyler D.J., Anderson J.S., and Triolo R.J. (2009). Chronic stability and selectivity of four-contact spiral nerve-cuff electrodes in stimulating the human femoral nerve. *J Neural Eng* 6:046010.

Frankel M.A., Dowden B.R., Mathews V.J., Normann R.A., Clark G.A., and Meek S.G. (2011). Multiple-input single-output closed-loop isometric force control using asynchronous intrafascicular multi-electrode stimulation. *IEEE Trans Neural Syst Rehabil Eng* 19:325–332.

Gaunt R.A., Hokanson J.A., and Weber D.J. (2009). Microstimulation of primary afferent neurons in the L7 dorsal root ganglia using multielectrode arrays in anesthetized cats: thresholds and recruitment properties. *J Neural Eng* 6:055009.

Gawad S., Giugliano M., Heuschkel M., Wessling B., Markram H., Schnakenberg U., Renaud P., and Morgan H. (2009). Substrate arrays of iridium oxide microelectrodes for in vitro neuronal interfacing. *Front Neuroeng* 2:7.

G'bbels K., Kuenzel T., van Ooyen A., Baumgartner W., Schnakenberg U., and Braunig P. (2010). Neuronal cell growth on iridium oxide. *Biomaterials* 31:1055–1067.

Goodall E.V. and Horch K.W. (1992). Separation of action potentials in multiunit intrafascicular recordings. *IEEE Trans Biomed Eng* 39:289–295.

Goodall E.V., Lefurge T.M., and Horch K.W. (1991). Information contained in sensory nerve recordings made with intrafascicular electrodes. *IEEE Trans Biomed Eng* 38:846–850.

Grill W.M. and Mortimer J.T. (1996). Quantification of recruitment properties of multiple contact cuff electrodes. *IEEE Trans Rehabil Eng* 4:49–62.

Gustafson K.J., Moffitt M.A., Wang X., Sun J., Snyder S., and Grill W.M. (2006). Topography of spinal neurons active during hindlimb withdrawal reflexes in the decerebrate cat. *Neuroscience* 141:1983–1994.

Helmholtz H.L.F. (1853). Ueber einige Gesetze der Vertheilung elektrischer Ströme in körperlichen Leitern mit Anwendung auf die thierishch-elektrischen Versuche. *Ann Physik und Chemie* 89:211–233, 354–377.

Hijjawi J.B., Kuiken T.A., Lipschutz R.D., Miller L.A., Stubblefield K.A., and Dumanian G.A. (2006). Improved myoelectric prosthesis control accomplished using multiple nerve transfers. *Plast Reconstr Surg* 118:1573–1578.

Hoffer J.A. and Loeb G.E. (1980). Implantable electrical and mechanical interfaces with nerve and muscle. *Ann Biomed Eng* 8:351–360.

Hoogerwerf A.C. and Wise K.D. (1994). A three-dimensional microelectrode array for chronic neural recording. *IEEE Trans Biomed Eng* 41:1136–1146.

Jan E., Hendricks J.L., Husaini V., Richardson-Burns S.M., Sereno A., Martin D.C., and Kotov N.A. (2009). Layered carbon nanotube–polyelectrolyte electrodes outperform traditional neural interface materials. *Nano Lett* 9:4012–4018.

Jensen W., Micera S., Navarro X., Stieglitz T., Guiraud D., Divoux J.L., Rossini P.M., and Yoshida K. (2010). Development of an implantable transverse intrafascicular multichannel electrode (TIME) system for relieving phantom limb pain. *Conf Proc IEEE Eng Med Biol Soc* 1:6214–6217.

Kamavuako E.N., Farina D., Yoshida K., and Jensen W. (2009). Relationship between grasping force and features of single-channel intramuscular EMG signals. *J Neurosci Methods* 185:143–150.

Keefer E.W., Botterman B.R., Romero M.I., Rossi A.F., and Gross G.W. (2008). Carbon nanotube coating improves neuronal recordings. *Nat Nanotechnol* 3:434–439.

Kipke D.R., Vetter R.J., Williams J.C., and Hetke J.F. (2003). Silicon-substrate intracortical microelectrode arrays for long-term recording of neuronal spike activity in cerebral cortex. *IEEE Trans Neural Syst Rehabil Eng* 11:151–155.

Kuiken T.A., Dumanian G.A., Lipschutz R.D., Miller L.A., and Stubblefield K.A. (2004). The use of targeted muscle reinnervation for improved myoelectric prosthesis control in a bilateral shoulder disarticulation amputee. *Prosthet Orthot Int* 28:245–253.

Kuiken T.A., Marasco P.D., Lock B.A., Harden R.N., and Dewald J.P. (2007a). Redirection of cutaneous sensation from the hand to the chest skin of human amputees with targeted reinnervation. *Proc Natl Acad Sci U S A* 104:20061–20066.

Kuiken, T.A., Miller L.A., Lipschutz R.D., Lock B.A., Stubblefield K., Marasco P.D., Zhou P., and Dumanian G.A. (2007b). Targeted reinnervation for enhanced prosthetic arm function in a woman with a proximal amputation: a case study. *Lancet* 369:371–380.

Kuiken T.A., Li G., Lock B.A., Lipschutz R.D., Miller L.A., Stubblefield K.A., and Englehart K.B. (2009). Targeted muscle reinnervation for real-time myoelectric control of multifunction artificial arms. *JAMA* 301:619–628.

Kundu A., Jensen W., Stieglitz T., Boretius T., and Yoshida K. (2010). Dependence of implantation angle of the transverse, intrafascicular electrode (TIME) on selective activation of pig forelimb muscles. 15th Annual Conference of International FES Society, Vienna, Austria.

Lago N., Yoshida K., Koch K.P., and Navarro X. (2007). Assessment of biocompatibility of chronically implanted polyimide and platinum intrafascicular electrodes. *IEEE Trans Biomed Eng* 54:281–290.

Lawrence S.M., Larsen J.O., Horch K.W., Riso R., and Sinkjaer T. (2002). Long-term biocompatibility of implanted polymer-based intrafascicular electrodes. *J Biomed Mater Res* 63:501–506.

Lawrence S.M., Dhillon G.S., Jensen W., Yoshida K., and Horch K.W. (2004). Acute peripheral nerve recording characteristics of polymer-based longitudinal intrafascicular electrodes. *IEEE Trans Neural Syst Rehabil Eng* 12:345–348.

Lefurge T., Goodall E., Horch K., Stensaas L., and Schoenberg A. (1991). Chronically implanted intrafascicular recording electrodes. *Ann Biomed Eng* 19:197–207.

Lertmanorat Z., Montague F.W., and Durand D.M. (2009). A flat interface nerve electrode with integrated multiplexer. *IEEE Trans Neural Syst Rehabil Eng* 17:176–182.

Leventhal D.K. and Durand D.M. (2003). Subfascicle stimulation selectivity with the flat interface nerve electrode. *Ann Biomed Eng* 31:643–652.

Leventhal D.K., Cohen M., and Durand D.M. (2006). Chronic histological effects of the flat interface nerve electrode. *J Neural Eng* 3:102–113.

Lin C.M., Lee Y.T., Yeh S.R., and Fang W. (2009). Flexible carbon nanotubes electrode for neural recording. *Biosens Bioelectron* 24:2791–2797.

Lovat V., Pantarotto D., Lagostena L., Cacciari B., Grandolfo M., Righi M., Spalluto G., Prato M., and Ballerini L. (2005). Carbon nanotube substrates boost neuronal electrical signaling. *Nano Lett* 5:1107–1110.

Ludwig K.A., Uram J.D., Yang J., Martin D.C., and Kipke D.R. (2006). Chronic neural recordings using silicon microelectrode arrays electrochemically deposited with a poly(3,4-ethylenedioxythiophene) (PEDOT) film. *J Neural Eng* 3:59–70.

Ludwig K.A., Langhals N.B., Joseph M.D., Richardson-Burns S.M., Hendricks J.L., and Kipke D.R. (2011). Poly(3,4-ethylenedioxythiophene) (PEDOT) polymer coatings facilitate smaller neural recording electrodes. *J Neural Eng* 8:014001.

Malagodi M.S., Horch K.W., and Schoenberg A.A. (1989). An intrafascicular electrode for recording of action potentials in peripheral nerves. *Ann Biomed Eng* 17:397–410.

Malmstrom J.A., McNaughton T.G., and Horch K.W. (1998). Recording properties and biocompatibility of chronically implanted polymer-based intrafascicular electrodes. *Ann Biomed Eng* 26:1055–1064.

Marasco P.D., Schultz A.E., and Kuiken T.A. (2009). Sensory capacity of reinnervated skin after redirection of amputated upper limb nerves to the chest. *Brain* 132:1441–1448.

Mazzatenta A., Giugliano M., Campidelli S., Gambazzi L., Businaro L., Markram H., Prato M., and Ballerini L. (2007). Interfacing neurons with carbon nanotubes: electrical signal transfer and synaptic stimulation in cultured brain circuits. *J Neurosci* 27:6931–6936.

McDonnall D., Clark G.A., and Normann R.A. (2004). Selective motor unit recruitment via intrafascicular multielectrode stimulation. *Can J Physiol Pharmacol* 82: 599–609.

McNaughton T.G. and Horch K.W. (1994). Action potential classification with dual channel intrafascicular electrodes. *IEEE Trans Biomed Eng* 41:609–616.

McNaughton T.G. and Horch K.W. (1996). Metallized polymer fibers as leadwires and intrafascicular microelectrodes. *J Neurosci Methods* 70:103–110.

Micera S., Navarro X., Carpaneto J., Citi L., Tonet O., Rossini P.M., Carrozza M.C., Hoffmann K.P., Vivo M., Yoshida K., and Dario P. (2008). On the use of longitudinal intrafascicular peripheral interfaces for the control of cybernetic hand prostheses in amputees. *IEEE Trans Neural Syst Rehabil Eng* 16:453–472.

Miller L.A., Lipschutz R.D., Stubblefield K.A., Lock B.A., Huang H., Williams T.W., Weir R.F., and Kuiken T.A. (2008a). Control of a six degree of freedom prosthetic arm after targeted muscle reinnervation surgery. *Arch Phys Med Rehabil* 89:2057–2065.

Miller L.A., Stubblefield K.A., Lipschutz R.D., Lock B.A., and Kuiken T.A. (2008b). Improved myoelectric prosthesis control using targeted reinnervation surgery: a case series. *IEEE Trans Neural Syst Rehabil Eng* 16:46–50.

Naples G.G., Mortimer J.T., Scheiner A., and Sweeney J.D. (1988). A spiral nerve cuff electrode for peripheral nerve stimulation. *IEEE Trans Biomed Eng* 35:905–916.

Navarro X., Krueger T.B., Lago N., Micera S., Stieglitz T., and Dario P. (2005). A critical review of interfaces with the peripheral nervous system for the control of neuroprostheses and hybrid bionic systems. *J Peripher Nerv Syst* 10:229–258.

Negi S., Bhandari R., Rieth L., and Solzbacher F. (2010). In vitro comparison of sputtered iridium oxide and platinum-coated neural implantable microelectrode arrays. *Biomed Mater* 5:15007.

Nordhausen C.T., Maynard E.M., and Normann R.A. (1996). Single unit recording capabilities of a 100 microelectrode array. *Brain Res* 726:129–140.

O'Shaughnessy K.D., Dumanian G.A., Lipschutz R.D., Miller L.A., Stubblefield K., and Kuiken T.A. (2008). Targeted reinnervation to improve prosthesis control in transhumeral amputees. A report of three cases. *J Bone Joint Surg Am* 90:393–400.

OttoBock. Michelangelo bionic hand. Available at http://www.living-with-michelangelo.com/gb/home/ (accessed April 8, 2013).

Pennisi C.P., Sevcencu C., Dolatshahi-Pirouz A., Foss M., Hansen J.L., Larsen A.N., Zachar V., Besenbacher F., and Yoshida K. (2009). Responses of fibroblasts and glial cells to nanostructured platinum surfaces. *Nanotechnology* 20:385103.

Pierce A.L., Sommakia S., Rickus J.L., and Otto K.J. (2009). Thin-film silica sol-gel coatings for neural microelectrodes. *J Neurosci Methods* 180:106–110.

Polasek K.H., Hoyen H.A., Keith M.W., Kirsch R.F., and Tyler D.J. (2009a). Stimulation stability and selectivity of chronically implanted multicontact nerve cuff electrodes in the human upper extremity. *IEEE Trans Neural Syst Rehabil Eng* 17:428–437.

Polasek K.H., Schiefer M.A., Pinault G.C., Triolo R.J., and Tyler D.J. (2009b). Intraoperative evaluation of the spiral nerve cuff electrode on the femoral nerve trunk. *J Neural Eng* 6:066005.

Popovic D.B., Stein R.B., Jovanovic K.L., Dai R., Kostov A., and Armstrong W.W. (1993). Sensory nerve recording for closed-loop control to restore motor functions. *IEEE Trans Biomed Eng* 40:1024–1031.

Prochazka A., Mushahwar V., and Yakovenko S. (2002). Activation and coordination of spinal motoneuron pools after spinal cord injury. *Prog Brain Res* 137:109–124.

Raspopovic S., Capogrosso M., and Micera S. (2011). A computational model for the stimulation of rat sciatic nerve using a transverse intrafascicular multichannel electrode. *IEEE Trans Neural Syst Rehabil Eng* 19:333–344.

Rigosa J., Weber D.J., Prochazka A., Stein R.B., and Micera S. (2011). Neuro-fuzzy decoding of sensory information from ensembles of simultaneously recorded dorsal root ganglion neurons for functional electrical stimulation applications. *J Neural Eng* 8:046019.

Rossini P.M., Micera S., Benvenuto A., Carpaneto J., Cavallo G., Citi L., Cipriani C., Denaro L., Denaro V., Di Pino G., Ferreri F., Guglielmelli E., Hoffmann K.P., Raspopovic S., Rigosa J., Rossini L., Tombini M., and Dario P. (2010). Double nerve intraneural interface implant on a human amputee for robotic hand control. *Clin Neurophysiol* 121:777–783.

Rousche P.J. and Normann R.A. (1998). Chronic recording capability of the Utah Intracortical Electrode Array in cat sensory cortex. *J Neurosci Methods* 82:1–15.

Rousche P.J., Pellinen D.S., Pivin D.P., Williams J.C., Vetter R.J., and Kipke D.R. (2001). Flexible polyimide-based intracortical electrode arrays with bioactive capability. *IEEE Trans Biomed Eng* 48:361–371.

Schiefer M.A., Triolo R.J., and Tyler D.J. (2006). Models of selective stimulation with a flat interface nerve electrode for standing neuroprosthetic systems. *Conf Proc IEEE Eng Med Biol Soc* 1:4639–4642.

Schiefer M.A., Triolo R.J., and Tyler D.J. (2008). A model of selective activation of the femoral nerve with a flat interface nerve electrode for a lower extremity neuroprosthesis. *IEEE Trans Neural Syst Rehabil Eng* 16:195–204.

Schiefer M.A., Polasek K.H., Triolo R.J., Pinault, G.C., and Tyler D.J. (2009). Intraoperative demonstration of selective stimulation of the common human femoral nerve with a FINE. *Conf Proc IEEE Eng Med Biol Soc* 2009:610–613.

Schiefer M.A., Polasek K.H., Triolo R.J., Pinault G.C., and Tyler D.J. (2010). Selective stimulation of the human femoral nerve with a flat interface nerve electrode. *J Neural Eng* 7:26006.

Schmidt C.E., Shastri V.R., Vacanti J.P., and Langer R. (1997). Stimulation of neurite outgrowth using an electrically conducting polymer. *Proc Natl Acad Sci U S A* 94:8948–8953.

Schultz A.E., Marasco P.D., and Kuiken T.A. (2009). Vibrotactile detection thresholds for chest skin of amputees following targeted reinnervation surgery. *Brain Res* 1251:121–129.

Seymour J.P. and Kipke D.R. (2007). Neural probe design for reduced tissue encapsulation in CNS. *Biomaterials* 28:3594–3607.

Stein R.B., Charles D., Davis L., Jhamandas J., Mannard A., and Nichols T.R. (1975). Principles underlying new methods for chronic neural recording. *Can J Neurol Sci* 2:235–244.

Stein R.B., Nichols T.R., Jhamandas J., Davis L., and Charles D. (1977). Stable long-term recordings from cat peripheral nerves. *Brain Res* 128:21–38.

Stein R.B., Aoyagi Y., Weber D.J., Shoham S., and Normann R.A. (2004a). Encoding mechanisms for sensory neurons studied with a multielectrode array in the cat dorsal root ganglion. *Can J Physiol Pharmacol* 82:757–768.

Stein R.B., Weber D.J., Aoyagi Y., Prochazka A., Wagenaar J.B., Shoham S., and Normann R.A. (2004b). Coding of position by simultaneously recorded sensory neurones in the cat dorsal root ganglion. *J Physiol* 560:883–889.

Sunderland S. (1978). *Nerves and Nerve Injuries*, 2nd ed. Edinburgh, U.K.: Churchill Livingstone.

Thanawala S., Palyvoda O., Georgiev D.G., Khan S.P., Al-Homoudi I.A., Newaz G., and Auner G. (2007). A neural cell culture study on thin film electrode materials. *J Mater Sci Mater Med* 18:1745–1752.

Tonazzini I., Bystrenova E., Chelli B., Greco P., Stoliar P., Calo A., Lazar A., Borgatti F., D'Angelo P., Martini C., and Biscarini F. (2010). Multiscale morphology of organic semiconductor thin films controls the adhesion and viability of human neural cells. *Biophys J* 98:2804–2812.

Tyler D.J. and Durand D.M. (2002). Functionally selective peripheral nerve stimulation with a flat interface nerve electrode. *IEEE Trans Neural Syst Rehabil Eng* 10:294–303.

Tyler D.J. and Durand D.M. (2003). Chronic response of the rat sciatic nerve to the flat interface nerve electrode. *Ann Biomed Eng* 31:633–642.

Veraart C., Grill W.M., and Mortimer J.T. (1993). Selective control of muscle activation with a multipolar nerve cuff electrode. *IEEE Trans Biomed Eng* 40:640–653.

Veraart C., Raftopoulos C., Mortimer J.T., Delbeke J., Pins D., Michaux G., Vanlierde A., Parrini S., and Wanet-Defalque M.C. (1998). Visual sensations produced by optic nerve stimulation using an implanted self-sizing spiral cuff electrode. *Brain Res* 813:181–186.

Veraart C., Wanet-Defalque M.C., Gerard B., Vanlierde A., and Delbeke J. (2003). Pattern recognition with the optic nerve visual prosthesis. *Artif Organs* 27: 996–1004.

Wagenaar J.B., Ventura V., and Weber D.J. (2011). State-space decoding of primary afferent neuron firing rates. *J Neural Eng* 8:016002.

Wang K., Fishman H.A., Dai H., and Harris J.S. (2006). Neural stimulation with a carbon nanotube microelectrode array. *Nano Lett* 6:2043–2048.

Warwick K., Gasson M., Hutt B., Goodhew I., Kyberd P., Andrews B., Teddy P., and Shad A. (2003). The application of implant technology for cybernetic systems. *Arch Neurol* 60:1369–1373.

Weber D.J., Stein R.B., Everaert D.G., and Prochazka A. (2006). Decoding sensory feedback from firing rates of afferent ensembles recorded in cat dorsal root ganglia in normal locomotion. *IEEE Trans Neural Syst Rehabil Eng* 14:240–243.

Weber D.J., Stein R.B., Everaert D.G., and Prochazka A. (2007). Limb-state feedback from ensembles of simultaneously recorded dorsal root ganglion neurons. *J Neural Eng* 4:S168-S180.

Weiland J.D. and Anderson D.J. (2000). Chronic neural stimulation with thin-film, iridium oxide electrodes. *IEEE Trans Biomed Eng* 47:911–918.

Weiland J.D., Anderson D.J., and Humayun M.S. (2002). In vitro electrical properties for iridium oxide versus titanium nitride stimulating electrodes. *IEEE Trans Biomed Eng* 49:1574–1579.

Wilks S.J., Richardson-Burns S.M., Hendricks J.L., Martin D.C., and Otto K.J. (2009). Poly(3,4-ethylenedioxythiophene) as a micro-neural interface material for electrostimulation. *Front Neuroeng* 2:7.

Yoo P.B. and Durand D.M. (2005). Selective recording of the canine hypoglossal nerve using a multicontact flat interface nerve electrode. *IEEE Trans Biomed Eng* 52:1461–1469.

Yoo P.B., Sahin M., and Durand D.M. (2004). Selective stimulation of the canine hypoglossal nerve using a multi-contact cuff electrode. *Ann Biomed Eng* 32:511–519.

Yoshida K. and Horch K. (1993a). Reduced fatigue in electrically stimulated muscle using dual channel intrafascicular electrodes with interleaved stimulation. *Ann Biomed Eng* 21:709–714.

Yoshida K. and Horch K. (1993b). Selective stimulation of peripheral nerve fibers using dual intrafascicular electrodes. *IEEE Trans Biomed Eng* 40:492–494.

Yoshida K. and Horch K. (1996). Closed-loop control of ankle position using muscle afferent feedback with functional neuromuscular stimulation. *IEEE Trans Biomed Eng* 43:167–176.

Yoshida K. and Riso R. (2004). Peripheral nerve recording electrodes and techniques. In: *Neuroprosthetics: Theory and Practic*, (Horch K.W. and Dhillon G.S., Eds.), pp. 683–744. Singapore: World Scientific.

Yoshida K. and Stein R. B. (1999). Characterization of signals and noise rejection with bipolar longitudinal intrafascicular electrodes. *IEEE Trans Biomed Eng* 46:226–234.

Yoshida K., Jovanovic K., and Stein R. B. (2000a). Intrafascicular electrodes for stimulation and recording from mudpuppy spinal roots. *J Neurosci Methods* 96:47–55.

Yoshida K., Pellinen D., Piven D., Rousche P. J., and Kipke D. R. (2000b). Development of the thin-film longitudinal intra-fascicular electrode. International Functional Electrical Stimulation Society (IFESS) Conference, Aalborg, Denmark, pp. 1497–1500.

Yoshida K., Hennings K., and Krammer S. (2006). Acute performance of the thin-film longitudinal intrafascicular electrode. IEEE Int Conf Biomed Robot Biomechatron, Pisa, Italy.

Yoshida K., Kurstjens M., Citi L., Koch K. P., and Micera S. (2007). Recording experience with the thin-film Longitudinal Intra-Fascicular Electrode, a multichannel peripheral nerve interface. IEEE 10th International Conference on Rehabilitation Robotics, Noordwijk.

Yoshida K., Farina D., Akay M., and Jensen W. (2010). Multichannel intraneural and intramuscular techniques for multiunit recording and use in active prostheses. *Proc. IEEE* 98:432–449.

5

FOCUSED CHAPTER

MULTIMODAL, MULTISITE NEURONAL RECORDINGS FOR BRAIN RESEARCH

ULRICH G. HOFMANN,[1,2] PETER DETEMPLE,[3] AND YIJING XIE[2]

[1]*Neuroelectronic Systems, Department for Neurosurgery, University Hospital Freiburg, Freiburg, Germany*

[2]*Graduate School for Computing in Medicine and Life Sciences, University of Lübeck, Lübeck, Germany*

[3]*Institut für Mikrotechnik Mainz GmbH, Mainz, Germany*

SUMMARY

Classical microelectrode recordings in animals' brains have come a very long way from already successful beginnings—from single metal microwires to bundled arrays, which may even include components other than electrical conducting leads. One or more wires of these bundles may possibly be replaced by glass fibers, thus providing the opportunity not only to record electrophysiologic data, but also to gain information from reflected optical parameters ahead of the multimodal probe.

To push the technological envelope even further, glass fibers can be detachably connected to polymeric and flexible multisite recording and stimulation probes, thus allowing visually controlled implantation of chronically stable flexible probes.

The combination of flexible probes and glass fibers will be found useful in deep brain navigation and optogenetics.

Introduction to Neural Engineering for Motor Rehabilitation, First Edition. Edited by Dario Farina, Winnie Jensen, and Metin Akay.

INTRODUCTION

The basically electrical nature of the brain's information processing opens a splendid way to learn about its details by using extracellular, microwire recordings to eavesdrop on the activity and information exchange among its neuronal constituents (Grundfest et al., 1950; Hubel, 1957). This method has the advantage over glass micropipettes of bringing quite a sturdy probe into the vicinity of a more or less randomly selected neuron. With this method an immense volume of knowledge was acquired with single- and multiple-electrode recordings from the living brain (Reitboeck, 1983; Mountcastle et al., 1991). Clearly, the drawback of intracortical recording is the way the exposed metal site of an insulated microwire or the thin pulled tip of a saline-filled glass pipette is brought to the area of interest. This has to be done surgically by performing skull opening (craniotomy) and punching the probe through the meninges into neuropil, a process discussed elsewhere (Jensen et al., 2003; Olamat et al., 2010).

The result of such a procedure is *one* recording site being carried into the brain by *one* mechanical structure, be it glass or metal. A quantitative measure to assess the traumatizing effect is the ratio of tissue displaced (by the probe) over the number of recording sites placed in it (Nicolelis, 1999). Quite logically, this ratio can be improved favorably by implanting probes with more than one recording site, thus being able to record from a greater number of cells at the cost of the same trauma (Gray et al., 1995). Again, even here another step may be taken, by using the recording site substrate as carrier for another type of measuring sensor, thus achieving with one penetration a truly multimodal record.

Thus, it is quite reasonable to try to place a number of recording sites into the same tissue where previously only one site was placed. However, some thought must be spent, in order to make the most of the effort. In the following discussion, the experiments undertaken are assumed to be acute, since little healing effect can take place in the brain (Polikov et al., 2005; Reichert, 2008) and the recordings have to be stable from the very beginning. There are at least two different positions at which multiple recording sites might be located on the carrier probe: at the very frontal tip of the carrying structure and at its circumference, with an orientation perpendicular to the main axis. Both positions have value on their own: the frontal geometry has a good chance to record from almost spherical areas of less than 100 μm diameter in front of the probe (Gray et al., 1995; Gritsun et al., 2007); the radial or side position offers the option to record from the layered brain structure (e.g., Wise et al., 1970; BeMent et al., 1986; Najafi and Wise, 1986; Drake et al., 1988; Hofmann et al., 1998b; Van Hooser et al., 2000; Csicsvari et al., 2003; Wise et al., 2004; Hofmann et al., 2006). So, where the frontal sites are suited for passing through and sequentially recording from several brain layers (e.g., Moll et al., 2005), the side position needs more thought in design with respect to the anatomical structure of the target tissue. Consultation with an anatomical atlas (Paxinos

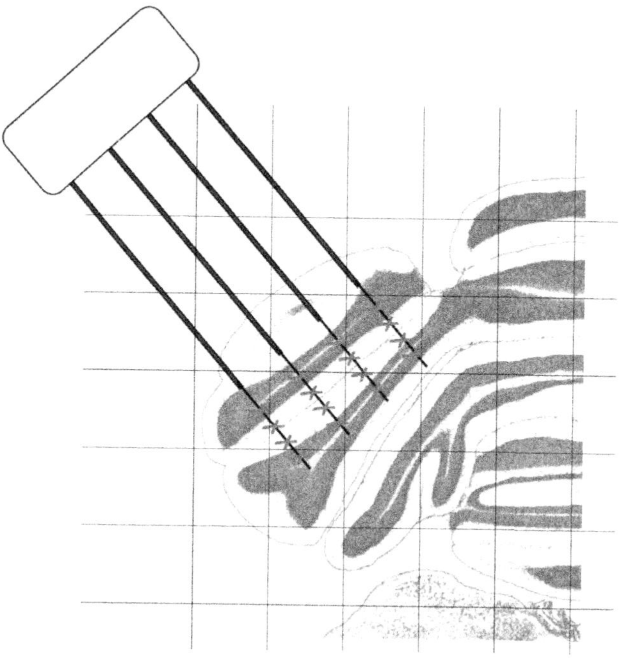

Figure 5.1. Planning step to decide in which layers and anatomical structures recording sites need to be placed according to a particular research paradigm (recordings from cerebellar cortex are desired here).

and Watson, 1998) and appropriate design of site location are needed here (see Fig. 5.1; Hofmann et al., 2000).

Another aspect not to be forgotten is the process of implantation itself. In rat experiments at least, the dura mater is easily removed during the craniotomy; the underlying pia, however, is not easily removed due to its intimate connection with blood vessels and brain tissue (Hofmann et al., 1998a). Consequently, forcing a probe of any kind through the pia will need to rupture this tough, yet thin tissue first (Olamat et al., 2010). In connection with the underlying very soft brain tissue, the latter will give way first and the tough pia will not (Knopp et al., 2004). The effect, called *dimpling*, is very picturesque and will affect underlying brain tissue to a certain depth. The effect is easily illustrated by trying to punch a hole into a semi-inflated rubber balloon—an experiment worth trying.

Simulations show that even a very sharp tip (which is hard to achieve with a microwire tip; Hubel, 1957) may cause dimpling, given that the full diameter of the probe will need to slip through the primary hole in the pia. Once the probe tip is in the brain, dimpling may happen another time: If geometrical considerations require the probe to widen further away from the penetrating tip, dimpling will be caused by delayed lateral rupturing of pial tissue. Here, so-called ball-tip probes are said to have shown a remarkable performance (Parker et al., 1973). Of course a small incision will rupture the pia without the dimpling, but again, this is recommended for deep recording positions only, not for shallow ones.

In many ways, the smallest possible probe cross section is to be preferred, optimally less than some 10,000 μm^2 for rodlike structures (i.e., a diameter below 110 μm).

MULTISITE MICROELECTRODES FOR ELECTROPHYSIOLOGY

Microwire-Based Multisite Probes

The simplest way of getting multiple recording sites in an area of interest is by twisting or intertwining insulated microwires, thus forming a quite stable bundle (e.g., McNaughton et al., 1983; Gray et al., 1995; Gritsun et al., 2007). The production process can even be sped up by bundling microwires with polytetrafluoroethylene (PTFE, or Teflon) shrinking tubes (Mankodiya et al., 2009).

Metal microwires from nickel–chrome alloy (Nichrome), platinum–iridium (PtIr), or plain platinum (Pt) are usually delivered on a roll, so they should be stretched slightly in order to straighten them out before mounting (Tsai and Yen, 2003). Twisting can then be done manually or on a concentric jig, not unlike how ropes were made in former times. Tissue-safe epoxy glue is applied to stabilize the strands and improve rigidity of the wire bundle as desired. It is a matter of geometry to decide which cross section the wire bundle has, but in any case one has to experiment with the most appropriate way to expose the conductive wires under the insulation and glue. To cut the wires, we cannot recommend using a shear cutter such as scissors, since the result is at least ill defined and poorly reproducible (see Fig. 5.2 left panel). The cutting results improve with sharp side cutter pliers; however, we achieved the best result with the blade-on-anvil geometry found in garden pliers—or just a single-use

Figure 5.2. Microwire bundle electrodes, from left: (a) stereotrode (McNaughton et al., 1983, with permission); (b) tetrode (courtesy of T. Davidson, Stanford); and (c) Niotrode bundle awaiting finishing process steps (Mankodiya et al., 2009).

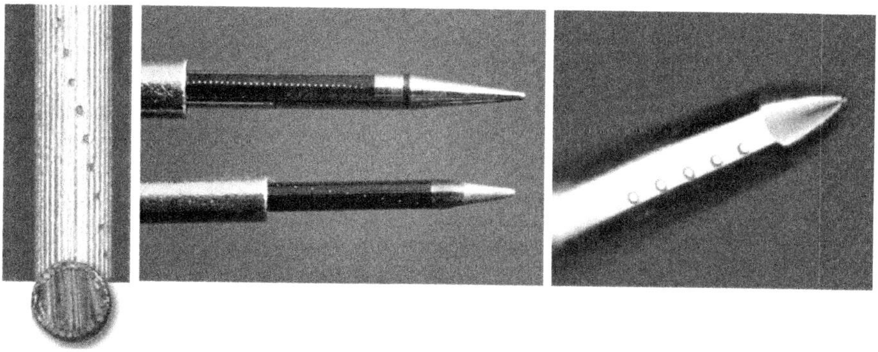

Figure 5.3. Precision-machined microwire probes. (left and center bottom) Insulated wires are glued on a carrier rod under microscope control. Recording sites are opened by targeted laser ablation. (center top and right) Microwires are threaded inside a carrying tube, thus limiting the number of potential channels. By precision-machined fine grinding, several tip options are possible.

scalpel on a glass surface. The final result may need some finishing touches on a polishing plate (Hammad et al., 2010).

Even though the starting point for radial multisite probes is again suitable, insulated microwires, the crucial complication here lies in precisely removing the wires' insulation to expose conductive material to the extracellular medium. Depending on the desired position of the microwires, they can be glued (Fig. 5.3) on a carrier structure or threaded in a tube. Their desired position defines the way to free contact areas: A targeted removal of insulation can be done most suitably by laser ablation followed by galvanic refilling of the developed crater on the external guided wires, whereas internal guided wires need to be run through holes in the carrier tube (Hofmann et al., 2006).

Radial metal-based probes are commercially available from sources in the United States and Germany.

Microstructured Multisite Probes

With the advent of lithographical microtechnologies, originating in the microelectronics industry, semiconductor-based, forklike multisite probes became available (e.g., Wise et al., 1970; Najafi et al., 1985; Wise and Najafi, 1991). Their main advantage is at the same time their main limitation: their lithographical (i.e., planar) production method provides the liberty to distribute high-quality recording sites only along the shanks of the fork in a radial-like geometry (see Fig. 5.4). Consequently almost any thinkable shank design and electrode arrangement can be fabricated in high quality by established processes (Kewley et al., 1997; Norlin et al., 2002; Kindlundh et al., 2004; Seidl et al., 2009). Due to the close connection with microelectronics, probes have been devised that

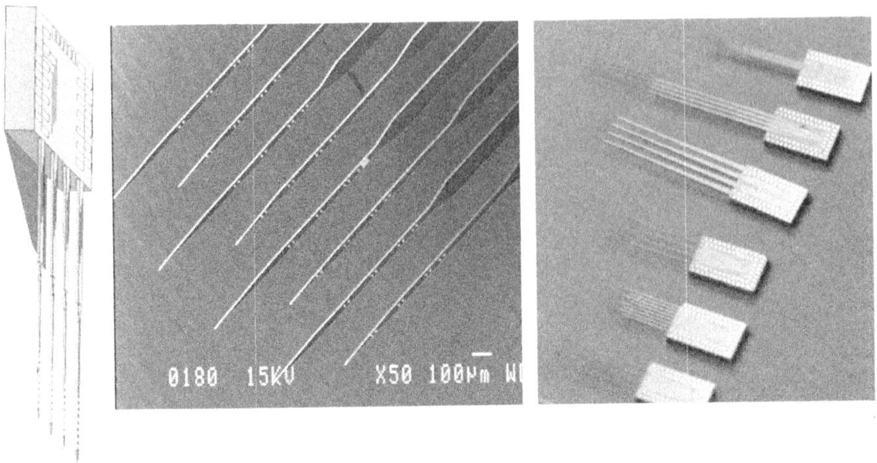

Figure 5.4. (from left) Schematic sketch of a forklike silicon-based multisite probe and electron microscope close-up of a 64-site array on eight staggered shanks. Recording sites here are arranged in a flat stereotrode design. Several multisite probes of differing shank design built by silicon-on-insulator lithographical processes (Norlin et al., 2002).

even contain signal amplification and processing units at the front end—although they are not in widespread use, yet (Ji et al., 1990, 1991; Ji and Wise, 1992; Hoogerwerf and Wise, 1994). All in all, microstructured semiconductor multisite probes in both radial and frontal design (Campbell et al., 1991; Nordhausen et al., 1996) have proven their value in many research settings and even in humans (e.g., Drake et al., 1988; Buzsaki et al., 1989; Blum et al., 1991; Williams et al., 1999; Csicsvari et al., 2003; Hochberg et al., 2006). Unfortunately, their rigid, yet fragile, base substrate (silicon, Si) cautions against their use in a wider clinical context and at the same time is thought to be the source of a strong foreign-body reaction when implanted for a long period, that is, chronic use (Turner et al., 1999; Szarowski et al., 2003; Leach et al., 2010).

Consequently, current research aims at combining the advantages given by lithographical precision with flexible and thus potentially more biocompatible substrates like polyimide (PI, e.g., PI-2555, HD Microsystems, Parlin, NJ, USA) (Stieglitz and Meyer, 1999; Rousche et al., 2001; Rubehn, 2010; Rubehn et al., 2010; Xie et al., 2010b). Differing from a standard silicon process, a silicon wafer is here only used to carry a thin spin-coated sacrificial layer followed by an approximately 10-µm PI layer. The surface of the PI layer is then structured very much like its silicon cousin leading in the end to thin and flexible carriers with many recording sites (see Fig. 5.5). These are then mechanically lifted off the silicon wafer and—after disinfection—used for implantation.

Electrical contacts to an external amplifier are established for both rigid and flexible probes by wire-bonding or appropriate zero-insertion-force (ZIF) multichannel plugs (Norlin et al., 2002). However, one major difficulty has to

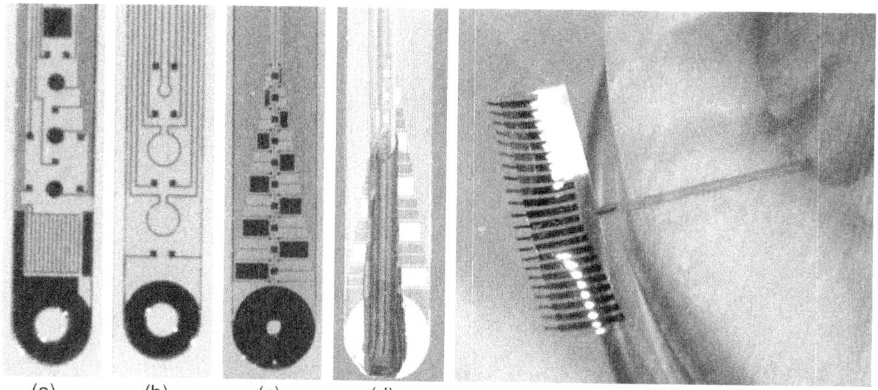

Figure 5.5. (from left) Different surface topologies of 300-μm-wide flexible polyimide (PI) probes with gold surface finish: (a) recording and stimulating sites with interdigitating electrochemical structures; (b) recording sites and magnetic loops; (c) recording, stimulation, and huge ground contacts; (d) a glass fiber inserted into the back side of a PI bearing; (e) a probe inserted with mechanical support into a boiled egg. Note the big contact tag on the left side of the probe, meant for ZIF electrical contacts.

be overcome in the case of flexible probes: how to implant such floppy structures intracortically? In general, the buckling force a probe needs to withstand when inserted through the pia into a rat brain lies in the range of 2–5 mN (Jensen et al., 2006; Olamat et al., 2010). As such, a flat PI probe 10 μm thick will not be able to deliver recording sites inside the neuropil due to early buckling. Clearly, this problem can only be solved with some type of minimal traumatic support. Learning from initial experiments with 150-μm metal rods, cheap glass fibers can be used for the same purpose. A 125-μm glass fiber is inserted into a short PI bushing glued to the back side of a flexible probe (Fig. 5.5d). Figure 5.5e shows the implantation of a flexible probe into a mockup brain given by a boiled egg. Part of the solid egg white is removed after penetration to show the probe's full length. The supporting fiber is then easily removed in case of chronic implantation; however, it can be utilized in many more ways than just a mechanical support.

Glass-Fiber-Based Microimaging

It is evident that the glass fibers' primary purpose is not mechanical support during brain probe implantation; instead (besides many more purposes) they are meant to deliver light at areas distant from a light source. They have been used for optogenetics experiments (Deisseroth et al., 2006; Aravanis et al., 2007; Zhang et al., 2007, 2009), photostimulation-assisted identification of neuronal populations (PINP; Lima et al., 2009), and fibered confocal microscopy (Perchant et al., 2004; Lledo and Saghatelyan, 2005) or *in vivo* anatomical

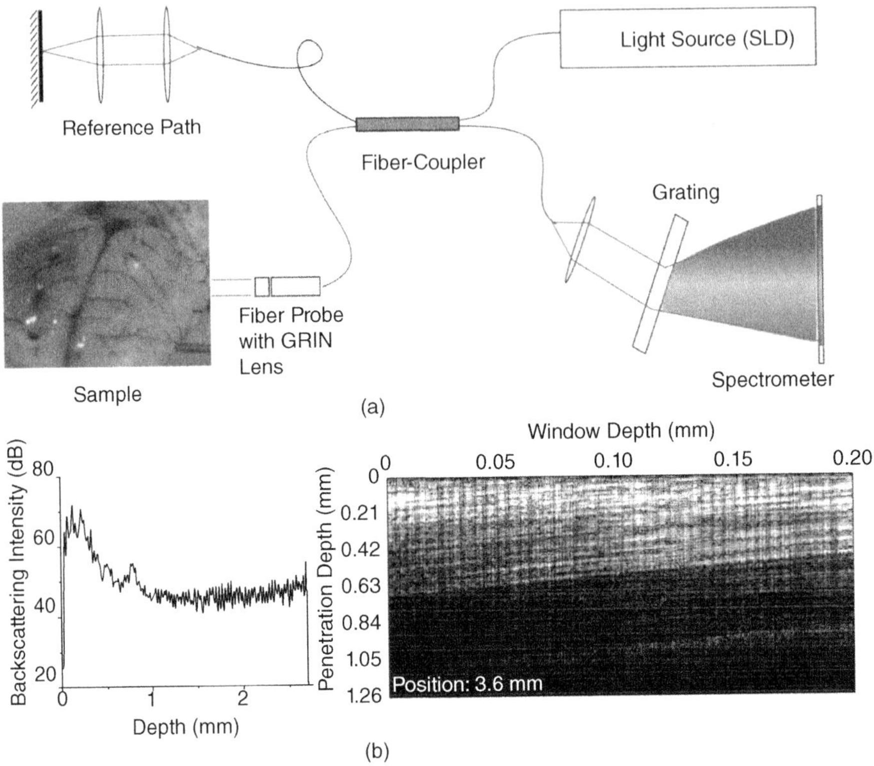

Figure 5.6. (a) Sketch of the essential elements of a fiber-optic OCT system illuminating the surface of a rat's brain. (b) Plot of backscattered intensity with respect to the distance from the tip and an A-scan-type endoscopic OCT picture calculated from 1000 reflectance measurements per picture column at brain depth 3.6 mm, known to be the hippocampus area of a rat. Due to the tip structure of the probing graded-index (GRIN) fiber and the optical properties of brain tissue, the A-scan shows reflecting layers up to 3.5 mm ahead of the probe.

imaging (Ramrath et al., 2007, 2008; Xie et al., 2010a). The first two use the transfection of neurons *in vivo* and their successive exprimation of light-sensitive channels to investigate neuronal network functionalities. The latter uses a method called optical coherence tomography (OCT), which in short provides signals resembling ultrasound pictures but is based on low-coherence light and interferometric measurement systems (Tearney et al., 1997; Bouma and Tearney, 2002; Fujimoto, 2003).

Figure 5.6 shows a schematic setup of an according measuring system: light from a superluminescence diode (SLD) is split by a fiber-coupler into a sample arm and a reference arm. In the reference arm, the light passes through a prism with an optical length corresponding to the length of the catheter of the sample arm and is then reflected back by a movable mirror. In the sample arm, the light passes through a fiber catheter (our glass-fiber probe) into the tissue and is backscattered by optical inhomogeneities. In spectral domain OCT, interference patterns of backscattered and reference light are identified with a spectrometer and are used to determine optical parameters and distance to the reflecting object (Xie et al., 2010a). A quarter-pitch graded-index (GRIN) fiber with numerical aperture NA = 0.275 is fused to the monomode fiber as working end, to yield a required spot size and penetrating depth of 3.5 mm in brain tissue, corresponding to 512 pixels in Figure 5.6b (right panel).

COMBINING OPTICAL AND ELECTROPHYSIOLOGICAL PROBES

Taking the previously stated goal of minimal penetration trauma seriously, it is more than reasonable not to punch two separate holes in the brain's delicate tissue, just to apply two different modalities. Instead, one unavoidable penetration should be used to gain as much insight from the brain's depths as possible. Consequently, several attempts have been made to combine electrophysiologic multisite recordings with light guides on the same probe.

Cho and coworkers (Cho et al., 2010) integrated glass fibers with a polymer waveguide on silicon multisite probes; even more complicated optical components such as optical mixers are now feasible (see Fig. 5.7; Im et al., 2011). Rubehn (Rubehn and Stieglitz, 2010) even succeeded in integrating an SU-8 waveguide onto a flexible PI-probe (see Fig. 5.7c). It remains to be proven in all these cases whether sufficient light intensity at a required wavelength can be carried into the tissue to trigger optogenetic neuronal activity. Moreover, SU-8 waveguides still need to be qualified to efficiently image biological tissue.

Instead of (as shown in Fig. 5.7b) sacrificing light intensity in the transition region between the optical fiber and a waveguide or even worse in the less-than-optimal light transfer parameters of the SU-8 polymer, we decided to use

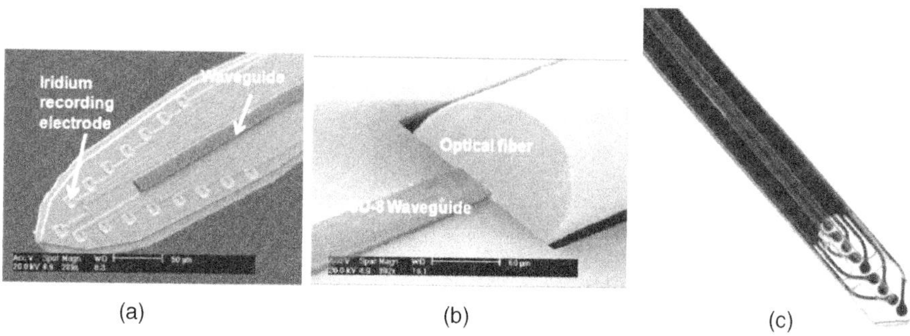

Figure 5.7. (a) and (b) Electron microscopy close-ups of an optical waveguide integrated on a Michigan-style silicon probe (Cho et al.. 2010, with permission). (c) Flexible and transparent shaft multisite probe with integrated waveguide (Rubehn, 2010) (probe width is 200 μm; picture courtesy of B. Rubehn, IMTEK).

optimal glass fibers in combination with multisite microprobes. SU-8 has a light attenuation in the range of 1–5 dB/cm, compared with a glass fiber's 5-dB/km attenuation. The core of this development was a commercial monomode glass fiber from which we removed the plastic coating, leaving a 125-μm glass rod with a GRIN lens fused to its working end (Xie et al., 2010a). We used this rod to replace the Niotrode bundle's standard tungsten core (Gritsun et al., 2007; Hammad et al., 2010). As starting point, a simple wrapping procedure was used to glue eight conductive microwires on the glass-fiber core (Krapohl et al., 2008). Instead of wrapping the wires around the core, we improved this procedure by threading the wires and the glass fiber under position control into a PTFE shrinking tube (Mankodiya et al., 2009). On heating the tube, wires and core are fixed inside in a predefined configuration and can, in principle, withstand the subsequent cleaning and polishing procedures. Unfortunately, the polishing process deteriorates the GRIN-lens' optical properties neccessary for imaging and this is thus a less than optimal procedure for our application (Hammad et al., 2010). It may, though, be an easy and versatile process for optogenetics applications with a lower optical demand at the front end.

Figure 5.8 depicts the two different microwire-based processes and a complete optoelectronic sensor assembly including electrical multichannel and optical fiber connections.

Since our imaging application demands a high-quality finish, which is difficult to maintain in a posteriori assembly steps, we decided to avoid postprocessing by using flexible multisite microprobes instead. Here, the only process needed is to glue a 160-μm PI-tube with bioinert glue as guide sleeve under the flexible probe and insert the delicate GRIN-ended fiber into it. Having the

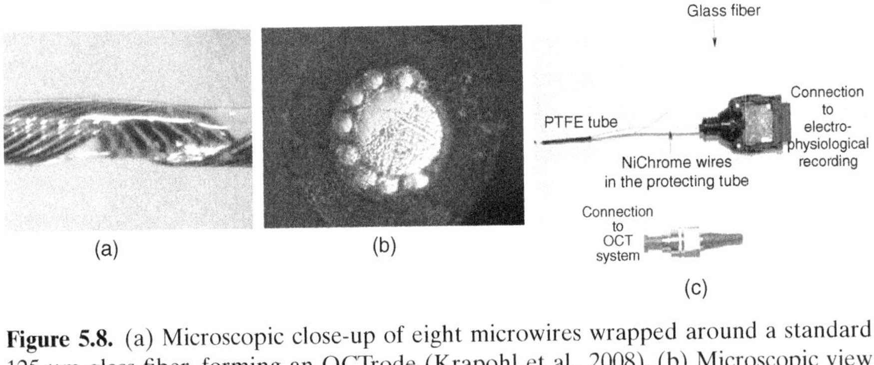

Figure 5.8. (a) Microscopic close-up of eight microwires wrapped around a standard 125-μm glass fiber, forming an OCTrode (Krapohl et al., 2008). (b) Microscopic view of the polished front tip of a similar glass fiber embedded with eight microwires in a PTFE shrinking tube (Hammad et al., 2010). (c) Overview of a complete OCTrode assembly (like that in panel b) with connections to both a multisite recording system and an OCT system (Mankodiya et al., 2009).

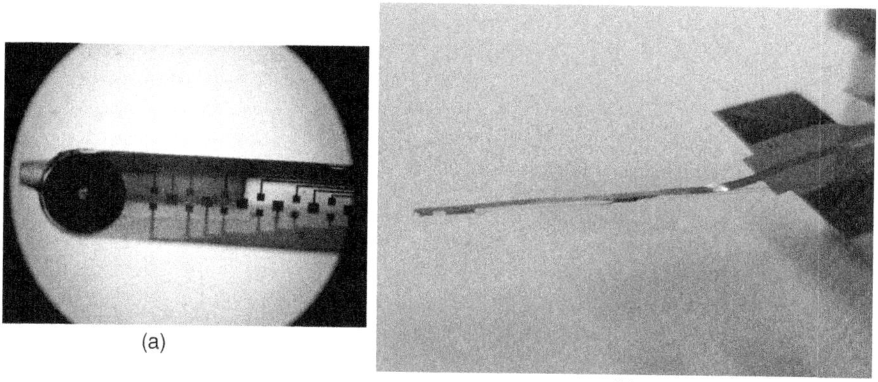

(a) (b)

Figure 5.9. (a) Microscopic close-up of an OCT glass fiber in the guiding sleeve on the back side of a 300-μm-wide, 10-μm-thick, flexible, PI multisite probe. (b) Detachable OCTrode assembly, based on flexible multisite probe inserted in a ZIF connector and a glass-fiber coupler (to right). Note the big PI flaps close to the connecting pad. They are intended to foster fixation of probes during chronic implantations.

coating removed to an appropriate distance from the working end defines the excess length of the glass fiber over the recording site positions (see Fig. 5.9). Another advantage is provided by this type of assembly: the glass fiber is reusable since it can be removed from the guiding sleeve even *in vivo*, given that the sleeve provides sufficient slack.

As stated previously, with the optical fiber connected to an OCT imaging system (e.g., Thorlabs Callisto, Luebeck, Germany), it is possible to acquire A-scans of the tissue lying ahead of the OCTrode probe assembly. Acquired pictures have both high lateral and axial resolution; however, missing a scanning opportunity due to the required space limitations makes the A-scans difficult to interpret. Nevertheless, some type of orientation is presented due to the identification of layered structures ahead of the probe that correspond to known anatomical tissue boundaries in the brain. That way the co-inserted multisite probe allows for high-resolution electrophysiological recording from anatomical structures validated *in vivo*. Figure 5.10 shows a sketch of such a detachable OCTrode together with data (image and neuronal signals) recorded from a brain layer close to the working tip.

CONCLUSION AND OUTLOOK

The examples discussed here illustrate that the technology of electrophysiological microrecordings—and for that matter stimulation as well—has come a long way toward multisite and minimally traumatic brain interfacing devices. However, the future of these technologies is even more exciting than the past,

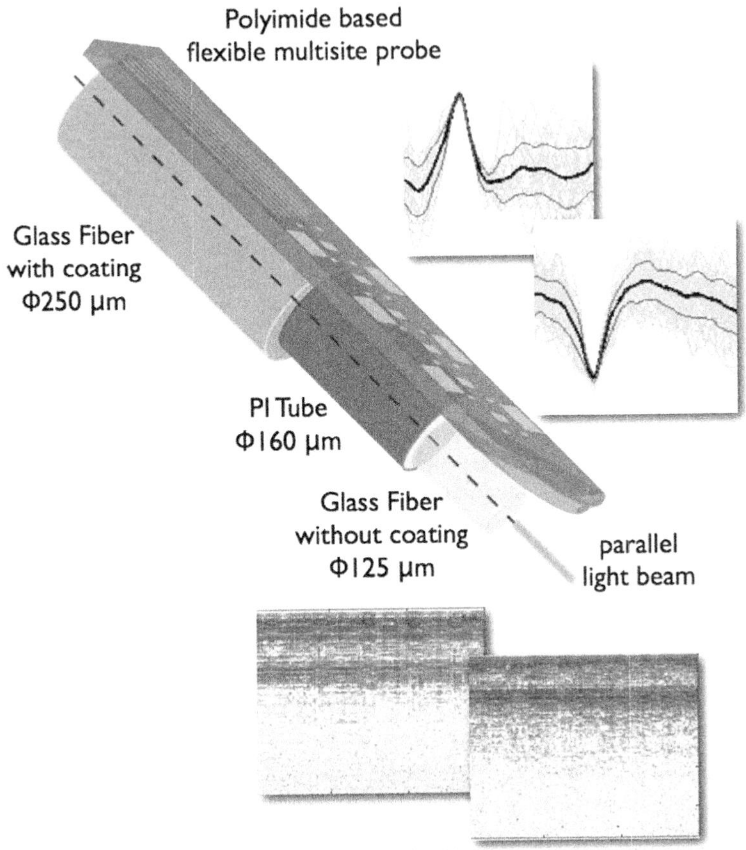

Figure 5.10. Sketch of a detachable OCTrode front end with a monomode glass fiber threaded in a guide sleeve as far as the fiber's coating permits. Curves depict spike recordings from two electrode sites clustered to identify units. The insert shows reconstructed OCT images of the tissue lying ahead at that particular depth in the brain, clearly indicating layered structures.

since the combination of electrical with optical modalities puts new procedures at the hands of neuroscientists and perhaps even clinical practitioners. A human-use combination of OCT-type imaging with electrophysiological multisite data analysis (Moll et al., 2005) may very well fulfill that neurosurgeon's dream of "looking into that burr-hole" and significantly improve stereotactic brain surgeries.

It remains to be mentioned that several approaches to coupling fluidic channels—and thus adding another modality—with multisite microelectrode probes have been successful (Chen et al., 1997; Metz et al., 2001; Takeuchi et al., 2004; Rubehn, 2010). Unfortunately, so far this has only meant that drug delivery at the site of recording was feasible, whereas microdialysis, the neurochemical standard procedure for acquiring *in situ* fluid samples (Ungerstedt,

1991), has not yet been implemented at the single-probe level. Admittedly, surrogating a co-localized structure for a minimally invasive single device with both microdialysis and multisite probes is not quite a satisfactory solution (Gritsun et al., 2006; Hiller et al., 2007).

ACKNOWLEDGMENTS

We are indebted to T. Davidson and B. Rubehn for providing us Figure 5.2b,c and Figure 5.8c, respectively. We gratefully acknowledge Peter Norlin, ACREO AB, Sweden, who provided the pictures in Figure 5.4. Silicon probes were developed for EU project VSAMUEL. Rigid and flexible probes were developed under German Research Ministry Grant BiCIRTS. We are grateful to the members of both research consortia.

REFERENCES

Aravanis A.M., Wang L.-P., et al. (2007). An optical neural interface: in vivo control of rodent motor cortex with integrated fiberoptic and optogenetic technology. *J Neural Eng* 4(3):143–156.

BeMent S.L., Wise K.D., et al. (1986). Solid state electrodes for multichannel multiplexed intracortical neuronal recording. *IEEE Trans Biomed Eng* 33(2):230–241.

Blum N., Carkhuff B., et al. (1991). Multisite microprobes for neural recordings. *IEEE Trans Biomed Eng* 38(1):68–74.

Bouma B.E. and Tearney G.J. (2002). *Handbook of Optical Coherence Tomography*. Marcel Dekker.

Buzsaki G., Bickford R.G., et al. (1989). Multisite recording of brain field potentials and unit-activity in freely moving rats. *J Neurosci Meth* 28(3):209–217.

Campbell P.K., Jones K.E., et al. (1991). A silicon based three dimensional neural interface: manufacturing processes for an intracortical electrode array. *IEEE Trans Biomed Eng* 38:758–768.

Chen J.K., Wise K.D., et al. (1997). A multichannel neural probe for selective chemical delivery at the cellular level. *IEEE Trans Biomed Eng* 44(8):760–769.

Cho I.-J., Baac H.-W., et al. (2010). A 16-site neural probe integrated with a waveguide for optical stimulation. *23rd IEEE International Conference on Micro Electro Mechanical Systems (MEMS 2010)*, Hong Kong.

Csicsvari J., Henze D.A., et al. (2003). Massively parallel recording of unit and local field potentials with silicon-based electrodes. *J Neurophysiol* 90:1314–1323.

Deisseroth K., Feng G., et al. (2006). Next-generation optical technologies for illuminating genetically targeted brain circuits. *J Neurosci* 26(41):10380–10386.

Drake K.L., Wise K.D., et al. (1988). Performance of planar multisite microprobes in recording single unit intracortical activity. *IEEE Trans Biomed Eng* BME-35: 719–732.

Fujimoto J.G. (2003). Optical coherence tomography for ultrahigh resolution in vivo imaging. *Nat Biotech* 21:1361–1367.

Gray C., Maldonado P., et al. (1995). Tetrodes markedly improve the reliability and yield of multiple single-unit isolation from multi-unit recordings in cat striate cortex. *J Neurosci Methods* 63(1–2):43–54.

Gritsun T., Litza M., et al. (2006). Semichronic, collocated deep brain stimulation and multisite recording in rats. *International Conference on Microtechnologies in Medicine and Biology, Bankoku-Shinryokan, Okinawa, Japan*, IEEE Publishing.

Gritsun T., Engler G., et al. (2007). A simple microelectrode bundle for deep brain recordings. *3rd International Conference on Neural Engineering, Hawaii, USA*, IEEE.

Grundfest H., Sengstaken R.W., et al. (1950). Stainless steel micro-needle electrodes made by electrolytic pointing. *Rev Sci Instruments* 21(4):360–361.

Hammad S., Löffler S., et al. (2010). Niotrode array for rodent brain recording. *Biomed Tech O Dössel Rostock, Walter de Gruyter* 55(Suppl 1):4.

Hiller A., Loeffler S., et al. (2007). Electrical high frequency stimulation induces GABA outflow in freely moving rats. *J Neurosci Methods* 159(2):286–290.

Hochberg L.R., Serruya M.D., et al. (2006). Neuronal ensemble control of prosthetic devices by a human with tetraplegia. *Nature* 442:164–171.

Hofmann U.G., Kewley D.T., et al. (1998a). Factors affecting brain dimpling during microelectrode insertion. *Soc Neurosci Abstr*, Los Angeles, Society for Neuroscience.

Hofmann U.G., Kewley D.T., et al. (1998b). Depth profiling of neuronal activity by multisite microelectrode arrays. *Europ J Neurosci* 10(Suppl 10):431.

Hofmann U.G., De Schutter E., et al. (2000). On the design of multi-site microelectrodes for neuronal recordings. *MICRO.tec 2000* 1:283–288.

Hofmann U.G., Moll C.K.E., et al. (2006). Multisite microelectrodes for use in human deep brain stimulation. *International Conference on Microtechnologies in Medicine and Biology, Bankoku-Shinryokan, Okinawa, Japan*.

Hoogerwerf A.C. and Wise K.D. (1994). A three-dimensional microelectrode array for chronic neural recording. *IEEE Trans Biomed Eng* 41(12):1136–1146.

Hubel D.H. (1957). Tungsten microelectrode for recording from single units. *Science* 125(3247):549–550.

Im M., Cho I.-J., et al. (2011). Neural probes integrated with optical mixer/splitter waveguides and multiple stimulation sites. MEMS 2011, Cancun, IEEE.

Jensen W., Hofmann U., et al. (2003). Assessment of subdural insertion force of single-tine microelectrodes in rat cerebral cortex. *25th Ann. Int. Conf of the IEEE Engineering in Medicine & Biology Society, Cancun, Mexico*, IEEE.

Jensen W., Yoshida K., et al. (2006). Implant mechanics of ACREO silicon electrodes in rat cerebral cortex. *IEEE Trans Biomed Eng* 53(5):934–940.

Ji J. and Wise K.D. (1992). An implantable CMOS circuit interface for multiplexed microelectrode recording arrays. *IEEE J Solid-State Circuits* 27(3):433–443.

Ji J., Najafi K., et al. (1990). A scaled electronically-configurable multichannel recording array. *Sensors and Actuators A—Physical* 22(1–3):589–591.

Ji J., Najafi K., et al. (1991). A low-noise demultiplexing system for active multichannel microelctrode arrays. *IEEE Trans Biomed Eng* 38(1):75–81.

Kewley D.T., Hills M.D., et al. (1997). Plasma-etched neural probes. *Sensors and Actuators A* 58:27–35.

Kindlundh M., Norlin P., et al. (2004). A neural probe process enabling variable electrode configurations. *Sensors and Actuators B: Chemical* 102(1):51–58.

Knopp U., Jensen W., et al. (2004). Compression of the cerebral cortex during corticotomy: a study of dimpling in rats. *55. Ann' Jahrestagung der Deutschen Gesellschaft für Neurochirurgie, Köln*, German Medical Sciences.

Krapohl D., Bonin T., et al. (2008). A new integrated optical and electrophysiological sensor. *Biomed Technik / Biomed Eng* 53(Suppl 1):126–128.

Leach J., Achyuta A.H., et al. (2010). Bridging the divide between neuroprosthetic design, tissue engineering and neurobiology. *Front Neuroeng*.

Lima S.Q., Hromádka T., et al. (2009). PINP: a new method of tagging neuronal populations for identification during in vivo electrophysiological recording. *PLos ONE* 4(7).

Lledo P.-M. and Saghatelyan A. (2005). Integrating new neurons into the adult olfactory bulb: joining the network, life–death decisions, and the effects of sensory experience. *Trends Neurosci* 28(5):249–254.

Mankodiya K., Krapohl D., et al. (2009). A simplified production method for multimode multisite neuroprobes. *IEEE—International Conference on Neural Engineering, Antalya, Turkey*, IEEE.

McNaughton B., O'Keefe J., et al. (1983). The stereotrode: a new technique for simultaneous isolation of several single units in the central nervous system from multiple unit recording. *J Neurosci Methods* 8:391–397.

Metz S., Holzer R., et al. (2001). Fabrication of flexible, implantable microelectrodes with embedded fluidic microchannels. *Transducers*.

Moll C.K.E., Struppler A., et al. (2005). Intraoperative Mikroelektrodenableitungen in den Basalganglien des Menschen. *Neuroforum* 1:11–20.

Mountcastle V., Reitboeck H., et al. (1991). Adaptation of the Reitboeck method of multiple microelectrode recording to the neocortex of the waking monkey. *J Neurosci Methods* 36(1):77–84.

Najafi, K. and K.D. Wise (1986). An implantable multielectrode array with on-chip signal processing. *IEEE J Solid-State Circ* 21:1035–1044.

Najafi K., Wise K.D., et al. (1985). A high-yield IC-compatible multichannel recording array. *IEEE Trans Electron Devices* 32(7):1206–1211.

Nicolelis M.A.L., Ed. (1999). *Methods for Neural Ensemble Recordings. CRC Methods in Neuroscience*. Boca Raton, FL: CRC Press.

Nordhausen C., Maynard E., et al. (1996). Single unit recording capabilities of a 100 microelectrode array. *Brain Res* 726(1–2):129–140.

Norlin P., Kindlundh M., et al. (2002). A 32-site neural recording probe fabricated by DRIE of SOI substrates. *J Micromech Microeng* 12(4):414–419.

Olamat A.E., Nkemasong N., et al. (2010). Physical and FEM simulation of microprobe insertion into rat brain. *European COMSOL Conference 2010*, Paris.

Parker T.D., Strachan D.D., et al. (1973). Tungsten ball microelectrodes for extracellular single-unit recording. *Electroencephalogr Clin Neurophysiol* 35:647–651.

Paxinos G. and Watson C. (1998). *The Rat Brain in Stereotaxic Coordinates*. Academic Press.

Perchant A., Le Goualher G., et al. (2004). An integrated fibered confocal microscopy system for in vivo and in situ fluorescence imaging: applications to endoscopy in small animal imaging. Proceedings of the IEEE International Symposium on Biomedical Imaging: Nano to Macro (ISBI'04), IEEE.

Polikov V.S., Tresco P.A., et al. (2005). Response of brain tissue to chronically implanted neural electrodes. *J Neurosci Methods* 148(1):1–18.

Ramrath L., Hofmann U.G., et al. (2007). Towards automated OCT-based identification of white brain matter. In: *Bildverarbeitung fuer die Medizin*. Munich: Springer, pp. 414–418.

Ramrath L., Pohl M., et al. (2008). Single fibre optical coherence tomography for brain imaging—a preliminary in vitro study. CURAC 2008, *Proceedings der 7. Jahrestagung der Deutschen Gesellschaft für Computer- und Roboterassistierte Chirurgie*, Leipzig.

Reichert W.M., Ed. (2008). *Indwelling Neural Implants. Frontiers in Neuroengineering*. Boca Raton, FL: CRC Press.

Reitboeck H.J.P. (1983). A 19-channel matrix drive with individually controllable fibre microelctrodes for neurophysiological applications. *IEEE Trans Syst Man Cybernet* 13(5):676–683.

Rousche P.J., Pellinen D.S., et al. (2001). Flexible polyimide-based intracortical electrode arrays with bioactive capability. *IEEE Trans Biomed Eng* 48:361–371.

Rubehn B. (2010). Polymer-based microimplants for the neurosciences. *IMTEK. Freiburg*, Albert-Ludwigs-University: 191.

Rubehn B. and Stieglitz T.A. (2010). A flexible neural shaft electrode with integrated optical waveguide. *Biomed Tech O Dössel Rostock, Walter de Gruyter* 55 (Supp 1):4.

Rubehn B., Lewis C., et al. (2010). Flexible shaft electrodes for transdural implantation and chronic recording. *Proceedings of the 15th Annual Conference of the IFESS. Vienna*, IFMBE. Abstr No 60:4.

Seidl K., Herwik S., et al. (2009). CMOS-based high-density silicon microprobe array for electronic depth control in neural recording. *Tech. Dig. 22nd IEEE International Conference on MEMS, Sorrento, Italy*.

Stieglitz T. and Meyer J.U. (1999). Implantable microsystems. Polyimide-based neuroprostheses for interfacing nerves. *Med Device Technol* 10:28–30.

Szarowski D.H., Andersen M.D., et al. (2003). Brain responses to micro-machined silicon devices. *Brain Res* 983(1–2):23–35.

Takeuchi S., Yoshida Y., et al. (2004). Parylene flexible neural probe with micro fluidic channel. IEEE MEMS 04, Maastricht.

Tearney G.J., Brezinski M.E., et al. (1997). In vivo endoscopic optical biopsy with optical coherence tomography. *Science* 276:2037–2039.

Tsai M.-L. and Yen C.-T. (2003). A simple method for fabricating horizontal and vertical microwire arrays. *J Neurosci Methods* 131:107–110.

Turner J., Shain W., et al. (1999). Cerebral astrocyte response to micromachined silicon implants. *Exp Neurol* 156(1):33–49.

Ungerstedt U. (1991). Microdialysis—principles and applications for studies in animals and man. *J Internal Med* 230(4):365–373.

Van Hooser S.D., Hofmann U.G., et al. (2000). Relationship between field potentials and spike activity in rat S1: multi-site cortical recordings and simulations. *Neurocomputing* 32–33:591–596.

Williams J., Rennaker R., et al. (1999). Stability of chronic multichannel neural recordings: implications for a long-term neural interface. *Neurocomputing* 26–27: 1069–1076.

Wise K.D. and Najafi K. (1991). Microfabrication techniques for integrated sensors and microsystems. *Science* 254:1335–1342.

Wise K.D. et al. (1970). An integrated circuit approach to extracellular microelectrodes. *IEEE Trans Biomed Eng* BME-17(3):238–247.

Wise K.D., Anderson D.J., et al. (2004). Wireless implantable microsystems: high density electronic interfaces to the nervous system. *Proc IEEE* 92(1):76–97.

Xie Y., Bonin T., et al. (2010a). Fiber spectral domain optical coherence tomography for in-vivo rat brain imaging. *Biophotonics: Photonics for Better Health Care. Brussels*, SPIE. 7715–7785.

Xie Y., Löffler S., et al. (2010b). A novel approach of flexible electrode implantation for deep brain recording. *Biomed Tech O Dössel Rostock, Walter de Gruyter* 55(Suppl 1):4.

Zhang F., Wang L.-P., et al. (2007). Multimodal fast optical interrogation of neural circuitry. *Nature* 446:633–639.

Zhang J., Laiwalla F., et al. (2009). Integrated device for optical stimulation and spatiotemporal electrical recording of neural activity in light-sensitized brain tissue. *J Neural Eng* 6:055007 (13p).

6

FOCUSED CHAPTER

SURFACE ELECTROMYOGRAM DETECTION

ALBERTO BOTTER, MARCO GAZZONI, AND ROBERTO MERLETTI

Laboratory for Engineering of the Neuromuscular System (LISiN), Politecnico di Torino, Torino, Italy

SUMMARY

The electromyogram (EMG) is a compound signal comprising the electrical activity of the motor units (MU) activated asynchronously during voluntary muscle contractions. The summation of action potentials of active MUs generates, on the skin surface, a bi-dimensional electrical field, the surface EMG (sEMG). Temporal and spatial evolution of this field might be sampled by surface electrodes appropriately positioned above active muscle regions. The properties of the detection system as well as the characteristics of the circuits for the conditioning of sEMG influence its quality and informative content.

The purpose of this chapter is to provide a basic overview of the technologies for the detection and conditioning of sEMG signals.

The first section focuses on electrode technology and the electrode–skin interface. The electrical model of the interface and the effect of electrode characteristics (e.g., size, constituent material) on the features of the detected signal are discussed.

Introduction to Neural Engineering for Motor Rehabilitation, First Edition. Edited by Dario Farina, Winnie Jensen, and Metin Akay.

The second section describes the most common solutions for the design of the front-end amplifier, signal filtering, and analog-to-digital (A/D) conversion.

The third section describes and discusses the use of different configurations of electrodes for the detection of sEMG signals. The concepts of spatial filtering and spatial sampling (mono- and bi-dimensional) are introduced and the effect of the detection system parameters on the sEMG characteristics is discussed.

ELECTRODE–SKIN INTERFACE

Equivalent Circuit, Electrode Impedance, Noise, and DC Voltage

A surface electrode for biopotential recordings can be defined either as a sensor of bioelectrical activity or as a transducer of an ionic current (flowing in the tissue) into an electrical current (flowing in the metallic connections between the electrode and the conditioning circuit).

The interface between the electrode and the skin is the first block of any system detecting bioelectric signals. It represents the boundary between two media where different charge carriers are involved (ions in the gel and skin; electrons in the metallic material). The electrical modeling of such an interface is very complex but can be roughly described by a nonlinear RC circuit whose components are frequency and current dependent (Hary et al., 1987; Ragheb and Geddes, 1990; Van Oosterom and Strackee, 1983). Figure 6.1a shows a possible model of the electrode–skin impedance. In this circuit R_p and C_p represent the resistive and capacitive behavior of the electrode–electrolyte junction and R_s is the resistance associated with the electrolyte gel. The epidermal layer can be modeled by a parallel RC circuit (R_e, C_e), whereas the dermis and subcutaneous layer under it are described as a pure resistance (R_{sub}).

The metal–electrolyte junction is intrinsically noisy because of the different charge carriers in the two phases. Thus, in addition to the passive components of the model, a noise generator must be taken into account (V_n in Fig. 6.1a). In the electromyogram (EMG) frequency band (10–400 Hz) the junction noise level is higher (a few μV_{RMS}) than the noise attributed to the front-end electronics (about 1 μV_{RMS}) and can be considered the main source of noise. In general silver–silver chloride (Ag–AgCl) electrodes present lower noise than other interfaces and are recommended for EMG detection (Freriks et al., 1997).

Another relevant issue to consider in biopotential recordings is the generation of DC voltages at the interfaces with the electrolyte gel. In the electrical model of Figure 6.1a, E_{hc} takes into account the half-cell potential due to the "battery effect" (Grimnes, 1982; Hary et al., 1987; Tam and Webster, 1977) generated at the metal–electrolyte junction; E_{ge} is the potential difference at the epidermis–electrolyte interface due to the gradient of ionic concentrations between the gel and the superficial layer of the skin. In differential recordings

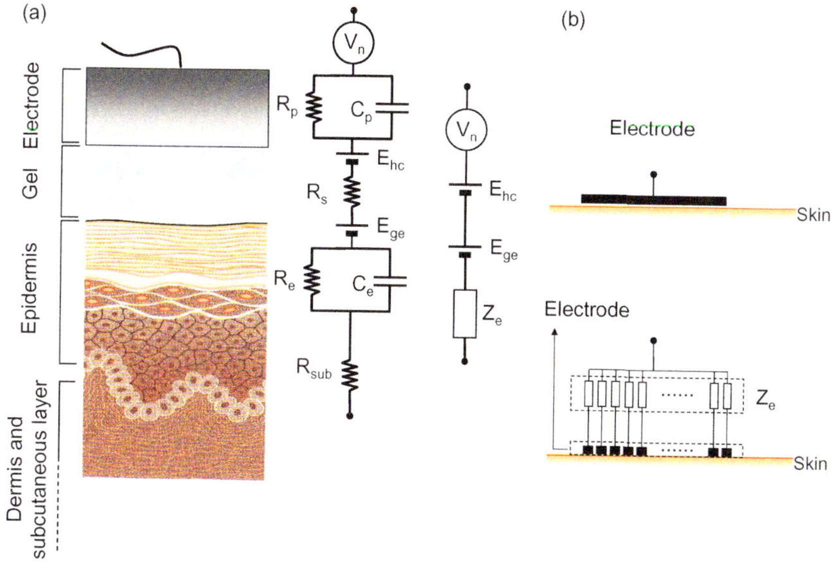

Figure 6.1. The electrode–skin interface. (a) Model of the electrode–skin impedance (Z_e): the parallel R_pC_p takes into account the capacitive and the resistive behavior of the electrode–gel junction; R_s describes the resistive behavior of the electrolyte gel. The parallel R_eC_e characterizes the gel–skin impedance; R_{sub} is the resistive component associated with the subcutaneous tissue layer; E_{hc} is the half-cell potential at the metal–electrolyte junction; E_{ge} is the DC potential due to differences in the ionic concentrations between the gel and the superficial layer of the skin; V_n is the voltage associated with the global noise component (equivalent noise generators are not indicated, for simplicity, but are present at all interfaces). (b) Model of electrode with finite area being approximated as a number of point electrodes connected in parallel.

the DC voltages of two electrodes with the same properties would theoretically cancel out. However, the intrinsic variability in the electrode–skin interface may cause a DC component of few hundred millivolts between two electrodes closely placed on the skin. This voltage could lead to a saturation of the front-end amplifier and must be considered in the design of the front-end electronics.

The electrode–skin impedance is one of the most important properties in relation to biopotential detection, especially when electrodes having small electrode–skin contact area are used. In general, the input impedance of an amplifier, Z_i, is much larger than the electrode impedance Z_e, and the attenuation of the signal is negligible. In the case of differential detection, if the two electrode impedances Z_{e1} and Z_{e2} are not identical, the common mode voltage (V_{CM}, e.g., power-line interference) will generate a differential voltage (V_d) at the input of the differential amplifier (Fig. 6.2a). For $Z_i >> Z_e$ this voltage value is approximately given by

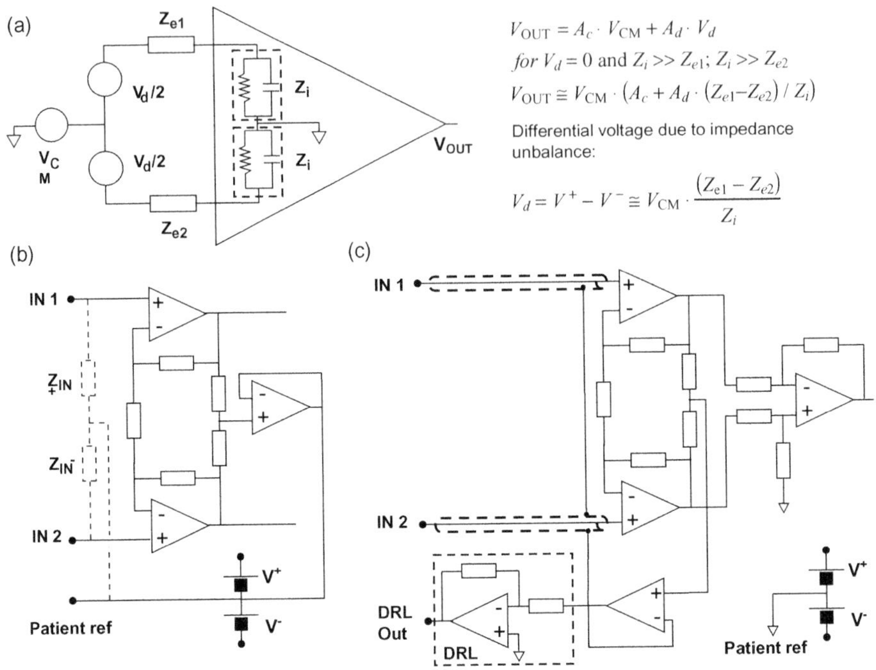

Figure 6.2. (a) Simplified model describing the presence of the common mode V_{CM} (e.g., power-line interference) and its conversion into a differential voltage due to the imbalance of the electrode–skin contact impedances (Z_{e1} and Z_{e2}). (b) Reduction of power-line interference by bootstrapping the isolated power supply; Z_{IN}^{+} and Z_{IN}^{-} are the internal input impedances of the amplifiers. The entire circuit floats at the common mode voltage and no common mode current theoretically flows into Z_{IN}^{+} and Z_{IN}^{-}; V^{+} and V^{-} are the power supply voltages for the operational amplifiers. (c) Instrumentation amplifier with guarding and common mode feedback (DRL = driven right leg circuit).

$$V_d \approx V_{CM}(Z_{e1} - Z_{e2})/Z_i \tag{6.1}$$

Since V_d is amplified with the differential gain (A_d) and V_{CM} is amplified with the common mode gain (A_c) of the differential amplifier, the output voltage will be

$$V_{out} = A_c V_{CM} + A_d V_d = V_{CM}(A_c + A_d(Z_{e1} - Z_{e2})/Z_i) \tag{6.2}$$

and the output referred to the input:

$$V_{out}/A_d = V_{CM}(A_c/A_d + (Z_{e1} - Z_{e2})/Z_i) \tag{6.3}$$

Equation 6.3 shows that the output signal due to the common mode input voltage is a function of the following: (a) the common mode rejection ratio

(CMRR = $20\log_{10}(A_d/A_c)$) of the amplifier, (b) the input impedance of the amplifier (Z_i), and (c) the unbalance of the two electrode impedances ($Z_{e1} - Z_{e2}$). The last contribution (c), often called the voltage divider effect, is usually the main source of power-line interference (Neuman, 1998b). Therefore, low and balanced impedances of the electrode–skin contacts are required to improve the rejection of common mode signals. Since the main factor affecting the impedance of the electrode–gel–skin interface is the outermost skin layer (the epidermis), the most common procedure to reduce, balance, and stabilize the impedances is to treat the skin in order to remove the body oils and flaky skin layers. The literature concerning skin treatments to improve the electrode–gel–skin interface is very heterogeneous: washing with soap, shaving, slight or strong abrasion, and rubbing with conductive gel, ether or alcohol, and solvents. Many authors mentioned the value of such impedance without indicating electrode size and the frequency at which measurements were made, so that comparison between observations of different authors is often impossible (Merletti et al., 2010). Evidence suggests that the treatment recommended by Hermens and co-workers (Hermens et al., 2000; rubbing with abrasive gel and then cleaning with tap water) is the best among those being tested (Merletti et al., 2010). Contrary to common practice, rubbing with alcohol or solvents could dry the skin and therefore increase the impedance.

Electrode Classification, Electrode Materials, and Gels

Surface electrodes are usually classified based on their constituent materials, on the technologies for their manufacturing, and on their physical properties (dimensions, shape). These characteristics may strongly influence the properties of the electrode–skin interface and therefore the quality of the detected signal.

The first electrode classification is based on the type of interface with the skin; on this basis we can distinguish between dry and nondry or wet electrodes. Wet electrodes include a layer of conductive gel, hydrogel, or sponge saturated with an electrolyte solution between the metal and the skin in order to reduce the electrode–skin impedance. The presence of gel also improves the electromechanical stability of the electrode–skin contact providing a flexible link between the skin and the metal that reduces the occurrence of movement artifacts. Dry electrodes do not require an electrolyte gel to establish and maintain a contact with the skin. Several types of dry electrodes exist: pin or bar electrodes, carbon electrodes, and sintered silver or silver chloride electrodes. Although dry electrodes offer the advantage of not requiring electrolyte gel, they embody certain disadvantages as well. The two major problems are the low contact stability and the high electrode–skin impedance. To overcome these limitations, dry disposable electrodes made of silver-coated silicon with etched spikes have been proposed (Griss et al., 2001). The spikes on the electrode surface reduce the contact impedance because they pierce the first layer of the skin (circumventing its high impedance) and increase the contact surface. Moreover, spiked electrodes should reduce noise as well as artifacts due to relative electrode–skin movement.

The constituent material of the electrode affects its electrochemical characteristics (Neuman, 1998a). Electrodes made of gold and platinum are characterized by a strong capacitive behavior due to a double layer of charges at the metal–electrolyte interface. For this reason they are usually called "polarizable electrodes." When a voltage is applied, there is no actual charge flow between the tissue and the electrode, but a change of charge distribution due to displacement currents. Although the use of capacitive electrodes is being investigated, their application is limited by their susceptibility to movement artifacts. In fact, a movement of the metal surface with respect to the skin can induce a change in the surface potential, leading to temporary instability of the contact and consequently to low-frequency artifacts. Conversely to capacitive electrodes, nonpolarizable electrodes have a mainly ohmic behavior, which allows a free flow of charges across the interface, and are currently preferred due to their higher stability. The electrode that better fits the requirements of nonpolarizability is the silver–silver chloride (Ag–AgCl) electrode (Janz and Ives, 1968; Neuman, 2000), which is highly stable and exhibits a lower noise level with respect the other metallic electrodes (Geddes and Baker, 1967).

Other important features affecting the properties of the detected signal are the geometrical characteristics of the electrode. Besides the effect of electrode size on the electrode–skin impedance (the greater the size of the electrode, the lower the contact impedance), the effect on the spatial filtering on the surface potential must be taken into account. The sources of electrical field generate a two-dimensional potential distribution on the skin surface. Such a potential is defined with respect to a reference point located far away and not influenced by the source. The ideal condition for measuring this potential distribution is to have a point electrode on the skin surface connected to an infinite input impedance voltmeter measuring the potential with respect to a reference electrode located at infinite distance (where the potential is zero). It should be noted that, in real conditions, the nonideality of the detection system (nonideal voltmeter) will add offset and noise to the signals. Moreover, the potential under the electrode will be modified by the physical dimension of the electrode. Assuming that the size of the electrode is small compared with the geometrical extension of the potential distribution (Dimitrov and Dimitrova, 1998; Dimitrova et al., 1999; Farina and Merletti, 2001), we can assume the finite area of the electrode to be made of a series of equally spaced point electrodes with the same impedance and the recorded potential to be the average of the potentials detected from these point electrodes (Fig. 6.1b). The averaging process implies a spatial smoothing on the surface potential that is a two-dimensional spatial filter whose smoothing effect increases with increasing electrode size. For EMG applications, in order to avoid loss of information, small electrodes (radius <5 mm for circular electrodes) are preferable (Dimitrova et al., 1999; Farina and Merletti, 2001; Helal and Bouissou, 1992).

CONDITIONING OF SURFACE ELECTROMYOGRAPHIC SIGNALS

Front-End Amplifiers

The input stage of the amplification chain plays a very important role in the amplification of sEMG signals (van der Locht et al., 1980) and of biopotentials in general (Silverman and Jenden, 1971). High-input impedance, high CMRR, and low noise are mandatory features of any biopotential amplifier.

In commercially available instrumentation amplifiers (IAs), the input impedance has a resistive and a capacitive component, usually quoted in the datasheet as a resistor (higher than 1 GΩ) in parallel with a capacitor (2–12 pF). To reduce the "voltage divider effect" described following Eq. 6.3, the input impedance for a surface EMG amplifier must be at least 100 times greater than the largest expected electrode–skin impedance. Considering a capacitive component of 10 pF and a resistive component of 1 GΩ, the input impedance of the IA at 100 Hz will be about 140 MΩ. Since the electrode–skin impedance can reach a few megohms for small and dry electrodes, a difference of 10% in the electrode–skin impedance between two electrodes of a differential recording would lead to a ratio $\Delta Z/Z_i$ of 10^{-3}. According to Eq. 6.3, this value would contribute 1 mV to the input referred voltage for each volt of common mode input voltage, regardless of the CMRR of the IA. Considering that the common mode voltage due to power-line interference can reach a few hundred millivolts and that the differential sEMG signal ranges between a few microvolts and a few millivolts, it becomes clear that the reduction (and the balancing) of the electrode–skin impedance is a key factor in the improvement of the signal-to-interference ratio.

To increase performance and power-line rejection, different front-end configurations have been used in the past and described in the literature (Accornero, 1981; Basmajian and Hudson, 1974; Bergey and Squires, 1971; Johnson et al., 1977; Metting van Rijn et al., 1991a,b; Nishimura et al., 1992). The circuit depicted in Figure 6.2b provides a bootstrap technique that strongly reduces the effect of $Z_{e1} - Z_{e2}$, Z_{IN}^{+}, and Z_{IN}^{-}. A buffer with unitary gain (e.g., an operational amplifier in voltage follower configuration) may be interposed between the electrodes and the IA of the input stage, as close as possible to the electrodes. This buffer acts as an impedance transformer, transforming a high impedance at the electrode side of the circuit to a low impedance at the cable side of the circuit, reducing the coupling of interfering sources to the EMG wires (Clancy et al., 2002). Alternatively, different shielding techniques have been proposed to reduce the interference currents in the measurement cables (Metting van Rijn et al., 1990). When the shield of the cable is driven by the signal at the inner wire (guarding technique) there is no cable capacitance and its contribution to the input impedance of the circuit is negligible (Morrison, 1977). For multichannel signal detection, guarding with the average of the input signals has been used (Metting van Rijn et al., 1990). According to this approach, if all the shields are driven by the average of the detected signals (the average is assumed to be the external common mode voltage), there is no potential difference between the inner wire and the shield for the common mode. Thereby, the equivalent input capacitance for the common mode is very small (theoretically zero).

Another possible approach that can be used in combination with the previously described methods is reduction of the common mode by a common mode feedback on the subject. This technique is widely used in ECG recording and is referred to as "driven right leg" (DRL).

The input noise of the IA affects the quality of the recorded signal for two reasons: the amplitude of the potential distribution on the skin surface is comparable to the noise amplitude, and both the signal and the input noise are amplified and filtered by the rest of the chain in the same way and can be discriminated off-line only by using sophisticated denoising techniques. It is therefore necessary to use front-end amplifiers with very low input noise.

The input noise introduced by the first stage has two contributions: (1) the input voltage noise, and (2) the input current noise flowing in the electrode–skin impedance. These noise sources are statistically independent; therefore, the power of the equivalent input noise can be estimated as the sum of the two noise contributions over the amplifier bandwidth:

$$P_n = B(e_n^2 + i_n^2 Z_e^2) \tag{6.4}$$

where e_n and i_n are the voltage and current noise densities of the amplifier, B is the amplifier bandwidth, and Z_e is the electrode–skin impedance. For high values of electrode–skin impedance, the input current noise may be relevant (e.g., for impedances greater than 1 MΩ, i.e., with the use of small dry electrodes). For example, an amplifier with $e_n = 20\ \text{nV/Hz}^{0.5}$ and $i_n = 50\ \text{fA/Hz}^{0.5}$ would produce a noise voltage of 1.2 μV_{RMS} (corresponding to about 7 μV peak-to-peak noise voltage) on an electrode with 1 MΩ impedance, in the bandwidth 10–500 Hz.

Signal Conditioning and Analog-to-Digital Conversion

The classical conditioning circuit that follows the front-end amplifier includes (a) band-pass filters to remove the components outside the sEMG bandwidth and to prevent aliasing, (b) amplifiers (usually with selectable gain) to exploit the entire input range of the analog-to-digital (A/D) converter, (c) A/D conversion block, and (d) optocouplers to guarantee the required high degree of safety for the patient, even in case of equipment malfunction.

The surface-detected signal often shows slow variations due to instability of the electrode–skin contact, usually referred to as movement artifact. The spectral content of these unwanted signals is generally in the frequency range 0–20 Hz and their amplitude can be one degree of magnitude higher than the peak amplitude of the surface EMG signal. For these reasons a high-pass filter (15–20 Hz) should be implemented directly in the first stage to remove low-frequency components, avoiding saturations and the consequent loss of information. A low-pass filter (400–500 Hz) is usually placed at the end of the amplifier chain to remove the high-frequency noise introduced by all the components of the chain. Additional high-pass filters can be placed along the chain to remove the offset introduced by the cascaded amplification stages.

The phase response of a filter is an important parameter to consider when choosing or designing the filter type. A filter with nonlinear phase will introduce different delays to the spectral components, resulting in a shape modification of the detected waveforms. Thereafter, filters with Bessel response are the most suitable for applications where waveform shapes are of interest, whereas this issue is not relevant when only EMG amplitude is of interest, such as in biofeedback applications. In some cases notch filters have been used to reduce power-line interference. These filters attenuate not only the interferences, but also spectral components of the surface EMG signal. In addition, notch filters introduce a nonlinear phase response below and above the central frequency.

The analog-to-digital conversion is performed in two steps: time sampling and amplitude quantization. The time sampling transforms the analog signal (defined for any time instants) in a set of values corresponding to the amplitude of the signal in time instants equally spaced in time. The time interval between two consecutive samples is the inverse of the sampling frequency. According to the Nyquist theorem, in order to avoid loss of information in the signal reconstruction, the sampling frequency must be at least twice the highest frequency of the harmonics of the sampled signal. For almost all muscles and most applications, a sampling frequency of 1000 Hz or 2000 Hz is suitable for sEMG signals. The amplitude quantization process converts the time-sampled voltages into discrete levels represented in binary code. The number of bits of the A/D converter (i.e., the number of digits of the output code, n) defines the number of levels ($2^n - 1$) into which the input range is subdivided. The input range divided by the number of levels is the resolution of the conversion. For example, an A/D converter with $n = 12$ bits has a resolution of 2.44 mV referred to the input of the A/D converter with an input range of ±5 V, resulting in a resolution of $2.44/G$ mV referred to the input of the front-end amplifier (where G is the global gain of the amplifier). Thus, the amplification of the conditioning chain and the input range of the A/D converter are tied by the desired resolution. For this reason, sEMG recording systems have a selectable gain amplifier that allows the user to modify the gain to match the amplitude of the amplified signal with the A/D converter input range. The gain is selected to prevent signal saturation (i.e., signal truncation due to exceeding the A/D input range) while using the full A/D converter range to obtain the highest resolution.

If the number of bits of the A/D converter is large enough (e.g., 24 bits), the conditioning of the signal can be performed by software. In this case the analog front end includes just a low-gain amplifier and a low-pass filter to prevent aliasing in the sampling process. This approach takes advantage of the high resolution of the A/D converter, which allows a low amplification of the signal and can thereby accommodate DC components, movement artifacts, and power-line interference without saturation. The signal can be stored as received from the amplifier, and filtering can be performed after acquisition.

DETECTION OF SEMG SIGNALS: ELECTRODE CONFIGURATION, DISTANCE, LOCATION

The importance of a standardization of electrode configuration and location for the reproducibility and correct interpretation of sEMG measurements has been widely recognized (Zuniga et al., 1970; Vitasalo and Komi, 1975; Rainoldi et al., 2000; Farina and Merletti, 2001; Farina et al., 2002a). The European concerted action SENIAM showed that a large variety of sensors, sensor placement procedures, and equipment is used in European laboratories to detect sEMG signals (Hermens et al., 1999) and this is still one of the major issues in clinical surface EMG.

Monopolar Detection

As discussed in the previous sections, the ideal configuration for the detection of the potential distribution on the skin is to move, on the skin, a point electrode measuring the voltage with respect to a remote reference where the potential is zero (monopolar detection).

Monopolar detection provides the whole information that can be recorded from the detection volume but it is mainly used in research applications because of its lack of spatial selectivity (recording of the contribution of sources that are near or far from the electrodes) and its sensitivity to common mode signals. As a consequence, only the information about the compound activity of a large number of MUs can be obtained form monopolar sEMG signals.

Spatial filtering techniques have been proposed to detect surface EMG signals, enhancing the spatial selectivity of surface recordings by limiting the detection volume (Broman et al., 1985; Disselhorst-Klug et al., 1997; Farina et al., 2002a; Rau and Disselhorst-Klug, 1997).

Spatial Filtering

Spatial filters in surface EMG detection are based on the linear combination of signals detected by a number of electrodes placed over the skin with a defined geometry with the purpose of attenuating specific spatial frequencies with respect to others. One of the most common goals is the attenuation of nonpropagating components of either physiological (such as the end-of-fiber effects and remote sources) or external (power-line interference) origin that arc present in the monopolar signals.

The effect of the spatial filter on the detected signal depends on the weights assigned to each electrode, on the geometry of the electrode configuration, and on the electrode shape and size.

Spatial Filters Composed of Point Electrodes. Consider a muscle fiber with infinite length aligned along the z-axis. The fiber generates a potential distribution $\phi(x, z)$ over the skin that moves along the fiber direction with a velocity v. The potential detected over the skin by a single point electrode with coordinates $x = x_i$ and $z = z_i$ at time zero is

$$V_i = [\phi(x,z) * (\delta(x + x_i)\delta(z + z_i))]_{\substack{x=0 \\ z=0}} \tag{6.5}$$

If we consider $M + 1$ detection points with weights a_i ($i = 0, \ldots, M$), the resulting spatial filter is

$$\begin{aligned} \sum_{i\_=0}^{M} a_i V_i &= \sum_{i=0}^{M} [\phi(x,z) * (a_i\delta(x + x_i)\delta(z + z_i))]_{\substack{x=0 \\ z=0}} \\ &= \left[\phi(x,z) * \sum_{i=0}^{M} (a_i\delta(x + x_i)\delta(z + z_i))\right]_{\substack{x=0 \\ z=0}} \end{aligned} \tag{6.6}$$

and, including the temporal variable,

$$\sum_{i\_=0}^{M} a_i V_i(t) = \left[\phi(x,z) * \sum_{i=0}^{M} a_i\delta(x + x_i)\delta(z + z_i)\right]_{\substack{x=0 \\ z=-vt}} = [\phi(x,z) * h(x,z)]_{\substack{x=0 \\ z=-vt}} \tag{6.7}$$

From Eq. 6.7 it appears that the operation performed by a spatial filter can be described by an equivalent two-dimensional spatial impulse response:

$$h(x,z) = \sum_{i=0}^{M} a_i\delta(x + x_i)\delta(z + z_i) \tag{6.8}$$

whose corresponding transfer function is given by the two-dimensional Fourier transform of Eq. 6.8:

$$H(f_x, f_z) = \sum_{i=0}^{M} a_i e^{j2\pi f_x x_i} e^{j2\pi f_z z_i} \tag{6.9}$$

The values assigned to the weights a_i allow the design of transfer functions with particular characteristics. Usually the spatial filters are designed to meet the following basic condition:

$$H(0,0) = \sum_{i=0}^{M} a_i = 0 \tag{6.10}$$

to obtain the rejection of the DC components in the two spatial directions.

The principle of spatial filtering can be applied to one-dimensional (Broman et al., 1985; Gydikov et al., 1986; Merletti et al., 1999) or two-dimensional electrode arrangements (Reucher et al., 1987a,b).

One-Dimensional Spatial Filters. The simplest and most widely used spatial filter is the bipolar or single differential (SD), which records the difference between the potentials detected by two electrodes placed at a fixed interelectrode distance (Fig. 6.3a).

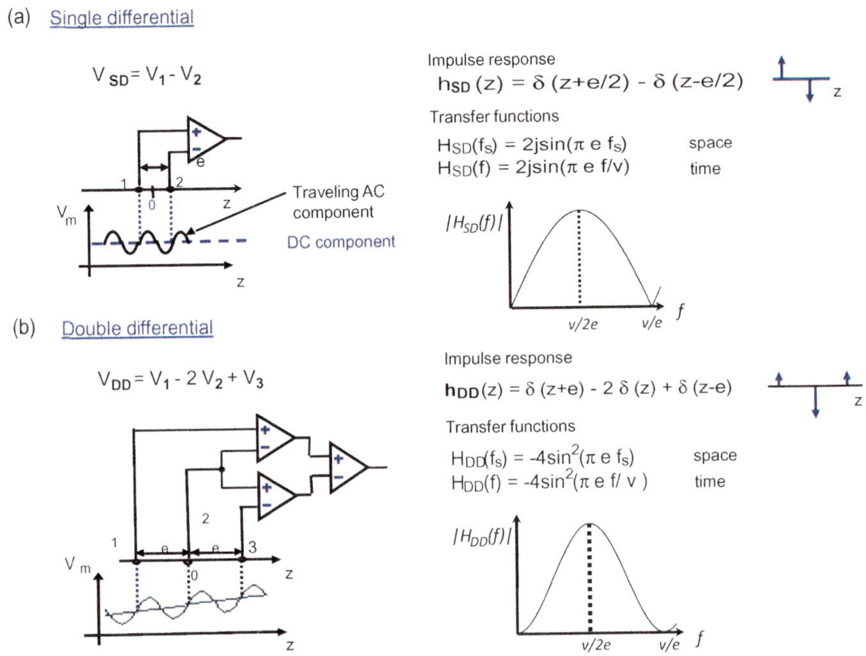

Figure 6.3. Impulse responses and transfer functions of the single and double differential configuration (a and b). (a) The transfer function of the differential system is a sinusoid: for $\lambda \ll e$ the system behaves as a differentiator, it shows a maximum for $\lambda/2 = e$, it approximates an integrator for $\lambda/2 < e < \lambda$, and it presents its first dip for $\lambda = e$. (b) The transfer function of a double differential system is a sinusoid raised to power 2: for $\lambda \ll e$ the system behaves as a second-order differentiator, its transfer function shows a maximum for $\lambda/2 = e$, and it presents its first dip for $\lambda = e$.

Despite the simplicity of the bipolar configuration, the effect of its transfer function as a spatial filter requires an accurate analysis for correct EMG detection and interpretation.

Consider a bipolar detection system with an interelectrode distance e, aligned along the fiber direction (z-axis whose zero is at the midpoint between the two electrodes). The detection system samples the potential distribution which changes in space and time.

For a given time instant, the SD voltage is $V(z)\delta(z + e/2) - V(z)\delta(z - e/2)$, where $\delta(\cdot)$ is the Dirac delta function (Fig. 6.3a). Power-line interference or contributions from distant sources whose field reaches both electrodes with the same intensity and at the same time are cancelled by the bipolar detection system because $V(-e/2) = V(e/2)$.

In the case of a potential distribution moving along z at a velocity v, $V(z)$ can be seen, in the spatial frequency domain, as a summation of sinusoidal harmonics. If we consider one harmonic at a time with wavelength λ and $e = n\lambda$, with $n = 1, 2, \ldots$, the detected voltage is zero, regardless of the propagation velocity v. If $e = n\lambda \pm \lambda/2$, the detected voltage is a sinusoid with the maximal possible peak value since the two electrodes detect the difference between the peaks of opposite phases of the harmonic. In the case of intermediate values of λ, the output of the system is a sinusoid whose amplitude is a function of λ. As a consequence, the differential detection system is a spatial filter whose output depends on the spatial frequency of the input. Recalling the relationships between the velocity v, the wavelength λ, the frequency in space f_s, the period T, and the frequency in time f_t ($v = \lambda/T = f_t/f_s$, where $\lambda = 1/f_s$ and $T = 1/f_t$), the impulse response and the transfer function in time and space are given by the equations in Figure 6.3a.

The preceding considerations hold for point electrodes and propagating potentials. In the case of nonpoint electrodes, the transfer function of an electrode system is the product of the electrode and of the spatial filter transfer functions. On the basis of these considerations, sEMG spectral parameters obtained with different interelectrode distances and electrode sizes cannot be compared (Fuglevand et al., 1992).

Single differential detection provides the rejection of common mode signals; a further enhancement of spatial selectivity can be achieved by using a more selective spatial filter.

One of the most important classes of spatial filters in sEMG recording is represented by the Laplace filters. The simplest Laplacian filter is the double differential (DD) filter (Fig. 6.3b) that is constituted by three equally spaced electrodes, the central electrode weighted with a factor −2 and the others with +1. As a consequence, the impulse response of the DD system is a set of three delta functions (with weights 1, −2, 1) and its transfer function is a sinusoid raised to the power 2 (Fig. 6.3b).

More complex detection configurations to provide selectivity high enough to separate single motor unit action potentials (MUAPs) from the interference EMG signal are represented by two-dimensional spatial filters.

Two-Dimensional Spatial Filters. The propagation of a MUAP along the muscle fibers results, on the skin surface, in a propagating two-dimensional distribution spatially low-pass filtered by the volume conductor. Since the optimal spatial filter is the one closest to the inverse of the volume conductor filter, two-dimensional spatial high-pass filters have been proposed (Disselhorst-Klug et al., 1997; Reucher et al., 1987a,b; Farina and Rainoldi, 1999).

Reucher and co-workers proposed the normal double differentiating (NDD) filter realized by five crosswise-arranged electrodes whereby the central electrode is weighted with the factor −4 and the surrounding electrodes with the factor +1. The NDD filter builds a two-dimensional spatial high-pass filter whose impulse response is

$$h(x,z)_{\text{NDD}} = \delta(x,z+d_z)+\delta(x+d_x,z)-4\delta(x,z)+\delta(x,z-d_z)+\delta(x-d_x,z) \quad (6.11)$$

where d_x and d_z are the interelectrode distances in the two spatial directions. The NDD filter improves the spatial selectivity in all directions (Fig. 6.4) and

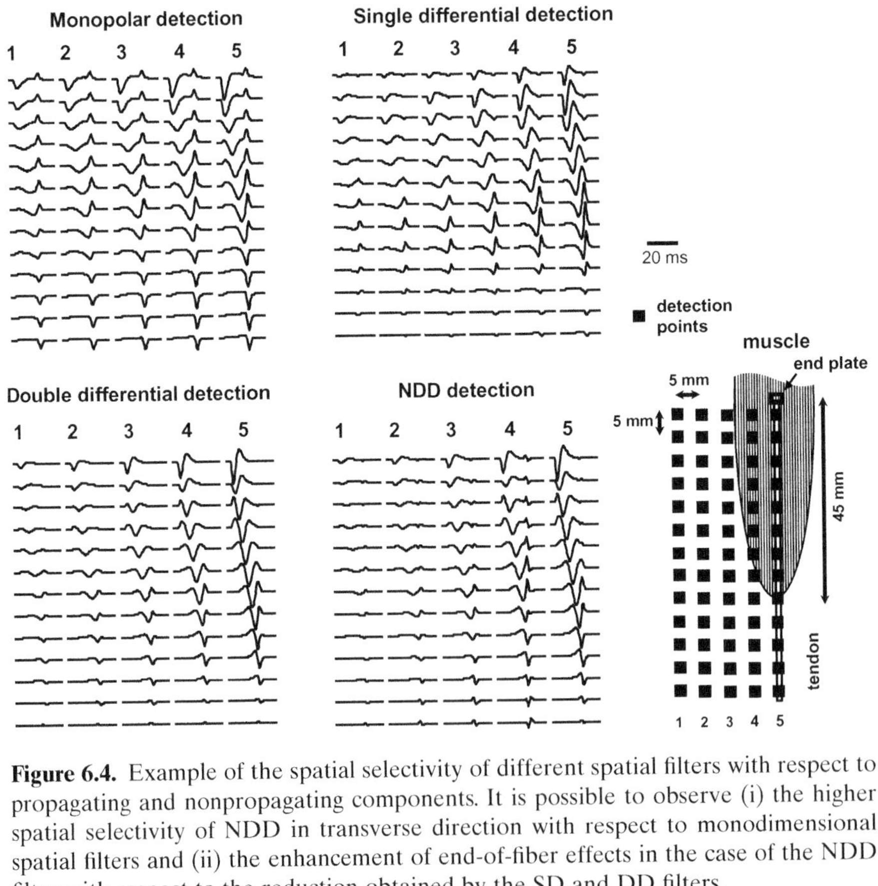

Figure 6.4. Example of the spatial selectivity of different spatial filters with respect to propagating and nonpropagating components. It is possible to observe (i) the higher spatial selectivity of NDD in transverse direction with respect to monodimensional spatial filters and (ii) the enhancement of end-of-fiber effects in the case of the NDD filter with respect to the reduction obtained by the SD and DD filters.

allows the separation of the activities of single MUs even at maximum voluntary contraction (Reucher et al., 1987a,b). Moreover the two-dimensional systems are more isotropic than one-dimensional systems: as a consequence, they are less sensitive to fiber orientation and the electrode placement is less critical with these types of filters.

Spatial Filtering and Interelectrode Distance. The interelectrode distance (IED) is regarded as one of the most relevant properties of the sEMG detection systems. Although it affects sEMG signal characteristics, a high variability and a wide range of values for IEDs (4–48 mm) can be found in the literature (Hermens et al., 1999).

It has often been suggested in the literature that a decrease of IED would increase the spatial selectivity of the detection system. By fitting experimental data, the relationship between the voltage detected by the electrodes and the distance between the motor unit and the recording electrodes has been described as $V = V_0 / (r / r_0)^D$, where D and V_0 are constants and r_0 is a reference distance from the electrical center (the location of the equivalent generator of the MUAP), where $V = V_0$. The exponent D is a function of IED and of the detection system. This would imply higher values of D upon decreasing IED. In literature no evidence can be found for this. Roeleveld (Roeleveld et al., 1997a) performed an experimental study investigating the effect of the IED variation (from 6 to 84 mm) in the bipolar detection on the contribution of motor unit potentials to the surface EMG. The contribution of superficial and deep motor units to the recorded sEMG signal was found to be unrelated to IED as long as IED < 40 mm while for IED exceeding 40 mm the contribution of deeper MUs to sEMG is greater than the contribution of superficial ones.

Detection System Orientation and Location on the Muscle. The *orientation* of a monodimensional detection system defines its direction with respect to the direction of the muscle fibers; the *location* defines the position of the detection system on the muscle.

The SD and DD detection systems are usually placed in the direction of the muscle fibers but they could also be placed in the transverse direction. When electrodes are arranged parallel to the muscle fibers, the spatial filters are referred to as longitudinal (LSD and LDD); when arranged transversely to the muscle fibers, the spatial filters are known as transverse (TSD and TDD). The LSD and LDD filters result in better longitudinal and poorer transverse selectivity with respect to TSD and TDD filters.

In the literature, the most common locations of the detection system on a muscle are the following: (a) in the center of the muscle; (b) on the muscle belly; (c) somewhere between the innervation zone and one tendon; (d) on the motor point.

Although the transfer function of the spatial filter is independent of the electrode location on the muscle, the motor end-plates (where MUAPs generate) and the muscle–tendon junctions (where MUAPs extinguish) are two positions that must be carefully considered.

As shown before, the signal generated by a single fiber and detected with a bipolar system placed along the fiber direction, symmetrically with respect to the end-plate of the fiber, provides a zero voltage. Since the innervations of motor units in a muscle are concentrated in one or two locations, this electrode location corresponds to signals that are small, noisy, and sensitive to small displacements between electrodes and muscle (Masuda et al. 1985) and must be avoided in practical applications. Similar observations apply to the muscle–tendon junctions. For this reason, the location of a pair of electrodes is critical and should be optimized by placing the electrodes between the innervation zone(s) and a muscle–tendon junction. Figure 6.5 depicts this situation and shows the importance of correct electrode placement.

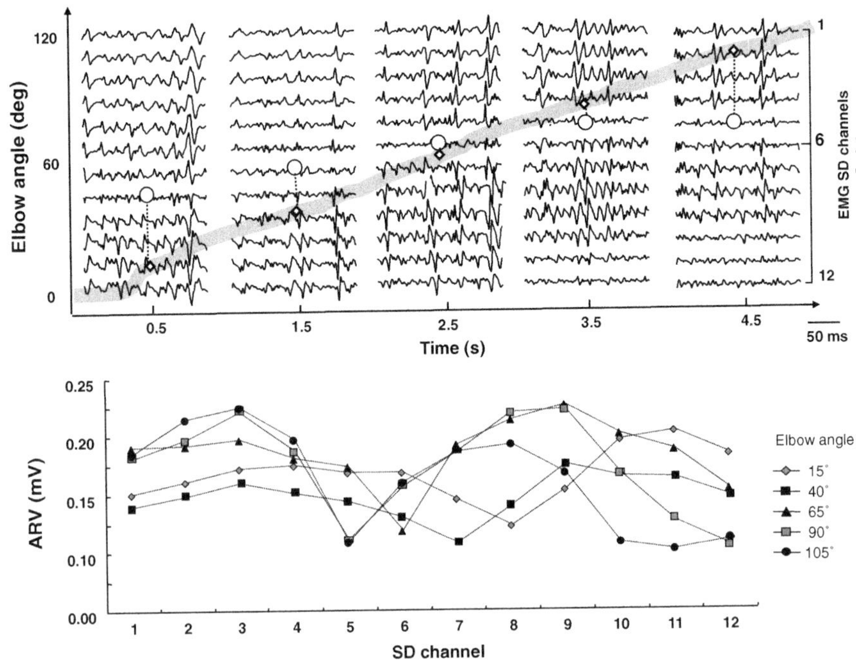

Figure 6.5. Example of the shift of innervation zone (IZ) and muscle–tendon junction during a progressive elbow flexion lasting five seconds (0° corresponds to maximum extension). Surface EMG signals have been recorded with 12 SD detection systems placed on the biceps brachii. (top) The thick gray line represents the elbow angle time course. Raw signals for the 250-ms epochs centered in the second 0.5, 1.5, 2.5, 3.5, and 4.5 are shown, and the IZ positions (expressed as SD channel) estimated on these epochs are reported (circles). (bottom) Averaged rectified value (ARV) estimated on each channel for different elbow angles. The influence of the shift of the muscle under the detection system on the amplitude estimation is evident. The minimum of the amplitude (corresponding to the IZ position) moves from channel 8 (for 15° of elbow flexion) to channel 5 (for 105° of elbow flexion).

Electromyographic global variables, such as amplitude, spectral characteristic frequencies, and estimates of conduction velocity, are heavily altered when electrode pairs are placed on or near the innervaton zone (Jensen et al., 1993; Lateva et al., 1993; Rainoldi et al., 2000; Roy et al., 1986).

These considerations imply the identification of the innervation zone(s) before the electrode pairs are applied: this task can be achieved by means of an electrode array (see section "Spatial sampling in one dimension: the linear arrays").

Spatial Filtering and Cross Talk. The signal detected on one muscle and generated by another active muscle is referred to as **cross talk.** Many applications of sEMG require the simultaneous detection of sEMG from many muscles to evaluate, for example, the muscular coordination pattern (Koh and Grabiner, 1993; Winter et al., 1994). In these cases, it is mandatory to reduce the cross talk from nearby muscles.

Spatial high-pass filters enhance the signals propagating along the fibers of MUs located close to the recording electrodes and suppress the contributions of more distant sources. At some distance from the sources, the contributions due to end-of-fiber effects become predominant with respect to the propagating components because the latter decay in space more slowly than the former. It could be concluded that high-pass spatial filtering would reduce cross talk, and different kinds of spatial filters have been applied for cross-talk reduction. However, the experimental results have shown that cross talk is not reduced by spatial high-pass filtering (van Vugt and van Dijk, 2001).

This discrepancy between theory and practice is justified by the fact that the model used for the description of spatial filter theory considered a potential distribution generated on the skin by a propagating MUAP, neglecting the MUAP generation and extinction phenomena. It is known that the generation and extinction phenomena produce a nonpropagating potential distribution on the skin surface, and spatial filters have different transfer functions with respect to propagating and nonpropagating signals (Dimitrov and Dimitrova, 1974). Simulations of the filter responses to nonpropagating potentials have shown that most spatial filters do not reduce them and in some cases even enhance them (Dimitrova et al., 2002; Farina et al., 2002b). Figure 6.4 shows an example of enhancement of end-of-fiber potentials by an NDD filter with respect to the reduction obtained by SD and DD filters.

Although different techniques have been tested to reduce cross talk, this issue is not yet satisfactorily solved.

SPATIAL SAMPLING

The surface EMG signal evolves in time and space, and it can be described as a three-dimensional signal with one temporal and two spatial (the skin plane) dimensions. Sampling the EMG potential distribution by placing a number of detection systems in different locations over the skin allows the study of how the surface EMG signal evolves in time and space.

If a spatial filter (one-dimensional (SD or DD) or two-dimensional (NDD or another type)) is applied to each detection point, the potential distribution is spatially filtered and also spatially sampled.

Spatial Sampling in One Dimension: the Linear Arrays

The first systems for performing spatial sampling of sEMG were proposed by Broman and Masuda (Broman et al., 1985; Masuda et al., 1985; Masuda and Sadoyama, 1986). They proposed linear arrays of electrodes placed along the fiber direction to estimate the velocity of propagation of action potentials, to identify some anatomical characteristics such as the innervation zone location and the muscle fiber length, and to investigate in detail the processes of generation, propagation, and extinction of the MUAPs along the muscle fibers.

Roeleveld and co-workers (Roeleveld et al., 1997b) proposed the use of two electrode arrays located both longitudinally and transversely with respect to the muscle fibers to estimate MU depth.

Linear electrode arrays have also been applied to obtain guidelines for the standardization of the sEMG recording when a global analysis of the signal is performed (Farina et al., 2002a; Farina and Merletti, 2001; Jensen et al., 1993; Rainoldi et al., 2000; Roy et al., 1986).

Spatial Sampling in Two Dimensions

The spatial distribution of the voltage on the skin above a muscle can be detected with a grid of electrodes that provides a two-dimensional (2D) sampling in space. If the grid covers a large part of the muscle, it provides spatial information that is largely independent of the temporal information. The time evolution of the voltage distribution on the skin can be tracked by sampling in time.

Multichannel sEMG is an interesting noninvasive methodology (1) to obtain muscle anatomical information (such as the location of innervation zones, tendon endings, and the direction of the muscle fibers), (2) to obtain a topographical representation of muscle activity, and (3) to decompose the surface EMG signal into the constituent single MU action potential trains if electrode grids with small electrode sizes and interelectrode spacing (high-density EMG, HD-EMG) are used (Zwarts and Stegeman, 2003; Drost et al., 2006; Holobar et al., 2009; Farina et al., 2010).

Anatomical Information and Topographical Representation of Muscle Activity. Knowledge of fiber direction and innervation zone location is relevant, for instance, for defining the optimal locations for estimating EMG variables in isometric and dynamic contractions. Topographical representation of muscle activity allows the study of the regional variations in the degree of

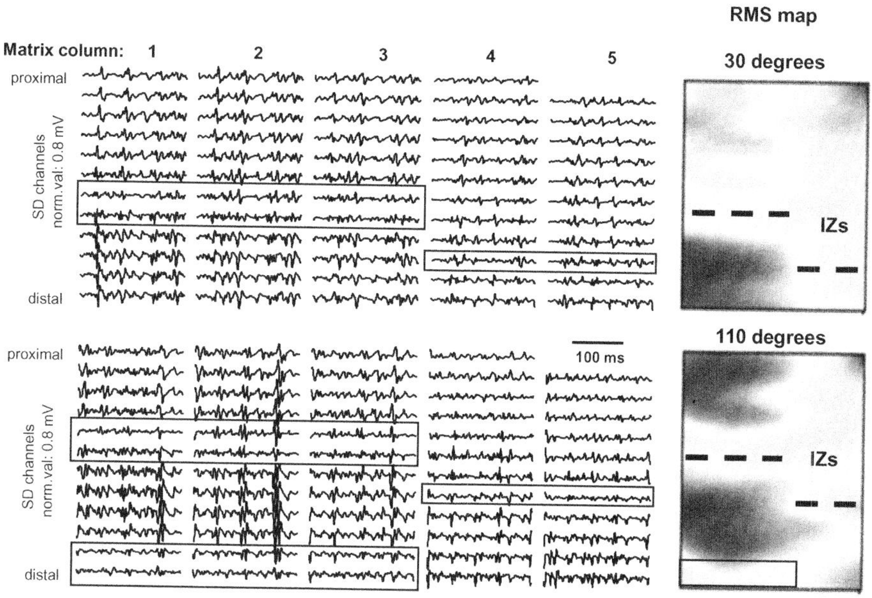

Figure 6.6. Example of topographical information obtained from multichannel detection systems. Surface EMG signals have been recorded during elbow flexion using a grid of electrodes (13 rows and 5 columns with one missed electrode, 8 mm IED) during a progressive elbow flexion. The SD sEMG signals (on the left) and the interpolated RMS distribution estimated on one epoch 250 ms long (on the right) are reported for two elbow angles (30° on the top and 120° on the bottom; 0° corresponds to maximum extension). The positions of the innervation zones and of the tendon are highlighted. From the images on the right it is possible to identify the two areas (left and right) of activity corresponding to the two heads of the biceps brachii and the different positions and shift of the innervation zones and tendon for the two biceps brachii heads.

muscle activation with time. This is particularly important in dynamic contractions. Figure 6.6 shows the distribution of single differential sEMG RMS at two different elbow angles of isometric contraction of the biceps brachii and demonstrates that a single sampling point is not representative of the spatially heterogeneous muscle activity and that the activity it detects depends, in a strong way, on the geometrical factors.

The 2D spatial sampling obtained using HD-sEMG detection systems results in a three-dimensional signal, which can be used to reconstruct the 2D potential distribution if the Nyquist limits are met in all three dimensions.

Aliasing in Space. To meet the Nyquist theorem in space, the interelectrode distance (IED) must be smaller than a threshold value. If we simplify the problem and we consider only propagating signals, the IED threshold value can be identified starting from the relationship $f_s = f_t/v$ between the spatial frequency (f_s, in cycles per meter), the temporal frequency (f_t, in cycles per second, or hertz), and the signal propagation velocity (v, in meters per second). If we consider 400 Hz as the highest temporal frequency of sEMG and a propagation velocity of about 4 m/s, the highest spatial frequency is 100 cycles per meter. For the Nyquist theorem, the spatial sampling frequency should be higher than 200 samples per meter, which means an IED less than 5 mm. Some commonly used values of IED (10 or 20 mm) imply aliasing in space but its consequences on the sEMG signals have not been investigated.

Decomposition of sEMG. Several methods have been developed to make use of the more detailed information provided by the HD-sEMG signal to identify individual MU discharge trains (Rau and Disselhorst-Klug, 1997; Kleine et al., 2007; DeLuca et al., 2006; Holobar and Zazula, 2007; Holobar et al., 2009). Contrary to decomposition of intramuscular EMG, only a few studies about the validation of decomposition of surface EMG techniques are present in the literature (Holobar et al., 2010); for this reason, although decomposition of the surface EMG has several potential applications, it is still not widely accepted and thus not extensively applied.

While the traditional one-electrode-pair technique remains the method of choice for movement analysis, HD-sEMG technology is opening new windows to look into the mechanisms associated with movement.

REFERENCES

Accornero N. (1981). Active surface EMG probe and contour follower. *Electroencephalogr Clin Neurophysiol* 51:331–332.

Basmajian J. and Hudson J. (1974). Miniature source attached differential amplifier for electromyography. *Am J Phys Med* 53:234–235.

Bergey G.E. and Squires R.D. (1971). Improved buffer amplifier for incorporation within a biopotential electrode. *IEEE Trans Biomed Eng* 18:430–431.

Broman H., Bilotto G., and De Luca C. (1985). A note on the non-invasive estimation of muscle fiber conduction velocity. *IEEE Trans BME* 32:341–343.

Clancy E.A., Morin F., and Merletti R. (2002). Sampling, noise-reduction and amplitude estimation issues in surface electromyography. *J Electromyogr Kinesiol* 12:1–16.

De Luca C.J., Adam A., Wotiz R., Gilmore L.D., and Nawab S.H. (2006). Decomposition of surface EMG signals. *J Neurophysiol* 96(3):1646–1657.

Dimitrov C.V. and Dimitrova N.A. (1974). Influence of the asymmetry in the distribution of the depolarization level on the extracellular potential field generated by an excitable fibre. *Electromyogr Clin Neurophysiol* 14(3):255–275.

Dimitrov C.V. and Dimitrova N.A. (1998). Precise and fast calculation of the motor unit potentials detected by a point and rectangular plate electrodes. *Med Eng Phys* 20:374–381.

Dimitrova N.A., Dimitrov C.V., and Chihman, V.N. (1999). Effect of electrode dimension on motor unit potentials. *Med Eng Phys* 21:479–485.

Dimitrova N.A., Dimitrov G.V., and Nikitin O.A. (2002). Neither high-pass filtering nor mathematical differentiation of the EMG signals can considerably reduce cross-talk. *J Electromyogr Kinesiol* (4):235–246.

Disselhorst-Klug C., Silny J., and Rau G. (1997). Improvement of spatial resolution in surface-EMG: a theoretical and experimental comparison of different spatial filters. *IEEE Trans Biomed Eng* 44:567–574.

Drost G., Stegeman D.F., van Engelen B.G., and Zwarts M.J. (2006). Clinical applications of high-density surface EMG: a systematic review. *J Electromyogr Kinesiol* 16:586–602.

Farina D. and Merletti R. (2001). Effect of electrode shape on spectral features of surface detected motor unit action potentials. *Acta Physiol Pharmacol Bulgarica* 26:63–66.

Farina D. and Rainoldi A. (1999). Compensation of the effect of sub-cutaneous tissue layers on surface EMG: a simulation study. *Med Eng Phys* 21(6–7):487–497.

Farina D., Cescon C., and Merletti R. (2002a). Influence of anatomical, physical and detection system parameters on surface EMG. *Biol Cybernet* 86:445–456.

Farina D., Merletti R., Indino B., Nazzaro M., and Pozzo M. (2002b). Cross-talk between knee extensor muscles. Experimental and model results. *Muscle Nerve* 26:681–695.

Farina D., Holobar A., Merletti R., and Enoka R.M. (2010). Decoding the neural drive to muscles from the surface electromyogram. *Clin Neurophysiol* 121:1616–1623.

Freriks B., Disselhorst-Klug C., Hermens H.J., and Rau G. (1997). Sensors and sensor placement procedures used in the European labs. In: *The State of the Art on Sensors and Sensor Placement Procedures for Surface Electromyography: A Proposal for Sensor Placement Procedures* (Hermens H. and Freriks B., Eds.), pp. 7–27. Enschede, NL: Roessingh Research and Development.

Fuglevand A.J., Winter D.A., Patla A.E., and Stashuk D. (1992). Detection of motor unit action potentials with surface electrodes: influence of electrode size and spacing. *Biol Cybern* 67(2):143–153.

Geddes L.A. and Baker L.E. (1967). Chlorided silver electrodes. *Med Res Eng* 6:33–34.

Grimnes S. (1982). Psychogalvanic reflex and changes in electrical parameters of dry skin. *Med Biol Eng Comput* 20(6):734–740.

Griss P., Enoksson P., Tolvanen-laakso, Merilainen P., Ollmar S., and Stemme G. (2001). Micromachined electrodes for biopotential measurements. *J Micromech Syst* 10:10–16.

Gydikov A., Kosarov D., Kossev A., Kostov K., Trayanova N., and Radicheva N. (1986). Motor unit potentials at high muscle activity recorded by selective electrodes. *Biomed Biochim Acta* 45(1–2):S63–S68.

Hary D., Bekey G., and Antonelli J. (1987). Circuit models and simulation analysis of electromyographic signal sources. I. The impedance of EMG electrodes. *IEEE Trans Biomed Eng* 34:91–97.

Helal J.N. and Bouissou P. (1992). The spatial integration effect of surface electrodes detecting myoelectric signal. *IEEE Trans Biomed Eng* 39:1161–1167.

Hermens H., Freriks B., Merletti R., Stegeman D., Blok J., Rau G., Disselhorst-Klug C., and Hägg G. (1999). *European Recommendations for Surface Electromyography*, ISBN 90-75452-15-2. Enschede, NL: Roessingh Research and Development.

Hermens H.J., Freriks B., Disselhorst-Klug C., and Rau G. (2000). Development of recommendations for SEMG sensors and sensor placement procedures. *J Electromyogr Kinesiol* 10(5):361–374.

Holobar A. and Zazula D. (2007). Multichannel blind source separation using convolution kernel compensation. *IEEE Trans Signal Process* 55:4487–4496.

Holobar A., Farina D., Gazzoni M., Merletti R., and Zazula D. (2009). Estimating motor unit discharge patterns from high-density surface electromyogram. *Clin Neurophysiol* 120:551–562.

Holobar A., Minetto M.A., Botter A., Negro F., and Farina D. (2010). Experimental analysis of accuracy in the identification of motor unit spike trains from high-density surface EMG. *IEEE Trans Neural Syst Rehabil Eng* 18:221–229.

Janz G.J. and Ives D.J.G. (1968). Silver–silver chloride electrodes. *Ann NY Acad Sci* 148:210.

Jensen C., Vasseljen O., and Westgaard R. (1993). The influence of electrode position on bipolar surface electromyogram recordings of the upper trapezius muscle. *Eur J Appl Physiol* 67:266–273.

Johnson S., Lynn P., Miller J., and Reed G. (1977). Miniature skin-mounted preamplifier for measurement of surface electromyographic potentials. *Med Biol Eng Comput* 15:710–711.

Kleine B.U., Van Dijk J.P., Lapatki B.G., Zwarts M.J., and Stegeman D.F. (2007). Using two-dimensional spatial information in decomposition of surface EMG signals. *J Electromyogr Kinesiol* 17:535–548.

Koh T.J. and Grabiner M.D. (1993). Evaluation of methods to minimize cross talk in surface electromyography. *J Biomech* 26(Suppl 1):151–157.

Lateva Z., Dimitrova N., and Dimitrov G. (1993). Effect of recording position along a muscle fiber on surface potential power spectrum. *J Electrom Kines* 3:195–204.

Masuda T., Myano H., and Sadoyama T. (1985). The position of innervation zones in the biceps brachii investigated by surface electromyography. *IEEE Trans BME* 32:36–42.

Masuda T. and Sadoyama T. (1986). The propagation of single motor unit action potentials detected by a surface electrode array. *Electroencephalogr Clin Neurophysiol.* 63(6):590–598.

Merletti R., Farina D., and Granata A. (1999). Non-invasive assessment of motor unit properties with linear electrode arrays. *Electroencephalogr Clin Neurophysiol* 50(Suppl):293–300.

Merletti R., Aventaggiato M., Botter A., Holobar A., Marateb H., and Vieira T.M. (2010). Advances in surface EMG: recent progress in detection and processing techniques. *Crit Rev Biomed Eng* 38(4):305–345.

Metting van Rijn A.C., Peper A., and Grimbergen C.A. (1990). High-quality recording of bioelectric events. Part 1. Interference reduction, theory and practice. *Med Biol Eng Comput* 28(5):389–397.

Metting van Rijn A.C., Peper A., and Grimbergen C.A. (1991a). High-quality recording of bioelectric events. Part 2. Low noise, low-power multichannel amplifier design. *Med Biol Eng Comput* 29:433–440.

Metting van Rijn A.C., Peper A., and Grimbergen C.A. (1991b). The isolation mode rejection ratio in bioelectric amplifiers. *IEEE Trans Biomed Eng* 38:1154–1157.

Morrison R. (1977). *Grounding and Shielding Techniques in Instrumentation*, 2nd ed., pp. 132–133. New York: Wiley.

Neuman M.R. (1998a). Biopotential electrodes. In: *Medical Instrumentation: Application and Design* (Webster J.G., Ed.), 3rd ed., pp. 183–232. Hoboken, NJ: Wiley.

Neuman M.R. (1998b). Biopotential amplifiers. In: *Medical Instrumentation: Application and Design* (Webster J.G., Ed.), 3rd ed., pp. 233–286. Hoboken, NJ: Wiley.

Neuman M.R. (2000). Biopotential electrodes. In: *The Biomedical Engineering Handbook* (Bronzino J.D., Ed.), 2nd ed., pp. 889–900. Boca Raton: CRC Press.

Nishimura S., Tomita Y., and Horiuchi T. (1992). Clinical application of an active electrode using an operational amplifier. *IEEE Trans Biomed Eng* 39:1096–1099.

Ragheb T. and Geddes L.A. (1990). Electrical properties of metallic electrodes. *Med Biol Eng Comput* 28:182–186.

Rainoldi A., Nazzaro M., Merletti R., Farina D., Caruso I., and Gaudenti S. (2000). Geometrical factors in surface EMG of the vasus medialis and lateralis muscles. *J od Electrom Kinesiol* 10:327–336.

Rau G. and Disselhorst-Klug C. (1997) Principles of high-spatial-resolution surface EMG (HSR-EMG): single motor unit detection and application in the diagnosis of neuromuscular disorders. *J Electromyogr Kinesiol* 7:233–239.

Reucher H., Rau G., and Silny J. (1987a). Spatial filtering of noninvasive multielectrode EMG: Part I—Introduction to measuring technique and applications. *IEEE Trans Biomed Eng* 34(2):98–105.

Reucher H., Silny J., and Rau G. (1987b). Spatial filtering of noninvasive multielectrode EMG: Part II—Filter performance in theory and modeling. *IEEE Trans Biomed Eng* 34(2):106–113.

Roeleveld K., Stegeman D.F., Vingerhoets H.M., and Van Oosterom A. (1997a). Motor unit potential contribution to surface electromyography. *Acta Physiol Scand* 160(2):175–183.

Roeleveld K., Stegeman D.F., Vingerhoets H.M., and Van Oosterom A. (1997b). The motor unit potential distribution over the skin surface and its use in estimating the motor unit location. *Acta Physiol Scand* 161(4):465–472.

Roy S., De Luca C., and Schneider J. (1986). Effects of electrode location on myoelectric conduction velocity and median frequency estimates. *J Appl Physiol* 61:1510–1517.

Silverman R.W. and Jenden D.J. (1971). A novel high-performance preamplifier for biological applications. *IEEE Trans Biomed Eng* 18:430.

Tam H.W. and Webster J.G. (1977). Minimizing electrode motion artifact by skin abrasion. *IEEE Trans Biomed Eng* 24(2):134–139.

van der Locht H.M., van der Straaten J.H., and Vredenbregt J. (1980). Hybrid amplifier–electrode module for measuring surface electromyographic potentials. *Med Biol Eng Comput* 18:119–122.

Van Oosterom A. and Strackee J. (1983). Computing the lead field of electrodes with axial symmetry. *Med Biol Eng Comput* 21:473–481.

van Vugt J.P. and van Dijk J.G. (2001). A convenient method to reduce crosstalk in surface EMG. *Clin Neurophysiol* 112:583–592.

Vitasalo J. and Komi P. (1975). Signal characteristics of EMG with special reference to reproducibility of measurements. *Acta Physiol Scand* 93:531–539.

Winter D.A., Fuglevand A.J., and Archer S.E. (1994). Crosstalk in surface electromyography: theoretical and practical estimates. *J Electromyogr Kinesiol* 4(1):15–26.

Zuniga E., Truong X., and Simons D. (1970). Effects of skin electrode position on averaged electromyographic potentials. *Arch Phys Med Rehab* 51:264–272.

Zwarts M.J. and Stegeman D.F. (2003). Multichannel surface EMG: basic aspects and clinical utility. *Muscle Nerve* 28(1):1–17.

7

FOCUSED CHAPTER

METHODS FOR NONINVASIVE ELECTROENCEPHALOGRAM DETECTION

CHRISTOPH GUGER AND GÜNTER EDLINGER

g.tec medical engineering GmbH/Guger Technologies OG, Schiedlberg, Austria

SUMMARY

The electroencephalogram (EEG) provides a noninvasive way to record and analyze brain functions in real time. For brain–computer interfaces (BCIs), slow waves, evoked potentials, and activation in the alpha, beta, and gamma bands are used. When a real-time biosignal processing system is configured, several influencing components have to be considered. This chapter discusses EEG electrodes and electrode placement, lists possible sources of noise and artifacts, and explains how to avoid them. Properties of biosignal amplifiers and analog-to-digital conversion are explained. Possibilities for host computers and processing environments are then discussed. Finally, examples of BCI systems are given for rehabilitation and for brain mapping.

INTRODUCTION

Real-time processing systems for neural engineering applications based on the electroencephalogram (EEG) have many different components influencing

Introduction to Neural Engineering for Motor Rehabilitation, First Edition. Edited by Dario Farina, Winnie Jensen, and Metin Akay.

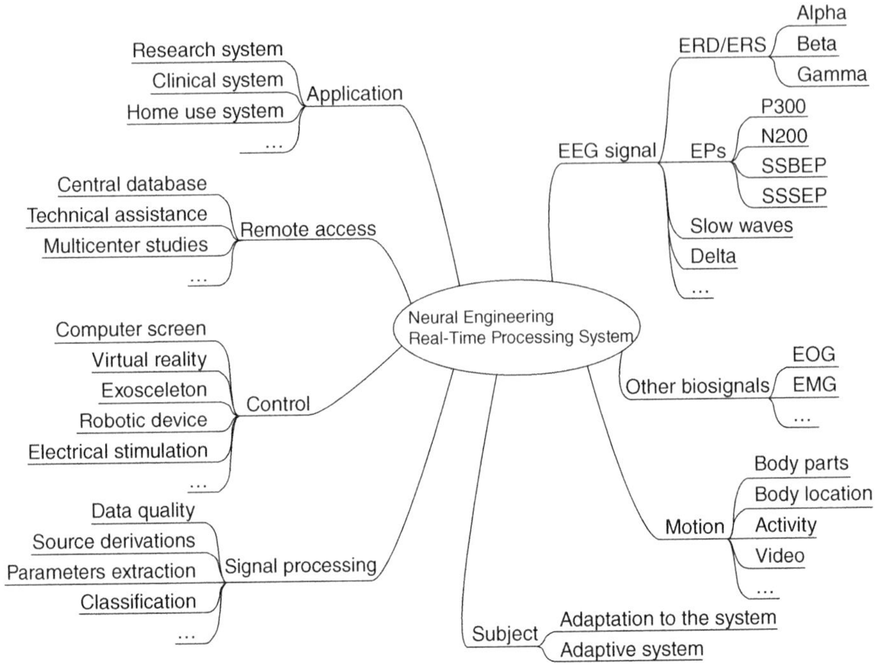

Figure 7.1. Influencing components for the design of a neural engineering real-time processing system.

the design (see Fig. 7.1; Guger et al., 2001; Schalk et al., 2004). Such brain–computer interfaces (BCIs) typically convert the electrical brain activity into control signals and use either slow waves, evoked potentials (EPs), event-related desynchronization–event-related synchronization (ERD/ERS), or delta–theta activity as input signals (Birbaumer et al., 2000; Neuper et al., 2000; Sellers et al., 2006; Friman et al., 2007; Bradberry et al., 2010). Systems based on ERD/ERS generally decode changes of EEG signals due to motor movement or imagination in the alpha (8–12 Hz), beta (14–30 Hz), and gamma (>40 Hz) frequency ranges. It has been shown that high gamma activity (80–100 Hz) can be acquired with the EEG (Darvas et al., 2010). The EEG signal power decreases exponentially as frequency increases. A typical alpha-activity amplitude is about ±40–60 μV, but amplitudes at high frequencies (gamma activity) are below ±1 μV. Besides the traditional alpha and beta bands, delta–theta activity has also been used to reconstruct three-dimensional hand movements (Bradberry et al., 2010). Systems based on EP rely heavily on P300 or N200 potentials occurring 300 or 200 ms after a rare stimulus (oddball paradigm, frequency >0.1 Hz) or on steady-state visual or somatosensory EPs that occur with stimulus frequencies of 5–40 Hz (Sellers et al., 2006; Friman et al., 2007; Fazel-Rezai et al., 2012). Typical amplitudes of such P300 and N200 responses are around 5–10 μV if several EEG epochs are averaged (Guger et al., 2009).

Besides the EEG signal, muscle activity (EMG; amplitude, ±several millivolts; frequency range, from 20 Hz to 1 kHz) or eye movements (EOG; amplitude, ±500 μV; frequency range, <5 Hz) can also be used as input signals for control applications. Both signals can also be used to suppress artifacts in the EEG with signal-processing algorithms. Furthermore, real muscle movements are sometimes recorded to detect movement onsets or explore correlations with the EEG data. Besides biological signals, motion-tracking systems based on acceleration sensors or videos can also be used for the latter task. In this case, it is important to synchronize the biosignal acquisition system with the motion-capture system.

Depending on the specific application, the design of such systems can be very different. Research systems need a very flexible setup to work with different numbers and types of electrodes. It must be possible to record different biosignals, work with various sensor systems, and easily adapt the signal processing, data storage, and visualization. On the contrary, clinical systems have less flexibility; they also need medical approval. Furthermore, home systems should be cheap, quite easy to handle, and dedicated to only a certain task.

Another important feature is remote access to systems for software updates, data management, or BCI system adaptation. For successful BCI operation, the BCI system should adapt itself as necessary to the user's EEG signal, ensuring sustained high performance. However, as the operation of a BCI system is a new skill, a user has to become familiar with it; hence the user must adapt himself or herself to the BCI system to learn effective control (Guger et al., 2001).

Furthermore, BCI systems can be connected to a central database, which could help log users' performance over time and react with customized medical treatment or BCI system adaptations.

For control purposes, the BCI system must be able to send its control signals to an external device. This could be a computer screen for the P300 spelling device (Sellers et al., 2006), a virtual reality system for navigation with the BCI (Guger et al., 2008), an exoskeleton or electrical stimulator for rehabilitation purposes, or robotic devices for supporting tasks. In all these cases, an appropriate fast communication channel between the device and the BCI system is required. A BCI control standard was proposed that allows devices to configure the BCI system accordingly (Putz et al., 2011).

For biosignal analysis systems, it is also essential to support different signal processing methods for data quality control, source derivations, parameter extraction, and classification.

The following sections explain some of the key components for EEG data acquisition and real-time analysis in more detail: electrodes, electrode placement, amplifier technology and artifacts, signal processing, and the programming environment.

ELECTRODES

In electrophysiology, the electrical contact between the amplification device and the brain or tissue from which electrical activity should be recorded is made with electrodes. The electrodes pick up the small electrical voltages generated by neural activity and pass that signal to the amplification unit. For EEG recordings, the electrodes are normally not in direct contact with the tissue, and conductive electrode gel is applied between the electrode and the skin. The metal–electrolyte and the electrolyte–skin conjunction can cause a DC offset voltage and electrical impedance that depends on the electrode material and its surface, the gel composition, and temperature. For example, stainless steel electrodes with a relatively high resistance can have high DC voltages. The amplitude of the DC offset voltage can be in the range of several hundred millivolts and can easily be larger than the source signal amplitudes (e.g., EEG ~ ±50 μV) on the amplifier input, which can saturate the amplifier input stages. Hence, biosignal amplifiers are designed such that, up to a specific DC offset amplitude, the performance of the amplifier is not affected. However, the same electrode material and same electrode gel should be used for all electrodes to minimize such polarization effects.

The electrode impedance of scalp electrodes depends on the skin–electrode junction, which depends on the condition and preparation of the skin, the properties of the gel, and the time elapsed after preparation. The electrode impedance consists of a resistive and a capacitive part and is time and frequency dependent (a function of $1/f^{1/2}$). Depending on the electrode material, polarization effects can be minimized. For example, silver–silver chloride (Ag–AgCl) electrodes show a lower polarization voltage and can be used for DC recordings. The electrode impedance also depends on the electrode size and contacting surface. Toxicity, scalp irritation, and mechanical strength are also important parameters. The electrode impedance normally has values between several kilohms and several hundred kilohms. The use of electrode gel with a NaCl concentration between 5% and 10% reduces impedance and thus facilitates stable recordings.

Before EEG electrodes are mounted on the head, the skin is typically cleaned with an abrasive gel to remove the outer dry layer of the skin to obtain lower skin–electrode impedance, ensuring high-quality EEG recordings. This procedure is performed because the outer layer of the skin can contribute up to several millivolts of DC potential because of small electrode position shifts. Tam and Webster showed that, with 20 strokes of fine sandpaper, the impedance of the epidermis layer can be reduced, which reduces motion artifacts produced, for example, by stretching the skin (Tam and Webster, 1977). The abrasion does not reach the capillary layers but does remove a barrier layer of dead skin cells that protects the skin. Furthermore, the skin–electrode impedance is influenced by temperature and humidity. The lower the impedance, the fewer the artifacts and the less the noise that are picked up. The disadvantage of skin preparation is that the procedure is time consuming and can cause some pain, especially if it is done every day over the same electrode location by an inexperienced person. Also, the drying of the gel and the skin regrowth after abrasion degrade the performance of EEG electrodes in long-term recordings. Therefore, so-called active electrodes were developed that have an amplifier in the electrode itself and can therefore accept much higher electrode–skin impedances. As a result, active electrodes pick up fewer artifacts, and no skin preparation using abrasive gel is necessary. However, the use of conductive gel ensuring a galvanic connection between the skin and the electrode is still needed. Active electrodes are also larger than passive electrodes, have additional electronic components, and are more expensive.

For EEG measurements, single disk electrodes made of gold or Ag–AgCl are normally used (see Fig. 7.2). Gold electrodes are maintenance free and have a good frequency response for EEG measurements. For DC derivations with EEG frequencies below 0.1 Hz, Ag–AgCl electrodes perform better than gold electrodes. Passive electrodes are connected with the electrode cable and a 1.5-mm medical safety connector to the biosignal amplifier; active electrodes have system connectors to supply the electronic components with power.

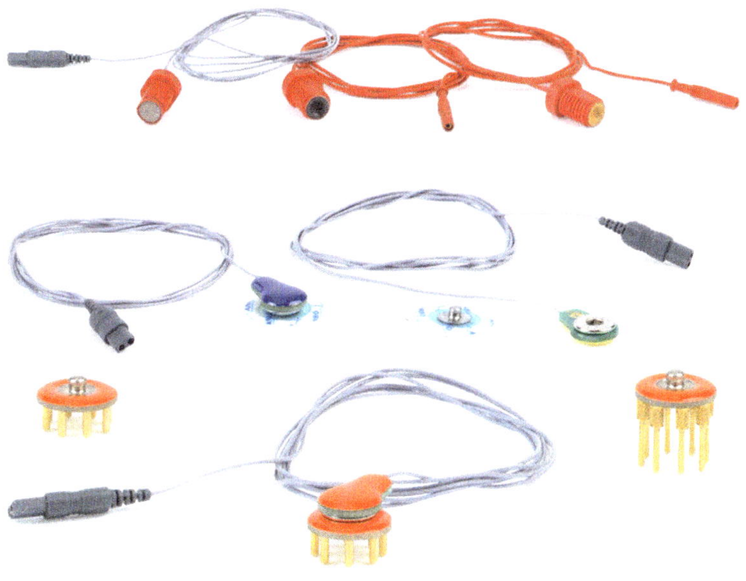

Figure 7.2. Different types of EEG electrodes. (top) Electrodes that can be inserted into an electrode cap and fixated at different depths—active sintered Ag–AgCl, passive Ag–AgCl, and gold electrodes. (middle) Active electrode clips with disposable electrodes for EEG/EMG/ECG. (bottom) Dry electrodes with different pin lengths and with active electrode clip. Passive electrodes have a 1.5-mm medical safety connector, and active electrodes have a two-pin safety connector.

Dry contact electrodes use either microneedles to penetrate the first layer of the skin and to get in contact with the conducting layers, or they penetrate the skin with mechanical springs that press the electrodes into the skin (Popescu et al., 2007; Taheri et al., 1994). Dry noncontact electrodes use capacitive electrical coupling through hair or clothing such as cotton. Early work focused on active and dry electrodes for recording electrocardiogram signals, which is easier than EEG because of the larger signal-to-noise ratio and easier montage on the thorax.

A dry contact electrode is shown in Figure 7.2; it was constructed by inserting seven gold-coated pins on a plate mounted in a circular arrangement (Guger et al., 2012). A small preamplifier is located in the electrode housing itself, ensuring that the electrode can work even with very high skin-to-electrode impedances. The length of the pins can be varied in order to reach the skin appropriately. The montage time of the dry electrodes is shorter than for gel-based electrodes. The P300 speller montage (8 channels) with passive electrodes is mounted with abrasive gel within about 10 minutes. If active electrodes are used, the montage time is reduced to about 1–3 minutes. With dry electrodes, the time is about 1 minute or less. But after the cap is attached, dry electrodes need a few minutes to adjust, and therefore the time span needed for preparation is comparable to active electrodes. The biggest advantage of dry electrodes is that there is no abrasive and no conductive gel to remain in the hair, eliminating the time-consuming need to wash the user's hair. Dry electrodes do not need to be washed afterward, which not only saves time but enhances their life span. However, dry electrodes pick up more noise from the surrounding environment.

Electrode Placement

For EEG recordings, electrode caps allow a fast montage of the electrodes in certain positions and are often assembled according to the international 10/20 system (Klem et al., 1999). Electrode caps (see Fig. 7.3, right panel) may use screwable single electrodes to adjust the depth and to optimize the electrode impedance. There are two main advantages of a single-electrode system: (i) if one electrode breaks down it can be removed immediately and (ii) different electrode montages can be implemented easily. The disadvantage is that every time a measurement is done, all electrodes must be inserted separately again. For this reason there are also caps available that are preconfigured and the electrodes remain in the cap (Fig. 7.3, left panel). The main disadvantage is that if one electrode breaks down it is more difficult to replace the electrode. The elastic material of the cap guarantees the proper fixation of the sensors, while arbitrary positions can still be chosen for optimal recordings depending on the experimental paradigm. Furthermore, the elastic material provides the necessary pressure to ensure that the electrode is in contact with the skin even for dry electrodes (as shown in Fig. 7.3, middle).

Figure 7.3. EEG electrode caps for active EEG electrodes (160 positions), for dry electrodes, and for screwable electrodes (65 positions).

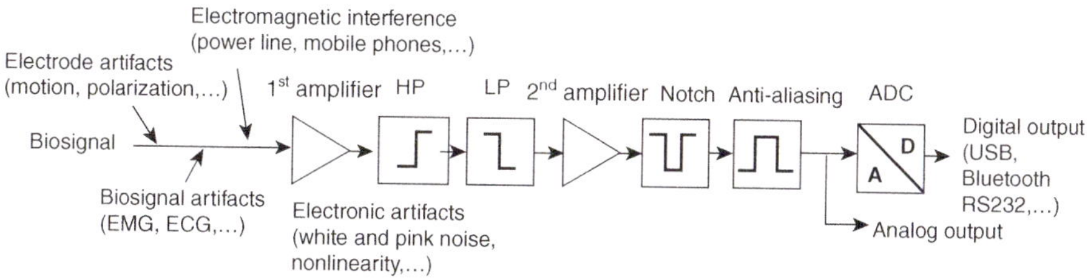

Figure 7.4. Biosignal amplification stages with artifacts.

Noise and Artifacts

Electroencephalographic recordings can be easily contaminated with biological artifacts (ECG, EMG, EOG, and other noise), electrode or sensor artifacts (movement, DC offset, noise, etc.), electromagnetic interference (capacitive coupling, inductive coupling, noise, etc.), amplifier noise, and digitization noise as shown in Figure 7.4. External electrical noise sources typically interfere at, for example, 50 or 60 Hz. However, these noise signals are picked up via all electrode leads, and such a common mode noise signal can be suppressed by the amplifier differential input stage. As the amplitudes of the noise signals will vary in practice due to, for example, different electrode impedances, the main frequency noise cannot be perfectly suppressed by the amplifier input stage and a noise signal at 50 or 60 Hz with amplitudes up to several millivolts can be present in the recordings. The ratio of the electrode impedance to the amplifier input impedance determines the signal attenuation that should be avoided. If the electrode impedance is sufficiently small compared with the amplifier and cable impedance, then the source signal is not attenuated. EEG electrodes have an impedance of several kilohms, and the electrode impedance is also frequency dependent. Thus, the amplifier input should be about 100 times larger, and this yields several megohms. Additional shielding, and proper grounding of the recording equipment and the subject, might be necessary to reduce the artifacts. Interference can also result from varying magnetic fields and depends on the areas defined by the electrodes' cables. Therefore, twisting the cables helps to reduce the area and therefore to reduce the induced voltages. Electronic components also add noise to the recordings that depends on both the electronic part itself and the layout of the amplifier, and it increases with wider bandwidth. Noise can be reduced by minimizing the amplifier bandwidth as much as possible. Typical values are around 1 μV peak-to-peak for 100-Hz bandwidth and 4 μV for 3-kHz bandwidth (Kamp et al., 2004). It is also important that the electrode itself be in contact only with the electrode gel and not with the skin, to minimize movement in the electrolyte–metal junction, and that the cap, electrodes, and electrode wires be kept lightweight to avoid movement.

AMPLIFIERS

Biosignal amplifiers are connected to the electrodes or sensors that pick up the very small source signals (EEG, 20–100 μV; EPs, a few microvolts) having a certain frequency range (EEG, DC–100 Hz). The task of the amplifier is to amplify the source signal without distortions and to suppress the noise as much as possible to increase the signal-to-noise ratio. A typical amplifier design including first and second amplifier stages, filters, and an analog-to-digital convertor (ADC) is shown in Figure 7.4. Amplifiers either have analog outputs or have an integrated ADC for digital signal transmission to the recording computer. A key aspect of biosignal amplifier design is also medical safety.

Amplifiers with bipolar inputs typically use instrumentation amplifiers as input units with a high common mode rejection ratio (CMRR). This means that, ideally, only signals are amplified that are different in the input channels, and common signals in amplitude and polarity are attenuated. The CMRR is typically between 60 and 110 dB, which is not only defined by the instrumentation amplifier at the input circuit, but also influenced by preamplifiers or additional electronic components at the input. If the preamplifiers have different gains, then the effective CMRR is also reduced. For bipolar recordings, the electrodes are connected to the differential inputs (the plus and the minus inputs) of the amplifier, and electrodes are mounted on the scalp at a distance of about 2.5–10 cm (see Fig. 7.5). The ground electrode is mounted, for example, on the forehead. Bipolar derivations have the advantage of suppressing noise and artifacts very well, so that only local brain activity nearby the electrodes is picked up.

Monopolar input amplifiers have in contrast a common reference electrode that is typically mounted at an ear lobe or a mastoid (Fig. 7.5) and is connected to the minus inputs of the amplifier. The amplifier ground electrode is also

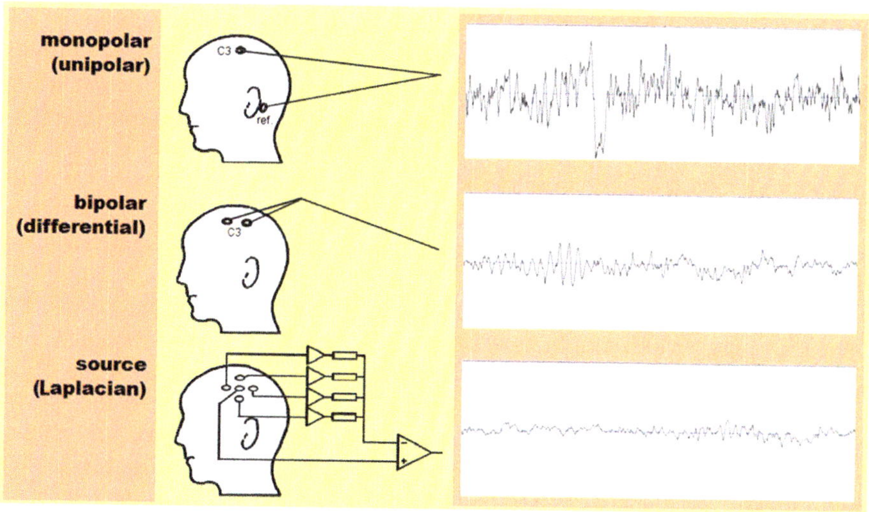

Figure 7.5. Electrode montage for different source derivations and corresponding raw EEG signals.

connected to the forehead. All other electrodes are connected to corresponding plus inputs of the amplifier. Monopolar recordings refer the measurements to the common reference electrode and are more sensitive to artifacts, but allow for re-referencing the recordings by calculating bipolar, small or large Laplacian, Hjorth's, or common average reference (CAR) derivations afterward (McFarland et al., 2003). Figure 7.5 shows also the setup of a Laplacian source derivation that subtracts the voltages of the surrounding electrodes from the middle electrode to increase the signal-to-noise ratio. This procedure results in a much smaller, but cleaner, EEG.

Typically, bipolar derivations are preferred if only a few channels are to be used, whereas monopolar channels are used for recordings with many electrodes such as when, for example, spatial filters are applied (Ramoser et al., 2000).

One of the most important features is the number of electrodes used. For slow-wave approaches, oscillations in the alpha and beta range, and P300 systems, a total of 1–8 EEG channels are sufficient (Pfurtscheller et al., 1997; Birbaumer et al., 2000; Krusienski et al., 2006; Guger et al., 2003). For realizations with spatial filtering such as common spatial pattern (CSP), a high number of channels is required (16–128) (Guger et al., 2000; Ramoser et al., 2000; McFarland et al., 2003). Therefore stackable systems might be designed to extend the functionality in future applications. A stackable 64-channel system, for example, can also be split into four 16-channels systems if required for some experiments.

The signal type (EEG, EP, EMG, EOG, etc.) influences the necessary sampling frequency and bandwidth of the amplifier. The conversion from analog to digital is performed with an ADC that works at a certain sampling frequency. An ADC with a sampling frequency of 256 Hz acquires 256 samples per second, which corresponds to one sample about every 4 ms. The sampling frequency must be at least two times larger than the highest frequency occurring in the signal, according to the Nyquist criterion. This is first guaranteed by selecting an adequate sampling frequency and second by applying an anti-aliasing filter before the ADC. The anti-aliasing filter is in the simplest case a low-pass filter that properly attenuates frequencies above the Nyquist frequency.

High-pass (HP) and low-pass (LP) filters allow users to limit the bandwidth of the biosignal to extract only the desired signal and to attenuate noise. If the bandwidth is too narrow, then signal components are lost; if is too broad, then too much noise is recorded. If the frequency ranges of different biosignals overlap, then of course both will be recorded. By using a high-pass and a low-pass filter, a so-called band-pass filter is realized. Filters are characterized by their cutoff frequency, order, type, and phase shift (time displacement).

Notch filters are used to suppress the 50- or 60-Hz power-line interference. A notch filter is typically a narrow band-stop filter having a very high order. High-pass and low-pass filters remove frequency components below or above the cutoff frequency. At the cutoff frequency, the signal is attenuated by –3 dB, which corresponds to a reduction to 70.7% of the signal amplitude. Filters are present in the analog and digital parts of the amplifier and can have different characteristics and filter orders. For example, a higher order low-pass filter has a higher attenuation of signal components above the cutoff frequency compared with a low-order low-pass filter but results in large phase shifts of the signal. Filters with high order can also be unstable. It is important to carefully select the filter characteristic (Bessel, Butterworth, etc.) to avoid unwanted distortions of the signal. The main advantage of digital filters is that filtering properties can easily be changed and that accurate BODE and phase diagrams can be generated. Analog filters inside the amplifier are predefined and therefore cannot be changed.

For EEG signals, sampling frequencies of 256 Hz with a bandwidth of 0.5–100 Hz are typically used (Guger et al., 2000). Slow waves may use a lower cutoff frequency of 0.01 Hz (Birbaumer et al., 2000). For P300-based systems, a bandwidth of 0.1–30 Hz is typically used (Sellers et al., 2006).

The input sensitivity of a biosignal amplifier must be suitable for the corresponding signal. For bipolar EEG measurements, a sensitivity of ±50 μV is sufficient, but it is important that the amplifier does not reach saturation if higher DC voltages occur. This must be considered in the input stage of the amplifier. The input signal is then amplified to optimally fit to the input range of the ADC. The ADC must convert the analog signals into digital format that can be transmitted to the host computer for further processing. For biosignal amplifiers, either sigma–delta (SD) or successive approximation (SA) converters are used. Sigma–delta converters have a higher resolution up to 24 bit, while SA converters typically have a resolution only up to 18 bit (mostly 16 bit). The resolution of the converter indicates the number of discrete values it can produce over the range of analog values. The advantage of SA converters is the higher sampling rate of several hundred kilohertz, while SD converters reach only several tens of kilohertz. For this reason, it is also possible to multiplex channels with SA converters to have more channels with a single converter. This also reduces the power consumption.

A 16-bit ADC has a resolution of 65,536 quantization steps. Assume a biosignal amplifier with an input range of ±250 mV. This yields to a resolution of 7.6 μV, which would be too low for EEG measurements. With a 24-bit converter, the resolution is 29 nV and therefore easily sufficient for EEG recordings. For an amplifier with an input range of ±100 μV and a 16-bit ADC, the resolution would be 3 nV, which is again enough for EEG recordings. This means that the bit rate and input range must be considered together to interpret the amplifier properties. A big advantage of a high input range is of course that the amplifier can be used for all biosignals without changing the amplification. Also, DC offset voltages and distortions do not saturate the amplifier. In practice, it is advisable to visualize all channels of the same biosignal type with the same amplitude to make comparisons easier. For biosignal recording devices, a linear relationship between the input and output signals is important within a specific amplitude range to avoid distortion of the signal.

Electroencephalographic recordings with up to 256 channels yield a spatiotemporal representation of the brain activity. Therefore, the signals must be simultaneously recorded, and it is important that ADCs guarantee a perfect nondelayed acquisition of all connected channels. This is especially important for EP recordings with short-lasting potentials or for recordings with many EEG channels. If only one ADC with a specific conversion time (time to convert an analog value to digital) is used for many channels, then a time lag between the first channel and the last channel would be the result (e.g., 100 channels × 10 μs = 1 ms). A solution is to use either an amplifier with simultaneous sampling-and-hold or an ADC for each channel. For recordings with many channels, it is also possible to daisy chain multiple devices, but in this case the amplifiers must be perfectly synchronized to avoid time shifts.

Biosignal acquisition systems must also provide trigger inputs and outputs to log external events in synchrony with the EEG data, or send trigger information to other external devices such as a visual flash lamp. Digital outputs can also be used to control external devices such as a prosthetic hand or a wheelchair (Neuper et al., 2000). An advantage here is to scan the digital inputs together with the biosignals to avoid time shifts between events and physiological data.

A medical power supply that works with 220 and 110 V is required for BCI systems that are used mainly in the lab. For mobile applications such as controlling a wheelchair, amplifiers that run on battery are also useful but require charging.

Biosignal amplifiers must fulfill the safety requirements for medical devices. Production, maintenance, and safety conditions are regulated and are approved by appropriate bodies. The regulations vary for every country, but a base standard for medical electrical equipment is defined by IEC 60601–1 (International Electrical Commission). It is important that the devices have the appropriate certifications depending on the country. In the European Union devices are required to be CE certified; in the United States devices have to be approved by the Food and Drug Administration (FDA).

Amplifier with Digital Signal Processing

Figure 7.6 shows a block diagram of a g.USBamp, which is a biosignal amplifier with integrated ADCs that sends the digitized data via USB to the processing unit. This device has 16 input channels, which are connected over software-controllable switches to the internal amplifier stages and anti-aliasing filters before the signals are digitized with sixteen 24-bit ADCs. The device is also equipped with digital-to-analog converters (DACs), enabling the generation of different signals such as sinusoidal waves, which can be sent to the inputs of the amplifiers for system testing and calibration. Additionally, the impedance of each electrode can be checked by applying a small current via the individual electrodes and measuring the voltage drops. All these components are part of the so-called applied part of the device, as a subject or patient is

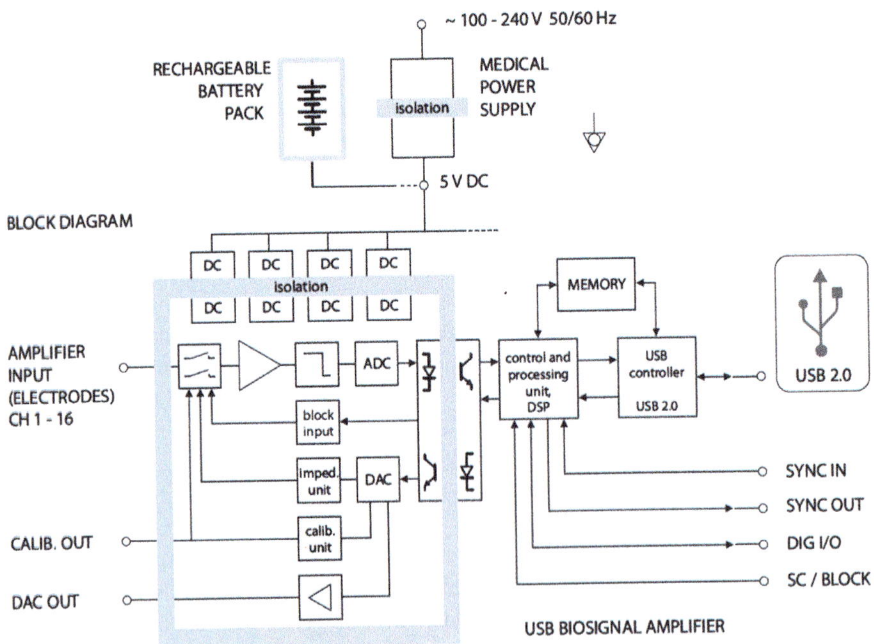

Figure 7.6. Block diagram of biosignal amplifier with medical power supply and battery, ADC conversion, and USB data transmission.

in galvanic contact with this part of the device via the electrodes. All subsequent parts of the device are separated via optical links from the subject or patient.

The digitized signals are passed to a digital signal processor (DSP) for further processing. The DSP performs an oversampling of the EEG data, band-pass filtering, and Notch filtering to suppress the power-line interference, and it calculates bipolar derivations. Then the preprocessed EEG data is sent to a controller that transmits the data via USB 2.0 to the PC. One important feature of the amplifier is the oversampling capability. Each ADC is sampling the data at 2.4 MHz. Then the samples are averaged to the desired sampling frequency of, for example, 128 Hz. Here, a total of 19,200 samples are averaged and this improves the signal-to-noise ratio by the square root of 19,200 = 138.6 times.

INTERFACING TO ELECTRODES, THE COMPUTER, AND PROGRAMMING ENVIRONMENT

Figure 7.7 shows three different amplifier concepts for the recording of 256 channels, 8 channels, and 16 channels. The 256-channel amplifier has 4 groups with 64 channels each, has electrode connector boxes for passive or active electrodes and sensors, and transmits the data via USB. Such a device is used for real-time brain mapping and high-resolution EEG. The 8-channel unit has electrode connector boxes for electrodes with 1.5-mm safety connectors and for disposable EMG/ECG electrodes and transmits the data via Bluetooth. This device is used, for example, in virtual environments and allows the user to move around. The 16-channel unit has 4 groups of 4 channels each, which allows developers to connect multiple users to the device or to record from different types of biosignals such as EEG and EMG without interference.

The computer system (PC, laptop, Pocket PC) must be able to read the biosignal data stream over the dedicated interface, such as USB, Bluetooth,

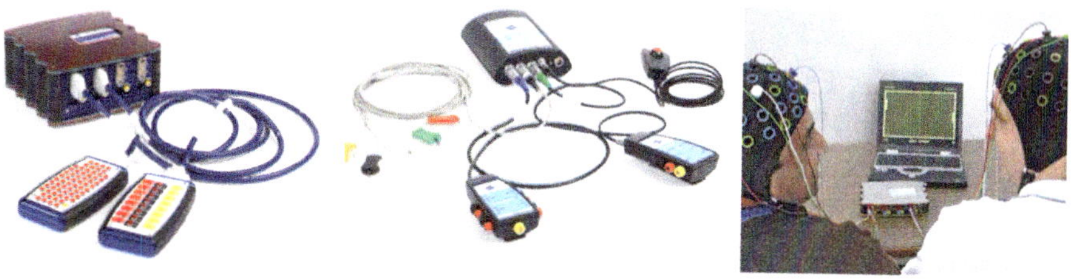

Figure 7.7. (left) 256-channel biosignal amplifier with connector boxes for passive, active EEG electrodes, and other sensors. (middle) Portable and wireless biosignal amplifier. (right) 16-channel amplifier connected to two subjects playing Pong.

or serial. A C application programming interface (API) is the most flexible choice, since it provides direct access to the recorded data and can handle all necessary amplifier configurations (sampling frequency, channel selection, low-pass filter, high-pass filter, Notch filter, source derivation, test signals, impedance check). It allows users to make their own applications that can be distributed and installed at other locations. Signal processing such as parameter extraction, classification, data visualization, and storage must be programmed in C. An example of such a BCI system is BCI2000 (Schalk et al., 2004).

A MATLAB API allows the recording of the biosignal data under MATLAB and allows the use of MATLAB-based toolboxes such as the signal-processing toolbox. Therefore the design cycle is much as that with the C API but has the disadvantage that a MATLAB license is required.

Simulink is a signal-flow-oriented toolbox running under MATLAB, and an example is given in Figure 7.8 that uses common spatial patterns for a BCI system (Guger et al., 2000). The Simulink model has a block that reads the biosignal data from the amplifier in real time into the computer system. In this case, the real-time clock of the amplifier is driving the Simulink model and guarantees the correct timing of the subsequent signal processing. The model then converts the data to double precision for optimal resolution and band-pass filters the data between 8 and 30 Hz. Then, four spatial filters are applied that were calculated with the common spatial patterns algorithm before the variance is estimated for each of the four channels. Finally, a linear discriminant analysis is applied to discriminate left from right hand movement imaginations, and this gives a control signal for controlling (in this case) hand movements of an avatar.

FUTURE DIRECTIONS

Figure 7.9 illustrates a complete real-time processing system able to read in biosignal data via the electrodes and the amplification unit. Then, the recording computer is capturing the data with the amplifier API and is performing the data storage, visualization, and signal processing. It is also controlling in real time the stimulator API, which gives access to a battery of devices such as eye and movement tracking, tone stimulation, haptic devices, electrical stimulators, and microphones to induce body functions or to record body functions. Additionally, the real-time system is connected to head-mounted devices or life-size virtual reality systems for highly immersive feedback or exoskeletons to capture or support body movements. Such a setup can be used in research to investigate brain functions linked to different body functions, but also in rehabilitation to drive external devices depending on brain functions.

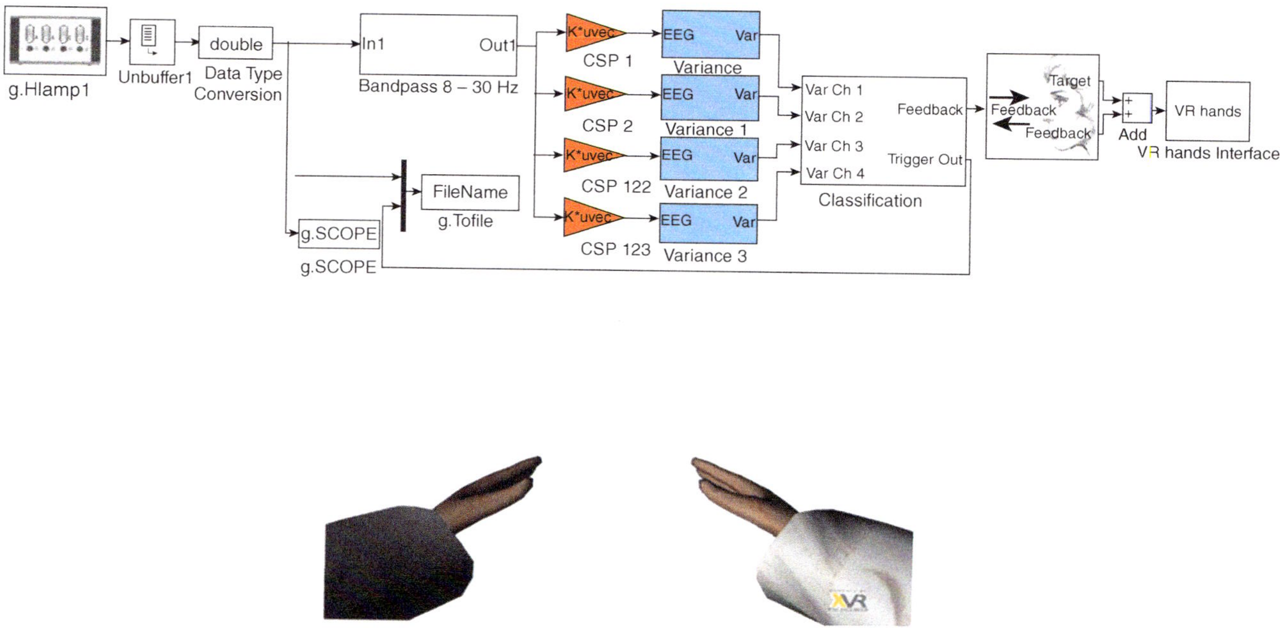

Figure 7.8. Simulink model running common spatial patterns implementation to detect motor movement imaginations. The model processes 128 EEG channels in order to control hand movements of a virtual avatar in real time (Bernhard Spanlang, University of Barcelona, and Mel Slater, ICREA–University of Barcelona).

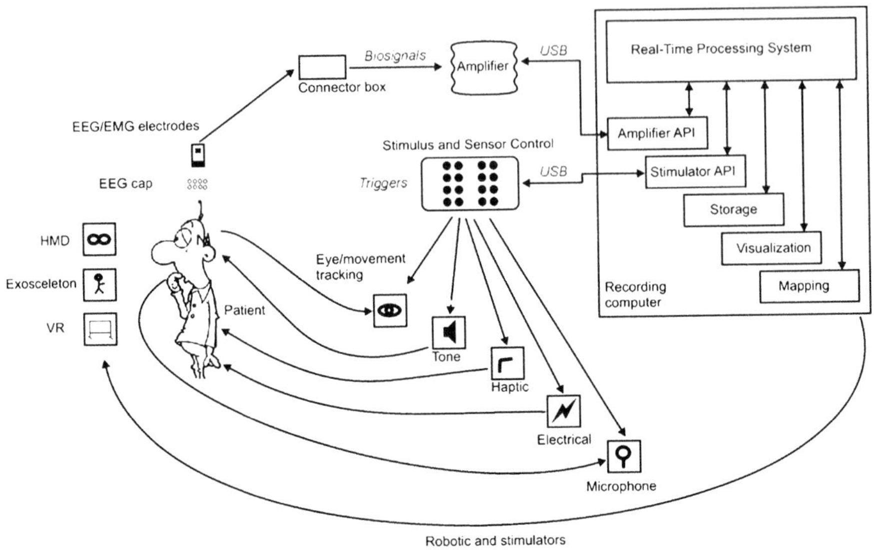

Figure 7.9. Components of a neurorehabilitation real-time processing system.

REFERENCES

Birbaumer N., Kubler A., Ghanayim N., Hinterberger T., Perelmouter J., Kaiser J., Iversen I., Kotchoubey B., Neumann N., and Flor H. (2000). The thought translation device (TTD) for completely paralyzed patients. *IEEE Trans Rehabil Eng* 8(2): 190–193.

Bradberry T.J., Gentili R.J. and Contreras-Vidal J.L. (2010). Reconstructing three-dimensional hand movements from noninvasive electroencephalographic signals. *J Neurosci* 30(9):3432–3437.

Darvas F., Scherer R., Ojemann J.G., Rao R.P., Miller K.J., and Sorensen L.B. (2010). High gamma mapping using EEG. *Neuroimage* 49(1):930–938.

Fazel-Rezai R., Allison B.Z., Sellers E., Guger C., Kleih S., and Kübler A. (2012). P300 brain computer interface: current challenges and future directions. *Front Neuroeng* 5:14.

Friman O., Volosyak I., and Graser A. (2007). Multiple channel detection of steady-state visual evoked potentials for brain–computer interfaces. *IEEE Trans Biomed Eng* 54:742–750.

Guger C., Ramoser H., and Pfurtscheller G. (2000). Real-time EEG analysis with subject-specific spatial patterns for a brain computer interface (BCI). *IEEE Trans Neural Syst Rehabil Eng* 8:447–456.

Guger C., Schlögl A., Neuper C., Walterspacher D., Strein T., and Pfurtscheller G. (2001). Rapid prototyping of an EEG-based brain–computer interface (BCI). *IEEE Trans Rehab Eng* 9(1):49–58.

Guger C., Edlinger G., Harkam W., Niedermayer I., and Pfurtscheller G. (2003) How many people are able to operate an EEG-based brain computer interface? *IEEE Trans Rehab Eng* 11:145–147.

Guger C., Groenegress C., Holzner C., Edlinger G., and Slater M. (2008). Brain–computer interface for controlling virtual environments. 2nd International Conference on Applied Human Factors and Ergonomics. Las Vegas, USA.

Guger C., Daban S., Sellers E., Holzner C., Krausz G., and Carabalona R. (2009). How many people are able to control a P300-based brain–computer interface (BCI)? *Neurosci Lett* 462(1):94–98.

Guger C., Krausz G., Allison B.Z., and Edlinger G. (2012). A comparison of dry and gel-based electrodes for P300 BCIs. *Front Neurosci* 6:60.

Kamp A., Pfurtscheller G., Edlinger G., and Lopes da Silva F. (2004). Technological basis of EEG recording. In: *Electroencephalography, Basic Principles, Clinical Applications and Related Fields* (Niedermeyer E. and Lopes da Silva F., Eds.), 5th ed. Philadelphia: Lippincott Williams & Wilkins.

Klem G., Lüders H., Jasper H., and Elger C. (1999). The ten–twenty electrode system of the International Federation. The International Federation of Clinical Neurophysiology. Cleveland Clinic Foundation, *EEG Clin Neurophysiol* 52(Suppl): 3–6.

Krusienski D.J., Sellers E.W., Cabestaing F., Bayoudh S., McFarland D.J., Vaughan T.M., and Wolpaw J.R. (2006). A comparison of classification techniques for the P300 Speller. *J Neural Eng* 3(4):299–305.

McFarland D.J., Sarnacki W.A., and Wolpaw J.R. (2003). Brain–computer interface (BCI) operation: optimizing information transfer rates. *Biol Psychol* 63(3): 237–251.

Neuper C., Pfurtscheller G., Guger C., Obermaier B., Pregenzer M., Ramoser H., and Schlögl A. (2000). Current trends in Graz brain–computer interface (BCI) research. *IEEE Trans Rehabil Eng* 8:216–219.

Pfurtscheller G., Neuper C., Flotzinger D., and Pregenzer M. (1997). EEG-based discrimination between imagination of right and left hand movement. *Electroenceph Clin Neurophysiol* 103:642–651.

Popescu F., Fazli S., Badower Y., Blankertz B., and Muller K.R. (2007). Single trial classification of motor imagination using 6 dry EEG electrodes. *PLoS One* 2(7):e637.

Putz V., Guger C., Holzner C., Torrellas S., and Miralles F. (2011). A unified XML based description of the contents of brain computer interfaces. In: *Proceedings of the 5th International Brain–Computer Interface Conference 2011*, pp. 328–331.

Ramoser H., Muller-Gerking J., and Pfurtscheller G. (2000). Optimal spatial filtering of single trial EEG during imagined hand movement. *IEEE Trans Neural Syst Rehabil Eng* 8(4):441–446.

Schalk G., McFarland D.J., Hinterberger T., Birbaumer N., and Wolpaw J.R. (2004). BCI2000: a general-purpose brain–computer interface (BCI) system. *IEEE Trans Biomed Eng* 51(6):1034–1043.

Sellers E.W., Krusienski D.J., McFarland D.J., Vaughan T.M., and Wolpaw J.R. (2006). A P300 event-related potential brain–computer interface (BCI): the effects of matrix size and interstimulus interval on performance. *Biol Psychol* 73(3):242–252.

Taheri B.A., Knight R.T., and Smith R.L. (1994). A dry electrode for EEG recording. *Electroencephalogr Clin Neurophysiol* 90(5):376–383.

Tam H.W. and Webster J.G. (1977). Minimizing electrode motion artifact by skin abrasion. *IEEE Trans Biomed* 24(2):134–139.

8

REVIEW CHAPTER

SPIKE SORTING

DI GE[1] AND DARIO FARINA[2]

[1]*Glaizer Groupe, Malakoff, France*

[2]*Department of Neurorehabilitation Engineering, Bernstein Focus Neurotechnology Göttingen, Bernstein Center for Computational Neuroscience, University Medical Center Göttingen, Georg-August University, Göttingen, Germany*

SUMMARY

Excitable cells, such as neurons, produce action potentials (APs) that in extracellular recordings are often referred to as **spikes**. In these recordings, APs indeed appear as sharp spikes deviated from the noisy baseline. The contributions of each cell must be isolated from the background noise and from those of the other cells. The process of estimating one or more AP waveforms and their discharge instances (firing patterns) from multiunit recordings is known as **spike sorting**. Resolving overlapped APs is a major difficulty in automatic spike sorting methods. Therefore, this chapter is in large part dedicated to discussing this issue; we focus on state-of-the-art techniques addressing the problem of spike sorting, including the resolution of overlapped APs. In the section "Signal model", we propose the mathematical modeling of multiunit recordings and the complexity in the resolution of overlapped APs. Then, in section "State of the art," we summarize state-of-the-art spike sorting algorithms and discuss the advantages and limitations of each and the applicability of these methods for different types of experimental demands. The final section provides the conclusions.

Introduction to Neural Engineering for Motor Rehabilitation, First Edition. Edited by Dario Farina, Winnie Jensen, and Metin Akay.

INTRODUCTION

Excitable cells, such as neurons, produce action potentials (APs) that in extracellular recordings are often referred to as **spikes**. In these recordings, APs indeed appear as sharp spikes deviated from the noisy baseline. Each cell generates a spike with a characteristic waveform, mainly determined by the electrode position relative to the recorded cell (Gold et al., 2006). The shape of the APs also depends on intrinsic characteristics of the cell, such as the composition of ionic currents (Gold et al., 2006).

In most cases, the recording electrode measures APs from multiple cells. For example, invasive electrodes inserted into the brain, nerves, or muscles provide multiunit recordings consisting of the activity of neural cells or muscle fibers. The contributions of each cell must be isolated from the background noise and from those of the other cells. The process of estimating one or more AP waveforms and their discharge instances (firing patterns) from multiunit recordings is known as **spike sorting**.

Spike sorting is a critical step in processing neural and muscular signals, both in basic neuroscience research and in neurotechnologies, such as brain–computer interfaces (BCIs). This processing stage is necessary when it is desirable to study local populations of cells that are too close to allow isolated recordings by specific acquisition techniques. Examples of signals for which this procedure is necessary are neural recordings from intrafascicular nerve electrodes (Djilas et al., 2010; Citi et al., 2008) and intramuscular recordings (Parsaei et al., 2010).

Several spike sorting techniques are reviewed in Lewicki (1998). The first stage of spike sorting is usually the segmentation achieved by thresholding a filtered version of the original signal. The classification of the segments can then be performed by various clustering approaches (Lewicki, 1998). The decision of the number of AP classes (neurons in the vicinity of electrodes) is a typical model selection problem and can be determined in a Bayesian context by comparing different model probabilities. In such clustering procedures, the temporally overlapped APs are generally considered as outliers (Quian Quiroga et al., 2004) for practical reasons. Indeed, overlapped waveforms yield too many AP classes, while many among them include only a few instances of unrecognized overlapped waveforms.

Excluding overlapped APs from the classification may provide an acceptable amount of information when only a few neural cells are active and they fire with relatively low rates; that is, the percentage of overlapped APs over the total number of APs is small enough. However, it is not possible to exclude overlapped APs when the full discharge patterns are needed. Bankman (Bankman et al., 1993) provided the expression of the average overlap rate as a function of the number of neural cells I, their average firing frequency f (Hz), and the spike duration d (second):

$$f^2 \cdot d \cdot I(I-1)$$

For example, if three units fire, all at 50 Hz, with AP waves of 1-ms duration, then the average overlap rate would be 15 s^{-1}.

Resolving overlapped APs is a major difficulty in automatic spike sorting methods. Therefore, this chapter is in large part dedicated to discussing this issue. In the following, we focus on state-of-the-art techniques addressing the problem of spike sorting, including the resolution of overlapped APs. The chapter outline is as follows. In the next section ("Signal model"), we propose the mathematical modeling of multiunit recordings and the complexity in the resolution of overlapped APs. Then, in section "State of the art," we summarize state-of-the-art spike sorting algorithms and discuss the advantages and limitations of each and the applicability of these methods for different types of experimental demands. The final section provides the conclusions.

SIGNAL MODEL

Basic Model of Multiunit Recordings

The generation of a multiunit single-channel recording is schematically represented in Figure 8.1. Each impulse train is convolved by a characteristic action potential whose shape is assumed to vary minimally during the recording.

The direct problem model can thus be expressed mathematically as a discrete convolution mixture:

$$z = \sum_{i=1}^{I} h_i * s_i + \varepsilon \qquad (8.1)$$

where z is the recorded signal of length N; $(s_i)_i$ and $(h_i)_i$ are the impulse trains (the discharge patterns) and their linear filters (action potentials), respectively;

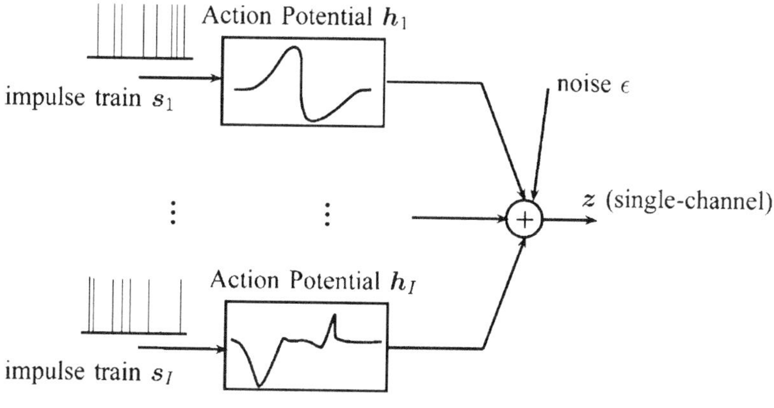

Figure 8.1. Multiple-unit single-channel direct model.

$(\bullet)_i$ denotes the set $\{\bullet, i = 1, \ldots, I\}$; for example, $(s_i)_i = \{s_i, i = 1, \ldots, I\}$ gathers the discharge patterns of I impulse trains. The number of firing patterns (neuron units) is not known a priori. The aim of spike sorting methods is to restore the shapes of the APs $(h_i)_i$ and the discharge patterns $(s_i)_i$.

For each impulse train to be restored, it is usually assumed that the $(s_i)_i$ are mutually independent discrete time impulse trains (this assumption will be discussed in subsection "Interunit dependence of impulse trains"). In the following, we assume that the coordinates of nonzero elements (discrete discharge instants) in each s_i are contained in a vector noted as x_i.

With the assumption of independence of impulse trains $(s_i)_i$ (and thus their discharge instants x_i), we obtain the conditional probability for the discharge time instants:

$$P((x_i)_i|(m_i,\sigma_i^2)_i) = \prod_{i=1}^{I} P(x_i|m_i,\sigma_i^2), \tag{8.2}$$

while the impulse train can be written as $s_i = 1_{x_i}$ by neglecting the AP shape variations during the recording.

Modeling of the sparsity in each impulse train will be discussed in the subsection "Sparsity of impulse trains," as well as the integration of physiological constraints that are application-specific to further impose sparsity characteristics. In the subsection "Interunit dependence of impulse trains," we discuss interunit dependence and its impact on the results of the spike sorting techniques that are based on the mutual independence assumption. Finally, in the subsection "Complexity of overlap decomposition," we deduce the decomposition complexity of the model in the presence of overlapped APs, without prior knowledge of the number or of the shape of the units involved.

Sparsity of Impulse Trains

We note that the sparsity in each impulse train is modeled by the nonzero impulse instants contained in the vector x_i, of considerably lower dimension than that of s_i. Similar models have been proposed by De Luca (De Luca, 1979) and, in a multiple-channel context, by Holobar and co-workers (Holobar and Zazula, 2007) for the case of EMG decomposition.

For restoration of sparse impulse trains, the Bernoulli–Gaussian (BG) model has been successfully used in several applications (Kormylo, 1983; Cheng et al., 1996; Champagnat et al., 1996; Bourguignon and Carfantan, 2005). It is worthwhile to introduce the model and its extensions since various efficient numerical tools exist for dealing with the BG model. The BG model first defines the nonzero impulse coordinates (the discharge instants x_i) for the ith impulse train using the Bernoulli law:

$$P(x_i) = \lambda^{n_i}(1-\lambda)^{N-n_i}$$

where $n_i = \dim(x_i)$ denotes the number of discharges (priorly unknown) in the total length ($N = \dim(s_i)$) of the impulse train, and $\lambda \in (0, 1)$ the Bernoulli parameter that controls the average density of nonzero elements in s_i. A Gaussian random variable is then associated with each nonzero impulse.

Such a model, however, needs some adaptations in order to model the neural spike train generation process, particularly with respect to some physiological constraints:

(A1) The **interspike interval** (ISI) $T_{ij} = x_{i,j+1} - x_{i,j}$ measures the temporal distance between two consecutive impulses of the same source and should be larger than a threshold value T_R, which expresses the refractory period of the firing cells.

(A2) In most applications, the firing pattern for each impulse train is more likely to be regular, that is, to have similar consecutive ISI values, than to be completely random.

The threshold value T_R represents the **refractory period** of the neural cells and thus depends on the specific application. For example, it is the absolute refractory period of the active motor neurons in the case of intramuscular EMG decomposition. The firing pattern's regularity in (A2) may be met more in some applications and less in others. For example, alpha motor neurons discharge very regularly during sustained muscle contractions (McGill et al., 2005).

We integrate the preceding assumptions for multiunit recordings into the BG model. First, the admissible solution space C for the ith neural unit spike train is written as $C = \{x_i \mid x_{i,j+1} - x_{i,j} > T_R \text{ for all } j\}$ and 1_C denotes the indicator function of C, such that

$$1_c(x_i) = \prod_{j=1}^{n_i-1} 1_{\{x_{i,j+1}-x_{i,j}>T_R\}}$$

Contrary to the BG case, a spike train s_i is no longer an independent and identically distributed (i.i.d.) sequence, because the discharge instants x_i are not independent. In the BG deconvolution framework, a model called a **modified Bernoulli process** has been proposed to impose the minimum distance constraint on the impulse train (Kail et al., 2009). It is noteworthy that the modified Bernoulli model reduces research complexity in solving overlapped AP.

Second, the firing pattern regularity can be modeled by a Gaussian-shaped distribution on the variables $T_{ij} - T_R$ (Ge et al., 2010; see Fig. 8.2 for histogram of the variable for a motor neuron). Fee et al. (1996) were the first to integrate the firing pattern regularity properties in their spike sorting algorithms. For each neural cell, the resulting law of x_i (in a discrete, regularly sampled time framework) can be written as follows:

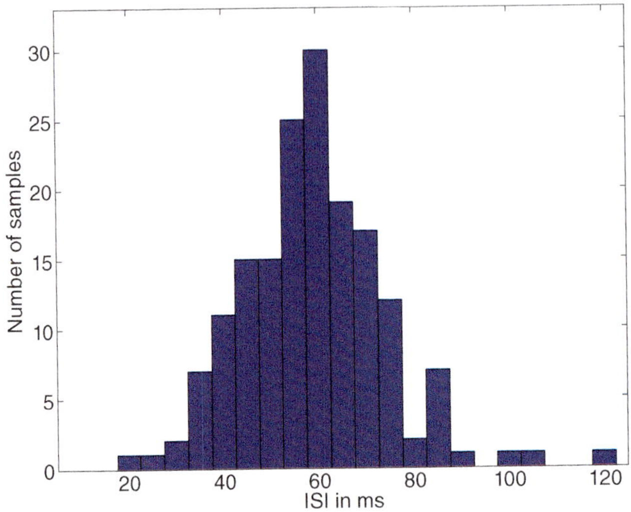

Figure 8.2. Representative histogram of ISI values within an impulse train of 10 seconds for a single alpha motor neuron.

$$P(x_i|m_i,\sigma_i^2) \propto 1_C(x_i) f_{m_{i,\sigma_i^2}}^{-(n_i-1)} \exp\left\{-\frac{1}{2\sigma_i^2}\sum_{j=1}^{n_i-1}(x_{i,j+1}-x_{i,j}-m_i-T_R)^2\right\} \quad (8.3)$$

where

$$f_{m_i,\sigma_i^2} = \sum_{k=-m_i}^{+\infty} \exp\left(-\frac{k^2}{2\sigma_i^2}\right)$$

In general, the integration constant f_{m_i,σ_i^2} depends on both m_i and σ_i^2. The following approximation can be adopted:

$$f_{m_i,\sigma_i^2} \approx \int_{-m_i}^{+\infty} \exp\left(-\frac{k^2}{2\sigma_i^2}\right) dk \approx \int_{-\infty}^{+\infty} \exp\left(-\frac{k^2}{2\sigma_i^2}\right) dk = \sqrt{2\pi\sigma_i^2} \quad (8.4)$$

under the following assumptions:

- σ_i is large enough to ignore the discretization error; for example, for $\sigma_i = 10$ ms and a sampling frequency of 10 kHz, the 3σ Gaussian lobe is discretized by 600 samples.
- σ_i/m_i is small enough to ignore the integral truncation at $-m_i$; for example, if $\sigma_i/m_i < 1/3$, then the truncation error is controlled at 0.135%.

To define a proper probability, Eq. 8.3 should be modified to incorporate a prior law on the first discharge instant x_{i1}, which can be supposed uniform for the sake of simplicity.

A further refinement in the model implies a more complex discrete random variable model of ISI for each unit, for example, a negative binomial (NB) distribution:

$$P(X = k|r, p) = C_{k+r-1}^{r-1}(1-p)^r p^k, \quad for\ k = 0,1,2\ldots \tag{8.5}$$

in which X stands for $T_{ij} - T_R$ as in the Gaussian-shaped model. Two degrees of freedom, that is, both the mean and variance of the probability density function, can be parameterized by (r, p). This extension is especially useful for discrete random variables over an unbounded positive range ($k \geq 0$) whose variance might exceed the mean. Typically for a Poisson distribution, the mean and the variance are equal. Observed ISI distributions can be overdispersed with respect to the Poisson law, while in the NB distribution model the second parameter can be used to adjust the variance independently of the mean. One advantage over the Gaussian-like distribution is the suppression of approximation in the formulation.

Interunit Dependence of Impulse Trains

The preceding discussion on modeling multiunit signals is based on the assumption of the independence of interunit firing patterns. Nevertheless, it is well known that in neural and muscular recordings this assumption is not fully met. For example, alpha motor neurons exhibit correlation both at low frequencies (usually called common drive; De Luca et al., 1982) and at higher frequencies, as near-simultaneous discharges of two or more units (short-time synchronization; Dietz et al., 1976; De Luca, 1979). It has to be noted that as a consequence of the dependence among discharge patterns, the actual overlap rates observed in multiunit recordings are generally higher than the average estimation given in Bankman et al. (1993), for which correlations are not considered. Models exist (e.g., see Jiang et al., 2007) to simulate the interdependent discharge patterns.

However, we note that the interunit independence assumption does not constitute a limitation in the search for the solution but rather implies that the potential correlation between discharge patterns is not taken into account in the prior laws. Such an assumption is usually motivated by model simplicity, since the complete discharge patterns as described in Jiang et al. (2007) would be extremely difficult to implement numerically in spike sorting methods. In cases with high correlation between neural firing patterns, such as tremor, the correlation information is simply not exploited in the statistical model.

Complexity of Overlap Decomposition

For the Gaussian-shaped ISI model, the posterior distribution $P(\Theta|z)$ for

$$\Theta = \{(x_i, m_i, \sigma_i^2, h_i)_i, \sigma_i^2\}$$

can be expressed as follows:

$$P(\Theta|z) \propto P(z|(x_i,h_i)_i,\sigma_\varepsilon^2)P(\sigma_\varepsilon^2)\prod_i P(x_i|m_i,\sigma_i^2)P(m_i)P(h_i) \quad (8.6)$$

This is a typical hierarchical Bayesian model. The NB model of ISI yields a similar posterior distribution for $\Theta = \{(x_i, r_i, p_i, h_i)_i, \sigma_\varepsilon^2\}$.

From a Bayesian perspective, the spike sorting task consists of deriving an estimator $\hat{\Theta}$ from the posterior distribution Eq. 8.6, containing both continuous and discrete parameters. The difficulty lies in the resolution of overlapped spikes, since the combinatorial nature of $(x_i)_i$ precludes the exhaustive exploration method (Atiya, 1992). Although analytic expressions of its complexity are difficult to derive due to the refractory period constraint, in the following we give examples of its cardinality to illustrate the order of magnitude.

We note variables with superscript (k) and $(-k)$ for those in the kth segment and in all other segments, respectively. Thus C_k denotes the set of eligible firing instants $(x_i)_i^{(k)}$ in the kth segment that satisfy the refractory period condition, in particular with respect to $(x_i)_i^{(-k)}$.

In a very favorable case for which at most I units are active in a segment of duration of L samples, the overall search space cardinality is bounded by $(L + 1)^I < |C_k| < 2^{I \times L}$. The inferior bound corresponds to a subspace containing at most one impulse for each unit in the segment, while the superior bound corresponds to the cardinality of the superclass neglecting the refractory constraint. For segment lengths that vary between 8 and 60 ms, sampled at 10 kHz, and number of units that varies between 4 and 8, the cardinality of $|C_k|$ is at least 4.30×10^7 and 1.68×10^{22} in each case.

The cardinality alone, however, does not sufficiently explain the spike sorting complexity. Another important phenomenon is illustrated in Figure 8.3, which shows the instability of overlapped AP shapes with respect to minor changes of firing instants $(x_i)_i$, for an extremely simple case of two overlapped APs. A similar problem exists when different sets of APs with different temporal alignments might produce very similar waveform shapes when overlapping. In these circumstances, spike sorting techniques uniquely based on a likelihood cost function

$$P(z^{(k)}|(x_i^{(k)}, h_i^{(k)})_i)$$

(using the Euclidean distance or the Minkowski distance in a more general sense) are bound to fail. This is the reason why a unit-by-unit maximization of the likelihood function with respect to x_i is not feasible, but instead the firing patterns of all units should be jointly optimized in each segment.

In conclusion, finding the $(x)_i^{(k)}$ by maximizing its posterior probability in C_k is the key problem of spike sorting, and exhaustive exploration methods yield unrealistic computational load, while the instability of overlapped AP shapes forbids any fast convergence methods without exploring the entire combinatorial space.

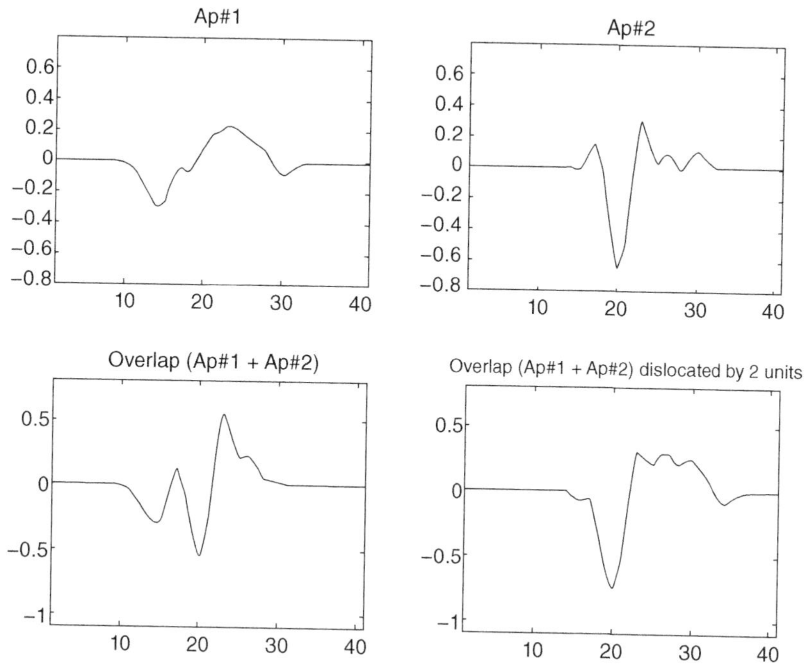

Figure 8.3. A simulation result illustrating drastic changes in overlapped APs due to minor temporal alignment changes. The upper panels show two isolated AP templates, while the lower panels show considerable shape changes with slight temporal alignments of the overlap. Each AP has a duration of 41 samples. *X*-axis is in samples, *y*-axis is in normalized units.

STATE OF THE ART

In this section, we first give a summary on the preprocessing methods that constitute the first step of most spike sorting algorithms. The fast one-pass algorithms allow segmentation of neural activities from noisy background and classification of neural units based on isolated AP waveforms.

Four classes of spike sorting methods are presented, with an emphasis on the ability of solving the complex AP overlaps.

Preprocessing

The identification of isolated spikes is usually performed by (1) filtering the signal to enhance spike activities, (2) segmenting the filtered signal into temporal intervals containing spikes (McGill et al., 2005), and (3) classifying isolated individual spikes (those not overlapped temporally with others) to determine the number of units I and the approximate spike shape for each class $h_i^{(0)}$ and the incomplete discharge patterns.

1. *Filtering.* Band-pass filtering is used to eliminate two sources of noise:
   (i) The potentials of neural units located relatively away from the electrodes that contain mainly low-frequency components due to the low-pass filtering effect of the tissues (LeFever and DeLuca, 1982)
   (ii) Thermal noises inherent to recording apparatus that contain mainly high-frequency components. In addition to band-pass filtering, various denoising techniques have been proposed for neural recordings, as in Kamavuako et al. (2003).
2. *Segmentation by Thresholding.* The goal of segmentation of spikes is to reduce the computational cost of the decomposition by estimating firing patterns only in selected temporal segments. Segments are selected to contain either isolated or overlapped APs, while pure noise intervals are ignored (see Fig. 8.4). Admissible solutions of the firing patterns $(1_{x_i})_i$ are thus limited to the selected segments.

Various ways to automatically determine an appropriate threshold have been proposed. As an example, for intramuscular recordings, a method for setting the threshold value T has been proposed in McGill et al. (1985):

$$T = c_1 \underbrace{\sqrt{\sum_{n=1}^{N} z_n^2 I(T,n) \bigg/ \sum_{n=1}^{N} I(T,n)}}_{\widehat{\sigma_\varepsilon}} \tag{8.7}$$

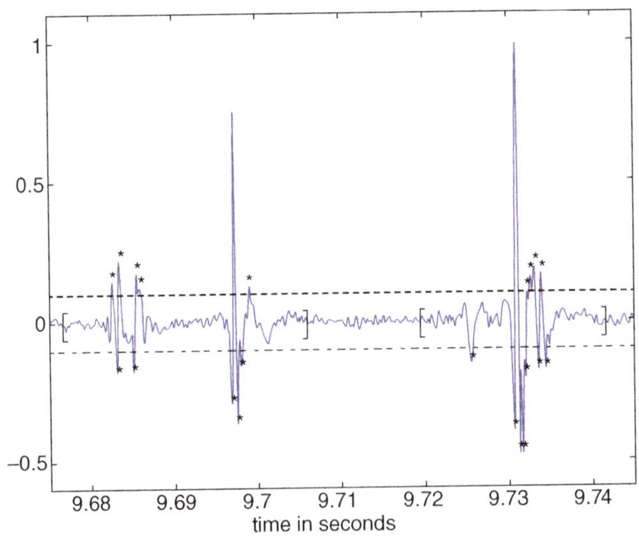

Figure 8.4. Two segments of an intramuscular EMG signal delimited by square brackets. The spikes in this signal indicate the activity of alpha motor neurons.

where z designates the filtered signal, N the total length of the signal, and $I(T, n)$ a binary function such that:

$$I(T,n) = \begin{cases} 1, & if\ |z_n| < T \\ 0, & otherwise \end{cases} \tag{8.8}$$

The brace under Eq. 8.7 shows the part that estimates the noise standard deviation $\widehat{\sigma_\varepsilon}$ by calculating the mean square, given the threshold value T. The threshold value S is fixed to be c_1 times $\widehat{\sigma_\varepsilon}$. The c_1 value is fixed empirically and the solution of T can be obtained in $O(N)$ using the algorithm in McGill and Dorfman (1984) and McGill et al. (1985).

Local extrema of z with absolute values greater than T are considered as AP peaks, as indicated in Figure 8.4, in which the two dashed line levels correspond to $\pm T$ and the AP peaks are marked by the stars. A new segment is created when the distance between two consecutive AP peaks is greater than the AP duration P. Figure 8.4 shows an example of segmentation for an EMG signal.

3. *Determination of the Spike Template* $(h_i^{(0)})_i$ *and the Number of Neural Units.* The goal is to determine an approximate waveform $h_i^{(0)}$ for each neural unit from the segmentation results. These approximate waveforms will be used for overlap decomposition as initialization values.

Automatic classification methods can be applied to isolated APs. The following steps can be performed:

(i) Select the 20% shortest segments.
(ii) Align the selected segments with respect to the peaks.
(iii) Classify the selected segments automatically.

Various automatic classification methods can be applied to isolated APs. Here we provide two examples of these methods, which well represent the problem at hand.

The first is a nonparametric Bayesian (NPB) method proposed by Wood (Wood and Black, 2008). The number of classes, a typical problem of model choice, is integrated analytically within an **infinite Gaussian mixture model** (IGMM; Rasmussen, 2000), as an alternative to the **reversible-jump Markov chain Monte Carlo** (RJMCMC) proposed by Nguyen et al. (2003) in AP classification. The method (Wood and Black, 2008) of stochastic simulation consists of generating a Markov chain in which each sample is a label vector, representing the class ID attributed to isolated APs. The dimension of these APs is reduced by using only a few coefficients extracted by principal component analysis (PCA).

Figure 8.5a,b shows typical results of the NPB method (Wood and Black, 2008). The selected segments initially have a duration between 5 and 6 ms. The

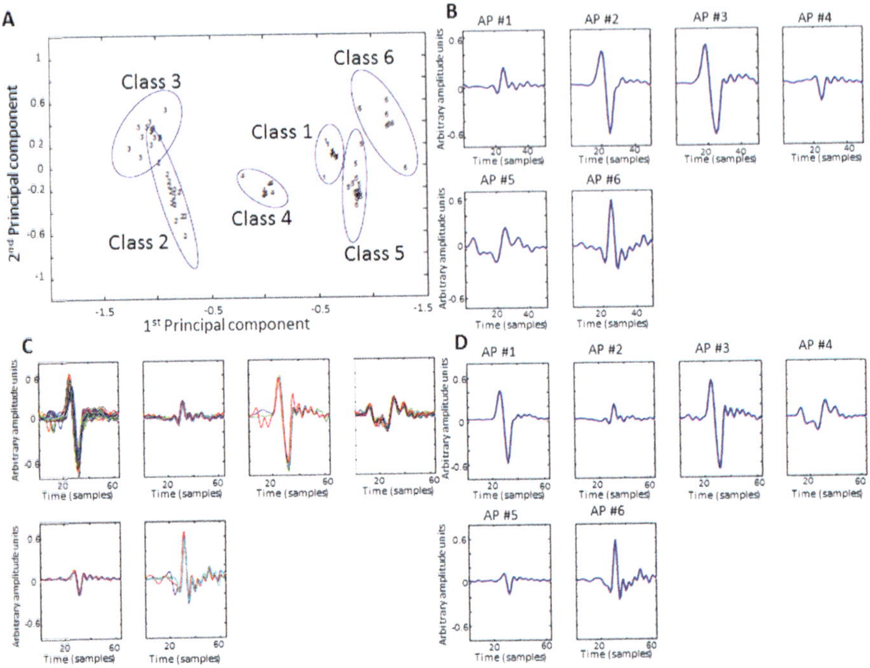

Figure 8.5. Comparison of two classification methods (NPB and CRDFT) applied to isolated APs.

coordinates in Figure 8.5a represent the two principal directions of an AP from the PCA, and each number is the class label attributed by the classification method. The equiprobability levels for each AP class (two-dimensional elliptical contours for a Gaussian distribution) are calculated from the covariance matrix of each class. Figure 8.5b illustrates representative AP waveforms $h_i^{(0)}$ obtained by averaging within each class.

The second classification method is based on the Euclidean distances between the **canonically registered discrete Fourier transform** (CRDFT) coefficients (McGill and Dorfman, 1984; McGill et al., 1985). This method has the advantage of better temporal alignment resolution for APs by interpolation. The idea is simple: a temporal shift of a signal can be represented by a phase multiplicative term in the Fourier transform domain. The minimization of the Euclidean distance with respect to the phase thus yields a finer resolution than the sample for temporal alignment. The APs are processed one after another, first by attempting to attribute it to an existing class if the distance is lower than a certain threshold (McGill and Dorfman, 1984); otherwise a new class of a single instance is generated.

The results of McGill and co-workers (McGill and Dorfman, 1984; McGill et al., 1985) are illustrated in Figure 8.5c,d. All instances of the same class obtained by McGill and Dorfman (1984) are given in Figure 8.5c and their averages $(h_i^{(0)})_i$ in Figure 8.5d. The same number of classes are identified by both the methods described. Moreover, representative AP waveforms $h_i^{(0)}$ are obtained up to a label permutation difference. However, determining the number of neural units remains an open problem in general settings, especially in cases of less frequently isolated APs when the level of neural activity is high.

Resolution of Overlapped Action Potentials

Once isolated action potentials are classified and incomplete discharge patterns obtained, the overlapped potentials need to be disentangled. In this subsection we describe some of the approaches that have been developed for this purpose. As stated in the Introduction, the resolution of overlapped APs is the most challenging task for complete spike sorting.

Among the spike sorting methods, the matched template filtering (MTF) method (also known as template matching) is by far the most popular and simple-to-implement sorter (D'Hollander and Orban, 2009; Bankman et al., 1993). It is a time-domain technique in which one template is generated in the learning phase to represent each neural unit AP waveform (similar to the preprocessing step) and a linear filtering is then performed to determine the firing pattern from the temporal correlation results. Faster MTF variants have also been proposed by extracting lower dimensional features of the templates, such as height, width, and onset time (Lewicki, 1998).

We introduce here an MTF sorter that solves the AP overlap problem. Bankman and co-workers (Bankman et al., 1993) proposed to perform an exhaustive exploration in a restrained space to find the best fit with respect to the observed signals. The performance gain was studied for the resolution of overlapped templates of up to two neural units. The optimal alignment of two templates was obtained by minimizing the residual errors (Euclidean distances). To do so, all configurations of temporal alignment by any two APs are exhaustively explored. Segments for which the minimum residual errors are higher than a given threshold are then considered as outliers. Despite its simple implementation, the method is representative in the temporal resolution of AP overlaps and illustrates the limit of exhaustive exploration methods when the number of neural units increases because of the very large cardinality of the solution (see "Complexity of overlap decomposition").

As an extension of the MTF approach for resolution of superpositions, Chandra and co-workers (Chandra and Optican, 1997) attempted the combination of MTF linear filtering and nonlinear classification within an artificial neural network (ANN) framework. The algorithm's highly parallel structure allows online real-time multiunit sorting if implemented by appropriate digital processors. The input and output training data are set to force the network to respond positively only when the spike is temporally aligned in the data buffer (first layer input nodes), and to reject shifted templates. Ideally, it will also respond when overlapped APs are n samples apart. In that case, the ANN would respond to the first spike and then n samples later to the second spike. The difficulty lies, however, in the training phase. Indeed, exhaustive training of all combinations of template overlaps is unrealistic, just as in MTF methods. In particular, Chandra and co-workers trained the network with up to two overlapped templates and tested on sequences with two firing neural units. The ANN hardly solves any overlapped waveforms that are separated by a small number of samples.

Alternative to the previous approaches, the idea of exploiting firing pattern statistics to further help overlap decomposition can be integrated in a Bayesian framework. Lewicki has given concrete examples (Lewicki, 1994) to illustrate that the best-fitting overlap solutions are not necessarily the most probable solutions in the posterior distribution sense by taking into account the firing pattern statistics (frequency, interspike intervals). It is shown that the best-fitting overlap solution actually increases the classification error, especially the false positive rate for neural units with smaller AP energies. The phenomenon is called **overfitting** (Lewicki, 1994). Thus, to minimize classification error, it is necessary to find the most probable overlap solution by considering deduced firing pattern statistics.

The dynamic programming method is proposed in Lewicki (1994) to solve overlaps in the spike sorting by a fast exploration of a k-d tree. Numerically the algorithm is also limited by the memory space and exploration complexity necessary to generate such a data structure, and in practice this results in an equally restrained search space of up to two overlapping action potentials.

Another method based on the Bayesian model is paired with either a maximum a posteriori (MAP) estimator (Ge et al., 2010) or a minimum mean square error (MMSE) estimator using stochastic simulation methods (Ge et al., 2011) for fully automatic spike sorting. The Bayesian model was based on the firing pattern regularity assumption, as formulated in Eq. 8.3 for a Gaussian-like ISI distribution. It is noteworthy that the search space is not restrained to two-AP overlaps, thus fully allowing decomposition of segments that contain extremely complex overlap situations. Figure 8.6 shows an example of raw intramuscular EMG signals and the decomposition of two adjacent segments with overlapped potentials using the proposed method. In this example, the proposed fully automatic method provided the same result as that obtained by a reference decomposition tool used by an expert operator.

Other approaches are not limited in the number of overlapping sources but require an interaction with an operator. For example, an algorithm for decomposing intramuscular EMG signals into the constituent motor unit spike trains recursively matches the templates and reevaluates the residual errors, until it is able to prove that it has found the global optimum (McGill, 2002). To avoid endless recursion in cases of difficult spike superpositions, the algorithm stops

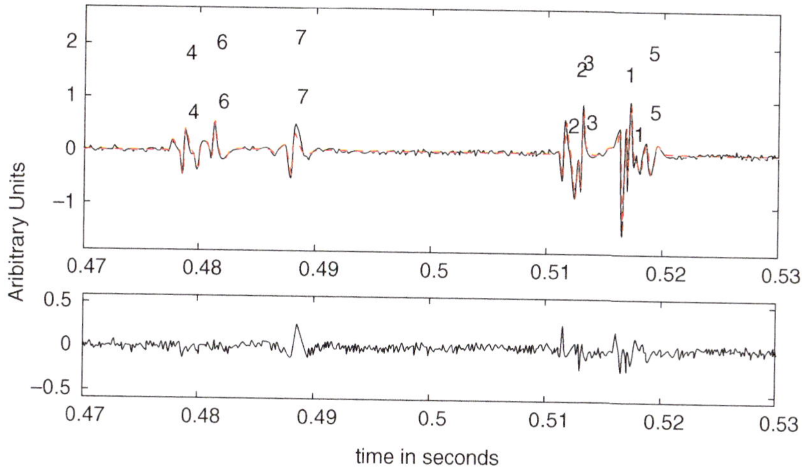

Figure 8.6. Example of spike sorting of two adjacent segments using the approach described in Ge et al. (2011), compared with the reference manual approach. In these two segments a total of seven sources are active. The numbers above the raw signal represent the source labels as identified by the automatic decomposition (top numbers) and the decomposition with interaction of an expert operator (lower numbers). The labeling of sources in this example led to perfect matching with the reference result. The raw (solid line) and the reconstructed (dashed line) signals are shown in the upper panel; the residual error is plotted in the lower panel.

and requires an expert intervention after a number of trials without reaching the lower-bound value.

CONCLUSIONS

In this chapter, we presented the spike sorting problem in the context of multiunit single-channel applications. We particularly addressed the issues of its utility, the main difficulties, and the state-of-the-art approaches.

Each cell generates spikes of a characteristic shape, and the identification of the time instants of cell activity is a challenging problem. Its main difficulty lies in the resolution of overlapped spikes. Isolated spikes can indeed be detected and clustered with robust techniques by high-performance classification methods. On the contrary, the complete sorting of overlapped action potentials is a nondeterministic polynomial-type (NP) hard problem and thus cannot be solved by polynomial complexity algorithms. Therefore, methods for full spike sorting differ mainly in their performance when separating overlapped action potentials.

Several approaches proposed for the resolution of overlapped spikes are limited to only a few overlapped APs. Bayesian inference methods provide a richer framework that allows the integration of additional information, such as firing pattern statistics, in order to limit the search space.

REFERENCES

Atiya F.A. (1992). Recognition of multiunit neural signals. *IEEE Trans Biomed Eng* 39:723–729.

Bankman I., Johnson K., and Schneider W. (1993). Optimal detection, classification, and superposition resolution in neural waveform recordings. *IEEE Trans Biomed Eng* 8(40):836–841.

Bourguignon S. and Carfantan H. (2005). Bernoulli–Gaussian spectral analysis of unevenly spaced astrophysical data. In IEEE Workshop Statist Signal Process, Bordeaux, France, pp. 811–816.

Champagnat F., Goussard Y., and Idier J. (1996). Unsupervised deconvolution of sparse spike trains using stochastic approximation. *IEEE Trans Signal Process* 44(12):2988–2998.

Chandra R. and Optican L. (1997). Detection, classification, and superposition resolution of action potentials in multiunit singlechannel recordings by an on-line real-time neural network. *IEEE Trans Biomed Eng* 44:403–412.

Cheng Q., Chen R., and Li T.-H. (1996). Simultaneous wavelet estimation and deconvolution of reflection seismic signals. *IEEE Trans Geosci Remote Sensing* 34: 377–384.

Citi L., Carpaneto J., Yoshida K., Hoffmann K.P., Koch K.P., Dario P., and Micera S. (2008). On the use of wavelet denoising and spike sorting techniques to process electroneurographic signals recorded using intraneural electrodes. *J Neurosci Methods* 172(2):294–302.

D'Hollander E.H. and Orban G.A. (2009). Spike recognition and on-line classification by unsupervised learning system. *IEEE Trans Biomed Eng* BME-26(5):279–284.

De Luca C.J. (1979). Physiology and mathematics of myoelectric signals. *IEEE Trans Biomed Eng* 26(6):313–325.

De Luca C.J., Le Fever R.S., McCue M.P., and Xenakis A.P. (1982). Control scheme governing concurrently active human motor units during voluntary contractions. *J Physiol* 329:129–142.

Dietz V., Bischofberger E., Wita C., and Freund H.J. (1976). Correlation between the discharges of two simultaneously recorded motor units and physiological tremor. *Electroencephalogr Clin Neurophysiol* 40:97–105.

Djilas M., Azevedo-Coste C., Guiraud D., and Yoshida K. (2010). Spike sorting of muscle spindle afferent nerve activity recorded with thin-film intrafascicular electrodes. *Intell Neurosci*. DOI: 836346.

Fee M.S., Mitra P.P., and Kleinfeld D. (1996). Automatic sorting of multiple unit neuronal signals in the presence of anisotropic and non-gaussian variability. *J Neurosci Methods* 69(2):175–188.

Ge D., Le Carpentier E., and Farina D. (2010). Unsupervised Bayesian decomposition of multi-unit EMG recordings using Tabu Search. *IEEE Trans Biomed Eng* 57(3):561–517.

Ge D., Le Carpentier E., Idier J., and Farina D. (2011). Spike sorting by stochastic simulation. *IEEE Trans Neural Syst Rehabil Eng* 19:249–259.

Gold C., Henze D.A., Koch C., and Buzsáki G. (2006). On the origin of the extracellular action potential waveform: a modeling study. *J Neurophysiol* 95:3113–3128.

Holobar A. and Zazula D. (2007). Multichannel blind source separation using convolution kernel compensation. *IEEE Trans Sig Process* 55(9):4487–4496.

Jiang N., Englehart K.B., and Parker P.A. (2007). A simulation method for the firing sequences of motor units. *J Electromyogr Kinesiol* 17(5):527–534.

Kail G., Novak C., Hofer B., and Hlawatsch F. (2009). A blind Monte Carlo detection–estimation method for optical coherence tomograph. In *IEEE International Conference on Acoustics, Speech, and Signal Processing (ICASSP)*, Taipei, Taiwan, R.O.C., pp. 493–496.

Kamavuako E.N., Jensen W., Yoshida K., Kurstjens M., and Farina D. (2003). A criterion for signal-based selection of wavelets for denoising intrafascicular nerve recordings. *J Neurosci Methods* 186(2):274–280.

Kormylo J.J. and Mendel J.M. (1983). Maximum-likelihood seismic deconvolution. *IEEE Trans Geosci Remote Sensing* GE-21(1):72–82.

LeFever R. and De Luca C. (1982). A procedure for decomposing the myoelectric signal into its constituent action potentials. Part I. Technique, theory, and implementation. *IEEE Trans Biomed Eng* 29:149–157.

Lewicki M.S. (1994). Bayesian modeling and classification of neural signals. *Neural Computation* 6:1005–1030.

Lewicki M.S. (1998). A review of methods for spike sorting: the detection and classification of neural action potentials. *Network: Computation Neural Syst* 9(4):R53–R78.

McGill K.C. (2002). Optimal resolution of superimposed action potentials. *IEEE Trans Biomed Eng* 49(7):640–650.

McGill K.C. and Dorfman L.J. (1984). High-resolution alignment of sampled waveforms. *IEEE Trans Biomed Eng* BME-31(6):462–468.

McGill K.C., Cummins K.L., and Dorfman L.J. (1985). Automatic decomposition of the clinical electromyogram. *IEEE Trans Biomed Eng* BME-32(7):470–477.

McGill K.C., Lateva Z.C., and Marateb H.R. (2005). EMGLAB: an interactive EMG decomposition program. *J Neurosci Methods* 149(2):121–133.

Nguyen D., Frank L., and Brown E. (2003). An application of reversible-jump Markov chain Monte Carlo to spike classification of multi-unit extracellular recordings. *Network* 14:61–82.

Parsaei H., Stashuk D., Rasheed S., Farkas C., and Hamilton-Wright A. (2010). Intramuscular EMG signal decomposition. *Crit Rev Biomed Eng* 38(5):435–465.

Quian Quiroga R., Nadasdy Z., and Ben-Shaul Y. (2004). Unsupervised spike detection and sorting with wavelets and superparamagnetic clustering. *Neural Computation* 16:1661–1687.

Rasmussen C. (2000). The infinite gaussian mixture model. *Adv Neural Inf Process Syst* 12:554–560.

Wood F. and Black M.J. (2008). A nonparametric bayesian alternative to spike sorting. *J Neurosci Methods* 173(1):1–12.

9

FOCUSED CHAPTER

WAVELET DENOISING AND CONDITIONING OF NEURAL RECORDINGS

LUCA CITI[1,2] AND SILVESTRO MICERA[3,4]

[1]*Department of Anesthesia, Critical Care and Pain Medicine, Massachusetts General Hospital, Harvard Medical School, Boston, Massachusetts, USA*
[2]*Department of Brain and Cognitive Sciences, Massachusetts Institute of Technology, Cambridge, Massachusetts, USA*
[3]*BioRobotics Institute, Scuola Superiore Sant'Anna, Pisa, Italy*
[4]*Center for Neuroprosthetics, Swiss Federal Institute of Technology Lausanne (EPFL), Lausanne, Switzerland*

SUMMARY

This chapter presents wavelet-based denoising algorithms as a preprocessing stage before spike detection and sorting. The first part of the chapter is an overview of wavelet-based denoising. The choices that the designer is required to make when implementing a wavelet-based denoising algorithm are presented in detail. In particular, the dyadic wavelet transform is compared with a time-invariant approach, showing that the latter is best suited to the denoising of neural signals.

The second part shows a sample application with eletroneurographic (ENG) signals recorded from the sciatic nerve of rabbits while the experimenter

Introduction to Neural Engineering for Motor Rehabilitation, First Edition. Edited by Dario Farina, Winnie Jensen, and Metin Akay.

stimulated the paw of the animal. A machine learning algorithm is trained to infer the stimulus applied based on the recorded ENG signals. The wavelet-based denoising is compared with a traditional band-pass filter in two cases: when followed by spike sorting and when followed by traditional rectified bin integration (RBI). The results illustrate the benefits of wavelet denoising over standard band-pass filtering and demonstrate that there is an even more marked improvement when the subsequent step requires signals with high signal-to-noise ratio (SNR), such as in the case of spike sorting.

BACKGROUND

The possibility of extracting action potentials (APs) from neural signal recordings is important for neuroscience research in general and for the development of neural human–machine interfaces in particular. In neural recordings, the APs of interest are often superimposed on instrument noise, environmental noise, and biological noise, such as electromyographic and electrocardiographic signals or spikes from distant neurons that are too small to be adequately discriminated. An ideal denoising algorithm should be able to identify and remove these distractors and isolate the clean APs while preserving the APs' waveforms. The shape information can thus be used to associate each spike with the correct neuron and reconstruct single-unit spike trains from multiunit recordings using spike sorting techniques (Lewicki, 1998; Quiroga, 2007).

Wavelet Denoising

Wavelet denoising is a set of techniques for removing noise from signals and images. This has been used in biomedical signal processing to reduce background noise that can be approximated to a Gaussian distributed random source (e.g., Tikkanen, 1999; Oweiss and Anderson, 2001; Kim and Kim, 2003; Diedrich et al., 2003; Wiltschko et al., 2008; Citi et al., 2008). The process involves the transformation of noisy data into an orthogonal time–frequency domain, application of thresholding to the resulting coefficients, and finally the transformation of the data back into the time domain. Within this framework, wavelet denoising can be performed with several distinct modalities, depending on the chosen type of decomposition, mother wavelet, thresholding function, and threshold selection.

Decomposition Schemes. A typical dyadic wavelet transform (DWT)–based denoising algorithm is made of three discrete phases: wavelet decomposition, thresholding, and reconstruction (Fig. 9.1). The decomposition is performed through separate iterative convolution with low-pass and high-pass wavelet decomposition filters. At each level the filter outputs are decimated by 2, thus

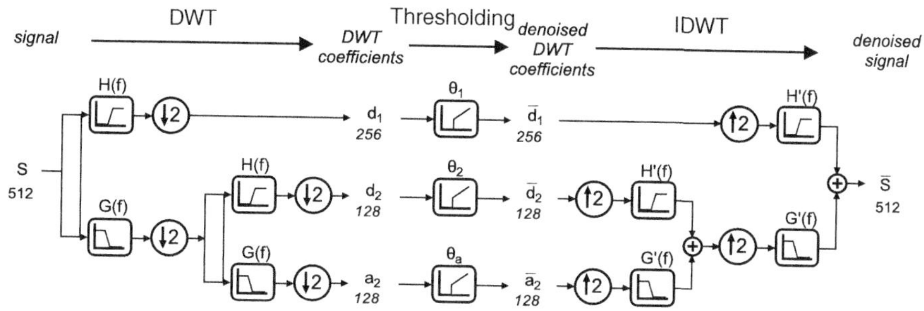

Figure 9.1. Example of denoising algorithm based on a two-level dyadic wavelet decomposition. The DWT decomposes the original signal into a nonredundant representation (e.g., a 512-sample signal is decomposed into 256+128+128 wavelet coefficients); then the coefficients are thresholded, maintaining only those above the noise level; finally, the inverse transform reconstructs the denoised signal.

obtaining a complete nonredundant representation of the original signal. The resulting wavelet coefficients are then processed by thresholding functions (more on this later) and finally the denoised signal is reconstructed.

Alternatively, a **time-invariant** decomposition scheme can be used. This technique is invariant to signal time shifts by an integer number of samples, an important property that makes it superior to the usual wavelet denoising performed via DWT when searching for abruptly changing signals such as spikes in our case.

In Brychta et al. (2007), a time-invariant approach outperformed the dyadic DWT in terms of spike detection. Of note, with a DWT-based denoising the same spike can be captured or missed depending on when the spike happens. Here we go one step further and show that, even when the DWT does not miss a spike, it can drastically change its shape depending on its time of occurrence, potentially hampering the subsequent spike sorting stage by artificially distributing a single unit into several arbitrary subunits. Figure 9.2 shows how the same spike can be captured, missed, or completely deformed by a DWT-based denoising algorithm, depending on the time the spike takes place.

A time-invariant denoising can be equivalently implemented using either the translation-invariant wavelet transform (Coifman and Donoho, 1995) or the stationary wavelet transform (Misiti et al., 2007). The latter is similar to the DWT scheme presented in Figure 9.1, with the difference that the signal is neither downsampled during decomposition nor upsampled during reconstruction. Also, it uses level-dependent filters that are upsampled versions of the corresponding DWT filters. Cascaded filters can then be combined, as depicted in Figure 9.3. An efficient algorithm can be used to process the data in the frequency domain by using an overlap-and-save approach such as in Citi (2010).

A third option is to use a continuous wavelet transform (CWT) that essentially performs a correlation analysis between the signal and the shifted and

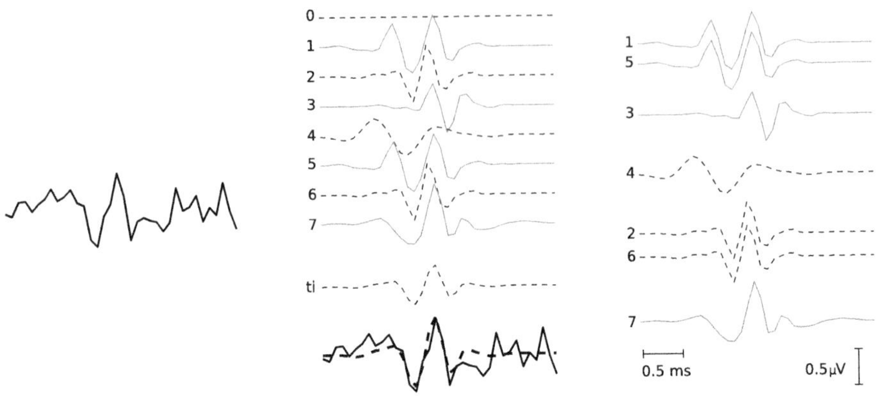

Figure 9.2. The figure shows that DWT-based denoising is not time-invariant. When processing the same noisy spike (on the left), an L-level DWT-based decomposition can produce up to 2^L different denoised spikes, depending on whether the spike happened at sample $n, n+1, \ldots, n+2^L-1$ (in the center, signals $0, 1, \ldots, 7$). This not only decreases the detection rate of the system but can also hamper the subsequent spike sorting stage by polluting it with several subunits that are an arbitrary partitioning of a single unit. Consider, for example, a signal made of several copies of the spike on the left, randomly scattered in time. The spike sorting algorithm can erroneously individuate 5 different units (on the right, the 5 different templates 1+5, 3, 4, 2+6, 7). Thus, the spikes produced by the original neuron are randomly assigned to these 5 false neurons, making inferences about the firing patterns of the real neuron impossible. Instead, with a time-invariant wavelet transform the denoised output (the signal labeled with "ti") does not depend on time. In the center bottom part of the figure, from the overlap between the original signal (solid line) and a scaled version of "ti" (dashed line), it is possible to observe the improved shape-preserving property of the time-invariant approach.

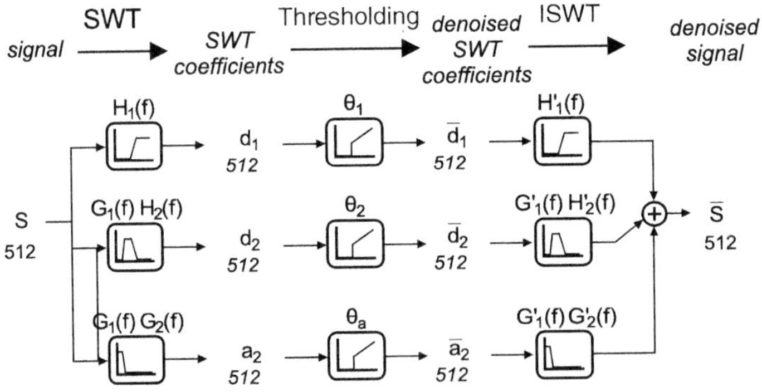

Figure 9.3. Example of a denoising algorithm based on a two-level stationary wavelet decomposition. The filters $H_l(f)$ and $G_l(f)$ used at each level l are the original wavelet filters $H(f)$ and $G(f)$ upsampled in time by a factor 2^{l-1}, that is, introducing $2^{l-1}-1$ zeros between each coefficient of the impulse response.

dilated versions of a mother wavelet. As opposed to the methods just discussed, in CWT the dilation and the shift factors can be any real value. This approach has been successfully tested on recordings from peripheral nerve interfaces (Djilas et al., 2010).

Mother Wavelet. The mother wavelet determines the transfer function of the decomposition ($H(f)$ and ($G(f)$)) and reconstruction ($H'(f)$ and ($G'(f)$)) filters used in the denoising procedure. The wavelet decomposition can be considered a matched filter between the signal and scaled versions of the mother wavelet. For this reason, the mother wavelet is often chosen by visual inspection among a catalog of wavelets by selecting the one that more closely resembles the shape of a typical action potential. For example, Diedrich et al. (2003) used the *Symmlet 7* wavelet, whereas Wiltschko et al. (2008) used *Daubechies 4*.

Alternatively, a signal-dependent mother wavelet can be designed by parameterizing the filters and then choosing the parameters that optimize some figure of merit. For example, Kamavuako et al. (2010) parameterized the wavelet filters using a single parameter whose optimal value was selected by maximizing the root mean square of the synchronized average (RMSSA) of the detected APs. This led to the selection of an optimized mother wavelet which performed similarly to the a posteriori chosen best wavelet.

Thresholding. The most common forms of thresholding are **hard-thresholding** and **soft-thresholding**, whose mathematical expressions are (respectively)

$$\mathrm{Th_H}(y)=\begin{cases} y & |y|\geq\theta \\ 0 & |y|<\theta \end{cases} \qquad \mathrm{Th_S}(y)=\begin{cases} y-\mathrm{sign}(y)\theta & |y|\geq\theta \\ 0 & |y|<\theta \end{cases}$$

The selection of the threshold is important because it determines how aggressive the denoising should be. Among the different approaches, the most frequently used are those proposed by Donoho and Johnstone (1994), such as the **universal threshold** and the **minimax threshold**, whose mathematical expressions are (respectively)

$$\theta_{\mathrm{U}}=\sigma\sqrt{2\ln(N)} \qquad \theta_{\mathrm{M}}=\sigma(0.3936+0.1829\log_2(N))$$

In statistics the minimax principle is often used to design estimators. The denoised signal can be thought of as an estimator of the unknown regression function. The minimax threshold minimizes the maximum mean square error over a given set of functions (Donoho and Johnstone, 1994). The minimax formulation thus leads to smaller thresholds than the corresponding universal threshold, making the denoising more conservative. For this reason, it is more appropriate when the most significant details of the signal lie near the noise range, such as with the low signal-to-noise ratio typical of peripheral neural signals.

When the noise may not be uniformly white, a different threshold can be adopted for each level, using a level-dependent estimation of the noise. The standard deviation of the noise, σ, should be evaluated using a robust estimator such as a trimmed standard deviation or the median absolute deviation:

$$\sigma_l = \frac{\text{median}_k(|d_l(k) - \text{median}_k(d_l(k))|)}{0.6745}$$

where $d_l(k)$ are the wavelet coefficients at level l.

Reconstruction. The reconstruction can be performed by using only selected levels of the decomposition and discarding the others (or equivalently, setting the corresponding threshold to $+\infty$). As each level approximately corresponds to a frequency band, this has a similar effect to filtering the signal. Each detail level $l = 1, \ldots, L$ roughly corresponds to the frequency band between $2^{-l}f_n$ and $2^{-l+1}f_n$, where f_n is the Nyquist frequency, that is, half the sampling frequency. The approximation level contains frequencies below approximately $2^{-L}f_n$, where L is the maximum level of decomposition.

In the case of neural recordings, if the decomposition level L is appropriately chosen, the approximation contains low-frequency components that are outside the band of interest and can be safely discarded. Some approaches avoid the reconstruction step altogether by performing the subsequent processing steps (e.g., spike sorting) directly in the wavelet domain (Kim and Kim, 2003).

APPLICATION

In this section, a real case application of wavelet denoising is presented. Wavelet denoising was applied to eletroneurographic (ENG) signals recorded from the sciatic nerve of rabbits while the experimenter stimulated the paw of the animal. The question we want to answer is whether wavelet denoising improves the probability of inferring the nature of the stimulus applied based on the recorded ENG signals.

Methods

Experimental Setup. A new version of longitudinal intrafascicular electrodes (LIFEs), named the thin-film LIFE (tfLIFE) was used in the experiments (Yoshida et al., 2006; Hoffmann and Koch, 2005). Thin-film LIFEs, which allow multiunit peripheral nerve recordings at eight recording sites per structure, were implanted through a lateral access to the sciatic nerve of four adult rabbits. During each session, various sensory stimuli were applied to the hind limb of the rabbit, and the elicited signals were recorded using the tfLIFEs. Stimuli were, for example, ankle flexion–extension, flexion–extension of one or more toes, and stroking cutaneous receptive fields. The various stimuli were selected as a means to obtain the thresholds for activation of the mechanoreceptors on the paw and, further, to localize the activity. Details of procedures for the animal preparation, the stimulation protocol, and the acquisition chain can be found in Citi et al. (2008). Recordings acquired at 48 kHz were later downsampled to 12 kHz.

Wavelet Denoising. The wavelet denoising technique just described was used in this experiment. Specifically, we adopted a three-level time-invariant wavelet decomposition scheme using the *Symmlet 7* mother wavelet because of its similarity to typical AP waveforms. *Hard-thresholding* was used with level-dependent thresholds estimated with the *minimax method*. The noise standard deviation was estimated at each decomposition level as the median absolute deviation of the corresponding wavelet coefficients during a 45-s quiescent epoch. The approximation was completely discarded because it contained components below approximately 750 Hz. Figure 9.4 shows how the wavelet denoising algorithm is able to reduce the noise in the recordings.

Spike Sorting. The denoised signal was later processed by an automated spike sorting algorithm to isolate single-unit action potential (spike sorting techniques are discussed in detail in Chapter 8, "Spike Sorting"). Our spike sorting algorithm consisted of two phases. During the first phase, a set of spike templates were created. During the second phase, the spikes in the signal were compared with each template and labeled as belonging to its best match. To assess similarity between the spike and the template, the spike was first aligned

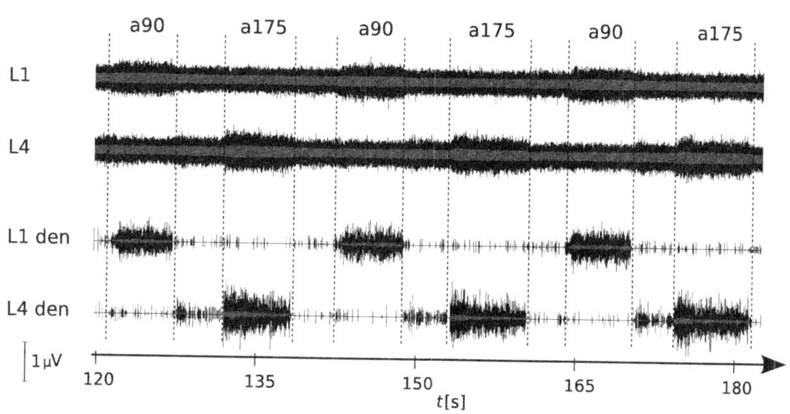

Figure 9.4. Example of applying the wavelet denoising algorithm to neural signals recorded from a tfLIFE inserted in the sciatic nerve of a rabbit while the experimenter performed leg extension at 175° (a175) or flexion at 90° (a90). The signals L1 and L4 are the raw signals recorded by two recording sites; L1den and L4den are the corresponding denoised signals.

to the template using the lag maximizing the cross-correlation, and then the following two criteria were checked: (a) the correlation coefficient between the two was greater than 0.9; (b) the ratio between the mean square difference of the two and the power of the template was less than 0.5. The first criterion is used to assess shape similarity regardless of the amplitude, while the second does the opposite.

Classification. For each rabbit, the recordings were annotated with the types of stimulus applied in each phase of the experiment. For each stimulus type, 6–12 epochs, during 3–6 seconds, were labeled. Each epoch served as an example that was used to train the classifier or to test its generalization skills. The feature vector was made from the ratios of the number of spikes matching each template to the total number of spikes in the epoch, that is, $\mathbf{F} = [f_1, \ldots, f_M]$, where M is the number of templates found during the template-creation phase and each f_i is given by $f_i = n_i/\sum_j n_j$, where n_i is the number of spikes matching the spike template i. The reason for using relative spikes rates, instead of absolute ones, was to prevent classification of the epochs based on the "quantity of activity" and favor the use of the "quality of activity" intended in terms of different spike waveforms (i.e., units) for different stimuli. To infer the type of stimulus applied during a given epoch from the feature vector $\mathbf{F}$, a classifier based on **support vector machines** (Cortes and Vapnik, 1995) was used.

Results

We used a Monte Carlo approach to assess the classification accuracy of the system. Each simulation consisted of assessing the classification performance on a test set formed by randomly choosing a single epoch per class (i.e., stimulus type) after training the system on a training set formed by the remaining epochs. After repeating this procedure several times for different randomly formed test sets, the overall classification accuracy could then be estimated.

Two preprocessing methods were compared: the wavelet denoising stage previously introduced (abbreviated as WD) and a band-pass filter (finite impulse response, FIR). In the latter case, the input data were processed with a 90-tap equiripple FIR filter with cutoff at 700 and 2000 Hz.

These processing methods were tested under two different conditions. The first (abbreviated as Srt) is when followed by the spike sorting algorithm described previously. The second (abbreviated as RBI) is when followed by traditional RBI (rectified and bin-integrated) algorithm evaluated as the mean over the epoch of the RBI computed on 50-ms windows.

The performance of the system for each of the four experimental sessions was assessed. Table 9.1 presents measures of performance, in terms of percentage of correct classifications and channel capacity, for the different processing configurations. The first column gives the results of using wavelet-based denoising followed by spike sorting (WD/Srt). The last column reports the results of

TABLE 9.1. Performance of Sessions with Four (A, C, E) or Five (D) Stimuli[a]

Session	Measure	WD/Srt	FIR/Srt	WD/RBI	FIR/RBI
A	*PC* [%]	92.6	66.7	68.6	52.8
	C [bit/symb]	1.59	0.87	1.01	0.84
C	*PC* [%]	99.1	95.2	88.8	69.0
	C [bit/symb]	1.92	1.72	1.50	1.28
D	*PC* [%]	94.5	80.4	76.0	66.5
	C [bit/symb]	2.01	1.64	1.61	1.54
E	*PC* [%]	90.1	60.7	92.6	89.1
	C [bit/symb]	1.51	0.64	1.71	1.51

[a]Performance is measured in terms of percentage of correct classifications (*PC*) and channel capacity (*C*); WD/Srt is wavelet-based denoising followed by spike sorting; FIR/RBI is a typical ENG approach using a FIR band-pass filter and rectified-bin integration; the others are hybrid configurations.

a typical ENG approach (FIR/RBI). The other columns present the performance of hybrid configurations.

A logistic regression was performed in order to assess the statistical significance of the possible improvement introduced by the use of WD over FIR, and Srt over RBI. The detailed results are reported in Citi et al. (2008). The main finding is that the use of wavelet denoising improves the classification accuracy with respect to band-pass filtering both when followed by spike sorting and when followed by RBI. Remarkably, the improvement is higher in the former case. This implies that even if the wavelet denoising step is useful per se, it becomes even more important when the following step requires high-SNR signals, such as in the case of spike sorting.

CONCLUSIONS

Wavelet-based denoising algorithms constitute a powerful preprocessing stage before spike detection and sorting. The different implementations and the several choices that the designer is required to make when implementing a wavelet-based denoising algorithm were presented in this chapter. A sample application demonstrated the improvement obtained by using this approach over a standard band-pass filter approach. In particular, we have concluded that the use of wavelet denoising is especially indicated when the following step requires high-SNR signals, such as in the case of spike sorting.

REFERENCES

Brychta R.J., Shiavi R., Robertson D., and Diedrich A. (2007). Spike detection in human muscle sympathetic nerve activity using the kurtosis of stationary wavelet transform coefficients. *J Neurosci Methods* 160(2):359–367.

Citi L. (2010). WavDen software for wavelet denoising of eng signals. Available at http://users.neurostat.mit.edu/lciti/software (accessed April 4, 2013).

Citi L., Carpaneto J., Yoshida K., Hoffmann K.P., Koch K.P., Dario P., and Micera S. (2008). On the use of wavelet denoising and spike sorting techniques to process electroneurographic signals recorded using intraneural electrodes. *J Neurosci Methods* 172(2):294–302.

Coifman R.R. and Donoho D.L. (1995). Translation-invariant de-noising. Technical report, Department of Statistics, Yale University and Stanford University.

Cortes C. and Vapnik V. (1995). Support-vector networks. *Machine Learning* 20(3):273–297.

Diedrich A., Charoensuk W., Brychta R.J., Ertl A.C., and Shiavi R. (2003). Analysis of raw microneurographic recordings based on wavelet de-noising technique and classification algorithm: wavelet analysis in microneurography. *IEEE Trans Biomed Eng* 50(1):41–50.

Djilas M., Azevedo-Coste C., Guiraud D., and Yoshida K. (2010). Spike sorting of muscle spindle afferent nerve activity recorded with thin-film intrafascicular electrodes. *Comput Intell Neurosci* 2010: 836346.

Donoho D.L. and Johnstone I.M. (1994). Ideal spatial adaptation by wavelet shrinkage. *Biometrika* 81(3):425–455.

Hoffmann K.P. and Koch K.P. (2005). Final report on design consideration of tLIFE2. Technical report, Fraunhofer Institute, Biomedical Engineering, IBMT, St. Ingbert, Germany.

Kamavuako E.N., Jensen W., Yoshida K., Kurstjens M., and Farina D. (2010). A criterion for signal-based selection of wavelets for denoising intrafascicular nerve recordings. *J Neurosci Methods* 186(2):274–280.

Kim K.H. and Kim S.J. (2003). A wavelet-based method for action potential detection from extracellular neural signal recording with low signal-to-noise ratio. *IEEE Trans Biomed Eng* 50(8):999–1011.

Lewicki M.S. (1998). A review of methods for spike sorting: the detection and classification of neural action potentials. *Network* 9(4):R53–R78.

Misiti M., Misiti Y., Oppenheim G. and Poggi J.M., Eds. (2007). *Wavelets and Their Applications*. Wiley–ISTE.

Oweiss K.G. and Anderson D.J. (2001). Noise reduction in multichannel neural recordings using a new array wavelet denoising algorithm. *Neurocomputing* 38–40: 1687–1693.

Quiroga R.Q. (2007). Spike sorting. *Scholarpedia* 2(12):3583.

Tikkanen P.E. (1999). Nonlinear wavelet and wavelet packet denoising of electrocardiogram signal. *Biol Cybern* 80(4):259–267.

Wiltschko A.B., Gage G.J., and Berke J.D. (2008). Wavelet filtering before spike detection preserves waveform shape and enhances single-unit discrimination. *J Neurosci Methods* 173(1):34–40.

Yoshida K., Hennings K., and Kammer S. (2006). Acute performance of the thin-film longitudinal intra-fascicular electrode. In: *First IEEE/RAS-EMBS International Conference BioRob*, Pisa, pp. 296–300.

10

FOCUSED CHAPTER

INSTANTANEOUS CROSS-CORRELATION ANALYSIS OF NEURAL ENSEMBLES WITH HIGH TEMPORAL RESOLUTION

ANTÓNIO R.C. PAIVA,[1,2] IL PARK,[3] JOSÉ C. PRÍNCIPE,[4] AND JUSTIN C. SANCHEZ[5]

[1]*Scientific Computing and Imaging Institute, University of Utah, Salt Lake City, Utah, USA*
[2]*ExxonMobil Upstream Research Company, Houston, Texas, USA*
[3]*Center for Perceptual Systems and Institute for Neuroscience, University of Texas, Austin, Texas, USA*
[4]*Department of Electrical and Computer Engineering, University of Florida, Gainesville, Florida, USA*
[5]*Department of Biomedical Engineering, University of Miami, Coral Gables, Florida, USA*

SUMMARY

One of the fundamental difficulties in neural assembly studies is the lack of an effective, high-resolution measure of the spatiotemporal structure in spike trains obtained from a single realization. In this chapter a systematic approach to estimate the cross-correlation (CC) of spike trains, over time and in only one realization, is proposed. The solution lies in an alternate definition of cross-correlation

Introduction to Neural Engineering for Motor Rehabilitation, First Edition. Edited by Dario Farina, Winnie Jensen, and Metin Akay.

which suggests that, rather than time averaging as is current practice, we should use ensemble averaging. This observation suggests a natural instantaneous CC estimator as required for high temporal resolution and real-time ensemble analysis and decoding. Results are shown in simulated data sets and neural activity of rat motor cortical neurons during a behavioral task.

INTRODUCTION

In multichannel neural studies, it is often necessary to quantify the spatiotemporal structure of spike train ensembles. How spike trains covary over time and at different scales is highly relevant to detecting and analyzing neural ensembles, studying their functional connectivity, and studying their role in information encoding (Buzsáki, 2010; Freiwald et al., 2001; Strangman, 1997; Gerstein and Turner, 1990; Gerstein et al., 1989). However, methods proposed in the literature to measure the covariation of spike trains neglect or smear heavily the evolution of this structure over time. However, we propose instantaneous cross-correlation (ICC) as a high temporal resolution measure for multiple spike trains obtained from a single realization.

Correlation-based measures are widely used tools to analyze spike train covariation (Brown et al., 2004). In particular, the cross-correlation (CC) is often used to assess associations between neurons (Perkel et al., 1967; Brody, 1999). However, because of the nature of spike train data, the estimation of cross-correlation is not straightforward due to the lack of the concept of signal amplitude. To deal with this problem, CC is commonly computed on the binned spike trains. In effect, binning is a transformation of the uncertainty in time onto amplitude variability and creates a form of averaging over time. In addition, binning limits the time resolution to the bin size due to the implicit quantization of the spike times. Moreover, to improve the estimation, local stationarity and ergodicity are tacitly assumed to allow averaging over a window in time. Although this averaging improves the estimation accuracy, this is obtained at the expense of temporal resolution since the filtering smears the temporal dynamic of the coupling.

Many other methods for analysis of spike train covariation have been proposed in the literature, including the Joint Peri-Stimulus Time Histogram (JPSTH) (Gerstein and Perkel, 1972; Aertsen et al., 1989), unitary events (Grün et al., 2002a,b), and frequency methods (Baccalá and Sameshima, 1999; Sameshima and Baccalá, 1999), among others. In any of these methods, however, the time structure of the analysis is smeared by averaging, over time or over trials, for gathering of statistics, or through the transformation involved.

To obtain an estimator of spike train cross-correlation with high temporal resolution, we must avoid averaging over time. The solution to this problem lies in the definition of a spatial cross-correlation. The crucial observation is that, rather than time averaging as is commonly done using the binned spike train CC estimator, we should use ensemble averaging. This observation suggests a natural ICC estimator as required for high temporal resolution and real-time ensemble analysis and decoding.

For spike train analysis we are particularly interested in measuring temporal structure in the coupling between neurons. Inevitably, this is often associated with the formation of neural assemblies and for which the most widely used descriptor is synchrony in the neuron firings (Gerstein and Turner, 1990; Gerstein et al., 1989). Due to the importance of precise spike times, synchrony studies in neural assemblies arise naturally as an obvious application for the ICC. Yet, the use of the ICC is not limited to this situation, as will be clear in our presentation. By selecting the analysis parameter appropriately it is possible to control the time scale of analysis covering the full spectrum, from synchronous firings to covariation in firing rate modulation.

BACKGROUND

The cross-correlation function of spike trains is typically computed on binned spike trains (Brown et al., 2001). The practical reasons for this approach are easily understood. By binning the spike trains one obtains a discrete-time random process, and thus random process theory can be applied as usual. Hence, the CC function of binned spike trains is defined at lag l as

$$C_{AB}^{\text{bin}}[l] = E\{b_A[n]b_B[n+l]\} \tag{10.1}$$

where b_A and b_B denote the binned spike trains, with $b_A[n]$ and $b_B[n]$ corresponding to the number of spikes in the nth bin for spike trains A and B, respectively. Additionally, in practice, the expectation is approximated as an average over time (i.e., bins),

$$C_{AB}^{\text{bin}}[l] \approx \frac{1}{M}\sum_{n=1}^{M} b_A[n]b_B[n+l] \tag{10.2}$$

with M the number of bins. To observe the time evolution of the cross-correlation, the averaging is usually done using a sliding window in time.

As stated earlier, a fundamental problem with the use of this approach for analyzing temporal structure is that the averaging over time will hide any existing temporal coupling dynamics at the fine time scale. Statistically, averaging over time implies that ergodicity and stationarity are assumed, at least locally. Yet, these assumptions are contrary to the goal of estimating the temporal dynamics.

Another issue with the preceding definitions is the use of binned spike trains. It is important to note that binning spike trains imposes a quantization on the spike times. Put differently, any information that is expressed at higher resolutions (i.e., finer time scales) than the bin size will be lost through binning.

This is especially concerning when one is interested in analysis of fine temporal structure, in the form of synchrony, for example (Grün et al., 1999).

INSTANTANEOUS CROSS-CORRELATION

Preserving Temporal Resolution

The goal here is to derive a CC estimator with high temporal resolution. We formalize this requirement in the CC definition in Eq. 10.1 by explicitly showing the dependence on the time instant n, yielding the **instantaneous cross-correlation** (ICC):

$$C_{ij}^{\text{bin}}[n,l] = E\{b_i[n]b_j[n+l]\} \tag{10.3}$$

Written in this way, the definition reinforces the idea that the expectation must be estimated over realizations rather than over time in order to preserve the time dynamics.

Nevertheless, one must still decide on how to define a realization to approximate the expectation. In a multitrial experimental paradigm one could approximate the expectation by averaging over trials, as done using the PSTH (Gerstein and Kiang, 1960) and JPSTH (Gerstein and Perkel, 1972; Aertsen et al., 1989). Averaging over trials preserves the trial temporal dynamics *only if the trials are stationary*. However, this assumption is hard to verify in practice and the results of the analysis can deviate significantly if the assumption does not hold (Brody, 1999). In a single-trial paradigm, and to avoid the limiting assumptions of stationarity in general, here it is proposed to approximate the expectation by the mean over all pairs of spike trains. Consider spike trains $\{s_1, \ldots, s_N\}$; then the ICC can be estimated as

$$\hat{C}^{\text{bin}}[n,l] = \frac{2}{N(N-1)} \sum_{i=1}^{N} \sum_{j=i+1}^{N} b_i[n]b_j[n+l] \tag{10.4}$$

Note that, in a sense, this approach is actually statistically more efficient than time averaging because the number of spike train pairs increases quadratically with the number of spike trains rather than linearly.

The ensemble-averaged ICC is a spatiotemporal measure of the ensemble covariation over time. In this form, and due to the exchange of time for ensemble averaging, the ICC is capable of detecting the presence of temporal dynamics within the ensemble with high temporal resolution. However, it raises the problem of how to select the subset of averaging neurons comprising an ensemble. A potential limitation in the use of the ICC is that spatial and neuron specificity is lost in the process of ensemble averaging. This is the trade-off incurred to reduce the estimation noise and still preserve the high temporal resolution.

Avoiding Time Quantization

The previous section introduced the ICC as an estimator of ensemble cross-correlation over time, but it did not address the issue of time quantization imposed by using binning. As stated earlier, any information expressed in the spike times at a finer time scale than the bin size will be lost through binning. Moreover, as pointed out by Richmond et al. (1990), the binning boundaries create an arbitrary assignment error. One might think that the error associated with binning is negligible, but recall that time quantization limits the resolution and significance of the analysis, which is precisely the goal here.

One should then ask, what is binning actually doing? Correspondingly, is there a better way to do it? As noted by Dayan and Abbott (2001), binning is an estimator of the instantaneous firing rate (apart from normalization by the bin size). Hence, the cross-correlation between two spike trains in Eq. 10.3 can be written in the more general form

$$C_{ij}(t,\theta) = E\left\{\hat{\lambda}_i(t)\hat{\lambda}_j(t+\theta)\right\} \tag{10.5}$$

where $\hat{\lambda}_i(t)$ and $\hat{\lambda}_j(t)$ denote the intensity functions estimated from spike trains i and j. This definition has the advantage that any intensity function estimator can be used. Specifically, in the statistical literature the conventional approach to intensity function estimation of point processes is kernel smoothing (Reiss, 1993), with clear advantages in the estimation (Kass et al., 2003; Richmond et al., 1990). Denoting the spike times in the time interval $(0,T)$ of the ith spike train as $\{t_n^i : n = 1,\ldots,N_i\}$, where N_i is the number of spikes of A in the interval. The kernel estimated intensity function is given by

$$\hat{\lambda}_i(t) = \sum_{n=1}^{N_i} h(t - t_n^i) \tag{10.6}$$

where h is the smoothing kernel function, which must integrate to 1. Note that, unlike binning, kernel intensity estimation does not introduce time-quantization errors.

Equation 10.6 can be interpreted as the convolution of the spike train with a window given by the smoothing function h. By controlling the smoothing introduced by h, one can regulate the scale at which the intensity function is estimated. Therefore, this controls the time scale of the CC coupling between two extremes: synchrony in neuron firings (for small smoothing) or firing rate (for large smoothing).

Nevertheless, it must be emphasized that the smoothing in Eq. 10.6 is fundamentally different from the time averaging in the estimation of CC. The smoothing in the estimation of the intensity function defines the time scale at which the spike times are analyzed for any information they may express. The CC time averaging, on the other hand, is used to reduce the estimation noise in order to improve the statistical significance of the result. Of course, if the smoothing introduced by h is large, there is a reduced need for time averaging, but this is simply because changes in firing rate must inherently vary slower than changes in synchrony.

Although many different smoothing functions h can be used, estimation methodologies that are causal and can be applied online are of particular interest. A smoothing function with these properties is the exponential function:

$$h(t) = (1/\tau)\exp(-t/\tau)u(t) \tag{10.7}$$

where $u(\cdot)$ is the step function. The exponential function provides both graded interactions and a time scale for the intensity estimation by controlling the time constant τ. In addition, the exponential function is biologically plausible since it can be interpreted as the evoked postsynaptic potential in a neuron (Dayan and Abbott, 2001). Accordingly, the estimated intensity function for the ith spike train is given by

$$\hat{\lambda}_i(t) = (1/\tau)\sum_{t_n^i \le t} \exp\left[-(t - t_n^i)/\tau\right]u(t - t_n^i) \tag{10.8}$$

which is simply the filtering of a spike train by a first-order infinite impulse response (IIR) filter and, therefore, it is computationally very simple and particularly suitable for hardware implementation.

Using the kernel intensity function estimator of Eq. 10.6, the ICC is defined as

$$\hat{C}(t,\theta) = \frac{2}{N(N-1)}\sum_{i=1}^{N}\sum_{j=i+1}^{N} \hat{\lambda}_i(t)\hat{\lambda}_j(t+\theta) \tag{10.9}$$

Hereafter, ICC refers to this formulation unless explicitly stated otherwise.

EXPERIMENTS

Synchronization of Pulse-Coupled Oscillators

In this example, we show that ICC can quantify the increase in coupling strength in a spiking neural network of leaky-integrate-and-fire (LIF) neurons designed according to Mirollo and Strogatz (1990)[1] and the ICC results compare favorably with the extended cross-correlation for multiple neurons.

[1]The following are the parameters for the simulation: 100 neurons; resting and reset membrane potential –60 mV; threshold –45 mV; membrane capacitance 300 nF; membrane resistance 1 MΩ; current injection 50 nA; synaptic weight 100 nV; synaptic time constant 0.1 ms; and all to all excitatory connection.

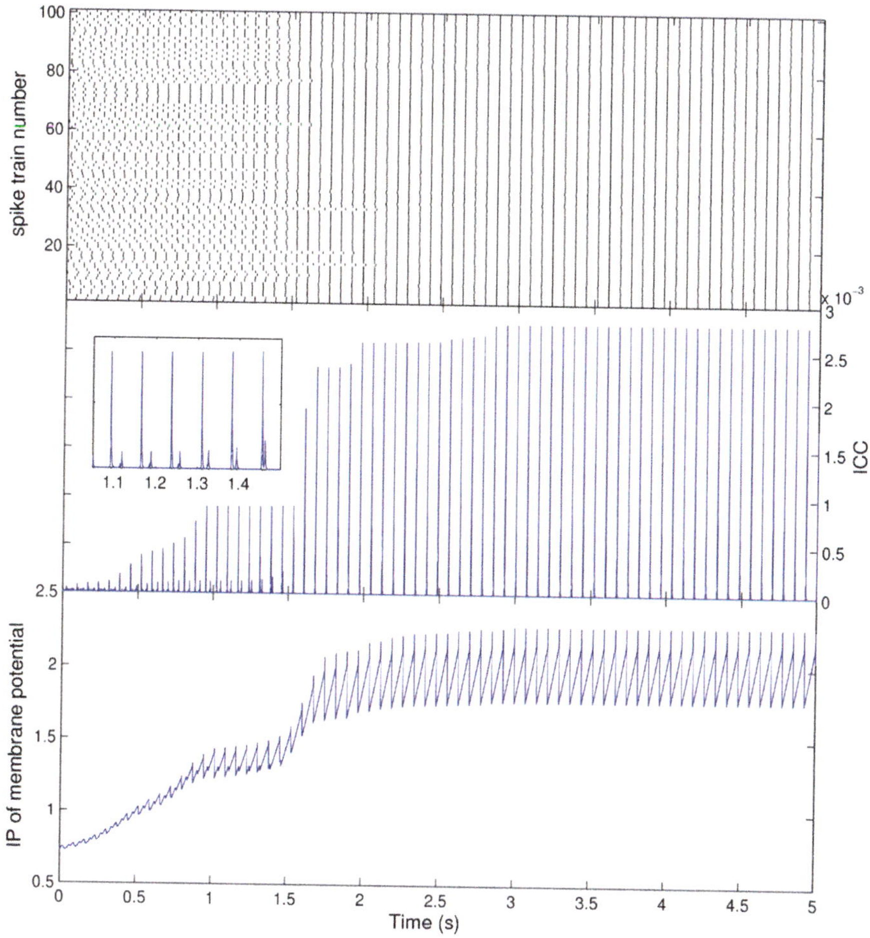

Figure 10.1. Evolution of synchrony in a spiking neural network. (top) Raster plot of the neuron firings. (middle) ICC over time. The inset highlights the merging of two synchronous groups. (bottom) Information potential of the membrane potentials. This is a macroscopic variable characterizing the entropy of the neurons' internal state.

The network is initialized in a random condition and is proven to synchronize over time. The synchronization is essentially due to leakiness and the weak coupling among the oscillatory neurons.

The raster plot of neuron firings is shown in Figure 10.1. There are two main observations: the progressive synchronization of the firings associated with the global oscillatory behavior of the network, and the local grouping that tends to preserve local synchronizations that either entrain the full network or wash out over time, as expected from theoretical studies of the network behavior (Mirollo and Strogatz, 1990). The ICC depicts this behavior precisely: the synchronization increases monotonically, with a period of fast increase in the first second followed by a plateau and slower increase as time advances. Moreover, it is possible to observe in the first 1.5 s the formation of a second group of synchronized neurons that slowly merges into the main group. In addition, the envelope of ICC reveals the coherence of the membrane potentials as quantified by the information potential (IP). The IP is an information-theoretic quantity inversely proportional to entropy (Principe, 2010). It was computed with

$$IP_\theta = \frac{1}{M^2}\sum_{i=1}^{M}\sum_{j=1}^{M}\exp(-d(\theta_i,\theta_j)/2\sigma^2)$$

with $\sigma = 75$ mV.[2] The IP measures the coherence of the neuron's internal state, which is only available in simulated networks. Yet, it is interesting to verify that the ICC was able to successfully and accurately characterize such information from the observed spike trains.

For comparison, the zero-lag cross-correlation over time, averaged through all pairwise combinations of neurons, is presented in Figure 10.2. The cross-correlation was computed with a sliding window 10 bins long, sliding 1 bin at a time. Results are shown for bin sizes of 1 ms and 1.1 ms. It is noteworthy that, although cross-correlation captures the general trends of synchrony, it masks the plateau and the final synchrony, and it is highly sensitive to the bin size as highlighted by the figure. These results highlight the importance of working in continuous time, as proposed in subsection "Avoiding time quantization," in contrast to the use of binning.

Analysis of Neural Synchronous Activity in Motor Neurons

In our second example, ICC is utilized to analyze the presence of synchronous activity in the motor cortex of a rat's brain. Throughout the literature, synchro-

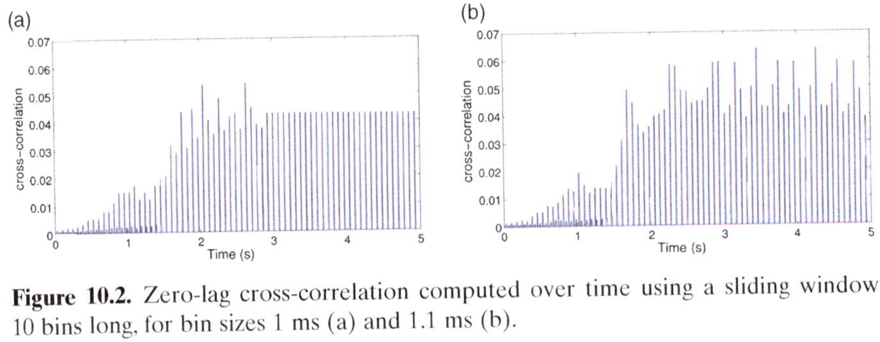

Figure 10.2. Zero-lag cross-correlation computed over time using a sliding window 10 bins long, for bin sizes 1 ms (a) and 1.1 ms (b).

[2]The distance used in the Gaussian kernel was $d(\theta_i, \theta_j) = \min(|\theta_i - \theta_j|, 15\text{ mV} - |\theta_i - \theta_j|)$ where θ_i is the membrane potential of the ith neuron. This wraparound effect expresses the phase proximity of the neurons before and after firing.

nous activity has been shown to provide additional information about motor movement when compared with firing rate modulation pattern analysis alone, including when no firing rate modulations are noticeable (Vaadia et al., 1995; Hatsopoulos et al., 1998; Riehle et al., 1997). Indeed, synchronous neural activity seems to be a widespread characteristic of the brain and can be found in a number of cortices, such as the auditory (Wagner et al., 2004; Carr and Konishi, 1990) and the visual cortices (Freiwald et al., 2001), for example.

Multichannel neuronal firing times from a male Sprague–Dawley rat were simultaneously recorded during a conditioned behavioral task at the University of Florida McKnight Brain Institute. The rat was chronically implanted with two 2×8 arrays of microelectrodes placed bilaterally in the forelimb region of the primary motor cortex (1.0 mm anterior, 2.5 mm lateral of bregma; Donoghue and Wise, 1982). Neuronal activity was collected with a Tucker–Davis recording rig with sampling frequency of 24,414.1 Hz and digitized to 16 bits of resolution. The firing times were recorded from individual neurons spike sorted with an online algorithm employing a combination of thresholding and template-based techniques. From sorting, a total of 44 single neurons were recorded, 24 neurons from the left hemisphere and 20 neurons from the right hemisphere. Simultaneously, the rat performed a go–no-go lever-pressing task in an operant conditioning cage (Med-Associates, St. Albans, VT, USA). The task consisted of choosing and pressing one out of two levers (left or right) depending on an LED visual stimulus to obtain a water reward. The queue and lever press signals were recorded simultaneously with the neural activity with sampling frequency 381.5 Hz. See Sanchez et al. (2005) for additional details on the experimental configuration.

Instantaneous cross-correlation was applied to this data set to quantify the synchronous neural activity across the ensemble. Figure 10.3 shows two trials

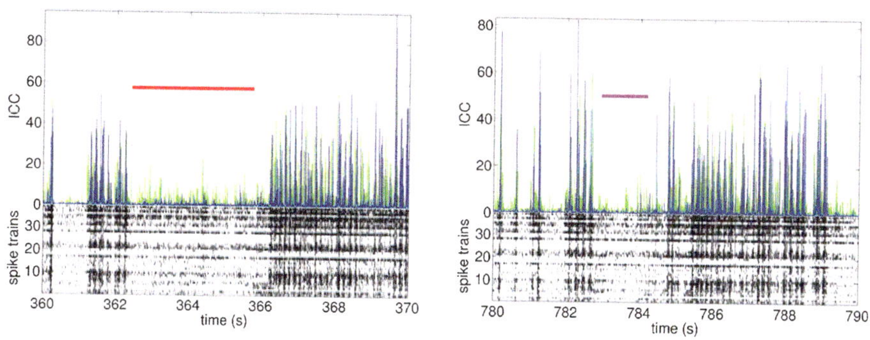

Figure 10.3. Instantaneous cross-correlation (ICC) and neuron firing raster plot on a single realization, showing the modulation of synchrony around the lever presses. The ICC was averaged throughout all neuron pairs separately for each hemisphere: left (blue) and right (green). The left plot shows a left lever press and the right plot shows a right lever press.

with the ICC at zero lag. As verified in other trials, the figure shows an increase in synchrony when the lever is released. Moreover, while the lever was being pressed the ensemble synchrony was observed to be significantly smaller than synchrony before and after the lever press. Actually, in this data set, visual inspection of the raster plots would yield such conclusions, but ICC provides a quantitative method to translate the visual evaluation. Furthermore, examining the raster plots we can verify the presence of ensemble synchronized activity reoccurring in a periodic manner after the lever is released. Notice that the ensemble ICC captures the presence of this oscillatory synchronized activity, seen in the envelope of ICC in Figure 10.3, directly from a single trial and with high temporal resolution.

To investigate either the observed synchrony modulation was widespread across all the neurons, we clustered the neural activity after the lever release was clustered. For that purpose, we used the spike train clustering algorithm described by Paiva and co-workers (Paiva, 2008; Paiva et al., 2009) with the memoryless cross-intensity (mCI) kernel. Figure 10.4 shows that all neurons contribute to the observed synchrony modulation, even though different clusters have slightly different rates and dynamics. In other words, neurons tend to form overlapping and time-evolving assemblies (Buzsáki, 2010; Strangman, 1997). Generally, clustering could be applied first to ensure that the ICC spatially averages only within an ensemble and, in this example, it could help to quantify subassembly dynamics.

For comparison with the commonly used methodologies, we show the cross-correlation computed at zero lag with a 200-ms sliding window over 1-ms bins, as is commonly done in other studies (see, e.g., Hatsopoulos et al., 1998). This is shown first for only some selected pairs of neurons (Fig. 10.5), and then spatially averaged (Fig. 10.6) as proposed in subsection "Preserving temporal resolution." Although the presence of synchrony is also successfully captured with cross-correlation, the existence of a periodic modulation in synchrony is not noticeable. This can be expected since the cross-correlation requires stationarity over time and uses time averaging to reduce the randomness of the

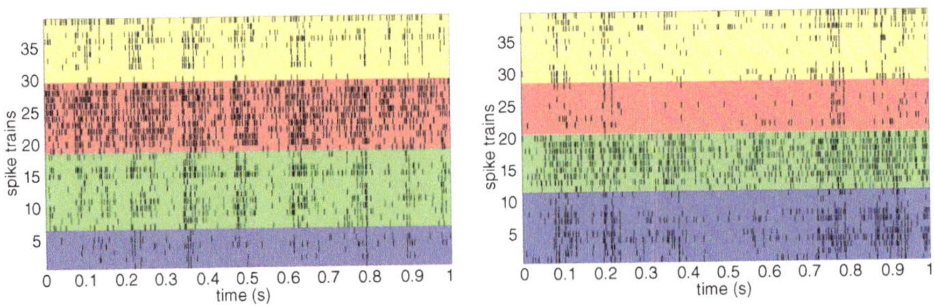

Figure 10.4. Clustering of neural activity following the lever release, considering four clusters, for the same trials shown in Figure 10.3.

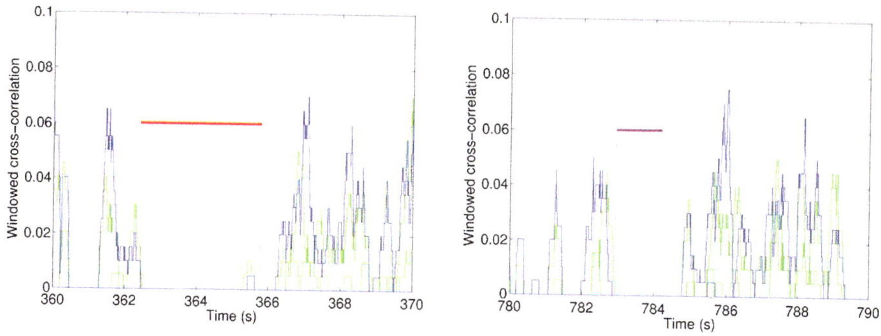

Figure 10.5 Windowed cross-correlation of selected six pairs of neurons, for the same segments shown in Figure 10.3. The cross-correlation was computed with a 200-ms sliding window over 1-ms bins. Four of the neuron pairs, two from each hemisphere, are known to synchronize and are shown with a solid line. The remaining two, one from each hemisphere, do not synchronize strongly and are shown with a dotted line.

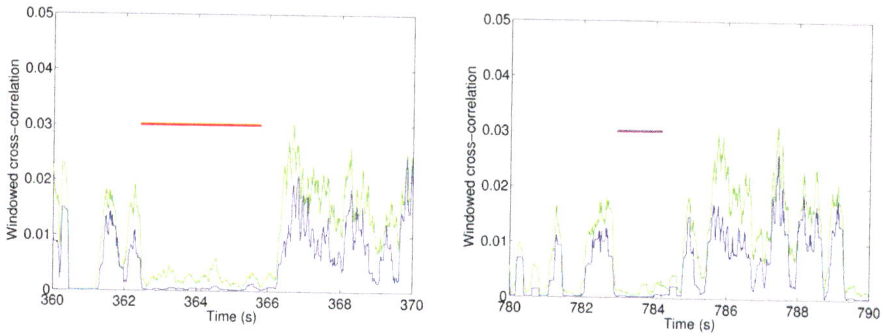

Figure 10.6. Ensemble-averaged windowed cross-correlation, for the same segments shown in Figure 10.3. The cross-correlation for each neuron pair was computed with a 200-ms sliding window over 1-ms bins. As proposed in subsection "Preserving temporal resolution," the spatial average was taken throughout all neuron pair combinations, separately for each hemisphere: left (blue) and right (green).

estimator. These two factors filter out potentially interesting features, such as the modulation of synchrony, on which information may represented. Put differently, time averaging imposes up front a lower bound on the frequencies that can be analyzed. This effect is most visible in Figure 10.6, where spatial averaging greatly improves the estimation as the variance of the estimator is reduced, but the time averaging masks the modulation in synchrony in favor of general trends. These figures highlight the importance of the spatial averaging proposed for ICC, in opposition to the time averaging employed in cross-correlation.

DISCUSSION

This chapter aims primarily (i) to highlight the limitations of cross-correlation as it is commonly estimated and (ii) to introduce the ICC as an alternative measure of cross-correlation with high temporal resolution. As reviewed in section "Background," the current practice for CC estimation utilizes time averaging and binning. Yet, this approach limits the temporal resolution of the analysis and the dynamics that can be characterized by the measure (cf. subsection "Synchronization of pulse-coupled oscillators"). Perhaps most important, the binned time-averaged CC estimator requires strong assumptions about the stationarity and ergodicity of the neural activity. However, if these assumptions do not hold, the outcome of the data analysis will likely have little or no significance (Brody, 1999). In contrast, the ICC uses ensemble averaging to obtain a measure of the neurons' cross-correlation with high temporal resolution. Using ensemble averaging has the added advantage that the approximation to the expectation has generally smaller variance because the number of spike train pairs increases quadratically. Thus, ICC can directly take full advantage of the increase in the channels in multichannel spike train recordings. In addition, binning artifacts are avoided by computing the ICC in continuous time, as defined by Eq. 10.9 and demonstrated in the experiments.

On the other hand, using ICC involves a change in analysis perspective because spatial and neuron specificity is lost as a result of ensemble averaging. However, this is a necessary trade-off because, as stated by the Heisenberg uncertainty principle, an increase in time resolution implies a decrease in spatial resolution. This is not a limitation, however, if the goal is to analyze the dynamic coupling behavior of the neural ensemble (Gerstein and Turner, 1990). In any case, it raises the question of how to divide the spike train recordings in neural ensembles. Although further tests still need to performed, the solution might involve clustering the spike trains (Paiva et al., 2009; Paiva, 2008), as illustrated in subsection "Analysis of neural synchronous activity in motor neurons."

The ICC is presented here as a data analysis tool for the practitioner, but it should be pointed out that the ICC also provides the framework for computation and adaptation in the spike train domain. Indeed, the ICC defines an inner product between spike trains at different times. This follows from the fact that, by definition, the ICC is a correlation function and therefore it is an inner product (Parzen, 1959). More explicitly, one can verify that the ICC is a symmetric and positive definite function and thus verifies the conditions of the Moore–Aronszajn theorem (Aronszajn, 1950). In other words, since the ICC is symmetric and positive definite, it induces a reproducing kernel Hilbert space (RKHS) for which the ICC evaluates the inner product. Having the mathematical structure of an RKHS means that there exists a complete vector space on which we can compute with spike trains directly and apply principal component analysis, clustering, Fisher discriminant analysis, and so on (Paiva et al., 2009; Paiva, 2008; Paiva et al., 2010). The inner product defined by the ICC is particularly interesting because it combines the pairwise cross-correlations to characterize the spike trains' joint distribution. Compared with existing statistical methods (Truccolo et al., 2005), this is a significant achievement considering that this information can be extracted directly, without knowledge of the data model, and in a single trial.

REFERENCES

Aertsen A.M., Gerstein G.L., Habib M.K., and Palm G. (1989). Dynamics of neuronal firing correlation: modulation of "effective connectivity." *J Neurophysiol* 61(5): 900–917.

Aronszajn N. (1950). Theory of reproducing kernels. *Trans Am Math Soc* 68(3): 337–404.

Baccalá L.A. and Sameshima K. (1999). Directed coherence: a tool for exploring functional interactions among brain structures. In *Methods for Neural Ensemble Recordings* (Nicolelis M.A.L., Ed.), pp. 179–192. Boca Raton, FL: CRC Press.

Brody C.D. (1999). Correlations without synchrony. *Neural Comp* 11(7):1537–1551.

Brown E.N., Nguyen D.P., Frank L.M., Wilson M.A., and Solo V. (2001). An analysis of neural receptive field plasticity by point process adaptive filtering. *Proc Natl Acad Sci* 98:12261–12266.

Brown E.N., Kass R.E., and Mitra P.P. (2004). Multiple neural spike train data analysis: state-of-the-art and future challenges. *Nature Neurosci* 7:456–461.

Buzsáki G. (2010). Neural syntax: cell assemblies, synapsembles, and reader. *Neuron* 68:362–385.

Carr C.E. and Konishi M. (1990). A circuit for detection of interaural time differences in the brain stem of the barn owl. *J Neurosci* 10(10):3227–3246.

Dayan P. and Abbott L.F. (2001). *Theoretical Neuroscience: Computational and Mathematical Modeling of Neural Systems*. Cambridge, MA: MIT Press.

Donoghue J.P. and Wise S.P. (1982). The motor cortex of the rat: cytoarchitecture and microstimulation mapping. *J Comp Neurol* 212(12):76–88.

Freiwald W.A., Kreiter A.K., and Singer W. (2001). Synchronization and assembly formation in the visual cortex. In *Progress in Brain Research* (Nicolelis M.A.L., Ed.), pp. 111–140. Elsevier Science.

Gerstein G.L. and Kiang N.Y.-S. (1960). An approach to the quantitative analysis of electrophysiological data from single neurons. *Biophys J* 1(1):15–28.

Gerstein G.L. and Perkel D.H. (1972). Mutual temporal relationships among neuronal spike trains. Statistical techniques for display and analysis. *Biophys J* 12(5):453–473.

Gerstein G.L. and Turner M.R. (1990). Neural assemblies as building blocks of cortical computation. In *Computational Neuroscience* (Schwartz E.L., Ed.), pp. 179–191. Cambridge, MA: MIT Press.

Gerstein G.L., Bedenbaugh P., and Aertsen A.M.H.J. (1989). Neuronal assemblies. *IEEE Trans Biomed Eng* 36(1):4–14.

Grün S., Diesmann M., Grammont F., Riehle A., and Aertsen A. (1999). Detecting unitary events without discretization of time. *J Neurosci Methods* 93(1):67–79. (10.1016/S0165-0270(99)00126-0.)

Grün S., Diesmann M., and Aertsen A. (2002a). Unitary events in multiple single-neuron activity. I. Detection and significance. *Neural Comp* 14(1):43–80.

Grün S., Diesmann M., and Aertsen A. (2002b). Unitary events in multiple single-neuron activity. II. Nonstationary data. *Neural Comp* 14(1):81–119.

Hatsopoulos N.G., Ojakangas C.L., Paninski L., and Donoghue J.P. (1998). Information about movement direction obtained from synchronous activity of motor cortical neurons. *Proc Natl Acad Sci* 95(26):15706–15711.

Kass R.E., Ventura V., and Cai C. (2003). Statistical smoothing of neuronal data. Networks: *Comp Neural Syst* 14:5–15.

Mirollo R.E. and Strogatz S.H. (1990). Synchronization of pulse-coupled biological oscillators. *SIAM J Appl Math* 50(6):1645–1662.

Paiva A.R.V. (2008). Reproducing kernel Hilbert spaces for point processes, with applications to neural activity analysis. PhD thesis, University of Florida.

Paiva A.R.C., Park I., and Prìncipe J.C. (2009). A reproducing kernel Hilbert space framework for spike train signal processing. *Neural Comp* 21(2):424–449.

Paiva A.R.C., Park I., and Prìncipe J.C. (2010). Inner products for representation and learning in the spike train domain. In *Statistical Signal Processing for Neuroscience and Neurotechnology* (Oweiss K.G., Ed.), chapter 8. Academic Press.

Parzen E. (1959). Statistical inference on time series by Hilbert space methods. Technical Report 23, Applied Mathematics and Statistics Laboratory, Stanford University, Stanford, CA.

Perkel D.H., Gerstein G.L., and Moore G.P. (1967). Neuronal spike trains and stochastic point processes. II. Simultaneous spike trains. *Biophys J* 7(4):419–440.

Principe J.C. (2010). *Information Theoretic Learning*. New York: Springer.

Reiss R.-D. (1993). *A Course on Point Processes*. New York: Springer-Verlag.

Richmond B.J., Optican L.M., and Spitzer H. (1990). Temporal encoding of two-dimensional patterns by single units in primate primary visual cortex. I. Stimulus–response relations. *J Neurophysiol* 64(2):351–369.

Riehle A., Grün S., Diesmann M., and Aertsen A. (1997). Spike synchronization and rate modulation differentially involved in motor cortical function. *Science* 278(5345):1950–1953.

Sameshima K. and Baccalá L.A. (1999). Using partial directed coherence to describe neuronal ensemble interactions. *J Neurosci Methods* 94(1):93–103.

Sanchez J.C., Principe J.C., and Carney P.R. (2005). Is neuron discrimination preprocessing necessary for linear and nonlinear brain machine interface models? In Proc. Intl. Conf. on Human–Computer Interaction.

Strangman G. (1997). Detecting synchronous cell assemblies with limited data and overlapping assemblies. *Neural Comp* 9(1):51–76.

Truccolo W., Eden U.T., Fellows M.R., Donoghue J.P., and Brown E.N. (2005). A point process framework for relating neural spiking activity to spiking history, neural ensemble, and extrinsic covariate effects. *J Neurophysiol* 93:1074–1089.

Vaadia E., Haalman I., Abeles M., Bergman H., Prut Y., Slovin H., and Aertsen A. (1995). Dynamics of neuronal interactions in monkey cortex in relation to behavioural events. *Nature* 373(6514):515–518.

Wagner H., Brill S., Kempter R., and Carr C.E. (2004). Microsecond precision of phase delay in the auditory system of the barn owl. *J Neurophysiol* 94(2):1655–1658.

11

FOCUSED CHAPTER

UNSUPERVISED DECOMPOSITION METHODS FOR ANALYSIS OF MULTIMODAL NEURAL DATA

FELIX BIESSMANN, FRANK C. MEINECKE, AND KLAUS-ROBERT MÜLLER

Machine Learning Group, Technische Universität Berlin, Berlin, Germany

INTRODUCTION

Technical advances in the field of noninvasive neuroimaging (Weiskopf et al., 2007; Blankertz et al., 2008) allow for innovative therapeutical strategies with application potential in neural rehabilitation (Sitaram et al., 2009; Krauledat et al., 2008; Tangermann et al., 2009; Schreuder et al., 2010). To improve these methods, combinations of multiple imaging modalities have become an important topic of research (Friston, 2009). Exploiting the advantages of each single modality, joint analysis can enhance the accuracy of unimodal analyses when it comes to decoding brain states. For instance, the spatial resolution of hemodynamic measurements acquired by functional magnetic resonance imaging (fMRI) can monitor certain brain areas much better than electroencephalography (EEG)—but the high temporal resolution of EEG can improve the analysis of hemodynamic activity (Eichele et al., 2005). In therapeutical real-time feedback systems it might be even more useful than in off-line analyses to have the spatial resolution of fMRI combined with the speed of EEG. With

Introduction to Neural Engineering for Motor Rehabilitation, First Edition. Edited by Dario Farina, Winnie Jensen, and Metin Akay.

the advent of new measurement devices, with more sensors and higher resolution, the overwhelming complexity of many data sets creates new challenges for experimenters and analysts. Therefore methods from the field of machine learning have become indispensable in many applications of neural data analysis (see, e.g., Norman et al., 2006; Haynes and Rees, 2006; Blankertz et al., 2008). Most of these methods need labeled data sets; they are called supervised algorithms. Not in every rehabilitation setting, however, can label information be provided for calibrating the learning algorithm. In those situations, unsupervised methods can be applied. In this chapter we review some of the most popular unsupervised statistical learning techniques used in the context of neuroscientific data analysis, and we place a special focus on multimodal neural data. All of the methods presented share a conceptual similarity—they can be cast in the form of a generalized eigenvalue problem. Using this formulation allows the implementation of various algorithms in a unified framework and offers a comprehensible view on many statistical learning techniques. We derive the solution for each algorithm and show how it can be formulated as an eigenvalue equation. For each method there are illustrative examples using practice data sets and an application example analyzing multimodal neural data.

We start with the well-known principal component analysis (PCA; Pearson, 1901). This method has become a workhorse of modern data analysis and is being used as a preprocessing step in many applications. First, we show how to derive the algorithm and give illustrative examples of the advantages and disadvantages of standard PCA. Several extensions are discussed, for instance, for high-dimensional data sets and for data with anisotropic noise covariances. Most often PCA is used for dimensionality reduction, that is, by keeping only the strongest principal components. The application example at the end of the first section shows how one can use PCA the other way round (discarding the strongest components) for cleaning neurophysiological data of artifacts induced by gradients of magnetic fields as used in fMRI.

The second method presented is canonical correlation analysis (CCA; Hotelling, 1936): a multivariate analysis method that reveals maximally correlated features of simultaneously acquired multiple data streams. While the derivation and computation of the CCA result is rather simple, it is a very powerful method if one has no prior knowledge about how two or more data sources are coupled. In an application example we show how to use CCA to recover patterns in fMRI data and weightings for different frequency bands of neural activity from simultaneously acquired fMRI signals and intracranial neurophysiology. The feature combinations found by CCA maximize the correlation between data streams and yield valuable insights into the coupling between data sources. This allows for a more accurate analysis of fMRI signals and the underlying neural activity for therapeutic and experimental purposes.

A major drawback of both of the aforementioned methods is that they do not take into account the temporal structure of the data. One example of how this can affect analysis results is the following: In the case of fMRI-EEG recordings, the hemodynamic response function (HRF) smoothes the neural activity reflected in the EEG and leads to a delay in the hemodynamic response to neural activation (Buxton et al., 2004). When looking at simultaneously recorded fMRI and EEG data, the correlation will therefore always be small. In contrast, when convolving the EEG with an HRF or when simply shifting the fMRI data back in time by the correct time lag, the correlation between the two signals will be larger. Any analysis of fMRI data has to take into account these hemodynamic response characteristics—and the correct HRF is not always known in advance. It has been shown that the HRF exhibits considerable variability, depending on subjects, brain area, or behavioral state (Aguirre et al., 1998), which can lead to inaccurate results compared with analyses that do not assume a fixed HRF (Lu et al., 2007). In the last part of this chapter we present a straightforward extension of CCA that estimates the correct solution even in the presence of noninstantaneous couplings, that is, temporal delays or convolutions between data sources. The application example will show how this method can be used to estimate neurovascular coupling parameters from simultaneous multimodal recordings (Bießmann et al., 2009; Murayama et al., 2010).

Eigenvalue Equations and Orthogonality

For the derivation of the methods presented we use a fundamental concept from linear algebra, namely *eigenvectors* and *eigenvalues* of square matrices. An **eigenvector** $w \in R^n$ of a square matrix $A \in R^{n \times n}$ is defined as a vector that is mapped to a multiple of itself when left-multiplied with A:

$$Aw = \lambda w \tag{11.1}$$

The corresponding factor λ is called an **eigenvalue** and Eq. 11.1 is called an **eigenvalue equation**. The scale of the eigenvectors is undefined, since for every eigenvector w a rescaled version αw with $\alpha \in R$ is also an eigenvector with the same eigenvalue. So it is useful to think of Eq. 11.1 as defining only eigendirections, that is, directions that do not change when projected on A.

From the eigenvalue equation we see that the matrix $A - \lambda I$ does not have full rank, since $(A - \lambda I)w = 0$, so we can find the eigenvalues of the matrix A by finding the zero set of the characteristic polynomial, that is, by solving

$$\det(A - \lambda I) = 0$$

for λ. Since this is an nth-order polynomial in λ, this implies that there are always n eigenvalue–eigenvector pairs. Note, however, that even for real matrices A the eigenvalues are not necessarily real, since in general even real polynomials have complex roots. If we write all the n eigenvectors as columns of a matrix W, we can write Eq. 11.1 as a matrix equation:

$$AW = W\Lambda \tag{11.2}$$

where Λ is a diagonal matrix with the eigenvalues λ on its diagonal. Equation 11.1 is a special form of the **generalized eigenvalue equation**:

$$Aw = \lambda Bw \tag{11.3}$$

with $A,B \in R^{n \times n}$ or, in matrix form,

$$AW = BW\Lambda \tag{11.4}$$

and the generalized eigenvalues can be found by solving $\det(A - \lambda B) = 0$. Generalized eigenvalue problems play a central role in the following sections. Throughout the entire chapter all matrices corresponding to A and B will be real and symmetric matrices computed from measured data. This not only leads to real eigenvalues and eigenvectors, but also to a certain orthogonality structure of the eigenvectors. Consider two different eigenvalue–eigenvector pairs (λ_i, w_i) and (λ_j, w_j). Multiplying the respective generalized eigenvalue equations with w_j^T and e_i^T from the left, we obtain

$$w_j^T Aw_i = \lambda_i w_j^T Bw_i \quad \text{and} \quad w_i^T Aw_j = \lambda_j w_i^T Bw_j$$

Symmetry of A and B means that either $\lambda_i = \lambda_j$ or

$$w_j^T Aw_i = w_j^T Bw_i = 0 \tag{11.5}$$

so that eigenvectors with different eigenvalues are orthogonal with respect to the scalar products defined by A and B. For ordinary eigenvalue problems, where $B = I$, the eigenvectors are orthogonal with respect to the standard Euclidean scalar product. In matrix notation this means that both matrices A and B are simultaneously diagonal with respect to the basis of the generalized eigenvectors:

$$W^T AW = D_A \quad \text{and} \quad W^T BW = D_B \tag{11.6}$$

where D_A and D_B are diagonal matrices. Therefore, solving the generalized eigenvalue problem for two symmetric matrices is sometimes also called **simultaneous diagonalization**. For a detailed introduction to linear algebra see, for instance, Strang (2009).

The advantage of eigendecompositions is that—depending on how one computes the matrices A and B—one can cast a large number of algorithms in this framework and apply standard eigenvalue solvers. This allows one to understand, implement, and extend many existing statistical learning algorithms using only Eq. 11.4. Generalized eigenvalue equations are a generic way of solving multivariate regression problems. Among the many applications are, for example, *t*-tests, analysis of variance (ANOVA; Hastie et al., 2003), partial least squares, ridge regression (Bie et al., 2005), (kernel) Fisher linear discriminant analysis (Mika et al., 1999), blind source separation (Diamantaras and Kung, 1996; Ziehe and Müller, 1998), and spectral clustering (von Luxburg, 2007). Despite the fact that eigenvalue equations are linear equations, they are not restricted to linear data analysis: many nonlinear methods can be solved efficiently in the form of Eq. 11.4 using nonlinear kernel functions (see, e.g., Schölkopf and Smola, 2001; Schölkopf et al., 1998; Shawe-Taylor and Cristianini, 2004). The intuition underlying all applications of eigendecompositions presented in this chapter is always the same: one wants to maximize the variance in some subspace of the data (expressed in how one computes A in Eq. 11.4), while minimizing or constraining the variance in another subspace of the data (expressed in how one computes B in Eq. 11.4). For a concise review of eigenvalue problems and their application to unsupervised and supervised analysis, such as regression, latent variable recovery, and clustering, the reader is referred to, for example, Bie et al. (2005) and Shawe-Taylor and Cristianini (2004).

PRINCIPAL COMPONENT ANALYSIS

In the past decades recording devices improved considerably and resulted in an ever increasing complexity of data sets. If a researcher gets a new data set that is too high dimensional to visualize, it is very useful to find a lower dimensional representation of the data that captures as much variance of the data as possible. This is what PCA (Pearson, 1901; Hotelling, 1933) does—it finds a linear combination of features (e.g., sensors, voxels) that maximizes the variance of the data. Inspecting the weights of each feature gives a good estimate of which features are important in the data-generating process and which combination of sensors captures most variance of the signal. Projecting the data onto those feature combinations allows a high-dimensional data set to be boiled down to only a few dimensions that can be easily visualized. PCA is probably one of the most often used analysis methods and, despite its age (it was proposed first in 1901, by Pearson), extensions of PCA are still the subject of active research (Schölkopf et al., 1998; Tipping and Bishop, 1999; Diamantaras, 2003; Mackey, 2009). The PCA algorithm is summarized in Box 11.1.

Derivation

We are given a data set stored in a data matrix $[x_1, x_2, \ldots, x_L] = X \in R^{D\times L}$, where D denotes dimensions or features of X, and L denotes the number of samples. Throughout the entire chapter we make two simplifying assumptions. First we assume that the mean has been subtracted from the data, that is,

$$\sum_i x_i = 0$$

BOX 11.1. PCA Algorithm

Input:

→ Data matrix $X = [x_1, x_2, \ldots, x_L]$, $x_i \in R^M$

1. Center data such that $\Sigma_i x_i = 0$.
2. Compute covariance matrix $C = XX^T$.
3. Compute eigenvalues and eigenvectors of C (Eq. 11.12).

Output:

← Principal components $W \in R^{M \times K}$; variances $\lambda_1, \lambda_2, \ldots, \lambda_k$

Second we omit the normalization constant when computing covariance matrices. PCA computes a linear projection w^* that maximizes the variance in the data set:

$$w^* = \text{argmax}_w w^T XX^T w \tag{11.7}$$

The matrix product XX^T is called the **covariance matrix** C of X. It contains in its ith row and jth column the covariance between the ith and jth dimension of X. To prevent PCA from finding a trivial solution, where $\|w\| \to \infty$ and $\|\|$ denotes the Euclidian norm,

$$\|w\| = \sqrt{\sum_i w_i^2}$$

we constrain the solution by requiring

$$w^T w = 1 \tag{11.8}$$

To maximize Eq. 11.7 with respect to Eq. 11.8, the constraint can be added to the objective function using a Lagrange multiplier λ (for an introduction to the Lagrangian multiplier technique see, e.g., Lang, 1973):

$$\mathcal{L} = w^T XX^T w + \lambda(1 - w^T w) \tag{11.9}$$

Setting the derivative with respect to w to 0 yields

$$\begin{aligned} \frac{\partial \mathcal{L}}{\partial w} &= XX^T w - \lambda w = 0 \\ &\Rightarrow XX^T w = \lambda w \end{aligned} \tag{11.10}$$

meaning that the optimal w has to be an eigenvector of XX^T. By left-multiplying with w^T and using Eq. 11.8, we see that the new variance is given by the eigenvalue λ:

$$w^T XX^T w = \lambda \tag{11.11}$$

Equation 11.10 is an example of an **ordinary eigenvalue equation**. The full spectrum of principal components can be obtained by solving Eq. 11.10 not only for the first eigenvector but for all k nonzero eigenvalues:

$$XX^T W = W\Lambda \tag{11.12}$$

We have seen in Eq. 11.5 that the eigenvectors of Eq. 11.12 are orthogonal. In geometric terms the space spanned by the principal components is an orthogonal subspace of the data. The basis vectors of W are aligned with the directions of maximal variance in the data, as depicted in Figure 11.1. Orthogonality implies that the data in the ith component $w_i^T X$ (X projected onto w_i) and $w_j^T X$ are **decorrelated**—their covariance is zero if $i \neq j$. The covariance along all principal components can be summed to give the total covariance captured by PCA:

$$\sum_{i=1}^{k} \Lambda_{i,i} = \sum_{i=1}^{k} w_i^T XX^T w_i \tag{11.13}$$

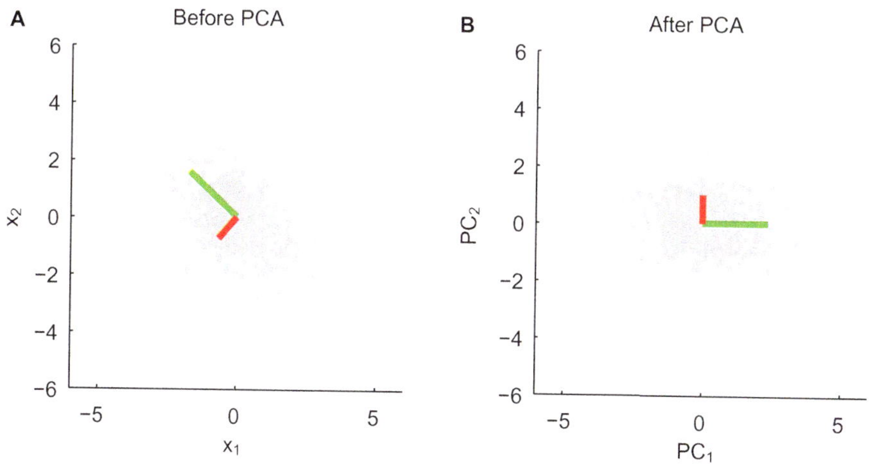

Figure 11.1. Two-dimensional practice-data example for PCA; in all panels, the first dimension is plotted on the x-axis, the second dimension along the y-axis. (A) Data was drawn from a two-dimensional Gaussian distribution; the first and second principal component directions are plotted in green and red, respectively. (B) The same data, projected onto the first two principal components; the direction of the data that exhibits most variance is now aligned with the x-axis and the direction of the second largest variance is parallel to the y-axis.

Looking at the eigenvalues in the diagonal matrix Λ helps to decide how many principal components should be considered when analyzing the data. A fast decay after i eigenvalues is a good sign that the data is well captured by the first i principal components. It is useful to divide each eigenvalue λ_i by the total variance $\sum_i^k \lambda_i$ to obtain the fraction of the variance explained by the ith component.

We can transform the data X into its new representation X_{PCA} in which all rows of the data matrix X are uncorrelated:

$$X_{PCA} = W^T X = \sum_{i=1}^{k} e_i w_i^T X \tag{11.14}$$

where the e_i are the basis vectors of the canonical Euclidean basis. An illustration using two-dimensional practice data is shown in Figure 11.1. The measurements x (in gray) are drawn from a two-dimensional Gaussian distribution. The directions of maximal variance, the principal components, are plotted on top of the data (Fig. 11.1A, red and green vectors). After projecting the data onto their first two principal components, the axes of maximal variances of X_{PCA} are rotated such that they fall on the main coordinates (Fig. 11.1B). Often PCA is used for dimensionality reduction. In this case, one is interested in a lower dimensional representation of the data, capturing as much variance as possible. This can easily be achieved by retaining only the first rows of X_{PCA}, that is, those that correspond to the highest eigenvalues. Since W is orthogonal we have $X = WW^T X$ and can therefore remove certain components also in the original basis of the data by

$$X' = \sum_{i \in \mathcal{J}} w_i w_i^T X \tag{11.15}$$

where the index set $\mathcal{J}$ contains all component indices that should be retained. Even though some components might be removed, the new data set X' is expressed in the same basis as X and therefore has the same number of rows.

PCA Extensions

Various extensions of PCA have been proposed. Popular extensions include **kernel PCA** (kPCA) and, as a special case, **nonlinear kernel** PCA (Schölkopf et al., 1998), which is useful in case the data is not well captured by linear feature combinations. Another important application of kPCA is in settings in which the data has more dimensions than samples. We begin this section by deriving **linear kernel PCA** and showing its relationship to standard PCA. Other PCA extensions such as **oriented PCA** (OPCA) have found applications in the field of **blind source separation** (Ziehe and Müller, 1998; Diamantaras and Papadimitriou, 2003). PCA has also been explored in a probabilistic framework (**probabilistic PCA**; Tipping and Bishop, 1999); for a review see, for instance, Roweis and Ghahramani (1999) or Bishop (2007). Many of these techniques can be cast as a generalized eigenvalue problem as in Eq. 11.4. Understanding how the PCA solution relates to the *generalized* eigenvalue problem (as opposed to the standard eigenvalue problem, where $B = I$) can be useful for better understanding the following sections. Before the application example at the end of this section we try to strengthen this intuition by deriving PCA as a generalized eigenvalue problem and give some illustrative examples using practice data sets.

PCA for High-Dimensional Data. If the dimensionality D of the data is higher than the number of samples L, it becomes more efficient to work not on covariance matrices but on **inner products** of the data, also known as the **kernel** or the **Gram matrix** of the data:

$$K_X = X^T X \tag{11.16}$$

The kernel K_X of the data is only of size $L \times L$, while the covariance matrix would be of size $D \times D$. Here we consider only linear kernels. An advantage of *linear* kernels is that the projections w can be easily recovered as a linear combination of data points. For the first principal component w_1 this means:

$$w_1 = X\alpha \tag{11.17}$$

where the ith entry of the vector α is the contribution of the ith data point (the ith column of X) to first principal component w_1. The intuition behind this is that if the number of dimensions is larger than the number of data points, then the solution has to lie in the space spanned by the data. Thus we can express the solution w_1 as a combination of data points, each weighted according to its respective coefficient in α. Plugging Eq. 11.17 into the eigenvalue equation in Eq. 11.10 yields

$$\begin{aligned} XX^T X\alpha &= \lambda X\alpha \\ XK_X\alpha &= X\lambda\alpha \end{aligned} \tag{11.18}$$

which can be equivalently solved (cf. Schölkopf et al., 1998) by computing the largest eigenvalue of

$$K_X\alpha = \lambda\alpha \tag{11.19}$$

So the solution of kernel PCA is just the same as Eq. 11.10, except that the eigenvalue decomposition is performed on $K_X = X^T X$ instead of $C_X = XX^T$; for a more detailed explanation see Schölkopf et al. (1998). One of the strengths of kernel methods is that nonlinear extensions are straightforward (Müller et al., 2001; Schölkopf and Smola, 2001). Besides being the more efficient computation, compared with diagonalizing the covariance matrix, the kernel version of PCA has another advantage: It is very simple to extend PCA to a nonlinear version by computing the covariance not in the input space but in a high-dimensional feature space defined by nonlinear **kernel functions** (Schölkopf et al., 1998). This can be helpful if the structure of the data is not well captured by a linear feature combination.

Relation between Standard PCA and Linear Kernel PCA. The relation between the standard PCA on the covariance matrix $C_X = XX^T$ and the linear kernel version on the kernel $K_X = X^TX$ can be understood best by performing a **singular value decomposition** (SVD; see, e.g., Strang, 2009) of the data matrix

$$X = ESF \tag{11.20}$$

where $E \in R^{D\times N}$ and $F \in R^{N\times D}$ are orthogonal matrices and S is a diagonal matrix containing the so-called singular values. It can be shown that $E \in R^{D\times N}$ and $F \in R^{N\times D}$ contain the eigenvectors of XX^T and X^TX, respectively. The eigenvalues of XX^T and X^TX are the squared singular values on the diagonal of S. It is straightforward to see this when computing C_X and K_X from the SVD solution:

$$C_X = XX^T = ESF(ESF)^T = ESFF^TS^TE^T = ESIS^TE^T = ES^2E^T \tag{11.21}$$

and

$$K_X = X^TX = FSE(FSE)^T = FSEE^TS^TF^T = FSIS^TF^T = FS^2F^T \tag{11.22}$$

where we used $EE^T = I = FF^T$. So, in conclusion, when $D < N$ (fewer dimensions than samples), it is more efficient to compute PCA on the covariance matrix $C_X = XX^T$. When $D < N$ (fewer samples than dimensions), the dimensionality of the principal component subspace is at most $(N - 1)$—thus it is more efficient to compute PCA on the Gram matrix $K_X = X^TX$.

PCA as a Generalized Eigenvalue Problem. The PCA solution in Eq. 11.12 can be cast in the form of a generalized eigenvalue equation as in Eq. 11.4 by setting $B = I$, where I is an identity matrix of the same size as A. PCA can be extended easily by different choices of $B \neq I$. To illustrate this it is helpful to have a look at a practice example. Let us consider measurements $x_i \in R^2, i = [1, 2, \ldots, L]$, in a data matrix $X = [x_1, x_2, \ldots, x_L]$. We assume that the measurement process is affected by noise $\varepsilon_i \in R^2$. So the model for the data-generating process would be:

$$\begin{aligned} N &= [\varepsilon_1, \ldots, \varepsilon_L], \qquad \varepsilon \sim \mathcal{N}(0, I) \\ X &= X_{\text{true}} + HN \end{aligned} \tag{11.23}$$

where the measurement noise $N = [\varepsilon_1, \ldots, \varepsilon_L]$ is drawn from an uncorrelated two-dimensional normal distribution with mean 0 and then mapped through

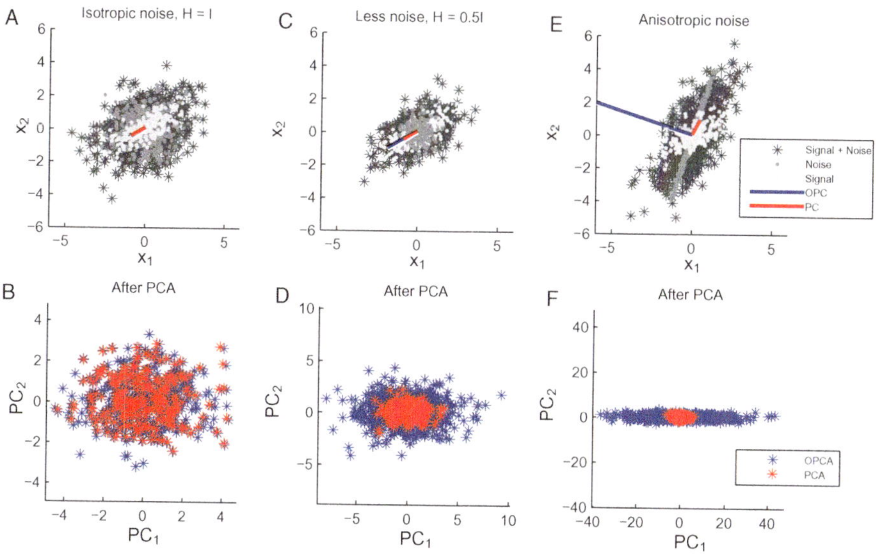

Figure 11.2. Effect of different signal and noise covariances; practice data set was generated according to Eq. 11.23; each column depicts a different noise covariance structure (cf. Table 11.1). (top row) Signal of interest X_{true} in light gray; noise HN in darker gray; measurements X (black); first PC of PCA (red) and OPCA (blue). (bottom row) Data projected on first PCs estimated by PCA (Eq. 11.10) and OPCA (Eq. 11.26). (A) data with isotropic noise; (B) measurements projected on solutions of PCA and OPCA are similar; (C,D) same as panels A and B with less noise, resulting in larger variance along first oriented principal component; (E,F) anisotropic noise covariance. The standard PCA solution aligns black measurement distribution with axes; the OPCA solution tries to squeeze the dark gray noise variance (panel E) while stretching apart the light gray signal variance; the resulting distribution (blue, panel F) has a larger variance than the PCA solution (red, panel F).

some linear mapping H. If H is not a multiple of I it stretches and rotates the uncorrelated noise distribution and distorts the original distribution of X_{true} (see Fig. 11.2E). Suppose we are interested in finding maximal variance directions of X_{true} and at the same time we want to constrain the noise variance. For the first principal component w this can be done by modifying the original constraint in Eq. 11.8 to

$$w^T HN(HN)^T w = w^T HH^T w = w^T C_\varepsilon w = 1 \tag{11.24}$$

where HH^T is the true noise covariance matrix (as $NN^T = I$) and C_ε denotes an empirical estimate of the noise covariance matrix. The corresponding Lagrangian is then

$$\mathcal{L} = w^T XX^T w + \lambda(1 - w^T C_\varepsilon w) \tag{11.25}$$

and the solution is found by setting the first derivative to zero, analogously to Eq. 11.7, Eq. 11.8, Eq. 11.9, and Eq. 11.10:

$$\frac{\partial \mathcal{L}}{\partial w} = XX^T w - \lambda C_\varepsilon w = 0$$
$$\Rightarrow XX^T w = \lambda C_\varepsilon w \qquad (11.26)$$

We refer to this generalized form of PCA as **oriented PCA** (OPCA; Diamantaras and Kung, 1996). Comparing the solution of standard PCA (Eq. 11.10) and the OPCA solution, we see that Eq. 11.26 simplifies to Eq. 11.10 if the noise covariance is the identity matrix. This assumption behind standard PCA implies that the noise is **isotropic**, it is uncorrelated between all dimensions x_i and x_j, where $i \neq j$, and has unit variance for $i = j$. If we do not know anything about the noise covariance structure, this is a reasonable assumption. But what happens if the noise is not isotropic? In the following examples we illustrate how the PCA and the OPCA solutions are affected by different noise covariance structures. Figure 11.2 shows data generated according to Eq. 11.23. The covariance structure of X_{true} was kept constant while the noise covariance imposed by H was changed in each column. Table 11.1 lists the true covariance matrix and three noise mappings H. In the top row (Fig. 11.2A,C,E), we plot the true data distribution X_{true} in light gray, the noise distribution HN in darker gray, and the measurements X in black. In the bottom row (Fig. 11.2B,D,F) we plot the data after projecting it on the principal components, computed with either standard PCA (red) or oriented PCA (blue). We varied the noise covariance starting in the leftmost panels with a noise mapping $H = I$, the identity matrix, leading to $C_\varepsilon = HN(HN)^T = \text{I}$. Isotropic noise is the underlying assumption in the standard PCA model in Eq. 11.12, and thus both methods, PCA and OPCA, find the same solution: the first principal component of PCA (red) and that of OPCA (blue) fall on top of each other. With less noise the projection found by OPCA has larger variance (Fig. 11.2C,D). In the setting of Figure 11.2C, the noise covariance is already uncorrelated (the variance is the same

TABLE 11.1. Practice-Data Settings for Data Shown in Figure 11.2

True Covariance	Noise Projection H		
$\begin{bmatrix} 0.7 & -0.3 \\ -0.3 & 0.7 \end{bmatrix}$	$\begin{bmatrix} 1 & 0 \\ 0 & 1 \end{bmatrix}$	$\begin{bmatrix} 0.5 & 0 \\ 0 & 0.5 \end{bmatrix}$	$\begin{bmatrix} 0.1 & 0.5 \\ 0.5 & 1.5 \end{bmatrix}$
$X_{\text{true}} X_{\text{true}}^T$	Isotropic noise (Fig. 11.2, panels A,B)	Less isotropic noise (Fig. 11.2, panels C,D)	Anisotropic noise (Fig. 11.2, panels E,F)

[a]The *left* column shows the true covariance matrix; the columns on the *right* show different noise mappings H for generating data according to Eq. 11.23; for all settings, we computed standard PCA or OPCA; data and results are plotted in Figure 11.2.

in all directions, the gray distribution is spherical), but the variance is smaller than 1. So OPCA tries to stretch the axis of maximal variance such that the dark gray distribution becomes a sphere as big as in Figure 11.2A. As result the OPCA solution in Figure 11.2D (blue) has a larger variance along its first principal component than the PCA solution (red). The right column (Fig. 11.2E,F) shows an example of anisotropic noise. The corresponding noise mapping H is shown in Table 11.1, rightmost column. Whereas standard PCA does not take the noise covariance into account and simply aligns the black distribution with the coordinate axes, OPCA maximizes the signal variance while keeping the noise variance small. Figure 11.2F shows how OPCA tries to keep the noise covariance small and maximize the signal variance by squeezing the dark gray noise distribution small while stretching apart the light gray signal distribution. Thus the resulting variance along the first principal component is much larger after OPCA, if the noise does not meet the isotropic noise assumption of standard PCA.

Application: Artifact Removal in Multimodal Recordings

One of the most difficult challenges in neuroscience is multimodal data acquisition (Friston, 2009). A very popular multimodal setting is the simultaneous measurement of neurophysiological activity, with electrodes, and the blood oxygenation level dependent signal. Recordings of neurophysiological activity with electrodes have been the methodological mainstay of neuroscientific research for many decades. However, they require placing the electrode close to the brain area under investigation, in order to obtain good signals. In contrast to these invasive procedures, noninvasive techniques such as functional magnetic resonance imaging (fMRI) do not require surgeries and thus have become increasingly popular—also in the field of neurorehabilitation (Sitaram et al., 2009). The advantages of electrode recordings are high temporal and spatial resolution. Hemodynamic activity measured with fMRI yields only an indirect view of brain activity with a lower temporal and spatial resolution—but fMRI can measure the hemodynamic response of the entire brain at once.[1] Combining these noninvasive measurements with direct neurophysiological measurements is crucial not only for the validation and improvement of fMRI. More important, multimodal measurements of neural and hemodynamic activity offer the possibility of exploiting the advantages of both modalities (high temporal resolution of electrode recordings and noninvasiveness of fMRI) while avoiding the drawbacks of the respective modality (small spatial focus of electrophysiological recordings and low spatiotemporal resolution of fMRI).

[1]The neurovascular coupling mechanism can be sketched as follows. Neural activity leads to changes in blood oxygenation in the brain. Changes in hemoglobin oxygenation are accompanied by changes in the magnetic properties of hemoglobin, which can be detected by fMRI scanners (Ogawa et al., 1992).

The problem with these multimodal setups is that fMRI scanners use very strong magnetic field gradients and thereby induce artifacts in the electrophysiological recordings. Even if the exact magnetic field gradient sequence is known, there is no trivial way of deriving from it the optimal filter for removing the artifactual voltage changes from the recorded electrophysiological signal. Therefore unsupervised methods are often used to clean the electrophysiological recordings by removing artifacts. In the following example we show how to use PCA to remove scanner-induced artifacts from intracranial electrophysiological recordings. Details of the experimental procedure are given in Rauch et al. (2008). In short, the electrophysiological data was recorded in the primary visual cortex during visual stimulation. The visual stimulus was a rotating checkerboard that was presented in 32-second blocks, interleaved with blocks of equal duration without stimulation. The simultaneous acquisition of fMRI data induced strong artifacts in the neurophysiological recordings (Fig. 11.3A,C E) that can be efficiently removed using PCA (Fig. 11.3A vs. B, and E vs. F). The steps involved in the cleaning process are given in the following subsections.

Splitting into Epochs. For artifact removal, the electrophysiological time series has to be split into epochs whose length matches the length of the aquisi-

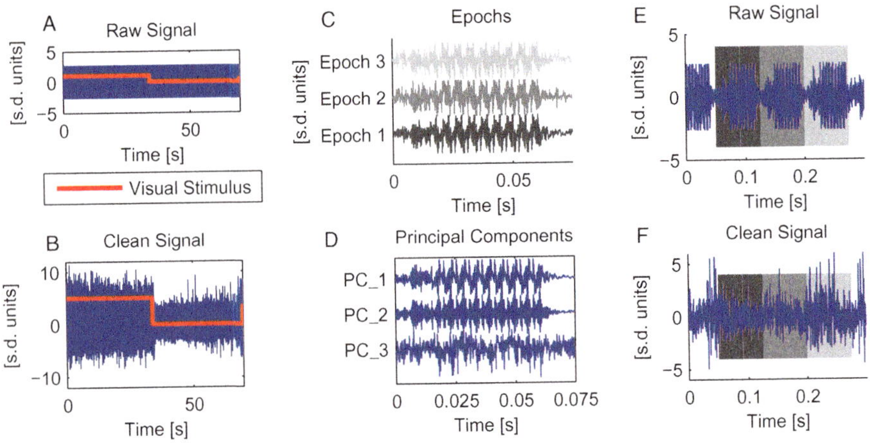

Figure 11.3. Example of PCA-based artifact removal from neurophysiological recordings made during fMRI data acquisition. (A) Example of raw contaminated time series (in blue) in units of standard deviation; neural responses to visual stimulation (in red) are much smaller than the artifacts and cannot be detected. (B) Clean signal after artifact removal; neural activity robustly follows visual stimulation. (C) Signal from panel A (magnified in panel E) split into single epochs (cf. shaded background in panel E); note the high-amplitude artifacts. (D) First three principal components computed from epochs in panel C. (E) Epochs from panel C concatenated before cleaning. (F) Same as panel E after subtraction of the principal components plotted in panel D.

tion period of one fMRI image (see Fig. 11.3, panels C and E). All epochs x_i, $i = [1, \ldots, L]$, L being the total number of fMRI volumes aquired, are then stacked as row vectors in a data matrix $X = [x_1^T, x_2^T, \ldots, x_L^T]^T$. It is crucial that the epochs be properly aligned, such that the artifact has the same phase in each epoch.

Projecting Out the Artifact. On the resulting data matrix we can compute PCA (or linear kPCA, in case the epoch size is larger than the number of data points). The first principal component will be the strongest artifact (cf. Fig. 11.3D). We then retain all but the i strongest (artifactual) components using Eq. 11.15. The number of components to reject has to be determined using empirical criteria (see the next subsection). The cleaning procedure relies on the assumption that the artifacts are temporally aligned in each epoch, as opposed to the neural signal. Temporal alignment of the artifacts means that there is a direction in data space where all the artifacts from the different epochs can sum up to a very large signal, whereas the unaligned neural activity cancels out. Therefore the most powerful principal components will capture the artifact. Figure 11.3D shows examples of the first three principal components reflecting the artifact. After projecting out the first three components, we can concatenate the epochs to recover a continuous time series (Fig. 11.3F). The effect of the cleaning procedure becomes evident when looking at stimulus-induced activity: Figure 11.3B shows the same data as in Figure 11.3A after cleaning. Neural activity in the stimulus-on periods is clearly stronger than in the stimulus-off periods.

How Many Components Should Be Discarded? When projecting out the artifactual principal components, it is important to know the number of artifactual components. Unfortunately in most cases one does not know exactly how the signal would look without the artifact. Also, exact noise measurements are often very difficult to obtain. However, one can use the decay of eigenvalues to decide how many principal components to reject. In addition most experimenters use empirical criteria, such as the shape of the power spectra of neural activity after cleaning, to decide how many components to discard. Figure 11.4A shows the eigenvalue spectrum of the application example. The first three eigenvalues (corresponding to the artifactual principal components in Fig. 11.3D) decay very fast. In panel B we plot the corresponding power spectra of continuous recordings after no cleaning (black) and cleaning using one and seven principal components. The threshold in the eigenvalue spectrum in Figure 11.4A is indicated by the color corresponding to the respective line in the power spectrum in Figure 11.4B. Inspection of the power spectra shows that, after removal of the first seven principal components, the peaks produced by the artifact are suppressed and the power spectrum approximately follows a $1/f$ distribution as expected for neurophysiological signals.

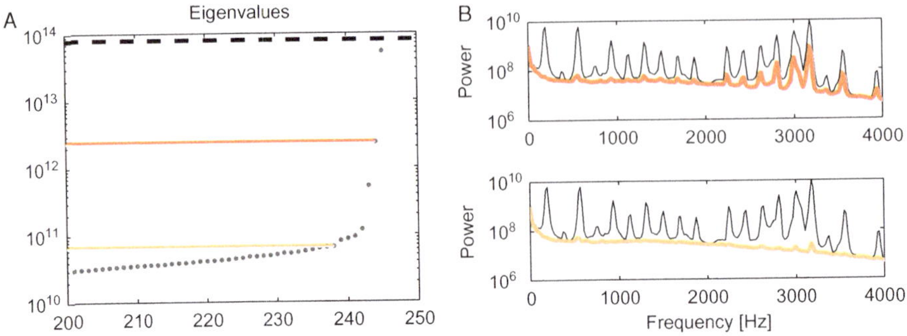

Figure 11.4. How many principal components should be subtracted? (A) Eigenvalues of the example in Figure 11.3—largest eigenvalues corresponding to artifactual components decay rapidly. (B) Power spectra of neurophysiological recordings without cleaning (black) and after subtraction of one (top) and seven (bottom) principal components; corresponding thresholds in the eigenvalue spectrum are plotted in respective colors in panel A.

Conclusion

To summarize this section, PCA finds feature combinations in the data that maximize the variance. Projecting the data onto the principal components results in a lower dimensional representation of the data that captures high-variance directions in the data. Removing principal components will decorrelate the data from these components. We showed in an application example how this can be applied in order to remove artifacts in multimodal recordings of neural activity.

CANONICAL CORRELATION ANALYSIS

Many data sets in modern neuroscientific studies are multimodal. Each recorded modality might have different dimensionalities, and often it is not clear how to select a set of features from the data that reflect common underlying processes. Let us consider two multivariate variables stored in matrices:

$$[x_1, x_2, \ldots, x_L] = X \in R^{U \times L}$$

$$[y_1, y_2, \ldots, y_L] = Y \in R^{V \times L}$$

where L is the number of samples and U,V the dimensionality of the respective modality. In the previous example of multimodal measurements, one of the variables could be electrophysiological measurements and the other variable could be hemodynamic measurements. Again we assume centered data such that $\sum_{i=1}^{L} x_i = 0$ and $\sum_{i=1}^{L} y_i = 0$ One could apply PCA on each modality or on a joint data set and continue with analyzing the data in their first principal components. In fact this is often done in multimodal analysis frameworks as a preprocessing step (Moosmann et al., 2008). However, the directions of maximal variance *within* one modality might not be what interests an experimenter. An example would be purely vascular signals in fMRI data, which are not due to changes in neural activity, and signals in EEG recordings that reflect muscle contractions instead of neural activity. Both would induce large variances in either modality (and thus would be considered by PCA as important), but neither reflects neural activity, the signal of interest. The PCA solution thus would lead astray interpretations of hemodynamic and neural activity. To select the best weighting of features, it is desirable to maximize the **canonical correlation** between the two variables X and Y:

$$\rho = \frac{w_x^T XY^T w_y}{\sqrt{w_x^T XX^T w_x w_y^T YY^T w_y}} \tag{11.27}$$

In this formulation we find the empirical **autocovariance matrices** of each data source, $C_{xx} = XX^T$ and $C_{yy} = YY^T$, the empirical estimate of the **cross-covariance matrix** $C_{xy} = XY^T$, and the **canonical variates** w_x and w_y of X and Y, respectively. The numerator contains the *cross-variance* of the data in its new representation, that is, after projecting the data onto their respective *canonical variates* w_x, w_y. It looks very similar to the objective of PCA in Eq. 11.7, except that in CCA we are now interested in *cross-covariances* not in *autocovariances*. In the denominator we have a product of the *autocovariances* within each data set, projected onto the canonical variates. Without this denominator, ρ would merely be a covariance. With the autocovariance terms, ρ will be scaled into the interval [−1 1]. This correlation coefficient is called *canonical* since ρ is invariant with respect to affine transformations of the data. In this section we present linear CCA and linear kernel CCA (Fyfe and Lai, 2000; Akaho, 2001), which is a slightly modified version of CCA that can be applied if the data is too high dimensional to compute covariance matrices. Besides these two versions many extensions of CCA have been proposed, including nonlinear versions of CCA (Leurgans et al., 1993; Leen et al., 2006; Bach and Jordan, 2002), CCA using stochastic processes (Fyfe et al., 2006), and sparse versions of CCA (Fyfe and Fyfe, 2006; Wiesel et al., 2008; Hardoon and Shawe-Taylor, 2011). It could also be shown that the CCA framework can be easily extended to more than two multivariate variables (Kettenring, 1971). CCA has found widespread application in multivariate statistics (Anderson, 1958), signal processing (Larimore, 1990; Bach and Jordan, 2002), data mining (Bie and Cristianini, 2004; Bießmann and Harth, 2010), artificial intelligence (Kim et al., 2007), and neuroscience (Friman et al., 2002; Hardoon et al., 2007; Macke et al., 2008), to name just a few examples.

BOX 11.2. CCA Algorithm

Input:

→ Data matrices $X = [x_1, x_2, \ldots, x_L] \in R^{U \times L}$, $Y = [y_1, y_2, \ldots, y_L] \in R^{V \times L}$

1. Center data such that $\sum_i x_i = \sum_i y_i = 0$.
2. Model selection to find right regularizer κ.:
   For each κ:
   a. Compute w_x, w_y (Eq. 11.40) using *training data*.
   b. Evaluate canonical correlation (Eq. 11.26) on *test data*.
3. Pick the κ that maximizes Eq. 11.26 on test data set.
4. Solve Eq. 11.40 again using the correct κ.

Output:

← Canonical variates $W_x \in R^{U \times \kappa}$, $W_y \in R^{V \times \kappa}$
← Canonical correlations $\rho_1, \rho_2, \ldots, \rho\kappa$

Derivation

The solution of CCA comprises sets of canonical variates $W_x = [w_x^1, w_x^2, \ldots, w_x^k]$ and $W_y = [w_y^1, w_y^2, \ldots, w_y^k]$, one for each variable. Each W_i spans a subspace in the respective data space of the ith modality that maximizes Eq. 11.27, the canonical correlation between the variables. We first derive the solution for the first pair of canonical variates w_x^1, w_y^1 using the Lagrange multiplier technique as in the previous section. Thereafter we show how to solve CCA using algebraic intuition about covariance matrices and eigenvalue equations. This allows the derivation of the solution not only for the first but for all k canonical variates, where k is the smallest rank of both matrices X, Y. The CCA algorithm is summarized in Box 11.2.

We want the first pair of canonical variates w_x, w_y to maximize the covariance between X and Y:

$$\operatorname{argmax}_{w_x, w_y} w_x^T C_{xy} w_y \tag{11.28}$$

To have a *correlation* instead of a *covariance* we need the right normalization for Eq. 11.28. This can be done by requiring the covariance *within* each variable X, Y to be 1; that is, we maximize Eq. 11.28 subject to

$$w_x^T C_{xx} w_x = 1, \qquad w_y^T C_{yy} w_y = 1 \tag{11.29}$$

Combining the objective function in Eq. 11.28 with the constraints in Eq. 11.29 yields the Lagrangian:

$$\mathcal{L} = w_x^T C_{xy} w_y - \frac{1}{2}\lambda_1\left(w_x^T C_{xx} w_x - 1\right) - \frac{1}{2}\lambda_2\left(w_y^T C_{yy} w_y - 1\right) \tag{11.30}$$

As we want to maximize Eq. 11.29, we calculate its first derivative with respect to w_x and set it to zero:

$$\begin{aligned} \frac{\partial \mathcal{L}}{\partial w_x} &= C_{xy} w_y - \lambda_1 C_{xx} w_x = 0 \\ &\rightarrow C_{xy} w_y = \lambda_1 C_{xx} w_x \end{aligned} \tag{11.31}$$

Analogously for w_y:

$$\begin{aligned} \frac{\partial \mathcal{L}}{\partial w_y} &= w_x^T C_{xy} - \lambda_2 w_y^T C_{yy} = 0 \\ &\rightarrow w_x^T C_{xy} = \lambda_2 C_{yy} w_y \end{aligned} \tag{11.32}$$

By left-multiplying Eq. 11.31 with w_x^T we obtain

$$w_x^T C_{xy}^T w_y = \lambda_1 w_x^T C_{xx} w_x \tag{11.33}$$

and by left-multiplying Eq. 11.32 with w_y^T:

$$\begin{aligned} w_y^T w_x^T C_{xy} &= \lambda_2 w_y^T C_{yy} w_y \\ w_y^T (C_{yx} w_x)^T &= \lambda_2 w_y^T C_{yy} w_y \\ w_x^T C_{xy} w_y &= \lambda_2 w_y^T C_{yy} w_y \end{aligned} \tag{11.34}$$

We see from Eq. 11.33 and Eq. 11.34 that $\lambda_1 = \lambda_2 = \rho$. Combining Eq. 11.31 and Eq. 11.32 into one equation results in the generalized eigenvalue equation (cf. Eq. 11.4):

$$\begin{bmatrix} 0 & C_{xy} \\ C_{yx} & 0 \end{bmatrix} \begin{bmatrix} w_x \\ w_y \end{bmatrix} = \rho \begin{bmatrix} C_{xx} & 0 \\ 0 & C_{yy} \end{bmatrix} \begin{bmatrix} w_x \\ w_y \end{bmatrix} \tag{11.35}$$

or, equivalently in matrix notation:

$$\begin{bmatrix} 0 & C_{xy} \\ C_{yx} & 0 \end{bmatrix} \begin{bmatrix} W_x \\ W_y \end{bmatrix} = \begin{bmatrix} C_{xx} & 0 \\ 0 & C_{yy} \end{bmatrix} \begin{bmatrix} W_x \\ W_y \end{bmatrix} \hat{C}_{xy} \tag{11.36}$$

where W_x, W_y are the matrices with the respective canonical variates in their columns, and

$$\hat{C}_{xy} = \begin{bmatrix} \rho_1 & \dots & 0 \\ \vdots & \ddots & \vdots \\ 0 & \dots & \rho_k \end{bmatrix} \tag{11.37}$$

is the diagonal matrix of all generalized eigenenvalues, that is, canonical correlations. Analogously to PCA, we can now project the data down to the subspace with the highest contribution to the canonical correlation by just multiplying the data with the j columns of W that correspond to the j highest eigenvalues:

$$X' = \sum_{i=1}^{j} e_i w_x^{iT} X = [w_x^1 w_x^2 \dots w_x^j]^T X = W_x' X \tag{11.38}$$

$$Y' = \sum_{i=1}^{j} e_i w_y^{iT} Y = [w_y^1 w_y^2 \dots w_y^j]^T Y = W_y' Y \tag{11.39}$$

One can think of CCA as a kind of generalized PCA: one wants to find a linear transformation $[W_x^T, W_y^T]^T$ that maximizes the left part of Eq. 11.36, the covariance *between* X and Y. The covariance between variables can be regarded as signal. At the same time one wants to constrain the variance *within* each variable, C_{xx}, C_{yy}. The covariance within variables can be regarded as noise. Just as in oriented PCA (cf. Eq. 11.26) these constraints are incorporated into the right-hand side of the eigenvalue equation.

CCA for High-Dimensional Data. When working with high-dimensional data, it can be infeasible to compute the covariance matrices C_{xx}, C_{yy}, and C_{xy}. In this case it is helpful to remember the **kernel trick** (Müller et al., 2001) mentioned in the PCA section: when the dimensionality U of the data is higher than the number of samples L it becomes more efficient to compute the solution not on covariance matrices but on the data *kernels*:

$$K_X = X^T X, \qquad K_Y = Y^T Y \tag{11.40}$$

The kernel of the data is only of size $L \times L$, while the covariance matrices would be of size $U \times U$ and $V \times V$. We will consider only linear kernels; for nonlinear kernel versions of CCA (also called **kernel CCA**, kCCA) see Fyfe and Lai (2000), Akaho (2001), and Bach and Jordan (2002). An advantage of linear kernels is that the canonical variates can be easily recovered as a linear combination of data points. For the first canonical component this means that

$$w_x = \alpha^T X, \qquad w_y = \beta^T Y \tag{11.41}$$

where the coefficients α can be obtained by solving an eigenvalue equation very similar to Eq. 11.36:

$$\begin{bmatrix} 0 & K_X K_Y \\ K_Y K_X & 0 \end{bmatrix} \begin{bmatrix} \alpha \\ \beta \end{bmatrix} = \rho \begin{bmatrix} K_X^2 & 0 \\ 0 & K_Y^2 \end{bmatrix} \begin{bmatrix} \alpha \\ \beta \end{bmatrix} \tag{11.42}$$

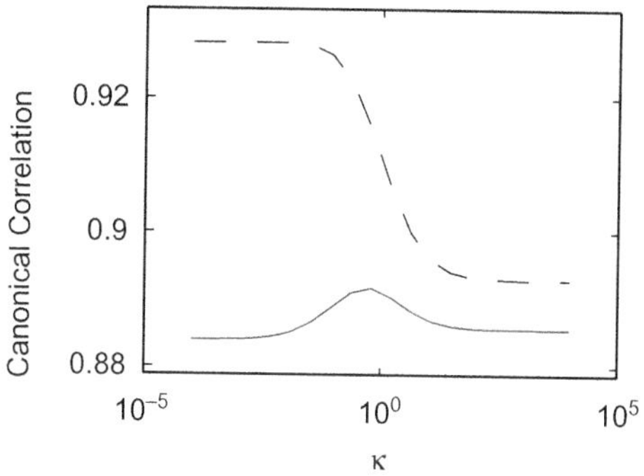

Figure 11.5. Examples of incorrect model selection for CCA from a 200-sample 10-dimensional practice-data set, generated similarly to Eq. 11.46: in black dashed lines the canonical correlation ρ on the training data set, in blue ρ on the test data set; ρ on test data is maximized for $\kappa \approx 1$; for $\kappa \ll 1$ the canonical correlation on the training data is high, but that on the test data is smaller than for $\kappa = 1$ (overfitting); for $\kappa \gg 1$, the correlation also decreases as the CCA solution is regularized too much.

This can be seen by plugging Eqs. 11.41 into the CCA solution obtained from Eq. 11.30.

Regularization

In the presence of noise or small data sets, it is important in CCA—and even more so in kCCA—to control the complexity of the solution, in order to make sure the variates found are generalizable to new data. Without this complexity control, CCA might *overfit* its variates to noise fluctuations. Given new data, on which it has not been trained, CCA would then result in nondiagonal covariance matrices and lower canonical correlations (see Fig. 11.5). In the machine-learning community, this complexity control is termed **regularization**. For standard CCA it was introduced by Vinod (1976). One way to control the complexity of the canonical variates w_x, w_y is by constraining the square of the Euclidian norm of the variates, $\|w_x\|^2$ and $\|w_y\|^2$. How small the norm should be is controlled by the **regularization parameter** K_i, where i denotes the variate (here, x, y). This constraint can be combined with the original CCA constraint such that Eq. 11.29 becomes

$$w_i^T C_{ii} w_i + \kappa_i \|w_i\|^2 = 1. \tag{11.43}$$

The larger κ_i, the smaller the norm of the variates and thus the smaller the coefficients of w_i, that is, the "simpler" the solution. When dealing with more than one canonical variate, imposing regularization to the objective corresponds to adding a diagonal matrix $\kappa_i I$ to the estimated autocovariance matrices C_{ii}, where I denotes an identity matrix of the same size as C_{xx}, and $\kappa_i I$ is a diagonal matrix that adds a **ridge** of height κ_i on the covariance matrix. The solution for w_y is derived analogously and both equations can again be solved jointly by solving the generalized eigenvalue problem

$$\begin{bmatrix} 0 & C_{xy} \\ C_{yx} & 0 \end{bmatrix} \begin{bmatrix} w_x \\ w_y \end{bmatrix} = \rho \begin{bmatrix} C_{xx} + \kappa_x I & 0 \\ 0 & C_{yy} + \kappa_y I \end{bmatrix} \begin{bmatrix} w_x \\ w_y \end{bmatrix} \tag{11.44}$$

It is helpful to think of what this means for the data-generating process. If the autocovariance matrices become more diagonal, this can be seen as adding more and more decorrelated variance to each dimension, as would be the case if decorrelated noise were added to the data. The solution for regularized kernel CCA can be derived analogously by replacing w_x and w_y in Eq. 11.36 by $\alpha^T X$ and $\beta^T Y$, respectively:

$$\begin{bmatrix} 0 & K_X K_Y \\ K_Y K_X & 0 \end{bmatrix} \begin{bmatrix} \alpha \\ \beta \end{bmatrix} = \rho \begin{bmatrix} K_X^2 + \kappa_x I & 0 \\ 0 & K_Y^2 + \kappa_y I \end{bmatrix} \begin{bmatrix} \alpha \\ \beta \end{bmatrix} \tag{11.45}$$

So regularizing the kCCA solution can also be achieved by simply adding a ridge on the denominator of the generalized eigenvalue equation (Eq. 11.42). This ridge on the autocovariance blocks effectively acts as a penalty on the squared norm of the α, β vectors, or directly on w_x, w_y in the CCA case (cf. Eq. 11.43).

Model Selection

The correct values for κ_X and κ_Y have to be estimated using classical model selection procedures. One way to estimate the correct regularization parameter is resampling-based methods as used in Bießmann et al. (2009). In this setting one produces surrogate data $\bar{X}$ that contains the same data as in X but shuffled in time. This will destroy the cross-covariance structure between X and Y while preserving the autocovariance structure within each variable. One then picks that value of κ that maximizes the difference between the canonical correlation on the true data set and the surrogate data. Another popular model selection procedure is **cross-validation**, as applied for instance in Hardoon et al. (2007). Here one splits the data into **training set** and **test set**. One computes CCA for all values of κ on the training set and uses the estimated covariates to compute canonical correlations on the test set. Maximizing the canonical correlation on the test data and not the training data will result in a solution that generalizes well to new data that CCA has not been trained on. A practice-data example of canonical correlations on test and training data is depicted in Figure 11.5. The optimal regularizer κ that maximizes the canonical correlation on the test data is close to 10^0, for smaller κ, the correlation on the test set is lower because CCA found a too-complex solution. For $\kappa > 1$ the regularization results in too-simple patterns and thus the correlation is also lower than with the optimal κ.

Practice-Data Example

Figure 11.6 illustrates the CCA solution on a two-dimensional practice data set, that is, $x \in R^2$, $y \in R^2$, generated according to the following:

$$\begin{aligned} & s \sim \mathcal{N}(0,1), \qquad \varepsilon_i \sim \mathcal{N}(0,1) \\ & x = \begin{bmatrix} 1 \\ 0.5 \end{bmatrix} \gamma s + \begin{bmatrix} 1 \\ 1 \end{bmatrix} (1-\gamma)\varepsilon_x \\ & y = \begin{bmatrix} 0 \\ -1 \end{bmatrix} \gamma s + \begin{bmatrix} 1 \\ 1 \end{bmatrix} (1-\gamma)\varepsilon_y \end{aligned} \tag{11.46}$$

where the underlying common signal s and the measurement noise ε_x, ε_y for each variable are each drawn from a normal distribution with mean 0 and standard deviation 1. The amount of the common signal s relative to the measurement noise is controlled by the parameter γ. The higher γ is, the stronger is the contribution of the common underlying signal to each variable x and y. In the following example we set $\gamma = 0.9$. Measurements in X are generated by projecting the signal in a two-dimensional space. For variable x the projection is such that s is present in both dimensions; the projection on y contains s in the second dimension only. The resulting data distributions are plotted in Figure 11.6A. We plotted each dimension of x against each dimension of y; all subplots have the same axis scaling. Above each subplot we show the corresponding univariate correlation coefficient r between x_i and y_i. The subplots are arranged in the form of a covariance matrix, blue blocks denote the autocovariance matrices C_{xx}, C_{yy}, and red blocks are cross-covariance matrices C_{xy}, C_{yx}. As expected from the data-generating process, both dimensions of x show strong autocovariances (blue frame, upper right corner), the cross-covariances (red frames) show large covariances between x_1 and y_2 as well as between x_2 and y_2. In this representation it is difficult to see how the signal of interest s is distributed across all dimensions of X and Y, respectively. The CCA transforms the data such that the common variance of s is reflected in one dimension only. In Figure 11.6B we show the same data after projecting them on their canonical variates. In this new representation x_1 is only correlated with y_1, the covariance matrix (in red) is diagonalized. The first eigenvectors found are $w_x = [10.5]^T$ and $w_y = [0 - 1]^T$. So after projecting the data onto their canonical variates the data is easily interpretable and the canonical variates found by CCA reflect how the underlying common signal s is reflected in X and Y.

Application: Analysis of Multimodal Data

We will use similar data as in the PCA section, simultaneous recordings of fMRI signals and electrophysiological recordings; experimental details are given in Murayama et al. (2010). This time we will include the fMRI data in the analysis and perform CCA between the fMRI images and spectrograms

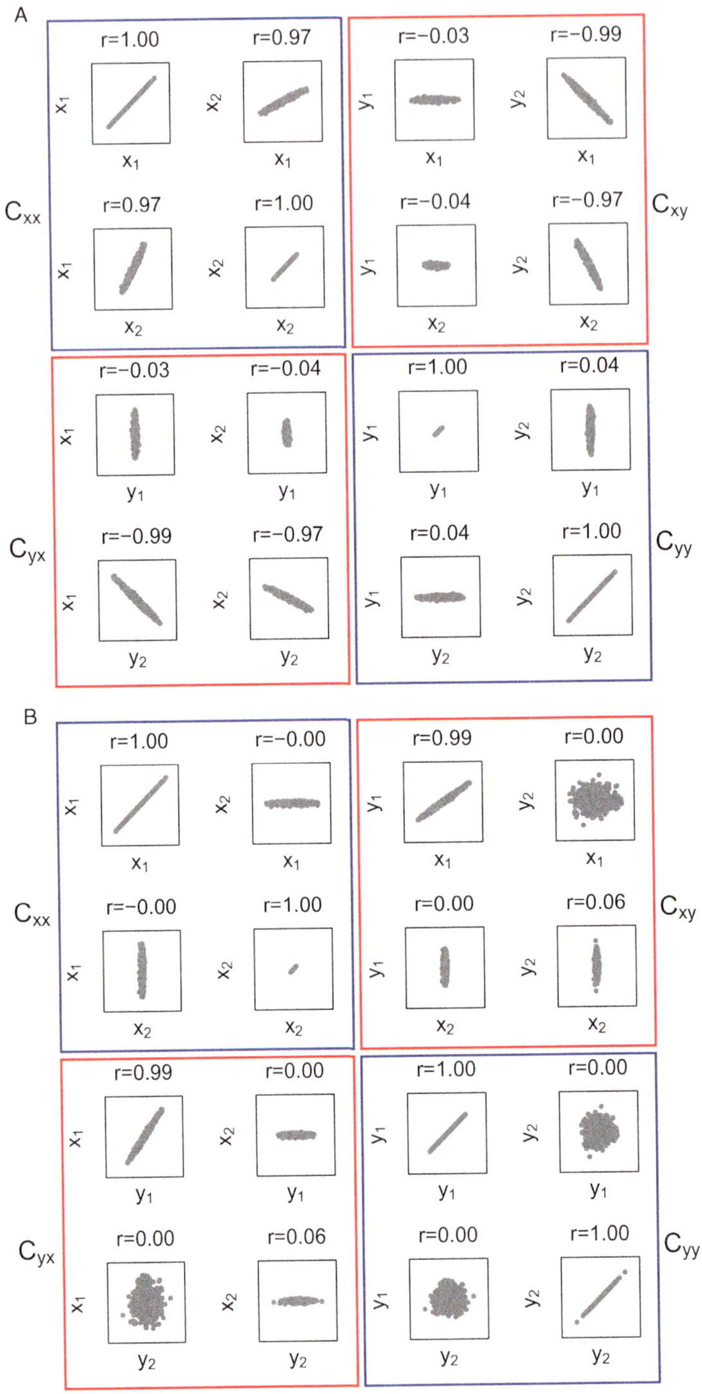

Figure 11.6. Pictorial representation of the joint covariance matrix $[X^T, Y^T]^T[X^T, Y^T]$ *before* (A) and *after* (B) CCA; in blue frames, the autocovariance matrices C_{xx}, C_{yy} are plotted; in red, the cross-covariance matrices C_{xy}, C_{yx}; for each plot, we plotted one dimension of X, Y against another dimension of X, Y (all axes have the same scaling) and the respective correlation coefficient r; after CCA, all variance of the underlying common signal is captured in the first dimension of x, y only (cf. red boxes in panel B).

of neural activity to find out which voxels are maximally correlated with which frequencies of neural bandpower. Neurophysiology spectrograms with seven frequency bands (0.1–8 Hz, 8–12 Hz, 12–24 Hz, 24–40 Hz, 60–100 Hz, 100–250 Hz, 1000–3000 Hz) are in the form of a data matrix $[x_1, x_2, \ldots, x_L] = X \in R^{7\times L}$, where L is the number of samples. The time bins of the spectrograms have been adjusted to match the time needed for acquisition of one fMRI image, one second in this case. The fMRI data consists of all voxels in an anatomically defined region of interest (ROI) comprising all voxels in the primary visual cortex in a radius of 6 mm around the recording electrode. All voxels (voxel size was 1×1×2 mm) in the ROI were extracted, vectorized, and concatenated in a data matrix $[y_1, y_2, \ldots, y_L] = Y \in R^{V\times L}$, V being the number of voxels in the ROI. Each frequency band was filtered with a neurovascular transfer function, modeled as a superposition of two Gamma functions (Friston et al., 1994). We computed the first canonical variates using regularized kCCA as in Eq. 11.45, and regularization parameters have been optimized as in Bießmann et al. (2009): For each regularizer we solve kCCA on the true data and on a surrogate data set, in which the temporal order of samples in one data set was shuffled. We then chose the regularizer that maximized the difference between the true and the shuffled data. The canonical variates for the fMRI data are three-dimensional images that indicate for each voxel how strongly it is correlated with the first canonical component of the neural signal. In Figure 11.7A we plot one coronal section of a three-dimensional fMRI volume, on which the canonical variate is superimposed in red. Voxels close to the electrode show strong correlations with the neural recordings. The canonical variates of the electrophysiology are simply a

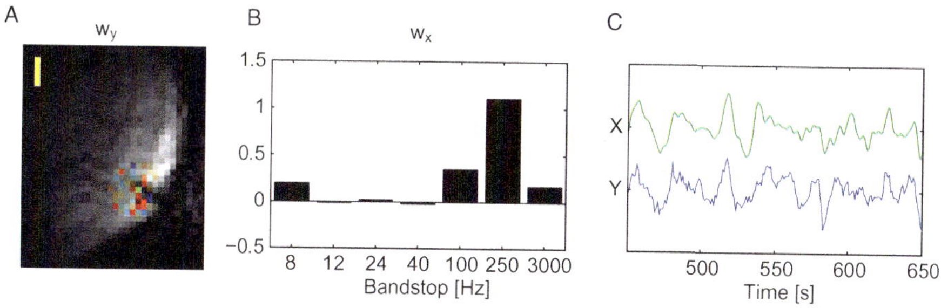

Figure 11.7. Application example of CCA: simultaneous recordings of fMRI and intracranial electrophysiology. (A) Canonical variate w_y of fMRI data, superimposed on primary visual cortex, with a 5-mm scale bar in yellow and recording position indicated by a green voxel (note the shadow around the electrode); voxels with high coefficients (red) are close to the recording electrode and extend only over a small part of the cortex. (B) Canonical variate w_x for the electrophysiological bandpower. (C) Time series of data projected on respective canonical variate, with electrophysiology in green and fMRI data in blue.

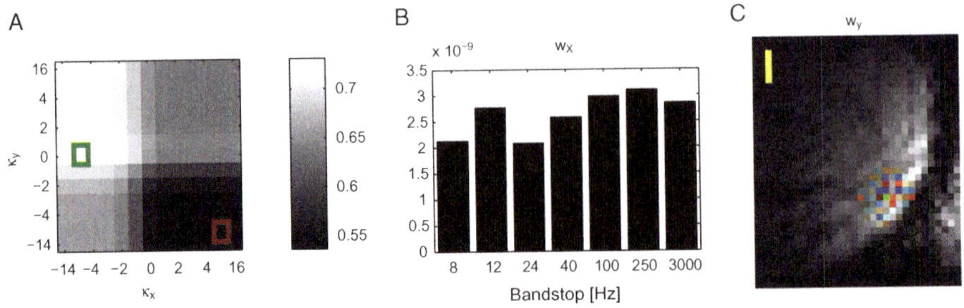

Figure 11.8. Examples of incorrect model selection on real data. (A) Correlation on test data for varying regularization parameter κ; optimal values (corresponding to Fig. 11.7) are $\kappa_x = 10^{-5}$, $\kappa_y = 1$, indicated by a green box; examples for κ too high or too low are taken from settings indicated by a red box. (B) When κ is too large, the solution has a small norm and coefficients become more similar (cf. Fig. 11.7B). (C) When there is not enough regularization the coefficients are lacking the spatial contiguity and are overfitted to random fluctuations (cf. Fig. 11.7C)

weighting vector for each frequency band of neural bandpower. Figure 11.7B shows the weighting coefficients as a bar plot. The results indicate that it is mainly neural oscillations in the gamma range (40–250 Hz) that are strongly coupled to the fMRI signal. This is in line with previous findings using univariate correlation coefficients (Logothetis et al., 2001; Goense and Logothetis, 2008). Figure 11.7C shows a typical time series of fMRI data and neural bandpower projected on their respective canonical variates. The neural canonical components exhibit faster dynamics than the fMRI canonical component; however, the peaks and troughs of the fMRI component are clearly correlated to the neural component. Figure 11.8 shows examples of regularizers that are too small or too large. Too little regularization leads to overfitting and incorrect canonical variates with strongly varying coefficients; too much regularization leads to the same coefficients across the entire variate and a small norm of the variate.

Conclusion

We have derived canonical correlation analysis (CCA) and drawn the connection to the previous section by formulating it as a generalized eigenvalue problem. Given multiple multivariate variables, CCA finds directions of common variance while constraining the variance within each variable. We have shown extensions for high-dimensional data and illustrated the problem of overfitting on practice data and neural data obtained in multimodal recordings of intracranial electrophysiology and high resolution fMRI data. An important assumption of CCA is that the variables are correlated *instanta-*

neously. The CCA will only maximize correlations between simultaneously acquired samples. If the underlying common signal is reflected in each variable with different temporal dynamics, one needs to compensate for this using prior knowledge about the respective temporal dynamics. This meant that for the application example of simultaneous electrophysiological and hemodynamic recordings we had to compensate for the hemodynamics by convolving the neural activity with a hemodynamic response function. In the next section we show an extension of CCA that learns the modality-specific temporal dynamics from multimodal data.

CONVOLUTIVE CCA

In CCA we optimized the canonical correlation (Eq. 11.27) between two signals. When the two signals are not *instantaneously* correlated but with some time lag τ, then classical CCA will fail to find sensible canonical variates. In general, however, we do not know the correct lag. And in many cases the modality-specific temporal dynamics are not well described by a simple delay but rather by a **convolution**, that is, a weighted combination of multiple time lags. A slight modification of CCA, called **temporal kernel canonical correlation analysis** (tkCCA) (Bießmann et al., 2009), can account for such non-instantaneous couplings. As we will consider not only the kernel version, but also a version operating on covariance matrices in this section, we use the more general term **convolutive canonical correlation analysis** (CCCA). This slight modification of standard CCA will not only compensate for modality-specific differences in temporal dynamics; CCCA will also reveal to what extent single features contribute to the coupling between two modalities and for which time lags they do so. The idea behind CCCA is simple: as the correlations between X and Y are not instantaneous, the canonical correlation coefficient (Eq. 11.28) will be small for simultaneous samples. By shifting one data source back and forth in time (relative to the other data source) by some relative time lag τ, the correlation will become larger for the correct time lag. Doing this for all (reasonable) time lags will yield time-lag-dependent weightings $w_x(\tau)$ for one data source (the one shifted in time) that can compensate for delays or convolutions between data sources. The convolution $w_x(\tau)$ should be such that there are high weights for those time lags and features that maximize the correlation with the time series in Y. If $w_x(\tau)$ is estimated correctly, the convolution of X with $w_x(\tau)$ will compensate for data-source-specific differences in temporal dynamics. In the neurovascular coupling example, the canonical convolution $w_x(\tau)$ will low-pass filter the neural activity and bring the signal in temporal alignment with the hemodynamic activity (Bießmann et al., 2009; Murayama et al., 2010). The CCCA algorithm is summarized in Box 11.3.

BOX 11.3. Convolutive CCA Algorithm

Input:

→ Data $X = [x_1, x_2, \ldots, x_L] \in R^{U\times L}$, $Y = [y_1, y_2, \ldots, y_L] \in R^{V\times L}$
→ Maximal time lag T

1. Center data such that $\sum_i x_i = \sum_i y_i = 0$
2. Model selection to find right regularizer κ:

   For each κ:

   a. Embed *one* variable in its temporal context

   $$\tilde{X} = [X_{-T}^T, X_{-T+1}^T, \ldots, X^T, \ldots, X_{T-1}^T, X_T^T]^T$$

   b. Compute $w_{\tilde{x}}, w_y$ (Eq. 11.40) using *training data.*

   c. Evaluate canonical correlation (Eq. 11.27) on *test data.*
3. Pick the κ that maximizes Eq. 11.26 on test data set.
4. Solve Eq. 11.40 again using the correct κ.
5. Reshape $w_{\tilde{x}} = [w_x(\tau = -T)^T, \ldots, w_x(\tau = T)^T]^T$

Output:

← Convolution $w_x(\tau) \in R^{U\times 2T+1}$
← Canonical variate $w_y \in R^V$
← Canonical correlogram $\rho(\tau) \in R^{2T+1}$

Derivation

In the following we assume that

$$\tau = [-T, -T+1, -T+2, \ldots, 0, \ldots, T=2, T-1, T]$$

We shift X back in time (relative to Y) by T samples and forward in time by the same amount. One might as well shift the time series in Y; the derivation is completely analogous to the following. Note that the shifting operation requires that the first and last T samples of X and Y be discarded. In CCCA the time-resolved canonical variate $w_x(\tau)$ and the canonical variate w_y are optimized such that

$$\text{argmax}_{w_x(\tau), w_y} \left(\frac{\sum_\tau (w_x(\tau)^T X_\tau), w_y^T Y}{\sqrt{\sum_\tau (w_x(\tau)^T X_\tau X_\tau^T w_x(\tau)) w_y^T Y Y^T w_y}} \right) \tag{11.47}$$

where X_τ denotes the data in X shifted by a time lag τ and $w_x(\tau)$ is the canonical variate of X at time lag τ. For $\tau = 0$ Eq. 11.51 reduces to the standard CCA (Eq. 11.28). This means that CCCA not only optimizes the canonical correlation for simultaneous samples, but over a short time window specified by the range of τ. To maximize Eq. 11.47, we embed the data set X into a higher dimensional space where the additional dimensions are occupied by copies of the original data, each shifted in time by a different τ:

$$\tilde{X} = \begin{bmatrix} X_{\tau=-T} \\ \vdots \\ X_{\tau=T} \end{bmatrix} \in R^{M(2T+1)\times L} \tag{11.48}$$

By using the same time-delay embedding for the $w_x(\tau)$,

$$\tilde{w}_x = \begin{bmatrix} w_x(\tau=-T) \\ \vdots \\ w_x(\tau=+T) \end{bmatrix} \in R^{M(2T+1)\times 1} \tag{11.49}$$

we reduce the optimization problem in Eq. 11.47 to an ordinary CCA problem in the embedded space that can be solved as in Eq. 11.36 by maximizing the CCA objective using the embedded data in $\tilde{X}$:

$$\text{argmax}_{\tilde{w}_x, w_y} \frac{\tilde{w}_x^T \tilde{X} Y^T w_y}{\sqrt{\tilde{w}_x^T \tilde{X}\tilde{X}^T \tilde{w}_x w_y^T Y Y^T w_y}} \tag{11.50}$$

The CCCA solution is thus the CCA solution in the embedded space. Analogously to a standard cross-correlogram for univariate time series, we can use the filter $w_x(\tau)$ to compute a time lag dependent canonical correlogram $\rho(\tau)$ by

$$\rho(\tau) = \frac{w_x(\tau)^T X_\tau, w_y^T Y}{\sqrt{w_x(\tau)^T X_\tau X_\tau^T w_x(\tau) w_y^T Y Y^T w_y}} \tag{11.51}$$

The canonical correlogram $\rho(\tau)$ reflects the temporal dynamics in the canonical correlations between the two variables and can be considered as a multivariate generalization of the univariate cross-correlogram. An example is shown in Figure 11.9D. It is worth mentioning that the canonical correlogram is different from a usual correlogram between two one-dimensional time signals in that it is always positive.[2] This is because CCA is ignorant with respect to the sign of the correlation; possible negative correlations between

[2]Since the $\rho(\tau)$ are not directly calculated by kCCA, but by Eq. 11.51, this is not strictly true, that is, only up to noise fluctuations.

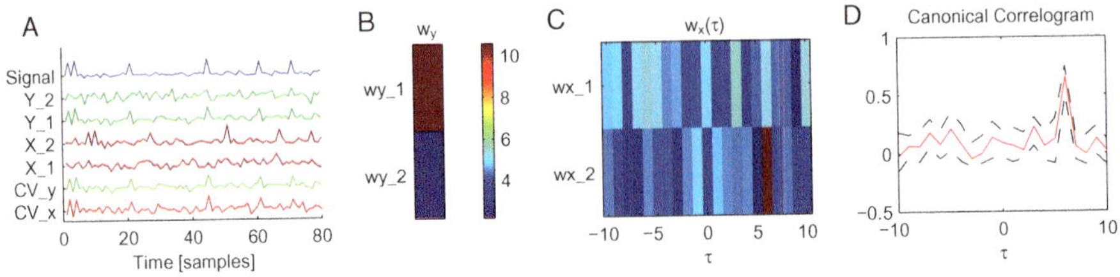

Figure 11.9. Practice-data example for data with noninstantaneous coupling. (A) Simulated data, where *blue* is the source signal; *green*, data stream Y and projection on canonical variate; Y_1 is strongly coupled to the source signal, while Y_2 contains mainly noise; *red* is data stream X and its projection on the first canonical variate; note that the source signal is reflected predominantly in X_2 but with a delay of six samples; the convolution with its canonical variate $w_x(\tau)$ compensates for this delay: both canonical variates are temporally aligned and capture the source signal well. (B) Canonical variate of Y CCCA found the right weighting. (C) Time-resolved canonical variate of X; all coefficients are close to 0, except those at the right time lag. (D) The canonical correlogram correctly reflects the coupling between X and Y at the right time lag τ.

the signals are encoded in the relative sign of the canonical variates. So, to interpret the results of CCCA, one always has to take into account both the canonical correlogram $\rho(\tau)$ and the corresponding canonical variates $w_x(\tau)$ and w_y.

CCCA for High-Dimensional Data

If the number of time lags and the dimensionality of X are large, one can compute the kernelized version of CCA as formulated in Eq. 11.42 from the kernel of $K_{\tilde{X}} = \tilde{X}^T\tilde{X}$. The temporal resolution of the correlogram is determined by the spacing of the τ-values; in particular, it can be higher than the temporal resolution of the signals in Y. There are some differences between solving CCCA on covariance matrices and tkCCA, which operates on kernels of the data. While the covariance matrices in the CCCA version can grow quickly in case the number of time lags or dimensions is large, the kernel computed from the embedded data $\tilde{X}$ is of the same size as the kernel K_X. Thus, after the kernel $\tilde{K}_X$ has been computed tkCCA is computationally not more demanding than normal kCCA. Another advantage of tkCCA is that cross-validation procedures or bootstrap resamplings can operate on the already-computed kernel, which can speed up the resamplings considerably. Since tkCCA just leads to a single eigenvalue equation of the type given by Eq. 11.45, neither the canonical correlation ρ nor the corresponding vectors α and β depend on τ. However, it is easy to recover the τ-dependent filter $w_x(\tau)$. Analogously to Eq. 11.41 we obtain from the solution α of kernel CCA on the embedded data $\tilde{X}$

$$\tilde{w}_x = \begin{bmatrix} w_x(\tau = -T) \\ \vdots \\ w_x(\tau = T) \end{bmatrix} = \tilde{X}\alpha$$
$$w_y = Y\beta. \tag{11.52}$$

Or, remembering the structure of $\tilde{X}$ in Eq. 11.48, the delay-dependent canonical variates $w_x(\tau)$ can be recovered equivalently as expansion of the α vector using the respective parts of $\tilde{X}$:

$$w_x(\tau) = X_\tau \alpha. \tag{11.53}$$

The canonical correlogram can now be calculated as in Eq. 11.51.

Practice-Data Example

Again we use two practice-data time series $x \in R^2$, $y \in R^2$. The data is generated from a univariate signal s consisting of Gaussian noise (mean 0 and standard deviation 0.1) interspersed by spikes occurring at intervals drawn from a univariate distribution. Multimodal data was then generated by

$$x = [0.1 \quad 0.9]^T s(t-6), \qquad y = [0.9 \quad 0.1]^T s(t) \tag{11.54}$$

where $s(t-6)$ means that s was shifted back in time; that is, the variable x reflects the source signal only six samples after variable y. Correlations between $w_x(\tau)^T x_\tau$ and $w_y^T Y$ at $\tau = 0$ will be close to zero, and only at the right delay of six samples will the canonical correlation be high. Figure 11.9A shows examples of the source signal, both variables and the canonical variate time courses. Panel B shows the estimated canonical variate for w_y with the right weighting for each dimension of y. Panel C shows the canonical convolution $w_x(\tau)$. Only at the correct time lag do the variates show high values and the weighting found is the same as in the data-generation process. In panel D the canonical correlogram (red) is plotted with 95th percentiles. All canonical correlations are close to 0 except the one at $\tau = 6$.

Application: Analysis of Multimodal Data with N\oninstantaneous Couplings

In Figure 11.10 we show an application example of CCCA on the same data as in the example of the previous section. This time we do not assume a hemodynamic lag but we do estimate a convolution $w_x(\tau)$ that will transform the neural bandpower signal x into a canonical representation that is maximally correlated with the hemodynamic response y. Panel A shows a section of the three-dimensional canonical variate w_y. The distribution of voxels with large

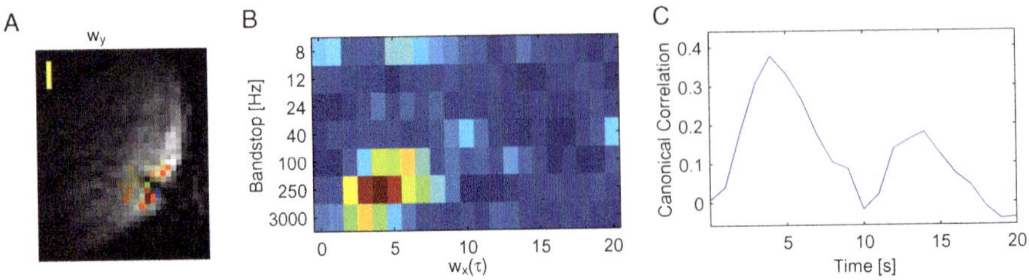

Figure 11.10. Neurovascular coupling dynamics estimated by CCCA. (A) Canonical variate w_y superimposed on a coronal section of primary visual cortex (in gray); electrode position is indicated by a green dot; close to the electrode voxels contribute strongly to the canonical variate (red), similar to Figure 11.7A. (B) Canonical convolution $w_x(\tau)$; the frequency weighting is the same as in Figure 11.7B, and the temporal delay of 4 s is compatible with established models of hemodynamic coupling. (C) Canonical correlogram $\rho(\tau)$ reflecting temporal coupling dynamics between X, Y; the peak corresponds to the peak in $w_x(\tau)$; the other maximum has to be interpreted with respect to the negative weights in $w_x(\tau = 15)$ as an undershoot.

positive contribution is similar to the one in Figure 11.7. The time-resolved canonical variate $w_x(\tau)$ is plotted in Figure 11.10B. High coefficients are at a physiologically plausible time lag of 4–5 seconds and the weighting of frequencies is the same as in the CCA solution (cf. Fig. 11.7B). Figure 11.10C shows the canonical correlogram obtained from the estimates of w_y and $w_x(\tau)$. The peak at $\tau = 5$ s corresponds to the peak in $w_x(\tau = 5)$; the peak at $w_x(\tau = 15)$ has a negative sign and can be interpreted as an undershoot, as expected from models of neurovascular coupling (Buxton et al., 2004).

Conclusion

In this section we have presented an extension of CCA, convolutional CCA (CCCA), which maximizes the correlation between multiple variables in the presence of noninstantaneous correlations. This is helpful in case one measures a signal of interest, such as neural activity, simultaneously with other modalities, such as hemodynamic activity, that reflect the signal of interest with different temporal dynamics. Applying the filters estimated by CCCA will remove modality-specific temporal delays or convolutions of the underlying signal. The kernel version of CCCA, temporal kernel CCA (tkCCA) allows the computation of convolutions between high-dimensional variables and renders nonlinear extensions straightforward. In an application example we showed how we can use tkCCA to estimate the temporal dynamics of the neurovascular transfer function from simultaneous recordings of hemodynamic and neurophysiological measurements. Despite the fact that the tkCCA results are completely data driven and do not rely on model assumptions that fMRI analyses typically make, the results are in line with well-established models of neurovascular coupling.

REFERENCES

Aguirre G.K., Zarahn E., and D'Esposito M. (1998). The variability of human, bold hemodynamic responses. *Neuroimage* 8:360–369.

Akaho S. (2001). A kernel method for canonical correlation analysis. In *Proceedings of the International Meeting of the Psychometric Society (IMPS2001)*, Springer-Verlag.

Anderson T. (1958). *An Introduction to Multivariate Statistical Analysis*. New York: Wiley.

Bach F.R. and Jordan M.I. (2002). Kernel independent component analysis. *J Machine Learning Res* 3:1–48.

Bie T.D.D. and Cristianini N. (2004). Kernel methods for exploratory pattern analysis: a demonstration on text data. *Springer Lecture Notes in Computer Science*. New York: Springer.

Bie T.D.D., Cristianini N., and Rosipal R. (2005). Eigenproblems in pattern recognition. In: *Handbook of Geometric Computing: Applications in Pattern Recognition, Computer Vision, Neural Computing, and Robotics* (Shawe-Taylor J., and Cristianini N., Eds.), pp. 129–170. Springer.

Bießmann F. and Harth A. (2010). Analysing dependency dynamics in web data. In *Proceedings of AAAI Spring Symposium, Stanford*, January.

Bießmann F., Meinecke F.C., Gretton A., Rauch A., Rainer G., Logothetis N.K., and Müller K.-R. (2009). Temporal kernel CCA and its application in multimodal neuronal data analysis. *Machine Learning J* 79(1–2):5–27.

Bishop C.M. (2007). *Pattern Recognition and Machine Learning*. New York: Springer.

Blankertz B., Losch F., Krauledat M., Dornhege G., Curio G., and Müller K.-R. (2008). The Berlin brain–computer interface: accurate performance from first-session in BCI-naive subjects. *IEEE Trans Biomed Eng* 55:2452–2462.

Buxton R.B., Uludag K., Dubowitz D.J., and Liu T.T. (2004). Modeling the hemodynamic response to brain activation. *Neuroimage* 23:220–233.

Diamantaras K.I. (2003). Temporal filtering and oriented PCA neural networks for blind source separation. *IEEE 13th Workshop on Neural Networks for Signal Processing, NNSP'03*, pp. 369–378.

Diamantaras K.I. and Kung S. (1996). *Principal Component Neural Networks: Theory and Applications*. Wiley.

Diamantaras K.I. and Papadimitriou T. (2003). Blind signal separation using oriented PCA neural models. Proceedings of the IEEE International Conference on Acoustics, Speech, and Signal Processing (ICASSP '03), Vol. 2, pp. II-733–II-736.

Eichele T., Specht K., Moosmann M., Jongsma M.L.A., Quiroga R.Q., Nordby H., and Hugdahl K. (2005). Assessing the spatiotemporal evolution of neuronal activation with single-trial event-related potentials and functional MRI. *Proc Natl Acad Sci U S A* 102:17798–17803.

Friman O., Borga M., Lundberg P., and Knutsson H. (2002). Exploratory fMRI analysis by autocorrelation maximization. *Neuroimage* 16:454–464.

Friston K.J. (2009). Modalities, modes, and models in functional neuroimaging. *Science* 326:399–403.

Friston K.J., Jezzard P., and Turner (1994). Analysis of functional MRI time-series. *Human Brain Mapping* 1:153–171.

Fyfe C. and Fyfe G. (2006). Two methods for sparsifying probabilistic canonical correlation analysis. *Springer Lecture Notes in Computer Science*. New York: Springer.

Fyfe C. and Lai P. (2000). ICA using kernel canonical correlation analysis. In *Proceedings of the International Workshop on Independent Component Analysis*.

Fyfe C. and Leen G. (2006). Stochastic processes for canonical correlation analysis. *Proceedings of the 14th European Symposium on Artificial Neural Networks*.

Goense J.B.M. and Logothetis N.K. (2008). Neurophysiology of the bold fMRI signal in awake monkeys. *Curr Biol* 18:631–640.

Hardoon D.R. and Shawe-Taylor J. (2011). Sparse canonical correlation analysis. *Machine Learning* 83(3):331–353.

Hardoon D.R., Mourão-Miranda J., Brammer M., and Shawe-Taylor J. (2007). Unsupervised analysis of fMRI data using kernel canonical correlation. *Neuroimage* 37:1250–1259.

Hastie T., Tibshirani R., and Friedman J.H. (2003). *The Elements of Statistical Learning*. Springer.

Haynes J.-D. and Rees G. (2006). Decoding mental states from brain activity in humans. *Nature Rev Neurosci* 7:523–534.

Hotelling H. (1933). Analysis of complex statistical variables into principal components. *J Educational Psychol* 24(6):417–441.

Hotelling H. (1936). Relations between two sets of variates. *Biometrika* 28(3):321–377.

Hyvarinen A., Karhunen J., and Oja E. (2001). *Independent Component Analysis*. John Wiley & Sons.

Kettenring J.R. (1971). Canonical analysis of several sets of variables. *Biometrika* 58(3):433–451.

Kim T.-K., Wong S.-F., and Cipolla R. (2007). Tensor canonical correlation analysis for action classification. In *Proceedings of the IEEE Conference on Computer Vision and Pattern Recognition*, pp. 1–8.

Krauledat M., Tangermann M., Blankertz B., and Müller K. (2008). Towards zero training for brain–computer interfacing. *PLoS ONE* 3(8):2967.

Lang S. (1973). *Calculus of Several Variables*. New York: Springer.

Larimore W. (1990). Canonical variate analysis in identification, filtering, and adaptive control. Proceedings of the 29th Conference on Decision and Control, pp. 596–604.

Leen G., Fyfe C., and Fyfe G. (2006). A gaussian process latent variable model formulation of canonical correlation analysis. Proceedings of the 14th European Symposium on Artificial Neural Networks.

Leurgans S., Moyeed R., and Silverman B. (1993). Canonical correlation analysis when the data are curves. *J Roy Statist Soc* 55(3):725–740.

Logothetis N.K., Pauls J., Augath M.A., Trinath T., and Oeltermann A. (2001). Neurophysiological investigation of the basis of the fMRI signal. *Nature* 412(6843): 150–157.

Lu Y., Grova C., Kobayashi E., Dubeau F., and Gotman J. (2007). Using voxel-specific hemodynamic response function in EEG–fMRI data analysis: an estimation and detection model. *Neuroimage* 34:195–203.

Macke J.H., Zeck G., and Bethge M. (2008). Receptive fields without spike-triggering. In *Advances in Neural Information Processing Systems 20: Proceedings of the 2007 Conference*, pp. 969–976.

Mackey L. (2009). Deflation methods for sparse PCA. In *Advances in Neural Information Processing Systems 22*, pp. 1–8.

Mika S., Schölkopf B., Smola A., Müller K.-R., Scholz M., and Rätsch G. (1999). Kernel PCA and de-noising in feature spaces. In *Advances in Neural Information Processing Systems* (Kearns, M.S., Solla, S.A., and Cohn, D.A., Eds.), pp. 536–542. MIT Press.

Moosmann M., Eichele T., Nordby H., Hugdahl K., and Calhoun V. D. (2008). Joint independent component analysis for simultaneous EEG–fMRI: principle and simulation. *Int J Psychophysiol* 67:212–221.

Müller K.-R., Mika S., Ratsch G., Tsuda K., and Scholkopf B. (2001). An introduction to kernel-based learning algorithms. *IEEE Trans Neural Networks* 12:181–201.

Murayama Y., Bießmann F., Meinecke F.C., Müller K.-R., Augath M.A., Oeltermann A., and Logothetis N.K. (2010). Relationship between neural and hemodynamic signals during spontaneous activity studied with temporal kernel CCA. *Magnetic Resonance Imaging* 28:1095–1103.

Norman K.A., Polyn S.M., Detre G.J., and Haxby J.V. (2006). Beyond mind-reading: multi-voxel pattern analysis of fMRI data. *Trends Cognitive Sci* 10:424–430.

Ogawa S., Tank D.W., Menon R., Ellermann J.M., Kim S.G., Merkle H., and Ugurbil K. (1992). Intrinsic signal changes accompanying sensory stimulation: functional brain mapping with magnetic resonance imaging. *Proc Natl Acad Sci U S A* 89:5951–5955.

Pearson V. (1901). On lines and planes of closest fit to systems of points in space. *Philos Mag* 2:559–572.

Rauch A., Rainer G., and Logothetis N.K. (2008). The effect of a serotonin-induced dissociation between spiking and perisynaptic activity on bold functional MRI. *Proc Natl Acad Sci U S A* 105:6759–6764.

Roweis S. and Ghahramani Z. (1999). A unifying review of linear Gaussian models. *Neural Computation* 11:305–345.

Schölkopf B. and Smola A.J. (2001). *Learning with Kernels: Support Vector Machines, Regularization, Optimization, and Beyond*. Cambridge, MA: MIT Press.

Schölkopf B., Smola A.J., and Müller K.-R. (1998). Nonlinear component analysis as a kernel eigenvalue problem. *Neural Computation* 10(6):1299–1319.

Schreuder M., Blankertz B., and Tangermann M. (2010). A new auditory multi-class brain–computer interface paradigm: spatial hearing as an informative cue. *PLoS ONE* 5:e9813.

Shawe-Taylor J. and Cristianini N. (2004). *Kernel Methods for Pattern Analysis*. Cambridge University Press.

Sitaram R., Caria A., and Birbaumer N. (2009). Hemodynamic brain–computer interfaces for communication and rehabilitation. *Neural Networks* 22:1320–1328.

Strang G. (2009). *Introduction to Linear Algebra*, 3rd ed. Wellesley-Cambridge Press.

Tangermann M., Krauledat M., Grzeska K., Sagebaum M., Blankertz B., Vidaurre C., and Müller K.-R. (2009). Playing pinball with non-invasive BCI. *Adv Neural Inform Process Syst* 22:1641–1648.

Tipping M. and Bishop C.M. (1999). Mixtures of probabilistic principal component analyzers. *Neural Computation* 11:435–474.

Vinod H. (1976). Canonical ridge and econometrics of joint production. *J Econometrics* 4(2):147–166.

von Luxburg U. (2007). A tutorial on spectral clustering. *Statist Comput* 17(4): 395–416.

Weiskopf N., Sitaram R., Josephs O., Veit R., Scharnowski F., Goebel R., Birbaumer N., Deichmann R., and Mathiak K. (2007). Real-time functional magnetic resonance imaging: methods and applications. *Magnetic Resonance Imaging* 25:989–1003.

Wiesel A., Kliger M., and Hero A. (2008). A greedy approach to sparse canonical correlation analysis. Available at http://arxiv.org/abs/0801.2748v1 (accessed January 17, 2008).

Ziehe A. and Müller K.-R. (1998). TDSEP—an efficient algorithm for blind separation using time structure. In *Proceedings of the 8th International Conference on Artificial Neural Networks*, pp. 675–680.

PART III

FUNCTION REPLACEMENT (PROSTHESES AND ORTHOSIS)

As discussed in the General Introduction, the three main approaches of neural engineering to the rehabilitation of impaired motor functions are restoration, replacement, and neuromodulation. The focus of Part III is on function replacement, where the aim is to substitute a lost part of the motor apparatus with an artificial system. These chapters present different interfacing modalities for function replacement, such as brain and muscle interfacing. As for all book parts, the topics are thus collected with the focus on the application (replacement of function in this section) rather than on the principle on which such application is exploited.

Part III includes six chapters. The first two provide an overview of the use of invasive and noninvasive brain–computer interfaces (BCIs), where the aim is to extract control signals directly from the brain and use them to control external devices (e.g., a cursor on a screen or a wheelchair). Chapter 12, by Millán, gives a review of the development and use of noninvasive BCIs. Chapter 13, by Dremstrup and collaborators, also focuses on noninvasive BCIs; however, these authors provide a more specific overview of the use of movement-related cortical potentials (MRCPs) and their possible applications in BCIs. They provide evidence that MRCPs may serve as an alternative or supplementary control signal for BCIs.

The following three chapters provide an overview of restoration of upper limb function through control of prosthetic devices. Prostheses may be passive,

Introduction to Neural Engineering for Motor Rehabilitation, First Edition. Edited by Dario Farina, Winnie Jensen, and Metin Akay.

body powered, or externally powered by a battery. The prosthetic devices may be controlled by signals extracted from invasive or noninvasive sensors in the body. In Chapter 14, Graimann and Dietl provide a review of commercial prosthetic devices for the upper limb that are controlled through information extracted from sensors located on the body surface. Chapter 15, by Hargrove and collaborators, extends the view to include more advanced methods of prosthetic control (i.e., pattern recognition) and experimental prosthetic devices controlled by the body's own muscles following repositioning of the peripheral nerves (i.e., targeted reinnervation). In Chapter 16, Carparneto and collaborators present a different approach to the control of prosthetic devices, where the control information is obtained directly from invasive neural implants.

In the final chapter of Part III, Chapter 17, Pons and collaborators introduce exoskeletal robots (i.e., robots worn by patients) as a promising technology for patients where no functional recovery is possible. The chapter reviews how these technologies are applied to control the flow of mechanical power and the flow of information between human and robot. The chapter also identifies the technological bottlenecks in the development of such robots.

12

REVIEW CHAPTER

BRAIN–COMPUTER INTERFACES

JOSÉ DEL R. MILLÁN

Center for Neuroprosthetics, School of Engineering, Ecole Polytechnique Fédérale de Lausanne (EPFL), Lausanne, Switzerland

SUMMARY

This chapter introduces the field of brain–computer interfaces (BCIs), also called brain–machine interfaces (BMIs), which has seen impressive achievements over the past few years. A BCI monitors the user's brain activity, extracts specific features from the brain signals that reflect the intent of the subject, and translates them into actions. BCI technology offers a natural way to augment human capabilities by providing a new interaction link with the outside world and, thus, it is particularly relevant as an aid for patients with severe neuromuscular disabilities. The chapter first reviews the different kinds of brain signals that can be recorded as input for a BCI. Then, it discusses a series of principles to build efficient BCIs that are independent of the particular signal of choice. There follows a short overview of BCI attempts to improve the outcome of neurorehabilitation, especially for motor control in stroke patients. The chapter concludes by pinpointing some future research directions in the field of BCI.

Introduction to Neural Engineering for Motor Rehabilitation, First Edition. Edited by Dario Farina, Winnie Jensen, and Metin Akay.

INTRODUCTION

The use of brain signals for interaction as well as for controlling robots and prosthetic devices has gained increasing attention over recent years. This is a rapidly emerging field of multidisciplinary research called brain–computer interfaces (BCI), or brain–machine interfaces (BMI), which has seen impressive achievements over the past few years: prototypes for writing messages with a virtual keyboard (Birbaumer et al., 1999; Donchin et al., 2000; Millán, 2003; Obermaier et al., 2003; Müller and Blankertz, 2006), playing brain games (Millán, 2003; Krepki et al., 2007; Nijholt et al., 2008; Tangermann et al., 2008), and even controlling robots or wheelchairs (Millán et al., 2004; Müller-Putz et al., 2005; Galán et al., 2008; Velliste et al., 2008; Iturrate et al., 2009; Millán et al., 2009; Tavella et al., 2010). A BCI monitors the user's brain activity, extracts specific features from the brain signals that reflect the intent of the subject, and translates their intentions into actions—such as closing a prosthetic hand or selecting a letter from a virtual keyboard—without using the activity of any muscle or peripheral nerve. The central tenet of a BCI is the capability to distinguish between different patterns of brain activity, each being associated with a particular intention or mental task. Hence, adaptation is a key component of a BCI, because, on the one side, users must learn to modulate their brainwaves so as to generate distinct brain patterns, while, on the other side, machine learning techniques ought to discover the individual brain patterns characterizing the mental tasks executed by the user. This chapter provides an introduction to the field of BCIs, with a particular focus on principles suitable for motor rehabilitation. For a more detailed coverage of BCI, interested readers can refer to Nicolelis (2001), Wolpaw et al. (2002), and Dornhege et al. (2007).

The technology of BCIs offers a natural way to augment human capabilities by providing a new interaction link with the outside world. In this respect, it is particularly relevant as an aid for patients with severe neuromuscular disabilities, although it also opens up new possibilities in human–machine interaction for able-bodied people. Figure 12.1 shows the general architecture of a brain-actuated device such as neuroprostheses for motor restoration and recovery. Brain activity, electroencephalogram (EEG) signals in this example, is recorded with a portable device. These raw signals are first processed in order to extract some relevant features that are then passed on to a mathematical model (e.g., statistical classifiers or regression, neural networks). This model computes, after some training process where it finds the prototypical patterns of brain activity associated with each mental command, the user's intention that is transformed into an appropriate action to control the device. Finally, visual feedback, and maybe other kinds such as haptic stimulation (Cincotti et al., 2007) or intracortical microstimulation (O'Doherty et al., 2011), informs the subject about the performance of the neuroprosthesis so that he or she can learn appropriate mental control strategies and make rapid changes to achieve the task.

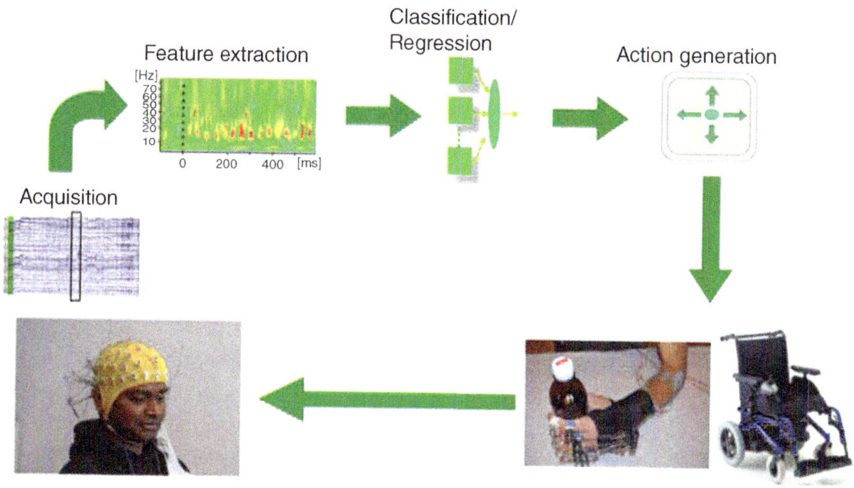

Figure 12.1. General architecture of a brain–computer interface (BCI) for controlling devices such as a hand orthosis or a motorized wheelchair. In this case the BCI measures electroencephalogram (EEG) signals recorded from electrodes placed on the subject's scalp.

This chapter is organized as follows. First, we review the different kind of brain signals that can be recorded as input for a BCI. Then, we discuss a series of principles to build efficient BCIs that are independent of the particular signal of choice. Next is a short overview of BCI attempts to improve the outcome of neurorehabilitation, especially for motor control in stroke patients. We conclude by pinpointing some future research directions in the field of BCI.

BCI SIGNALS

A BCI may monitor a variety of brain signals, such as electrical, magnetic, and metabolic. Magnetic fields can be recorded with magnetoencephalography (MEG), while brain metabolic activity—reflected in changes in blood flow—can be observed with positron emission tomography (PET), functional magnetic resonance imaging (fMRI), and near-infrared spectroscopy (NIRS). Unfortunately, MEG, PET, and fMRI require sophisticated devices that can be operated only in special facilities. Moreover, metabolic signals have long latencies and thus are less appropriate for interaction. Brain electrical activity, on the other hand, can be recorded with portable devices and can have an excellent time resolution—we can detect changes in brain activity at the millisecond range—and this makes it the natural candidate for a BCI. We can record the electrical brain activity invasively or noninvasively. The former technique employs microelectrode arrays implanted in the brain that record the activity of single neurons (single-unit activity, SUA) or from small neuronal populations (local field potentials, LFP). The overall concerted activity of neuronal populations can also be recorded (semi)invasively with electrodes placed on the surface of the brain (electrocorticography, ECoG). Noninvasive electroencephalographic (EEG) activity is recorded from electrodes placed on the scalp, measuring the synchronous activity of thousands of cortical neurons. LFP, ECoG and EEG signals are similar in nature, although at different levels—microscopic, mesoscopic, and macroscopic, respectively.

Invasive approaches (LFP and SUA) carry very detailed information, which may allow decoding of complex movement intentions (Velliste et al., 2008; O'Doherty et al., 2011; Carmena et al., 2003; Musallam et al., 2004; Hochberg et al., 2006; Santhanam et al., 2006; Ganguly and Carmena, 2009). However, we need to record from many electrodes and different areas to capture the global set of motor parameters, which are encoded in a rather distributed brain network. Also, implanted microelectrode arrays damage brain tissue, and so signals do not last long. Consequently, one area of active research is the design of safe biophysical interfaces that, in addition, should be ultralow power and wireless. On the other hand, being recorded on the scalp, noninvasive EEG signals suffer from a reduced spatial resolution and increased noise. In particular, EEG can be contaminated with muscular—such as subtle facial movements—and ocular artifacts that, ideally, must be filtered out. Alternatively, we must ensure that the selected control features are proper brain signals and not generated partially, or completely, by non-brain sources.

Regarding the spatial resolution of EEG, it can be greatly improved by means of spatial filters that estimate the electrical activity originated in radial sources immediately below each recording electrode. Such filters are known as surface Laplacian, which requires a large number of electrodes and a spherical spline model. Other examples of spatial filters are common average reference (CAR) and Laplacian, which compute an approximation of the surface Laplacian. The former removes, for each electrode, the average activity over all electrodes. The latter subtracts, for each electrode, the mean activity of its four nearest neighboring electrodes. The use of spatial filters should be an essential step in the design of an EEG-based BCI, as they have largely been shown to yield significant increases in performance. Another way to increase the spatial resolution of raw EEG is to estimate the cortical activity from the scalp EEG, thus unravelling the contributions of different small cortical areas that are picked up and mixed by a single scalp electrode (Cincotti et al., 2008; Grosse-Wentrup et al., 2009).

Analysis of human ECoG is attracting significant interest as, being semi-invasive, it blends the advantages of both approaches: high spatial resolution and simple surgery with minimal health risks. Although not largely used because of limited access to patients—who undergo neurosurgery for epilepsy or brain tumor—during short periods of time, initial results show high potential (Leuthardt et al., 2004; Schalk et al., 2008; Acharya et al., 2010; Pasley et al., 2012).

VOLUNTARY ACTIVITY VERSUS EVOKED POTENTIALS

A BCI can exploit signals associated with external sensory stimulation—such as visual flashes or auditory tones—or to endogenous voluntary decision processes—such as the onset of a movement. In the former case, the brain reacts with so-called evoked potentials. Two different evoked potentials have been widely explored in the field of EEG-based BCI, namely P300 (Donchin et al., 2000; Iturrate et al., 2009; Farwell and Donchin, 1988; Allison and Pineda, 2003; Schreuder et al., 2011) and steady-state visual evoked potential (SSVEP; Sutter, 1992; Gao et al., 2003). P300 is a potential evoked by an awaited infrequent event that appears at centroparietal locations along the midline of the scalp, independently of the sensory stimulation modality. As its name indicates, it is a positive wave peaking at around 300 ms after task-relevant stimuli. The amplitude of the P300 depends on the frequency of stimulus occurrence—less frequent stimuli produce larger responses—and task relevance. Visual evoked potentials (VEPs) reflect electrophysiological mechanisms underlying the processing of visual information in the brain and vary in response to changes in visual stimuli. Steady-state visual evoked potentials are VEPs induced by a stimulus repeated at a rate higher than 6 Hz and are composed of a series of components over the visual cortex whose frequencies are exact integer multiples of the stimulus frequency.

Evoked potentials are, in principle, easy to pick up. The necessity of external stimulation does, however, restrict the applicability of evoked potentials to a limited range of tasks. This is particularly the case when controlling robotics devices and, more important in the context of this book, for motor rehabilitation. In both cases, BCI has to rely upon brain correlates of voluntary mental activity, which users can modulate at different frequency ranges—or rhythms. Populations of neurons can form complex networks with feedback loops, which give rise to oscillatory activity. In general, the frequency of such oscillations becomes slower the larger the size of the synchronized neuronal assemblies. A particularly relevant rhythm can be recorded from the central region of the scalp overlying the sensorimotor cortex during the imagination of body movements. Correlates of imaginary movements can be recorded at any scale: microspopic, SUA and LFP; mesoscopic, ECoG; and macroscopic, EEG. Apart from their different degrees of spatial resolution, microscopic and mesoscopic signals also have a broader bandwith (up to 300–500 Hz) than macroscopic signals (normally, less than 100 Hz).

In the case of frequencies below 2 or 4 Hz, we can observe slow cortical potentials (SCPs). The analysis of SCP is usually done in the temporal domain. Notably, SCP measured with scalp electrodes was the basis of the first demonstration of a BCI for lock-in patients (Birbaumer et al., 1999), who learned to modulate its amplitude. The negative amplitude of SCP is related to the overall preparatory excitation level of a given cortical network—the more negative, the more active. Another example of SCP that can be observed in the EEG is the so-called Bereitschaftspotential (BP), or readiness potential—a slow negative shift over the contralateral motor cortical area starting around 400–500 ms before the onset of a movement. This makes the readiness potential particularly relevant for motor rehabilitation. However, being close to DC, detection of its presence in single trials seems to be elusive. Nevertheless, it has been shown that BP allows detection of self-paced reaching movement intention in single trials with humans, including stroke patients (Lew et al., 2012a). Readiness potentials are also present in intracranial LFP (Lew et al., 2012b) as well as in ECoG (Ball et al., 2009). In the former case, intracranial LFP recorded from the supplementary motor area (SMA) in an epileptic patient yielded high recognition rates in the detection of self-paced reaching movements well before onset. The intention to execute a self-paced movement has also been decoded from SUA in humans (Fried et al., 2011), where it was observed in progressive neuronal recruitment in the SMA over 1500 ms before subjects made the decision to move. As a last observation, experiments have shown that scalp SCP also carries enough information to decode three-dimensional arm trajectories (Bradberry et al., 2010).

Most spontaneous BCIs rely on variations of brain rhythms in higher frequencies than SCP. In the case of EEG and ECoG, imagination of limb movements gives rise to amplitude suppression (event-related desynchronization, ERD) of Rolandic μ (8–12 Hz) and β (13–28 Hz) rhythms over the contralateral limb motor cortical area (Pfurtscheller and Lopes da Silva, 1999). This imagination-related ERD shows different time courses in the two bands. In the μ band the ERD recovers to baseline level within a few seconds. On the other hand, the central β activity displays a short-lasting ERD followed by an amplitude increase (event-related synchronization, ERS). In particular, ERD has been largely exploited to build BCIs where subjects imagine movements of their left versus right hand or arm or of their upper limbs versus feet. References are too numerous to cite them all, but see Obermaier et al. (2003), Müller and Blankertz (2006), Millán et al. (2004), Müller-Putz et al. (2005), Tavella et al. (2010), Cincotti et al. (2007), Cincotti et al. (2008), Grosse-Wentrup et al. (2009), Leuthardt et al. (2004), Schalk et al. (2008), and Pfurtscheller and Neuper (2001). Apart from these "classical" imagined movements of hands, arms, and feet, researchers have also shown the feasibility of recognizing different wrist movements (Vuckovic and Sepulveda, 2008; Gu et al., 2009), which can be quite relevant for motor rehabilitation.

Finally, because of its higher spatial resolution and broader frequency range, ECoG seems also to carry information about different motor parameters such as movement kinematics (Acharya et al., 2010; Ball et al., 2009; Pistohl et al., 2008; Kubánek et al., 2009) that could provide finer control of neuroprostheses and could greatly help in motor rehabilitation. As mentioned before, decoding of complex movement intentions has been largely demonstrated with the use of intracranial recordings (LFP and SUA; Velliste et al., 2008; O'Doherty et al., 2011; Carmena et al., 2003; Musallam et al., 2004; Hochberg et al., 2006; Santhanam et al., 2006; Ganguly and Carmena, 2009).

MUTUAL LEARNING

A critical issue for the development of a BCI is training: that is, how users learn to operate the BCI. Some groups have demonstrated that some subjects can learn to control their brain activity through appropriate, but lengthy, training in order to generate fixed EEG patterns that the BCI transforms into external actions (Birbaumer et al., 1999; Wolpaw and McFarland, 2004). In this case the subject is trained over several months to modify the amplitude of a prespecified parameter of their brain signals, from EEG to fMRI. An alternative approach is to relieve the user from any training by using machine learning techniques to rapidly find individual patterns of brain activity associated with the mental commands the user wants to convey (Blankertz et al., 2006). The drawback of this approach is that the BCI must be recalibrated at the beginning of every session. Most BCI systems lie in between these two extreme approaches as they adopt a mutual learning principle, where the user and the BCI are coupled together and adapt to each other. Different groups have shown that, in this case, humans can learn to operate the brain-actuated device very rapidly, in a few hours normally split between a few days (Tangermann et al., 2008; Millán et al., 2004; Schalk et al., 2008; Bradberry et al., 2010).

Mutual learning facilitates and accelerates users' training process thanks to the use of statistical machine learning techniques that both select relevant, stable features and build optimal models to decode the user's intention. Feature selection yields user-specific brain components—normally spatiofrequency features for rhythmic activity or spatiotemporal for evoked potentials—that maximize the separability between mental commands (Millán et al., 2002; Galán et al., 2007). In addition, because of the nonstationary nature of brain signals, selected features must also be stable over time. These initial features represent those brain components that the user can naturally modulate and, via feedback received during online BCI training, learn to control quickly and voluntarily. Selection of stable features can be enhanced by removing nonstationary brain sources (von Bünau et al., 2010).

Although there are reports of users who keep a stable level of performance over months and even years, BCIs suffer from the natural variability of brain signals due to changes in background activity and learning. This calls for the use of online adaptation techniques to keep the BCI tuned to drifts in the signals (Millán, 2004; Buttfield et al., 2006; Shenoy et al., 2006; Vidaurre et al., 2006, 2011; Perdikis et al., 2011).

Adaptation should play a key role in neurorehabilitation. Indeed, since neurorehabilitation is based on plasticity—the capability of the brain to generate new connectivity patterns among neurons so as to encode new skills or recover from damage—optimal discriminant features (i.e., those encoding the parameters of the motor tasks we want to recover) evolve and it is necessary to track them. However, it remains an open question what is the optimal way to do adaptation in the framework of neurorehabilitation, since we need to trade off speed of adaptation and stability of the classifier to promote activity-dependent plasticity.

CONTEXT-AWARE BCI

Independently of the kind of BCI signal (SUA, LFP, ECoG, or EEG), paradigm (voluntary activity or evoked potential) and learning approach, users cannot sustain high levels of performance over long periods of time. This is mainly due to the natural variability of brain signals. Furthermore, EEG-based approaches yield low throughput. As a consequence, it seems that current state-of-the-art BCI technology is insufficient for full dexterous control of complex applications such as neuroprostheses. Nevertheless, we can cope with those limitations and achieve reliable mental control by designing context-aware BCIs (Chavarriaga and Millán, 2010). Such a BCI collects information about the state of the device (e.g., position and velocity of the cursor or neuroprosthesis), as well as its environment (e.g., icons on the screen or potential targets and obstacles perceived by the prosthesis' sensors), and combines it with the user's mental commands. In this approach, the smart brain-controlled device interprets and executes the mental commands with respect to the context, thus enabling the performance of complex tasks even with a reduced scattered number of commands. Furthermore, any eventual decrease in mental control is compensated, up to a certain degree, by the smart device. Examples of context-aware BCIs are neuroprostheses such as robots and wheelchairs (Millán et al., 2004; Galán et al., 2008; Millán et al., 2009; H.K. Kim et al., 2006; Tonin et al., 2010), as well as smart virtual keyboards (Wills and MacKay, 2006; Williamson et al., 2009) and other assistive-technology (AT) software with predictive capabilities. Remarkably, in the case of control of complex devices such as a telepresence robot, this approach yields performance similar to manual control despite the fact a BCI is not a perfect control channel. This was the case not only for expert BCI users, but also for novice BCI subjects (Tonin et al., 2010) and users with physical disabilities (Tonin et al., 2011).

A common form of context awareness is shared control, a technique widely used in robotics, where the user and the smart device share responsibilities in choosing the final action to be executed that best matches the user's intent. For example, it requires quite precise control to drive a wheelchair in a home environment (scattered with chairs, tables, doors, etc.). Also, it can take a long time to navigate between rooms. In a shared-control BCI, the user delivers high-level mental commands such as left, right, and forward, which the wheelchair interprets based on the contextual information from its sensors to compute the actual motor commands (speed of the motors) to travel smooth trajectories and avoid obstacles. In such a way, shared control can facilitate the operation of the device by inferring goals, reducing the cognitive workload (the user does not need to care about all low-level details), inhibiting pointless mental commands (e.g., driving zig-zag), preventing critical situations (e.g., collisions), and determining meaningful motion sequences (e.g., for an arm neuroprosthesis).

BCI-BASED MOTOR REHABILITATION

In addition to motor substitution, where a BCI bypasses a central nervous injury to control a neuroprosthesis, BCI technology can also facilitate motor rehabilitation. This is an emerging area of research and application, especially for stroke. The rationale for BCI-based motor rehabilitation is threefold. First, and perhaps more pragmatic, contrary to most other well-established rehabilitation paradigms requiring some degree of residual mobility, BCI technology can assist stroke patients even in the absence of residual motor activity (Wang et al., 2010). Second, motor rehearsal (or motor imagery) is often used in conventional therapies (Nilsen et al., 2010), and BCI is the right tool to assess patient compliance with therapist's instructions. Third, and most important, it can promote neuroplasticity—activity-dependent plasticity—as it provides online feedback about brain patterns associated with motor intention. In other words, BCI-based training provides patients with self-induced sensory feedback—whether visual only or even together with somatosensory and proprioceptive—for the paralyzed limb.

A critical component of a BCI for rehabilitation is the selection of the control features. Some studies follow standard principles where the selected features discriminate the most between motor imagery and rest, independently of their localization (Buch et al., 2008). However, if BCI is intended to guide neuroplasticity in order to recover "normal" motor activity of the paralyzed limb, it seems logical that the selected control features are constrained to those leading to "normal" motor cortical activity (Daly et al., 2009; Ang et al., 2010; Broetz et al., 2010; Mattia et al., 2012). An option is to select features only from the ipsilesional motor hemisphere (contralateral to the imagined movement of the affected limb).

Despite some positive preliminary results, there is no solid evidence that BCI-based rehabilitation yields significant improvements compared with control conditions (for a review see Mattia et al., 2012). In the European project TOBI (http://www.tobi-project.org), we are conducting such a comparative study. Importantly, the approach has been designed in close cooperation with rehabilitation professionals to make it operational in daily practice. Results with 10 unilateral stroke patients show that, after one month of training, EEG patterns in the motor areas of the ipsilesional hemisphere, and associated with the motor imagery task, are more prominent and are accompanied by an improvement of the functional motor scales (Fugl–Meyer score related to the affected upper limb). Finally, the BCI-based rehabilitation approach has received a high level of acceptability among the patients and the professionals. For more details about the approach and results, see Mattia et al. (2012).

FUTURE DIRECTIONS

Current BCI technology, in particular EEG-based, enables the operation of simple brain-actuated devices over short periods of time. Increasingly complex prototypes are being developed and demonstrated. No doubt, this will represent an important achievement for motor-disabled people. Yet, robust and natural brain interaction with more complex devices, such as neuroprostheses, over long periods of time is a major challenge.

A related issue is to develop practical BCI technologies that can be brought out of the lab and into real-world applications in order to improve the lives of countless disabled individuals. For this we need to confirm the benefit of BCI for disabled people outside laboratory conditions. Only a few studies have been conducted so far. In addition, we have to design novel hardware and sensors as well as principled approaches.

One of the main barriers to wide use of BCI technology is the limitation of today's electrodes, both invasive and noninvasive. In the former case, they damage brain tissue and do not last long before they lose signal. Consequently, one area of active research is the design of safe biophysical interfaces that, in addition, should be ultralow power and wireless. In the latter case, EEG signals suffer from a reduced spatial resolution and increased noise when measurements are taken on the scalp. Significant research is taking place worldwide on algorithms to improve EEG analysis and decoding. Another of the shortcomings of today's EEG technology, which is essentially the same as 30 to 50 years ago, is the need for conductive gel to reduce impedance. Success of BCI, although still modest, is creating a market for new sensors that will make the use of EEG-based systems much easier and robust over long periods of time. Examples of this new technology are dry electrodes that do not require any gel and can be integrated into aesthetic helmets (Guger et al., 2012), as well as skin sensors that can remain operational for months, if not years (D.-H. Kim et al., 2012). An associated challenge, however, is to develop new techniques to analyze the signals they will measure, which will probably differ from conventional electrodes.

Apart from addressing new sensors, and the associated signal processing techniques, we should also develop new principles to make BCI work effectively for disabled people. In this respect, it seems natural to combine BCI with existing assistive technologies (ATs; Millán et al., 2010). This is called a hybrid BCI: a combination of different signals including at least one BCI channel (Millán et al., 2010; Pfurtscheller et al., 2010). Thus, it could be a combination of two BCI channels but, more important, also a combination of BCI and other residual biosignals (such as electromyographic, EMG) or special AT input devices (e.g., joysticks, switches). More recently, we have introduced a general architecture and a common software framework for hybrid BCIs (Müller-Putz et al., 2011). Importantly, shared control (context-aware techniques) is a key component of the architecture of a hybrid BCI, as it will shape the closed-loop dynamics between the user and the brain-actuated device so that tasks can be performed as easily and effectively as possible. Also, it is worth noting that hybrid BCIs are particularly relevant for motor neurorehabilitation as the final goal is to recover motor functions. Therefore, patients should utilize both EEG and muscular activity.

REFERENCES

Acharya S., Fifer M.S., Benz H.L., Crone N.E., and Thakor N.V. (2010). Electrocorticographic amplitude predicts finger positions during slow grasping motions of the hand. *J Neural Eng* 7:046002.

Allison B.Z. and Pineda J.A. (2003). ERPs evoked by different matrix sizes: implications for a brain computer interface (BCI) system. *IEEE Trans Neural Syst Rehabil Eng* 11:110–113.

Ang K.K., Guan C., Chua K.S., Ang B.T., Kuah C., Wang C., Phua K.S., Chin Z.Y., and Zhang H. (2010). Clinical study of neurorehabilitation in stroke using EEG-based motor imagery brain–computer interface with robotic feedback. *Proc. 32nd Annual Int. Conf. IEEE Eng. Med. Biol. Soc.*

Ball T., Schulze-Bonhage A., Aertsen A., and Mehring C. (2009). Differential representation of arm movement direction in relation to cortical anatomy and function. *J Neural Eng* 6:016006.

Birbaumer N., Ghanayim N., Hinterberger T., Iversen I., Kotchoubey B., Kübler A., Perelmouter J., Taub E., and Flor H. (1999). A spelling device for the paralysed. *Nature* 398:297–298.

Blankertz B., Dornhege G., Krauledat M., Müller K.-R., Kunzmann V., Losch F., and Curio G. (2006). The Berlin brain–computer interface: EEG-based communication without subject training. *IEEE Trans Neural Syst Rehabil Eng* 14:147–152.

Bradberry T.J., Gentili R.J., and Contreras-Vidal J.L. (2010). Reconstructing three-dimensional hand movements from noninvasive electroencephalographic signals. *J Neurosci* 30:3432–3437.

Broetz D., Braun C., Weber C., Soekadar S.R., Caria A., and Birbaumer N. (2010). Combination of brain–computer interface training and goal-directed physical therapy in chronic stroke: a case report. *Neurorehabil Neural Repair* 24:674–679.

Buch E., Weber C., Cohen L.G., Braun C., Dimyan M.A., Ard T., Mellinger J., Caria A., Soekadar S., Fourkas A., and Birbaumer (2008). Think to move: a neuromagnetic brain–computer interface (BCI) system for chronic stroke. *Stroke* 39:910–917.

Buttfield A., Ferrez P.W., and Millán J.d.R. (2006). Towards a robust BCI: error recognition and online learning. *IEEE Trans Neural Syst Rehabil Eng* 14:164–168.

Carmena J.M., Lebedev M.A., Crist R.E., O'Doherty J.E., Santucci D.M., Dimitrov D.F., Patil P.G., Henriquez C.S., and Nicolelis M.A.L. (2003). Learning to control brain–machine interface for reaching and grasping by primates. *PLoS Biol* 1:192–208.

Chavarriaga R. and Millán J.d.R. (2010). Context-aware brain–computer interfaces. *PerAda Magazine* (DOI: 10.2417/2201006.003009).

Cincotti F., Kauhanen L., Aloise F., Palomäki T., Caporusso N., Jylänki P., Mattia D., Babiloni F., Vanacker G., Nuttin M., Marciani M.G., and Millán J.d.R. (2007). Vibrotactile feedback for brain–computer interface operation. *Comput Intell Neurosci* 2007:48937.

Cincotti F., Mattia D., Aloise F., Bufalari S., Astolfi L., De Vico Fallani F., Tocci A., Bianchi L., Marciani M.G., Gao S., Millán J.d.R., and Babiloni F. (2008). High-resolution EEG techniques for brain–computer interface applications. *J Neurosci Methods* 167:31–34.

Daly J.J., Cheng R., Rogers J., Litinas K., Hrovat K., and Dohring M. (2009). Feasibility of a new application of noninvasive Brain Computer Interface (BCI): a case study of training for recovery of volitional motor control after stroke. *J Neurol Phys Ther* 33:203–211.

Donchin E., Spencer K.M., and Wijesinghe R. (2000). The mental prosthesis: assessing the speed of a P300-based brain–computer interface. *IEEE Trans Neural Syst Rehabil Eng* 8:174–179.

Dornhege G., Millán J.d.R., Hinterberger T., McFarland D.J., and Müller K.-R. Eds. (2007). *Toward Brain–Computer Interfacing*. Cambridge, MA: MIT Press.

Farwell L.A. and Donchin E. (1988). Talking off the top of your head: toward a mental prosthesis utilizing event related brain potentials. *Electroenceph Clin Neurophysiol* 70:510–523.

Fried I., Mukamel R., and Kreiman G. (2011). Internally generated preactivation of single neurons in human medial frontal cortex predicts volition. *Neuron* 69:548–562.

Galán F., Ferrez P.W., Oliva F., Guardia J., and Millán J.d.R. (2007). Feature extraction for multi-class BCI using canonical variates analysis. *Proc IEEE Int Symp Int Sig Process*.

Galán F., Nuttin M., Lew E., Ferrez P.W., Vanacker G., Philips J., and Millán J.d.R. (2008). A brain-actuated wheelchair: asynchronous and noninvasive brain–computer interfaces for continuous control of robots. *Clin Neurophysiol* 119:2159–2169.

Ganguly K. and Carmena J.M. (2009). Emergence of a stable cortical map for neuroprosthetic control. *PloS Biol* 7:e1000153.

Gao X., Dingfeng X., Cheng M., and Gao S. (2003). A BCI-based environmental controller for the motion-disabled. *IEEE Trans Neural Syst Rehabil Eng* 11:137–140.

Grosse-Wentrup M., Liefhold C., Gramann K., and Buss M. (2009). Beamforming in noninvasive brain–computer interfaces. *IEEE Trans Biomed Eng* 56:1209–1219.

Gu Y., Dremstrup K., and Farina D. (2009). Single-trial discrimination of type and speed of wrist movements from EEG recordings. *Clin Neurophysiol* 120:1596–1600.

Guger C., Krausz G., Allison B.Z., and Edlinger G. (2012). Comparison of dry and gel based electrodes for P300 brain–computer interfaces. *Front Neurosci* 6:60 (DOI: 10.3389/fnins.2012.00060).

Hochberg L.R., Serruya M.D., Friehs G.M., Mukand J.A., Saleh M., Caplan A.H., Branner A., Chen D., Penn R.D., and Donoghue J.P. (2006). Neuronal ensemble control of prosthetic devices by a human with tetraplegia. *Nature* 442:164–171.

Iturrate I., Antelis J., Kübler A., and Minguez J. (2009). Non-invasive brain-actuated wheelchair based on a P300 neurophysiological protocol and automated navigation. *IEEE Trans Robotics* 25:614–627.

Kim D.-H., Lu N., Ma R. et al. (2012). Epidermal electronics. *Science* 333:838–843.

Kim H.K., Biggs S.J., Schloerb D.W., Carmena J.M., Lebedev M.A., Nicolelis M.A.L., and Srinivasan M.A. (2006). Continuous shared control for stabilizing reaching and grasping with brain–machine interfaces. *IEEE Trans Biomed Eng* 53:1164–1173.

Krepki R., Blankertz B., Curio G., and Müller K.-R. (2007). The Berlin brain–computer interface (BBCI)—towards a new communication channel for online control in gaming applications. *J Multimedia Tools Appl* 33:73–90.

Kubánek J., Miller K.J., Ojemann J.G., Wolpaw J.R., and Schalk G. (2009). Decoding flexion of individual fingers using electrocorticographic signals in humans. *J Neural Eng* 6:066001.

Leuthardt E.C., Schalk G., Wolpaw J.R., Ojemann J.G., and Moran D.W. (2004). A brain–computer interface using electrocorticographic signals in humans. *J Neural Eng* 1:63–71.

Lew E., Chavarriaga R., Silvoni S., and Millán J.d.R. (2012a). Detection of self-paced reaching movement intention from EEG signals. *Front Neuroeng* 5:13 (DOI: 10.3389/fneng.2012.00013).

Lew E., Chavarriaga R., Zhang H., Seeck M., and Millán J.d.R. (2012b). Self-paced movement intention detection from human brain signals: invasive and non-invasive EEG. *Proc 34th Annual Int Conf IEEE Eng Med Biol Soc.*

Mattia D., Pichiorri F., Molinari M., and Rupp R. (2012). Brain computer interface for hand motor function restoration and rehabilitation. In: *Towards Practical Brain–Computer Interfaces* (Allison B., Dunne S., Leeb R., Millán J.d.R., and Nijholt A., Eds.). New York: Springer-Verlag.

Millán J.d.R. (2003). Adaptive brain interfaces. *Commun ACM* 46:74–80.

Millán J.d.R. (2004). On the need for on-line learning in brain–computer interfaces. *Proc Int Joint Conf Neural Networks.*

Millán J.d.R., Franzé M., Mouriño J., Cincotti F., and Babiloni F. (2002). Relevant EEG features for the classification of spontaneous motor-related tasks. *Biol Cybern* 86:89–95.

Millán J.d.R., Renkens F., Mouriño J., and Gerstner W. (2004). Noninvasive brain-actuated control of a mobile robot by human EEG. *IEEE Trans Biomed Eng* 51:1026–1033.

Millán J.d.R., Galán F., Vanhooydonck D., Lew E., Philips J., and Nuttin M. (2009). Asynchronous non-invasive brain-actuated control of an intelligent wheelchair. *Proc 31st Annual Int Conf IEEE Eng Med Biol Soc.*

Millán J.d.R., Rupp R., Müller-Putz G.R., Murray-Smith R., Giugliemma C., Tangermann M., Vidaurre C., Cincotti F., Kübler A., Leeb R., Neuper C., Müller K.-R., and Mattia D. (2010). Combining brain–computer interfaces and assistive technologies: state-of-the-art and challenges. *Front Neurosci* 4:161 (DOI:10.3389/fnins.2010.00161).

Müller K-R. and Blankertz B. (2006). Toward noninvasive brain–computer interfaces. *IEEE Signal Process Mag* 23:125–128.

Müller-Putz G.R., Scherer R., Pfurtscheller G., and Rupp R. (2005). EEG-based neuroprosthesis control: a step towards clinical practice. *Neurosci Lett* 382:169–174.

Müller-Putz G.R., Breitwieser C., Cincotti F., Leeb R., Schreuder M., Leotta F., Tavella M., Bianchi L., Kreilinger A., Ramsay A., Rohm M., Sagebaum M., Tonin L., Neuper C., and Millán J.d.R. (2011). Tools for brain–computer interaction: a general concept for a hybrid BCI. *Front Neuroinform* 5:30 (DOI: 10.3389/fninf.2011.00030).

Musallam S., Corneil B.D., Greger B., Scherberger H., and Andersen R.A. (2004). Cognitive control signals for neural prosthetics. *Science* 305:162–163.

Nicolelis M.A.L. (2001). Actions from thoughts. *Nature* 409:403–407.

Nijholt A., Tan D., Allison B., Millán J.d.R., Graimann B., and Jackson M.M. (2008). Brain–computer interfaces for HCI and games. *Proc ACM CHI 2008*.

Nilsen D.M., Gillen G., and Gordon A.M. (2010). Use of mental practice to improve upper-limb recovery after stroke: a systematic review. *Am J Occup Ther* 64:695–708.

O'Doherty J.E., Lebedev M.A.,. Ifft P.J, Zhuang K.Z., Shokur S., Bleuler H., and Nicolelis M.A.L. (2011). Active tactile exploration using a brain–machine–brain interface. *Nature* 479:228–231.

Obermaier B., Müller G.R., and Pfurtscheller G. (2003). Virtual keyboard controlled by spontaneous EEG activity. *IEEE Trans Neural Syst Rehabil Eng* 11:422–426.

Pasley B.N., David S.V., Mesgarani N., Flinker A., Shamma S.A., Crone N.E., Knight R.T., and Chang E.F. (2012). Reconstructing speech from human auditory cortex. *PLoS Biol* 10:e1001251.

Perdikis S., Tavella M., Leeb R., Chavarriaga R., and Millán J.d.R. (2011). A supervised recalibration protocol for unbiased BCI. *Proc 5th Int Brain–Computer Interface Conf.*

Pfurtscheller G. and Lopes da Silva F.H. (1999). Event-related EEG/MEG synchronization and desynchronization: basic principles. *Electroenceph Clin Neurophysiol* 110:1842–1857.

Pfurtscheller G. and Neuper C. (2001). Motor imagery and direct brain–computer communication. *Proc IEEE* 89:1123–1134.

Pfurtscheller G., Allison B.Z., Bauernfeind G., Brunner C., Solis Escalante T., Scherer R., Zander T.O., Müller-Putz G., Neuper C., and Birbaumer N. (2010). The hybrid BCI. *Front Neurosci* 4:42 (doi:10.3389/fnpro.2010.00003).

Pistohl T., Ball T., Schulze-Bonhage A., Aertsen A., and Mehring C. (2008). Prediction of arm movement trajectories from ECoG-recordings in humans. *J Neurosci Methods*,167:105–114.

Santhanam G., Ryu S.I., Yu B.M., Afshar A., and Shenoy K.V. (2006). A high-performance brain–computer interface. *Nature* 442:195–198.

Schalk G., Miller K.J., Anderson N.R., Wilson J.A., Smyth M.D., Ojemann J.G., Moran D.W., Wolpaw J.R., and Leuthardt E.C. (2008). Two-dimensional movement control using electrocorticographic signals in humans. *J Neural Eng* 5:75–84.

Schreuder M., Rost T., and Tangermann M. (2011). Listen, you are writing! Speeding up online spelling with a dynamic auditory BCI. *Front Neurosci* 5:112.

Shenoy P., Krauledat M., Blankertz B., Rao R.P.N., and Müller K.-R. (2006). Towards adaptive classification for BCI. *J Neural Eng* 3:R13–R23.

Sutter E.E. (1992). The brain response interface: communication through visually-induced electrical brain response. *J Microcomput Appl* 15:31–45.

Tangermann M.W., Krauledat M., Grzeska K., Sagebaum M., Vidaurre C., Blankertz B., and Müller K.-R. (2008). Playing pinball with non-invasive BCI. *Proc NIPS*.

Tavella M., Leeb R., Rupp R., and Millán J.d.R. (2010). Towards natural non-invasive hand neuroprostheses for daily living. *Proc 32nd Annual Int Conf IEEE Eng Med Biol Soc*.

Tonin L., Leeb R., Tavella M., Perdikis S., and Millán J.d.R. (2010). The role of shared-control in BCI-based telepresence. *Proc 2010 IEEE Int Conf Systems, Man, Cybernetics*.

Tonin L., Carlson T., Leeb R., and Millán J.d.R. (2011). Brain-controlled telepresence robot by motor-disabled people. *Proc 33rd Annual Int Conf IEEE Eng Med Biol Soc*.

Velliste M., Perel S., Spalding M.C., Whitford A.S., and Schwartz A.B. (2008). Cortical control of a prosthetic arm for self-feeding. *Nature* 453:1098–1101.

Vidaurre C., Schögl A., Cabeza R., Scherer R., and Pfurtscheller G. (2006). A fully on-line adaptive BCI. *IEEE Trans Biomed Eng* 53:1214–1219.

Vidaurre C., Kawanabe M., von Bünau P., Blankertz B., and Müller K.-R. (2011). Toward unsupervised adaptation of LDA for brain–computer interfaces. *IEEE Trans Biomed Eng* 58:587–597.

von Bünau P., Meinecke F.C., Scholler S., and Müller K.-R. (2010). Finding stationary brain sources in EEG data. *Proc 32nd Annual Int Conf IEEE Eng Med Biol Soc*.

Vuckovic A. and Sepulveda F. (2008). Delta band contribution in cue based single trial classification of real and imaginary wrist movements. *Med Biol Eng Comput* 46:529–539.

Wang W., Collinger J.L., Perez M.A., et al. (2010). Neural interface technology for rehabilitation: exploiting and promoting neuroplasticity. *Phys Med Rehabil Clin N Am* 21:157–178.

Williamson J., Murray-Smith R., Blankertz B., Krauledat M., and Müller K.-R. (2009). Designing for uncertain, asymmetric control: interaction design for brain–computer interfaces. *Int J Hum Comput Stud* 67:827–841.

Wills S. and MacKay D. (2006). DASHER—an efficient writing system for brain–computer interfaces? *IEEE Trans Neural Syst Rehab Eng* 14:244–246.

Wolpaw J.R. and McFarland D.J. (2004). Control of a two-dimensional movement signal by a noninvasive brain–computer interface in humans. *PNAS* 101:17849–17854.

Wolpaw J.R., Birbaumer N., McFarland D.J., Pfurtscheller G., and Vaughan T.M. (2002). Brain–computer interfaces for communication and control. *Clin Neurophysiol* 113:767–791.

13

FOCUSED CHAPTER

MOVEMENT-RELATED CORTICAL POTENTIALS AND THEIR APPLICATION IN BRAIN–COMPUTER INTERFACING

KIM DREMSTRUP,[1] YING GU,[1] OMAR FEIX DO NASCIMENTO,[1] AND DARIO FARINA[1,2]

[1]*Center for Sensory–Motor Interaction, Department of Health Science and Technology, Aalborg University, Aalborg, Denmark*

[2]*Department of Neurorehabilitation Engineering, Bernstein Center for Computational Neuroscience, University Medical Center Göttingen, Georg-August University, Göttingen, Germany*

SUMMARY

We provide an overview of basic research on movement-related cortical potentials (MRCPs) and their potential applications in brain–computer interfacing (BCI). MRCPs have been studied for decades with the scope of understanding brain motor functions. More recently, it has been shown that the characteristics of the MRCPs, averaged over trials, are associated with movement-related parameters, such as the speed of a task, during both the execution and the imagination of movements. Based on these findings, a series of studies have tested the hypothesis that the identification of these characteristics could be done in single trials, thus allowing the use of MRCPs in BCI systems. The average single-trial

Introduction to Neural Engineering for Motor Rehabilitation, First Edition. Edited by Dario Farina, Winnie Jensen, and Metin Akay.

misclassification rate of MRCPs generated by imagination of the same task at two speeds (two-class problem) in healthy subjects has been reported to be ~20%. Moreover, it has been shown that the performance was only slightly degraded (error of ~30%) when MRCPs were classified in amputees and in patients suffering of amyotrophic lateral sclerosis (ALS). The results from these studies provide evidence that MRCPs could serve as an alternative or supplementary control signal for BCIs.

INTRODUCTION

Brain–computer interface (BCI) systems aim at providing a nonmuscular communication and control channel to patients with severe disabilities (Wolpaw et al., 2002) or at promoting neuroplasticity (Daly and Wolpaw, 2008). BCI technology is based on the ability of individuals to voluntarily and reliably produce brain signals in forms of electrical, magnetic, or metabolic changes. Currently, the large majority of BCI research and applications in humans are based on electroencephalogram (EEG) signals due to their high temporal resolution and the low cost, low risk, and acceptable portability of detection systems. Motor imagery is the most common approach to producing EEG changes in EEG-based BCI research. Extensive work has been devoted to distinguishing movements from different joints with the purpose of increasing the number of control commands in BCIs (McFarland et al., 2000; Neuper et al., 1999, 2003; Pfurtscheller et al., 1997, 2006). An alternative approach, which we discuss in this chapter, consists in distinguishing movement-related parameters, such as speed, for the same imagined task. Although this approach may seem doomed by an intrinsically greater challenge for the classification algorithms, it has the advantages of providing a more intuitive way of control for the user and of increasing the potential number of commands. Therefore, in a series of studies, we have analyzed the characteristics of movement-related cortical potentials (MRCPs) and explored their potential application to BCI.

Movement-related cortical potentials are the EEG evidence of motor cortical involvement during movement preparation and execution (Kornhuber and Deecke, 1965). They have been studied for decades, mostly in motor control physiology and psychophysiology (Libet et al., 1983a,b; Klein, 2002; Jahanshahi and Hallett, 2003; Eagleman, 2004). It is known that MRCPs occur in association with both executed and imaginary movements and that their magnitude and latency are modulated by movement parameters (Slobounov et al., 1998; Romero et al., 2000; do Nascimento et al., 2005, 2006; Nielsen et al., 2006). This modulation, especially with motor imagination, has led to development of single-trial classification methods of MRCPs (Farina et al., 2006, 2007; do Nascimento and Farina, 2006, 2008; Gu et al., 2006, 2009a–c; Gu and Dremstrup, 2011; Niazi et al., 2011, 2012).

This chapter describes the characteristics of MRCPs and how these characteristics are modulated by movement parameters, such as speed and torque.

These results are discussed from the viewpoint of applying MRCPs generated during imagined movements of the same joint as control signals for BCIs. Evidence for this possibility is provided with results obtained online and offline in several studies, including in patients with severe motor impairments.

CHARACTERISTICS OF MOVEMENT-RELATED CORTICAL POTENTIALS

Kornhuber and Deecke (1964) first observed a negative slow potential measured over the scalp in correspondence to the motor and premotor areas, appearing more than one second before a voluntary movement. This slow wave was called Bereitschaftspotential (BP) and was interpreted as the representation of the cortical activity related to the preparation for a movement. Further investigation revealed that the BP is actually part of a longer potential that consists of a negative and a positive phase (Fig. 13.1) and is now referred to as a MRCP.[1,2] The MRCP reflects not only the movement planning but also the monitoring of motor execution and performance (Deecke et al., 1976; Kristeva et al., 1979; Grünewald and Grünewald-Zuberbier, 1983). These processes are identified in subcomponents of the MRCP, called readiness potential (RP, corresponding to the original BP), motor potential (MP), and movement-monitoring potential (MMP) (Fig. 13.1, panel IV). Some other nomenclature and subdivision of MRCPs can be found in the literature, according to the specific motor-physiologic aspect being assessed; see Shibasaki and Hallet (2006) for a review.

Movement-related cortical potentials occur in connection with practically any movement that can be self-paced under voluntary control, such as hands (Ikeda and Shibasaki, 1992), fingers (Shibasaki et al., 1980; Slobounov et al., 1999), arms (Siemionow et al., 2000), legs, ankle (do Nascimento et al., 2005), feet (Shibasaki et al., 1981; Böcker et al., 1994; Terada et al., 1999), toes (Boschert and Deecke, 1986), mouth, lips (Yoshida et al., 1999), tongue (Ikeda et al., 1995), jaw (Nakajima et al., 1991), and eyes (Jentzsch and Leuthold, 2002). The main generators of MRCPs are the supplementary motor area (SMA) and primary motor cortex (M1). The early part of MRCP (i.e. RP) is

[1]In Figure 13.1, panel I, the *simple movement* is to pull down a toggle switch with the index finger. The *motor sequence* is to pull down a toggle switch with the index finger, move the hand and forearm to the right to turn successively on and off a button located 20 cm to the right side of the switch, and finally return to the starting position with an inverse movement of the hand and forearm.

[2]In Figure 13.1, panel II, the *discrete movement* is to apply index finger pressure on a load cell (either as accurately or as fast as possible) until the pressure is equal to 50% of the maximal voluntary force and hold it for 5 s. The *repetitive movement* is to vary index finger pressure on a load cell between 10% and 50% of the maximal voluntary force (either as accurately or as fast as possible) in a 5-s time period.

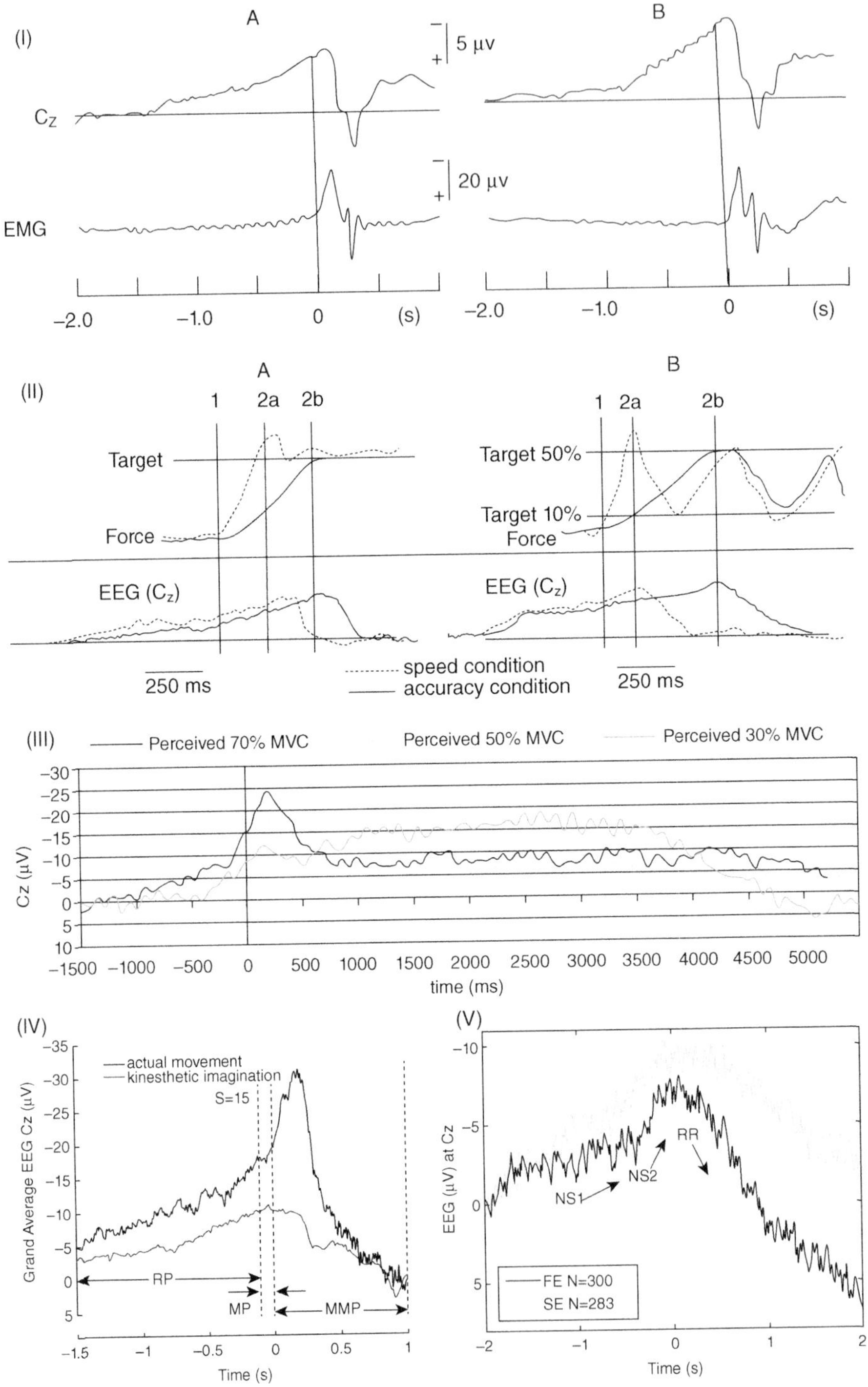
(I)
A
B
Cz
EMG
5 μv
20 μv
−2.0
−1.0
0
(s)
(II)
1
2a
2b
Target
Force
Target 50%
Target 10%
EEG (Cz)
250 ms
speed condition
accuracy condition
(III)
Perceived 70% MVC
Perceived 50% MVC
Perceived 30% MVC
Cz (μV)
time (ms)
(IV)
actual movement
kinesthetic imagination
S=15
Grand Average EEG Cz (μV)
RP
MP
MMP
Time (s)
(V)
EEG (μV) at Cz
NS1
NS2
RR
FE N=300
SE N=283
Time (s)

symmetrically distributed over the SMA, whereas MP and MMP evolve to a lateralized distribution over M1 if a unilateral movement occurs. Hence, RP reflects unspecific motor planning and its development has been shown to be an unconscious process, with the awareness of a planned movement appearing only at about 300 ms after RP initiation (start of negative deflection; Libet et al., 1983a). It has been claimed that an individual is not capable of controlling the planning of a motor action, as an unconscious process, but rather is capable of vetoing its the execution (at around MP, when the motor processes take place at M1; Libet et al., 1983b).

For hands and fingers, maximal MRCP amplitudes are detected in the portion of the scalp contralateral to the movement effector, whereas for distal portions of lower limbs (toes, ankles) maximal MRCP amplitudes are detected in the ipsilateral portion. The latter, called paradoxical ipsilaterality, is the result of a potential dipole projection caused by the neuronal activity of the toe–foot–ankle region in M1, which is somatotopically located in the medial brain fissure (Brunia and van den Bosch, 1984).

The latency and magnitude of MRCPs can be influenced by various factors, such as force exerted, speed, perceived effort, precision, training, repetition rate, complexity, anticipation, and concentration (Birbaumer et al., 1990; Shibasaki and Hallet, 2006); see Figure 13.1. Some of these factors reduce the

Figure 13.1. Average MRCPs influenced by various factors. Panels I–III are from actual movements (see footnotes 1 and 2 for definitions); panel IV is a comparison between actual movement and motor imagery; panel V is from motor imagery. Amplitudes are shown with negativity upward. Actual and imaginary movement onsets occur at time instance 0 s (except in panel II). (I) Grand average MRCPs from Cz in (A) a simple movement and (B) a motor sequence (modified from Simonetta et al., 1991); see footnote 1. The motor sequence elicited a BP with earlier onset and larger amplitude than in the simple movement. (II) Examples of force production and MRCP from Cz in speed and accuracy conditions during (A) discrete and (B) repetitive movements; see footnote 2. Vertical lines represent: 1, force output onset; 2a, target force reached at speed condition; 2b, target force reached at accuracy condition (modified from Slobounov and Ray, 1998). MMP amplitude was enhanced at higher rates of force development for both speed and accuracy tasks and consistently higher for fast than for accurate force output; MRCP began earlier for fast than for accurate force output. (III) Grand average MRCPs at Cz preceding and accompanying the force production tasks reaching perceived 30%, 50%, and 70% of the maximum voluntary contraction (MVC). The black vertical line indicates the force initiation (modified from Slobounov et al., 2004). The MRCP preceding the initiation of force proportionally increased as a function of perceived effort. (IV) Grand average of ballistic real and kinesthetic imaginary ankle plantar flexion, based on results by do Nascimento et al. (2006). The amplitude of MRCP was smaller for imaginary movement than for actual movement. (V) Grand average MRCP at the channel Cz, based on results by Gu et al. (2009b). Imaginary movements were fast right-wrist extension (FE) and slow right-wrist extension (SE). Rebound rate (RR) was stronger for fast than for slow wrist extension.

expression of MRCPs, whereas others have a modulatory character. Reports show, for instance, that there is a positive correlation in the nominal amplitude of MRCPs and the amount of force exerted in isometric motor tasks, for both upper (fingers, Slobounov et al., 1998; elbow, Siemionow et al., 2000) and lower (ankle, do Nascimento et al., 2005) limbs. The same modulation is observed in relation to the rate of torque/force development (that can be translated into speed in nonisometric motor tasks). For ankle plantar flexions, do Nascimento et al. (2005) have shown that RP is more influenced by variations in force level, whereas MP and MMP are more sensitive to variations in speed. Slobounov et al. (2004) suggested, however, that it is the perceived effort, and not necessarily the nominal amount of force or speed, that primarily modulates MRCPs. As shown in their EEG experiments, where the exerted force and the visual feedback presented to subjects were purposefully mismatched during finger pressing, greater MRCP amplitudes were elicited when the effort was perceived as high, with not necessarily high levels of force exertion (Fig. 13.1, panel III).

Since they are related to planning of movements, MRCPs are also expressed during kinesthetic imagination of movements and, interestingly, similar modulations according to variation of perceived kinesthetic parameters are observed. do Nascimento et al. (2006) have shown that imaginary ankle plantar flexions generate MRCPs that are somewhat similar to those identified in executed movements. The amplitude of MRCPs from imaginary movements is approximately two-thirds smaller than that observed in executed movements and their topographical distribution is more diffuse, with spread of activity mainly over SMA and parietal areas. MRCPs of imaginary movements are also more sensitive to speed than to force level, probably because perceiving differences in exerted force may be more complex than perceiving differences in speed.

A striking feature of MRCPs in imaginary movements is that their postmovement portion lacks the MMP peak (a plateau is observed instead; Fig. 13.1, panel IV), corroborating the hypothesis that MMP reflects control of performance, which is highly dependent on afferent response. Studies in amputees have shown that MRCPs associated with phantom movements also lack the MMP peak, but the MRCP amplitudes generally correspond to those observed in overtly executed movements (do Nascimento et al., 2008).

More recently, Gu et al. (2009b) have performed experiments on imaginary dynamic wrist extension and rotation under two speeds, extracting the premovement potentials and features NS1 (Negative Slope 1), NS2 (Negative Slope 2), peak RP, and the postmovement potential and feature rebound rate (RR). This experiment showed that stronger MRCP rebounds (RR) were associated with faster imaginary movements (Fig. 13.1, panel V). Those rebounds were, however, not affected by variations in movement type.

Movement-related cortical potentials have also been investigated in ALS patients (Gu et al., 2009c), who have their neuromuscular channels disrupted and therefore are incapable of communicating with and controlling its external environment. In ALS patients, the disease progresses from the first symptoms

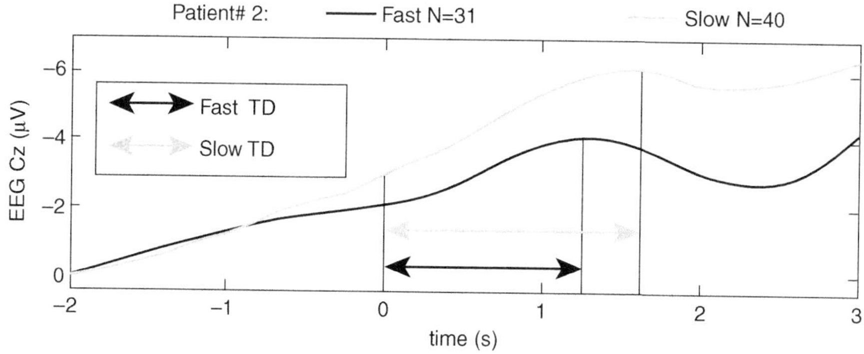

Figure 13.2. Averaged MRCPs from one patient at the channel Cz during fast and slow wrist extensions. The imaginary movement onset is represented by time 0 s; N is the number of averaged trials; TD, time delay; black double-arrowed line indicates fast TD; gray double-arrowed line indicates slow TD. Based on results by Gu et al. (2009c).

of muscular or respiratory weakness to the locked-in state;, sensory, emotional, and cognitive processing often remain largely intact despite extensive degeneration of the motor system (Kübler et al., 2005). Modern life-support technology allows most ALS individuals to live long lives, but paralysis greatly limits their independence and communication. ALS patients are among those who are in greatest need of BCI systems for control and communication purposes. Gu and co-workers (Gu et al., 2009c; Gu and Demstrup, 2011) performed experiments on ALS patients to investigate the characteristics of MRCPs by motor imagery. The components of MRCPs in four ALS patients, with different degrees of impairment, have shown great variability within and between patients. Despite this variability, MRCPs associated with imaginary movements were also identified in these patients and variations in imaginary movement speed also had an influence on those MRCPs, with faster movements associated with shorter time delays of peak negativity (Fig. 13.2). MRCPs seem to persist and be modulated by movement parameters even in serious motor degenerative diseases.

The association between MRCP features and variations of perceived kinesthetic parameters leads to the possibility of using these modulations as control signal for BCIs.

SINGLE-TRIAL CLASSIFICATION OF MRCPs

Here we report approaches that have been proposed for detecting MRCPs in single trials, recognizing its modulation according to perceived kinesthetic parameters, and its application in both healthy subjects and patients, for the generation of control outputs in BCIs.

The current classification methods being developed for BCI applications include quadratic discriminant analysis (QDA; Neuper et al., 1999), linear discriminant analysis (LDA; Donoghue et al., 2004), regression (McFarland and Wolpaw, 2005), Fisher discriminant analysis (Mika et al., 2001), support vector machine (SVM; Müller et al., 2001; Farina et al., 2007), linear programming machine (Bennett and Mangasarian 1992), and kernel methods (Müller et al., 2001), to mention a few.

The first attempted method for classifying single-trial MRCPs is described in the following subsections, as a representative of the possibly many pattern recognition alternatives available for tackling the challenge of classifying MRCPs.

Pattern Recognition for MRCP Classification

The pattern recognition method applied to MRCP classification is derived from that described by Farina et al. (2007) and is based on features extracted with optimized wavelet transform and classification performed with SVM. Figure 13.3 shows the structure of the algorithm.

The discrete wavelet transform (DWT) projects the signal into different scales with multiple resolutions by dilating a mother wavelet. The projection coefficients are obtained by multiresolution analysis (MRA) through the application of a bank of digital filters (Mallat, 1989). Wavelet coefficients form a multiscale representation of the signal, leading to localized frequency subbands with equal width on a logarithmic scale. Thus, resolution improves for subbands at lower frequencies. This property may be advantageous in classifying MRCPs, since the energy of MRCPs is mainly concentrated at low frequencies (approximately up to 1 Hz). The optimized wavelet transform provides the way of obtaining a feature space adapted to a specific classification problem. Wavelet optimization was chosen on the basis of a learning step (supervised classification), as described later.

The SVM is a robust classification method that has demonstrated good generalization properties in various applications, including BCI (Farina et al., 2007; Schlögl et al., 2005). In the general algorithm used for MRCP studies, a nonlinear SVM classifier with Gaussian kernel of parameterized width was

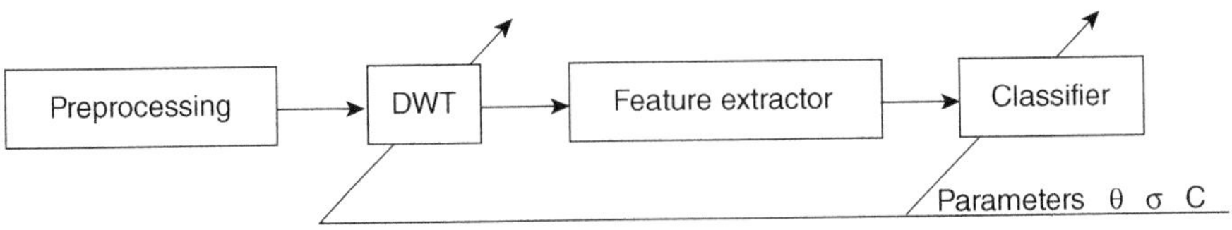

Figure 13.3. Block diagram of the algorithm used for single-trial classification; θ is the parameter for tuning the mother wavelet; σ and C are the parameters of the SVM classifier. See text for details.

used. Both the parameters that defined the mother wavelet and the parameters of the SVM were optimized on the basis of cross-validation.

Accuracy in MRCP Classification

In the following, studies aimed at the single-trial classification of MRCPs in healthy subjects and patients are reviewed.

Healthy Subjects. Do Nascimento and Farina (2008) showed that two rates of torque development (RTDs) in imaginary isometric plantar flexions can be classified on healthy subjects with minimum and average misclassification rates of 8.3% and ~20%, respectively. This result provided the first evidence that MRCPs could potentially be used for a new type of BCI that relies on kinesthetic parameters of a single limb rather than movements of opposite limbs. More recently, Gu et al. (2009a) investigated the accuracy in discriminating combinations of RTD and target torque (TT) of imaginary isometric plantar flexions and concluded that discrimination of RTD is preferable over discrimination of TT. MRCP modulations under imaginary dynamical wrist movements were also investigated, where movement speed was found to be better distinguishable than movement type at the same joint (wrist rotation vs. wrist extension; Gu et al., 2009b).

Patients. The classification approach just described has also been applied to recognize MRCP modulations in lower limb amputees and ALS patients. In amputees, MRCPs modulated by two speeds of plantar flexion with the phantom limb were classified with average misclassification rate of ~30% (do Nascimento et al., 2008). In ALS patients, MRCPs modulated by two speeds of imaginary wrist extension have been differentiated on four ALS patients. The classification error ranged from 25% to 34% after individual temporal–spatial optimization. The use of speed as the discriminating variable had advantages in this patient population. It was indeed simple to instruct the patients about the task either by showing them the movement or by performing the movement passively. The difference between fast and slow speed was quite obvious and intuitive for the patients. A possible explanation for the lower classification accuracy, on these patients compared with healthy subjects, is that the healthy subjects probably have better feeling of imagined movements due to the intact motor function pathway and the continuous performance of movements. Nevertheless, the accuracy obtained seems sufficient for clinical application of the technique.

CONCLUSION AND OUTLOOK

The studies reviewed in this chapter have focused on the influence of movement parameters on MRCPs and exploring the potential application of MRCPs in BCI. One of the outcomes of this research is that single-trial off-line classification of MRCPs is possible with reasonable accuracy, paving the way to building online BCI systems with feedback and training capabilities. With training and feedback, it may be expected that single-trial classification of MRCPs can further improve since previous studies have reported improved accuracy of EEG control after several training sessions and with continuous feedback during training (Curran and Stokes, 2003; Neuper et al., 1999). Identifying MRCPs tuned by movement parameters in single trials presents an important step in BCI research (Vuckovic, 2009), with the potential of increasing the number of separable classes. In addition, the tasks being classified are closely related to the natural control of complex movements. MRCP-based BCIs constitute a powerful tool for better understanding of brain motor functions, on both healthy subjects and clinical patients, with the potential of providing clues to explain the various movement disorders. MRCPs should be examined on larger patient populations to explore the full potential of this approach.

Detection of movement intention from neural signals combined with assistive technologies may be used for effective neurofeedback in rehabilitation. To promote plasticity, a causal relation between intended actions, detected for example from MRCP and the corresponding feedback, should be established. Application of MRCP-based BCI on neurological rehabilitation has been studied in our lab. Our present researches include the development of a system that couples functional electrical stimulation (FES) with self-paced BCI (SBCI, based on detection of MRCP) for the rehabilitation of stroke patients. In this system, FES will be driven by the user's intention, which will be estimated by decoding EEG signals, to move his or her upper or lower limb. The novelty is that we analyze the plasticity of MRCPs after training and learning processes and examine whether relevant features improve in parallel with improvement in motor activity. Finally, we will assess whether combining FES with SBCI might lead to better rehabilitation protocol because of active participation in training of subject compared with current existing physiotherapy technologies where they are trained passively. The results are promising for the application of this approach to provide patient-driven real-time neurofeedback (Niazi et al., 2011, 2012).

REFERENCES

Bennett K.P. and Mangasarian O.L. (1992). Robust linear programming discrimination of two linearly inseparable sets. *Optimization Methods and Software* 1:23–34.

Birbaumer N., Elbert T., Canavan A.G., and Rockstroh B. (1990). Slow potentials of the cerebral cortex and behavior. *Physiol Rev* 70(1):1–41.

Boschert J. and Deecke L. (1986). Handedness, footedness and finger and toe movement-related cerebral potentials. *Hum Neurobiol* 5:235–243.

Böcker K.B., Brunia C.H., and Cluitmans P.J. (1994). A spatio-temporal dipole model of the readiness potential in humans. II. Foot movement. *Electroencephalogr Clin Neurophysiol* 91:286–294.

Brunia C.H.M. and van den Bosch W.E. (1984). Movement related slow potentials. I. A contrast between finger and foot movement in right handed subject. *Electroencephalogr Clin Neurophysiol* 57:515–527.

Curran E.A. and Stokes M.J. (2003). Learning to control brain activity: a review of the production and control of EEG components for driving brain–computer interface (BCI) systems. *Brain and Cognition* 51:326–336.

Daly J.J. and Wolpaw J.R. (2008). Brain–computer interfaces in neurological rehabilitation. *Lancet Neurology* 7:1032–1043.

Deecke L., Grözinger B., and Kornhuber H.H. (1976). Voluntary finger movements in man: cerebral potentials and theory. *Biol Cybern* 23:99–119.

Donoghue J.P., Blankertz B., Curio G., and Muller K. (2004). Boosting bit rates in non-invasive EEG single trial classification by feature combination and multi class paradigm. *IEEE Trans Biomed Eng* 51(6):993–1002.

do Nascimento O.F. and Farina D. (2006). Classification of movement-related cortical potentials according to variations of rate of torque development during imaginary isometric plantar-flexion. In *Proceedings of the 3rd International Brain–Computer Interface Workshop and Training Course*, Graz, Austria, pp. 70–71.

do Nascimento O.F. and Farina D. (2008). Movement-related cortical potentials allow discrimination of rate of torque development in imaginary isometric plantar flexion. *IEEE Trans Biomed Eng* 55:2675–2678.

do Nascimento O.F., Nielsen K.D., and Voigt M. (2005). Relationship between plantar-flexor torque generation and the magnitude of the movement-related potentials. *Exp Brain Res* 160:154–165.

do Nascimento O.F., Nielsen K.D., and Voigt M. (2006). Movement-related parameters modulate cortical activity during imaginary isometric plantar-flexions. *Exp Brain Res* 171:78–90.

do Nascimento O.F., Andersen C.R., and Farina D. (2008). Single-trial movement-related cortical potentials are modulated by rate of torque development in lower extremity amputees. In *Proceedings of the 4th International Brain–Computer Interface Workshop and Training Course*, Graz, Austria, pp. 256–261.

Eagleman D.M. (2004). The where and when of intention. *Science* 303:1144–1146.

Farina D., do Nascimento O.F., Lucas M.F., and Doncarli C. (2006). Parameterization of wavelets for optimized signal representation in the classification of movement-related cortical potentials. In *Proceedings of the 3rd International Brain–Computer Interface Workshop and Training Course*, Graz, Austria, pp. 10–11.

Farina D., do Nascimento O.F., Lucas M.F., and Doncarli C. (2007). Optimization of wavelets for classification of movement-related cortical potentials generated by variation of force-related parameters. *J Neurosci Methods* 162:357–363.

Grünewald G. and Grünewald-Zuberbier E. (1983). Cerebral potentials during skilled slow positioning movements. *Biol Psychol* 31:71–78.

Gu Y. and Dremstrup K. (2011). Comparing MRCP of healthy subjects with that of ALS patients. *Proc IFMBE* 34:141–143.

Gu Y., do Nascimento O.F., Mazzaro T., Sinkjaer T., and Farina D. (2006). Classification of cortical potentials evoked by modulation of force in perceived and real wrist flexion in an amputee. In *Proceedings of the 3rd International Brain–Computer Interface Workshop and Training Course*, Graz, Austria, pp. 74–75.

Gu Y., do Nascimento O.F., Lucas M.-F., and Farina D. (2009a). Identification of task parameters from movement-related cortical potentials. *Med Biol Eng Computing* 47(12):1257–1264.

Gu Y., Dremstrup K., and Farina D. (2009b). Single-trial discrimination of type and speed of wrist movements from EEG recordings. *Clin Neurophysiol* 120:1596–1600.

Gu Y., Farina D., Murguialday A.R., Dremstrup K., Montoya P., and Birbaumer N. (2009c). Offline identification of imagined speed of wrist movements in paralyzed ALS patients from single-trial EEG. *Front Neurosci* 3, Article No. 62.

Ikeda A. and Shibasaki H. (1992). Invasive recording of movement-related cortical potentials in humans. *J Clin Neurophysiol* 9:509–520.

Ikeda A., Luders H.O., Burgess R.C., Sakamoto A., Klem G.H., Morris H.H. III, and Shibasaki H. (1995). Generator locations of movement-related potentials with tongue protrusions and vocalizations: subdural recording in human. *Electroencephalogr Clin Neurophysiol* 96:310–328.

Jahanshahi M., and Hallett M., Eds. (2003). *The Bereitschaftspotential. Movement Related Cortical Potentials*. New York: Kluwer Academic/Plenum Publishers.

Jentzsch I., and Leuthold H. (2002). Advance movement preparation of eye, foot, and hand: a comparative study using movement-related brain potentials. *Brain Res Cogn Brain Res* 14:201–217.

Klein S. (2002). Libet's research on the timing of conscious intention to act: a commentary. *Conscious Cogn* 11:273–279.

Kornhuber H.H. and Deecke L. (1964). Hirnpotentialänderungen beim Menschen vor und nach Willkürbewegungen, dargestellt mit Magnetband-Speicherung und Rückwärtsanalyse. *Pflugers Archiv* 281:52.

Kornhuber H.H. and Deecke L. (1965). Hirnpotentialänderungen bei Willkürbewegungen und passiven Bewegungen des Menschen: Bereitschaftspotential und reafferente Potentiale. *Pflugers Archiv* 284:1–17.

Kristeva R., Keller E., Deecke L., and Kornhuber H.H. (1979). Cerebral potentials preceding unilateral and simultaneous bilateral finger movements. *EEG Clin Neurophysiol* 47:229–238.

Kübler A., Nijboer F., Mellinger J., Vaughan T.M., Pawelzik H., Schalk G., McFarland D.J., Birbaumer N., and Wolpaw J.R. (2005). Patients with ALS can use sensorimotor rhythms to operate a brain–computer interface. *Neurology* 64:1775–1777.

Libet B., Gleason C.A., Wright E.W., and Pearl D.K. (1983a). Time of conscious intention to act in relation to onset of cerebral activity (readiness-potential). The unconscious initiation of a freely voluntary act. *Brain* 106(3):623–642.

Libet B., Wright E.W. Jr., and Gleason C.A. (1983b). Preparation- or intention-to-act, in relation to pre-event potentials recorded at the vertex. *Electroencephalogr Clin Neurophysiol* 56(4):367–72.

Mallat S.G. (1989). A theory for multiresolution signal decomposition: the wavelet representation. *IEEE Trans Pattern Anal Mach Intell* 11:674–693.

McFarland D.J. and Wolpaw J.R. (2005). Sensorimotor rhythm-based brain–computer interface (BCI): feature selection by regression improves performance. *IEEE Trans Rehabil Eng* 13:372–379.

McFarland D.J., Miner L.A., Vaughan T.M., and Wolpaw J.R. (2000). Mu and beta rhythm topographies during motor imagery and actual movement. *Brain Topogr* 3:177–186.

Mika S., Rätsch G., and Müller K.R. (2001). A mathematical programming approach to the kernel Fisher algorithm. In: *Advances in Neural Information Processing Systems (NIPS00)* (Lee T.K., Dietterich T.G., and Tresp V., Eds.), 13:591–597. Cambridge, MA: MIT Press.

Müller K.R., Mika S., Rätsch G., Tsuda K., and Schölkopf B. (2001). An introduction to kernel based learning algorithm. *IEEE Trans Neural Networks* 12(2):181–201.

Nakajima I., Tanaka Y., Uchida A., Sakai T., Akasaka M., Mori A., and Sumino R. (1991). Cortical potentials associated with voluntary biting movement in humans. *Neurosci Res* 10:285–289.

Neuper C., Schlögl A., and Pfurtscheller G. (1999). Enhancement of left–right sensorimotor EEG differences during feedback-regulated motor imagery. *Clin Neurophysiol* 16(4):373–382.

Neuper C., Müller G.R., Kübler A., Birbaumer N., and Pfurtscheller G. (2003). Clinical application of an EEG-based brain–computer interface: a case study in a patient with severe motor impairment. *Clin Neurophysiol* 114(3):399–409.

Niazi I.K., Jiang N., Tiberghien O., Nielsen J.F., Dremstrup K., and Farina D. (2011). Detection of movement intention from single-trial movement-related cortical potentials. *J Neural Eng* 8(6):2011.

Niazi I.K., Mrachacz-Kersting N., Jiang N., Dremstrup K., and Farina D. (2012). Peripheral electrical stimulation triggered by self-pace detection of motor intention enhances motor evoked potentials. *IEEE Trans Rehabil Eng* 20(4):595–604.

Nielsen K.D., Cabrera A.F., and do Nascimento O.F. (2006). EEG based BCI—towards a better control. Brain–computer interface research at Aalborg University. *IEEE Trans Rehabil Eng* 14:202–204.

Pfurtscheller G., Neuper C., Flotzinger D., and Pregenzer M. (1997). EEG-based discrimination between imagination of right and left hand movement. *Electroencephalogr Clin Neurophysiol* 103(6):642–651.

Pfurtscheller G., Brunner C., Schlögl A., and Lopes da Silva F.H. (2006). Mu rhythm desynchronization and EEG single-trial classification of different motor imagery tasks. *NeuroImage* 31:153–159.

Romero D.H., Lacourse M.G., Lawrence K.E., Schandler S., and Cohen M.J. (2000). Event-related potentials as a function of movement parameter variations during motor imagery and isometric action. *Behav Brain Res* 117:83–96.

Schlögl A., Lee F., Bischof H., and Pfurtscheller G. (2005). Characterization of four class motor imagery EEG data for the BCI-competition. *J Neural Eng* 2:14–22.

Shibasaki H. and Hallett M. (2006). What is the Bereitschaftspotential? *Clin Neurophysiol* 117:2341–2356.

Shibasaki H., Barrett G., Halliday E., and Halliday A.M. (1980). Components of the movement-related cortical potential and their scalp topography. *Electroencephalogr Clin Neurophysiol* 49:213–226.

Shibasaki H., Barrett G., Halliday E., and Halliday A.M. (1981). Cortical potentials associated with voluntary foot movement in man. *Electroencephalogr Clin Neurophysiol* 52:507–516.

Siemionow V., Yue G.H., Ranganathan V.K., Liu J.Z., and Sahgal V. (2000). Relationship between motor activity-related cortical potential and voluntary muscle activation. *Exp Brain Res* 133(3):303–311.

Simonetta M., Clanet M., and Rascol O. (1991). Bereitschaftspotential in a simple movement or in a motor sequence starting with the same simple movement. *Electroencephalogr Clin Neurophysiol* 81:129–134.

Slobounov S.M. and Ray W.J. (1998). Movement-related potentials with reference to isometric force output in discrete and repetitive tasks. *Exp Brain Res* 123:461–473.

Slobounov S.M., Ray W.J., and Simon R.F. (1998). Movement-related potentials accompanying unilateral finger movement with special reference to rate of force development. *Psychophysiology* 35:537–548.

Slobounov S., Tutwiler R., Rearick M., and Challis J.H. (1999). EEG correlates of finger movements with different inertial load conditions as revealed by averaging techniques. *Clin Neurophysiol* 110:1764–1773.

Slobounov S., Hallett M., and Newell K.M. (2004). Perceived effort in force production as reflected in motor-related cortical potentials. *Clin Neurophysiol* 115(10): 2391–2402.

Terada K., Ikeda A., Yazawa S., Nagamine T., and Shibasaki H. (1999). Movement-related cortical potentials associated with voluntary relaxation of foot muscles. *Clin Neurophysiol* 110:397–403.

Vuckovic A. (2009). Non-invasive BCI: how far can we get with motor imagination? *Clin Neurophysiol* 120(8):1422–1423.

Wolpaw J.R., Birbaumer N., McFarland D.J., Pfurtscheller G., and Vaughan T.M. (2002). Brain–computer interfaces for communication and control. *Clin Neurophysiol* 113:767–791.

Yoshida K., Kaji R., Hamano T., Kohara N., Kimura J., and Iizuka T. (1999). Cortical distribution of Bereitschaftspotential and negative slope potential preceding mouth-opening movements in humans. *Arch Oral Biol* 44:183–190.

14

REVIEW CHAPTER

INTRODUCTION TO UPPER LIMB PROSTHETICS

BERNHARD GRAIMANN[1] AND HANS DIETL[1,2]

[1]*Otto Bock HealthCare GmbH, Duderstadt, Germany*

[2]*Otto Bock HealthCare Products GmbH, Vienna, Austria*

SUMMARY

The progress in upper limb prostheses has been moderate for decades. Only recently have new commercial prosthetic developments become clinically available. This new generation of devices offers better functionality and new compliance components by improved mechatronic designs that utilize advances in materials, rechargeable batteries, and actuators. Compared with human limbs, however, the performance of these new prosthetic components is still very limited. Further progress is hampered by the limitations in the human–machine interfaces that control these devices. Their signal processing and control mechanisms, which are usually based on myoelectric activity, are still very similar to conventional systems. That is, control signals are still derived from a small number of myoelectric signals recorded from surface electrodes. Consequently, the information transfer rates are very limited, the control of various degrees of freedom is cumbersome and slow, and using surface electrodes causes a number of practical problems. Academic research has suggested various ways to overcome these problems and significantly improve the human–machine interface.

Introduction to Neural Engineering for Motor Rehabilitation, First Edition. Edited by Dario Farina, Winnie Jensen, and Metin Akay.

The possible solutions include shared control approaches, advanced signal processing, selective nerve transfer, new noninvasive and invasive signal acquisition methods, and methods for providing somatosensory feedback. However, many of these new technologies still require considerably more research before they can become clinically available.

INTRODUCTION

Prostheses are artificial substitutes that replace missing or lost parts of the body. There are many different artificial replacements for body parts available now. For instance, heart valves, teeth, arteries, and joints can be replaced by artificial parts, which are then called prostheses. However, the artificial components that are most commonly regarded as prostheses are those that replace limbs. To better differentiate the various forms of prostheses, the terms **endoprostheses** and **exoprostheses** are used. Prostheses that replace parts inside of the body are called endoprostheses. In contrast, exoprostheses are applied outside of the body. Typical exoprostheses are artificial replacements for the upper or lower extremities. Artificial substitutes for fingers, hands, wrists, forearms, elbows, upper arms, or shoulders are also termed **upper limb prostheses**. Correspondingly, exoprostheses for the lower extremities are also called **lower limb prostheses**.

Neuroprostheses are another form of prostheses that have been gaining considerable attention over the last decades. Neuroprostheses are devices that are direct or indirect interfaces to the central or peripheral nervous system and can substitute lost or impaired motor or sensory functions. Examples are cochlear and retinal implants as well as invasive and noninvasive brain–machine interfaces (BMIs). Although conventional artificial limbs have not much in common with neuroprostheses, new developments of exoprosthetic devices that use neural human–machine interfaces have an increasing similarity to motor neuroprostheses.

As is often the case in science and engineering, the terms used for specific applications or technologies are coined by the field in which they have been developed. *Prosthesis* is a term traditionally used in the field of orthopedics. A very closely related term but used in the field of robotics is *robotic arm* or *robotic hand*. In general, a robotic arm is a robotic manipulator with functions similar to the human arm (and hand). Robotic components can be large and heavy, can work in hazardous and hostile environments, and can perform tasks such as welding or assembling of parts in industrial environments. However, wearable robotic arms or hands are worn by human operators and can supplement the function of a limb or replace a limb completely (Pons et al., 2008). If such a robotic limb substitutes the human limb, the terms robotic arm and robotic hand are synonymous with prosthetic arm and prosthetic hand, respectively. Wearable robots may also embrace and support an already existing limb to restore functionality in case of an impaired limb or to enhance functionality beyond the capability (e.g., strength) of a normal healthy limb. Such wearable robots are also called **exoskeletons**. Exoskeletons in turn share many similarities with orthoses, which are orthopedic medical devices that support, correct, or restrict the function of body parts.

Most patients require exoprostheses as the result of amputations. Patients suffering from dysmelia, which is a congenital disorder that can result in missing limbs or malformations of limbs, may require prostheses as well. Larger statistical surveys about the number of amputations are rare. Dillingham and co-workers identified 52 amputations per 100,000 persons in the United States for the year 1996. Five of these amputations were related to upper limb, 47 to lower limb amputations (Dillingham et al., 1998). A similar ratio of upper to lower limb amputations can be found in the United Kingdom (U.K. Prosthetics Services, 2008). The major cause of lower limb amputations is related to dysvascularity, which affects mostly elderly people. Vascular problems requiring amputation are much less common in the upper extremities. The loss of an upper limb is most commonly caused by traumatic amputation, and amputees tend to be younger and in good health otherwise. The specific reasons of trauma-related amputations are diverse but the leading causes in industrialized countries include injuries involving machinery, powered tools and appliances, and motor vehicle accidents (Dillingham et al., 1998). Military conflicts are another source for traumatic amputations. Advances in the body armor of soldiers have increased the survival rate in armed conflicts, but also have increased the number of amputations due to severe injuries of the unprotected limbs (U.S.G.P.O., 2010). Cancer and infections are examples of further reasons for upper limb amputations but are not as frequent as trauma.

The length of the residual limb after amputation determines the options for prosthetic components and affects the potential for functional rehabilitation (Näder, 2011). The amputation levels for upper extremities include (Fig. 14.1): upper digits and partial hand, trascarpal, wrist disarticulation, transradial (below elbow), elbow disarticulation (transection through the elbow joint), transhumeral (above elbow), and shoulder disarticulation (transection through the shoulder joint). Finger and partial hand amputations, transradial amputations, and transhumeral amputations are the most frequent amputations. From the 215 upper limb referrals reported by the U.K. Prosthetics Services (2008), 44% were more distal or equal to wrist disarticulation, 18% were transradial, and 25% were transhumeral.

As a general rule, the residual limb should be as long as possible. The longer the residual limb, the better is the leverage for lifting, pushing, and pulling activities, and the more sensory information and residual muscles are still available for moving and controlling the prosthesis (Näder, 2011). Conversely, higher amputation levels require prostheses of additional complexity (more joints), but the control options are reduced. In the case of a distal transradial amputation (long transradial amputation), for instance, hand and wrist are replaced by prosthetic components, but a large part of the arm, including

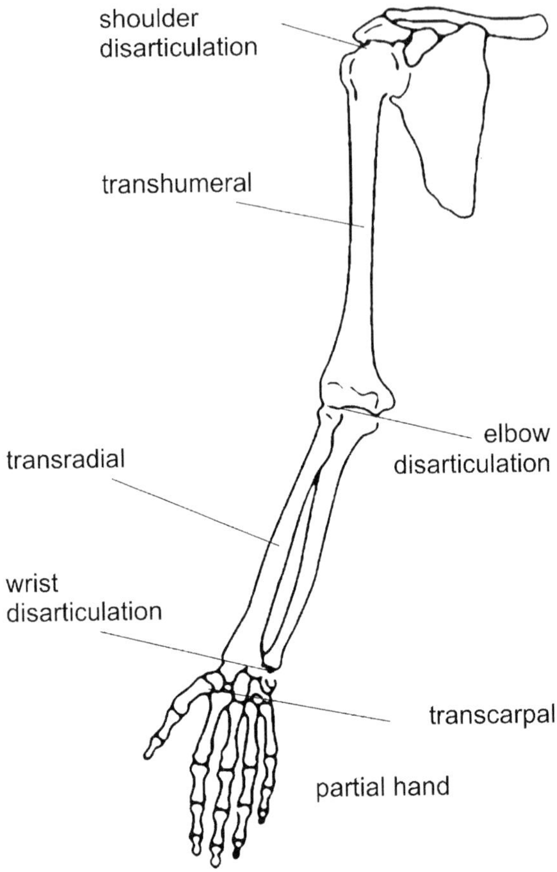

Figure 14.1. Upper limb amputation levels. (Modified from H.G. Näder, 2011.)

residual forearm muscles, is still available to use the prosthesis. Most of the wrist rotation capability (pronation and supination) is still preserved at this level. In a short transradial amputation, however, this capability is not available anymore. As another example, amputees with shoulder disarticulation would need a prosthesis that substitutes the entire arm including shoulder. The complexity of such a prosthesis is considerably higher than in, for instance, the transradial case, since all the structures, motors, and sensors (bones, muscles, and receptors) of the arm are lost. A prosthesis of this amputation level has to provide the functionality of a transradial prosthesis as well as elbow and shoulder functionality.

Most amputees suffer from a unilateral amputation. Those with a high-level bilateral amputation such as bilateral shoulder disarticulation amputees present the greatest challenges for functional restoration. The need for functional restoration is enormous, since many essential activities of daily living

(ADL) cannot be carried out independently. Even for the most basic and personal tasks such as dressing, eating, and performing personal hygiene, the bilateral amputee depends on the help of others when an appropriate rehabilitation by prostheses is not attained.

The purpose of a prosthesis is to improve the appearance and to provide the ability to perform tasks of daily living and increase quality of life. Amputees have to benefit from their prostheses. If function, comfort, or appearance is not sufficient, amputees will not use the device. According to several researchers (Biddiss and Chau, 2007; Schultz et al., 2007; Atkins et al., 1996), the two most common reasons for not using a prosthesis are the lack of functionality and discomfort. Unfortunately, function, comfort, and appearance often conflict with each other. Prostheses that have a very natural appearance are often limited in their functionality, while highly functional devices may look unattractive or are heavy and therefore uncomfortable.

Depending on the level of amputation, upper limb prostheses may comprise a terminal device, wrist unit, elbow system, shoulder joint. Of special importance is the **socket**, which is the interface between the amputee's body and the prosthesis. It has to remain attached (suspended) to the residual limb for the entire range of motion, and it has to allow the load required for performing ADL tasks. Depending on the amputation level and the residual limb condition, fitting a socket can be a challenging task. It is important to note that the most versatile and advanced prosthetic components are useless when the socket interface is not well fitted to the amputee. There are three broad categories of upper limb prosthesis: passive, body-powered, and externally powered. Figure 14.2 depicts examples of these three different types. Passive prostheses provide cosmetic restoration, but do not provide active grasping capabilities. Natural appearance and lightweight design are the main advantages. Modern materials allow texture, skin tone, superficial veins, and other structures to be individually optimized to the amputee's needs, making it very difficult to tell apart from the natural hand. Body-powered prostheses are controlled by body movements such as gross shoulder movements. A harness-and-cable system captures these movements and actuates one to three degrees of freedom (DOFs) of the prosthesis. This design is very durable and has only moderate weight. Another advantage is the feedback that is provided to the user by the cable tension. Through this feedback the user can get a sense of the position or state of the terminal device. The disadvantages include rather limited grip force, uncomfortable harness, required energy expenditure, unnatural appearance, and additional loading of the shoulder joint.

Externally powered prostheses contain actuators that are driven by electrical, hydraulic, or any other power source external to the body. Although the power source is external, it has to be wearable and contained in the prosthetic system. Almost all externally powered prostheses available today use electric motors and rechargeable batteries. Compared with body-powered prostheses, these prostheses are heavier, more costly, and do not provide any somatosensory feedback other than the very rudimentary feedback provided by the

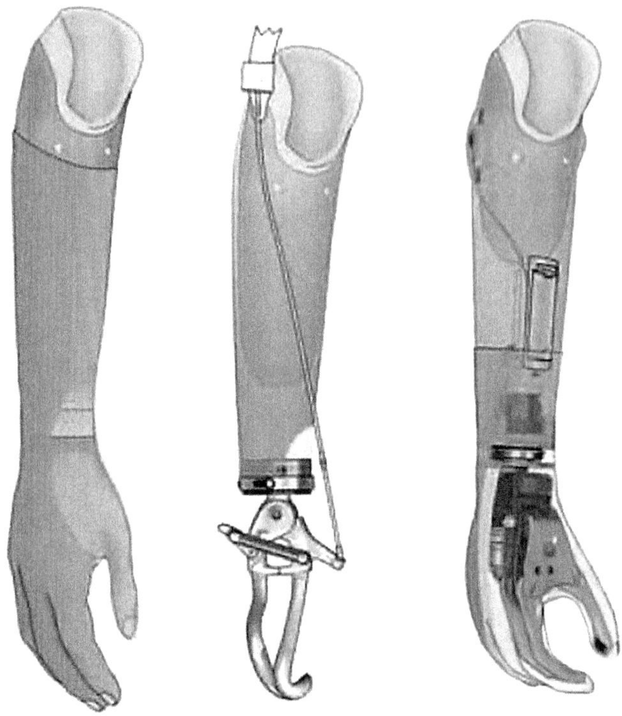

Figure 14.2. From left to right: passive (cosmetic) prosthesis, body-powered (cable-driven) prosthesis, and externally powered prosthesis with indicated cosmetic cover. (Modified from H.G. Näder, 2011.)

socket–stump connection. Externally powered prostheses, however, offer superior grasping force, responsiveness, speed, and, most important, a large variety of control options. Push-buttons and harnesses with switch or linear transducer mechanisms are simple forms of controlling such devices. The most common means of controlling this type of prosthesis is **myoelectric control**, which is derived from muscle signals generated by voluntary muscle contractions. These signals are typically recorded from surface electrodes located in the prosthetic socket and placed over the residual muscles of the amputee's stump. Current commercial systems allow the control of only one DOF at a time, for instance, hand opening or hand closing. However, advances in human–machine interfaces based on myoelectric signals or invasive approaches utilizing signals from the nervous system directly do offer considerable improvements.

In the following sections, we give an overview of the current state of the art of commercial externally powered prostheses and outline current developments in research that will most likely have an important impact on future commercial prosthetic systems. We present the requirements of advanced prostheses and discuss the rather large gap between commercial systems and systems in academic research. Since the prosthetic hand is a key component in upper limb prosthetics, we mainly focus on prosthetic hands.

STATE OF THE ART

A prosthesis consists of several parts depending on the amputation level. Similar to the human hand, the prosthetic hand or the terminal device is the key element of functional performance for the upper extremity. The main task of the other parts of the prosthetic arm is to bring the terminal device in a suitable position for grasping and carrying out vocational operations and ADL tasks (Muzumdar, 2004). For this reason, terminal devices have a central role in upper limb prosthetics, and it is interesting to compare their capabilities with the human hand.

The Human Hand

Natural evolution has formed the human hand over millions of years. The result is an extremely versatile grasping and manipulation device with unmatched dexterity, reliability, and durability. The hand is capable of performing a large number of activities ranging from simple to very complex tasks. Healthy people with natural hands can hold objects, change their position or orientation, and assess object features such as temperature or texture with a minimum of cognitive effort. These capabilities are based on a remarkably complex biomechanical system of bones, joints, ligaments, tendons, skeletal muscles, and sensors. Twenty-seven bones, more than 19 DOFs (depending on the mechanical model applied), and actuation provided by more than 30 muscles, many of which are not in the hand but in the forearm, form the hand–wrist complex (Changmok et al., 2008; Carrozza et al., 2006; Chappell et al., 2007). The large number of DOFs enables the hand to adapt to a wide range of objects that are different in size and shape. In addition, thousands of sensors measure a number of physical variables such as temperature, pressure, force, joint position, and sensations such as pain. These sensors together with the muscles are organized in a closed loop that is mainly controlled by the sensorimotor system of the brain (Bear et al., 2006). The computational complexity required is indicated by the large proportion of the somatotopic map on the cortex (homunculus) associated with the hand. In fact, it is the combination of the complex biomechanical system and the sophisticated control mechanisms that gives the human hand its capabilities.

Prosthetic Hands

Prosthetic hands that are currently on the market are unable to match even a small fraction of the abilities of the human hand. Most commercial systems provide only a very limited number of DOFs. Moreover, there is no system available that provides explicit somatosensory feedback to the amputee. Amputees can rely on the pressure on the residual limb exerted by the socket.

Experienced users may also utilize vibrations and mechanical noise produced by the prosthetic actuators. But these feedback mechanisms provide only a very poor substitute for the sensory capabilities of the natural hand. Therefore, amputees depend on visual feedback for positioning the limb and grasping objects. This has the consequence that amputees have to continuously monitor prosthetic activities, which requires much more selective attention and cognitive effort than when using natural limbs.

The task of a prosthetic hand is to substitute a lost hand. Therefore, one of the main requirements of a prosthetic hand is functionality. In addition to functionality, however, there are a number of basic requirements that prosthetic designs have to satisfy in order to be successful (Muzumdar, 2004; Biddiss, 2009; Pylatiuk et al., 2007). These requirements include compact and lightweight design, natural appearance, minimal power consumption, reliable and silent operation, and affordable costs. The prosthetic hand has to be an intrinsic design that also contains the actuators. Designs with actuators in other components of the prosthesis would limit the number of amputees who could use it. For instance, transradial amputees with an amputation level at the distal forearm could not use such a design. Moreover, all components of the hand have to fit into an anthropomorphic shape, because appearance is of great importance to amputees. Weight is another important feature. Since the weight of a prosthesis is concentrated at the socket interface, amputees prefer lightweight designs. An essential requirement, sometimes overlooked in research, is costs. The resources of healthcare systems are constantly decreasing. Insurance companies assess very carefully whether new expensive medical aids justify the additional costs compared with already existing less expensive solutions. Often the commercial viability of a medical aid depends on such cost units. When cost units are not willing to cover the expenses of a new advanced prosthesis, it may disappear from the market and consequently patients cannot benefit from the new development.

Current commercial prosthetic hands can be categorized roughly into two groups: Single-degree-of-freedom (1-DOF) actuator designs and multiple-DOF actuator designs. The 1-DOF actuator design has one DOF to open and close the hand with one grasping pattern. Since there is only one actuator, the actuator can be large, and therefore these hands are usually very robust, fast, and strong. In contrast, there is a new generation of commercial prosthetic hands that offer more than one actuated DOF and compliant grasping. Both groups, however, share the same control principles based on myoelectric signals, which will be explained later in this section.

The Otto Bock Sensorhand Speed (http://www.ottobock.com), shown in Figure 14.3, is a typical representative of the single-degree-of-freedom actuator designs to open and close the fingers and thumb. Many commercial prosthetic hands have such a design. Hands with a similar cinematic assembly are offered, for instance, by Motion Control Inc (http://www.utaharm.com) and RLSSteeper Inc. (http://www.rlssteeper.com). The grip pattern offered by the Sensorhand Speed is a tripod pinch grip with the thumb opposing the two

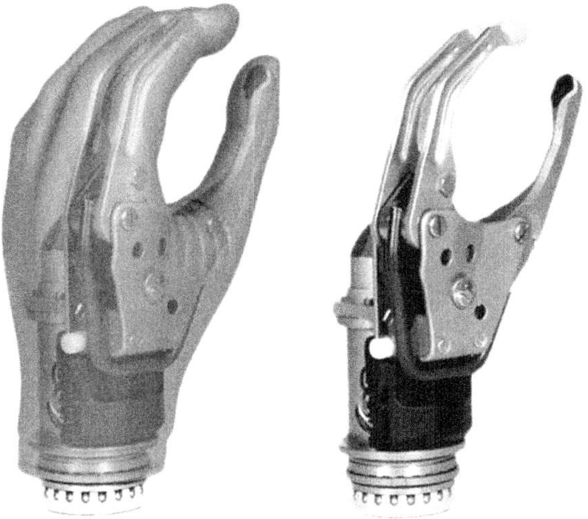

Figure 14.3. Otto Bock SensorHand Speed with an indicated cosmetic cover (left) and without cover (right). (Modified from H.G. Näder, 2011.)

fingers (see Fig. 14.3). The tripod pinch is one of the most common grasp patterns, and it also offers cylindrical grasp for wider objects (Muzumdar, 2004). The hand weighs 460 g and is very robust and durable. The maximum speed is 300 mm/s and the maximum grip force is 100 N. A number of myoelectric control options for discrete and proportional control using one or two bipolar electrodes are possible. To avoid allowing objects to slip, a sensor integrated into the thumb detects when a gripped object is about to slip, and the control system integrated into the prosthetic hand automatically adjusts grip strength so that the object remains securely in the hand. This intelligent behavior together with the speed and strength provided by this hand compensate to some extent the missing somatosensory feedback and the lack of a larger number of DOFs.

The Touch Bionics i-LIMB (http://www.touchbionic.com) is a new-generation prosthetic hand introduced just a few years ago. It was the first hand with articulated fingers on the market. Each of the five digits of the i-LIMB has two joints. Each finger is individually actuated by a micromotor contained in the finger. The thumb has an additional DOF for abduction and adduction. This motion, however, can only be performed passively. Due to the two DOFs per finger, the fingers can curve around an object. A control mechanism determines when each finger has achieved sufficient grip. Then the fingers lock in this position until the amputee triggers an open signal. The grip patterns offered by this hand include a key grip, a power grip, and a precision grip. Additionally, there is the finger-pointing grip with the index finger extended and all other fingers closed against the palm. To select a particular grip, the amputees have to manually rotate the thumb in different positions (Conolly et al., 2008). They can do that either with the healthy hand (in the case of a unilateral amputation) or by applying external forces such as moving the thumb against other objects. A similar prosthetic hand is the RSL Steeper, BeBionic hand (http://www.rslsteeper.com). This hand also offers fingers that are separately actuated and curl as they flex around an object.

The traditional 1-DOF prosthetic hands have rigid digits. Such digits cannot wrap around an object in a natural way. As a consequence, the contact areas between the objects and the prosthetic hand can be small. Due to the lack of friction and compliance the actuator has to provide stronger forces to ensure stable grasping. Hand designs with fingers that can wrap around an object and thus can spread the grip across a larger surface area may reduce the need for greater grip force (Carrozza et al., 2006).

Another new prosthetic hand that is about to enter the market (as of 2012) is the Otto Bock Michelangelo hand (http://www.ottobock.com). Its design is a trade-off between a fully articulated and a 1-DOF hand. It combines high strength and speed (grip force, 120 N; speed, 408 mm/s) with an increased number of DOFs and a set of additional compliance features. Each finger has its own motion axis (flexion——extension) defined by the joint that connects the finger to the metacarpus. The thumb has two perpendicular motion axes (abduction–adduction, flexion–extension). Thus the hand has 6 DOFs. Similar to the relaxed fingers of the human hand, all digits are slightly arch-shaped. This gives the hand a neutral look and appealing cosmetic appearance. The hand has two actuators. The main actuator provides the grasping motions. It drives flexion and extension of the thumb, index finger, and middle finger, which are mechanically linked. The second actuator, which is considerably smaller, drives the abduction and adduction of the thumb. These two separate motions of the thumb allow two different grasping patterns: tripod pinch and lateral pinch. Closing the hand while the thumb is in a palmar position simultaneously performs a finger adduction. This can be used to hold thin objects between the fingers, for instance, a paper or some bills. Moreover, the hand is equipped with compliance elements such as soft finger tips, passive finger movements in palmar direction (ring and little finger in all directions), and a force-balancing mechanism between the index and middle fingers. These elements may further increase the performance of this prosthetic hand in vocational and ADL tasks (Puchhammer, 2008). Figure 14.4 shows the Michelangelo hand and some of these compliance elements.

Although the new generation of commercial prosthetic hands offers more grasping options and compliance features, the performance of these hands is still very limited compared with the human hand. A number of intrinsic prosthetic hands developed in research provide a larger number of actuated DOFs, more grasping patterns, and various control mechanisms. Some of these hands even provide proprioceptive and exteroceptive sensory information. Examples are the Southampton-REMEDI hand (Light and Chappell, 2000), the RTR II hand (Massa et al., 2002), the MANUS hand (Pons et al., 2004), the Fluidhands (Gaiser et al., 2009; Schulz et al., 2005), the DLR hands (Butterfass et al., 2001; Liu et al., 2007), the Cyberhand (Carrozza et al., 2006),

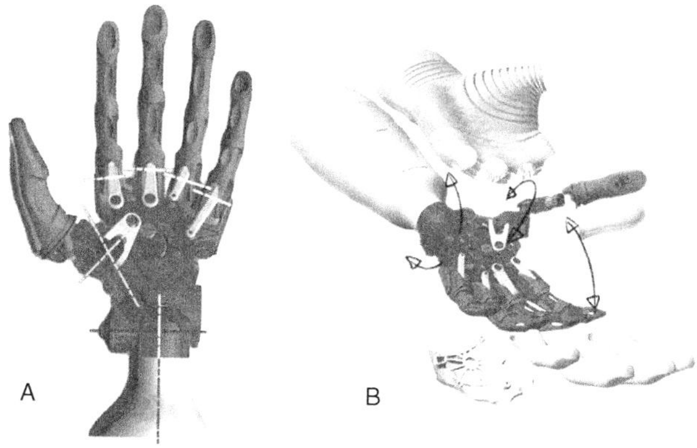

Figure 14.4. Otto Bock Michelangelo hand and wrist: (A) Dashed lines indicate joints and axes of motion. (B) The hand with compliance elements. Arrows indicate the possible motion directions of the digits. (Modified from H.G. Näder, 2011.)

and the Smarthand (Cipriani et al., 2009). A very recent hand is an 18-DOF intrinsic prototype hand developed jointly by the Rehabilitation Institute of Chicago, Otto Bock HealthCare, and Johns Hopkins University. It is designed to mimic the biomechatronic functionality of the human hand and to be robust enough to withstand the rigors of daily living (Weir et al., 2008). Although such multifunctional hands have the potential to provide amputees with improved prosthetic options, they depend on human–machine interfaces that can exploit the capabilities of these hands. Conventional myoelectric human–machine interfaces for prosthetic control, which will be explained in the next section, provide only very limited information transfer rates. That is, the number of commands that can be detected with sufficient accuracy and within a certain time frame is small. Controlling each actuated DOF directly with such an interface would be too cumbersome and unpractical.

Myoelectric Human–Machine Interfaces

Externally powered prostheses are typically controlled by myoelectric human–machine interfaces. Myoelectric signals (electromyographic signals, EMG signals) are the result of muscle contractions induced by motor commands generated in the central nervous system and sent to the muscles via the peripheral nervous system. For this reason, myoelectric human–machine interfaces can also be considered as neural human–machine interfaces (Ohnishi et al., 2007).

Conventional myoelectric control of prosthetic devices uses surface electrodes to record voluntary EMG signals. Although electrodes with electrode gel (wet electrodes) provide better signal quality, they are not an option in

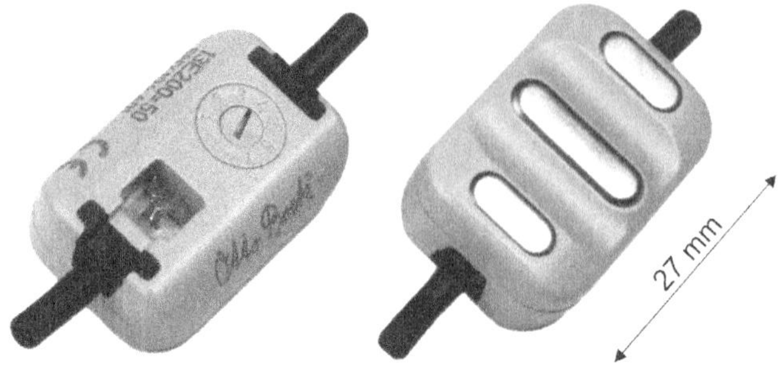

Figure 14.5. Active bipolar myoelectric electrodes. These electrodes provide adjustable amplification and do not require gel or any special skin preparation.

prosthetic control, because they may dry out and are inconvenient to set up. Dry bipolar electrodes are typically used instead (Fig. 14.5). To achieve an acceptable signal-to-noise ratio, such an electrode not only picks up the potential differences from the skin, but also already amplifies them at the recording site (active electrode). Integrated into a socket, the electrodes are automatically placed over the residual muscles during the donning process. No particular skin preparation or manual electrode placement is necessary. However, the prosthetist has to take special care in the design of the socket to ensure consistent electrode positioning as well as reliable contact between the electrodes and the skin. The actual placement of the electrodes depends on the residual muscles, the condition of the residual limb, and the control strategy.

Common electrode configurations in commercial systems employ one or two electrodes. The two-site system uses two bipolar electrodes placed over the antagonistic residual flexor and extensor muscles. In cases where two sufficiently independent signals cannot be derived, for instance, due to a difficult stump condition, a single-site system is used (Muzumdar, 2004). Regardless of whether one or two site systems are used, conventional systems employ very simple signal processing: The signal is filtered to eliminate or reduce the influence of the power line and other external artifacts (noise). To have a measure for the muscle activity level, the mean absolute value or a similar estimate of the amplitude of each myoelectric signal is calculated. This signal feature is usually extracted by averaging (low-pass filtering) the rectified signal. Thresholds determined heuristically during the fitting or rehabilitation process define when a muscle contraction is detected. In fact, either the amplitude or the changing rate of the signal feature is used to detect appropriate muscle contractions (Muzumdar, 2004). All commercial myoelectric systems perform such or similar signal processing. Currently, there are no commercial systems available that can automatically determine the thresholds in an optimal and adaptive way to account for changes in the properties of the myoelectric signal.

Using this signal processing scheme, various control strategies are possible. In a single-site system, a simple switch (on/off) control can trigger hand closing or hand opening. In a two-site system, the contraction of the residual flexor muscles closes the hand; the contraction of the residual extensor opens the hand. The simultaneous contraction of flexor and extensor muscles, called a cocontraction, switches to another DOF; for instance, from the hand to the wrist. Then the contraction of the flexor or extensor muscle results in pronation or supination, respectively. When the amputee is able to voluntarily modulate the strength of the required muscle contractions, the speed and the strength of a prosthetic action (e.g., grasping) can be controlled proportionally to the amplitude of the myoelectric signal.

The control schemes just described require one or two largely independent EMG signals to control one DOF. Since there is a one-to-one association between EMG activity of one muscle (or muscle group) and the prosthetic command, these are **direct control** schemes. These control schemes, however, are not very suitable for prosthetic systems with more than two DOFs, because the number of independent EMG signals available is very limited. Although cocontraction of two muscles can be used to sequentially select different joints to be controlled, with more than two DOFs such sequential control becomes cumbersome, slow, error prone, and difficult for the user to learn. New approaches based on advanced signal processing and pattern recognition promise more independent control signals for improved prosthetic functionality (Englehart et al., 2000). However, at the moment there is no commercial system that utilizes these new approaches.

Targeted Muscle Reinnervation

Selective nerve transfer or targeted muscle reinnervation (TMR) is a new surgical procedure that has been successfully used to develop a clinically viable neural machine interface for people with high-level amputations. TMR is based on three facts (Kuiken et al., 2007): (1) The residual nerves of the peripheral nervous system that once innervated the amputated limb still carry the motor control and sensory information. (2) These nerves can be transferred to other muscles or muscle groups, and they can reinnervate these muscle. (3) After a rehabilitation period these residual nerves control the newly innervated muscles. The reinnervated muscles then have the role of biological amplifiers of the motor control information transmitted by the transferred nerves. Therefore, TMR provides myoelectric signals that are closely related to the myoelectric activity of the lost limb. In patients with shoulder disarticulation, for instance, the pectoralis ipsilateral to the amputation is typically selected for TMR. This is because this muscle is no longer biomechanically functional after the arm has been removed. Often the targeted muscle is surgically divided into several segments. Then the residual nerves (e.g., median, radial, ulnar, and musculocutaneous) are transferred to these segments. The purpose of this segmentation is to increases the independence of the muscle activity invoked by the transferred nerves. After rehabilitation, the muscle segments are reinnervated with nerves that once innervated the lost limb, and

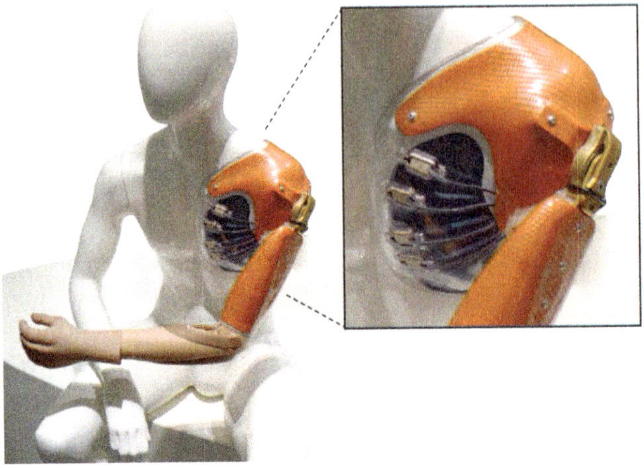

Figure 14.6. Example of a prosthetic fitting with a commercial TMR prosthesis (Otto Bock HealthCare Products GmbH, Duderstadt, Germany).

patients can produce muscle contraction patterns by attempting to move their missing limb. Advanced signal processing techniques are required to fully utilize these patterns and transform them into appropriate control signals for prostheses (Kuiken et al., 2007). However, TMR can provide enough independent muscle activity with sufficiently good signal-to-noise ratio so that even with simple direct control approaches a limited number of DOFs can be controlled simultaneously. In fact, TMR is already a clinically viable treatment for high-level amputees. It has already been successfully applied to more than 40 amputees with shoulder disarticulation and transhumeral amputation worldwide (Schultz and Kuiken, 2011; Kuiken et al., 2009; Dietl, 2008; Miller et al., 2008a,b). Figure 14.6 shows a prosthesis with three DOFs and direct control specifically designed for TMR patients. The selected nerve transfer is not limited to motor nerves. Targeted sensory reinnervation (TSR), which is the selected nerve transfer of sensory nerves, is also possible. It provides the patient with a sense of touch or vibration in the reinnervated area of the skin, for example, in the skin segments near or over the TSR site. Since the transferred sensory nerves are still mapped to the somatosensory cortical areas of the missing limb, the amputee perceives a stimulation of these areas as if the missing limb (phantom limb) were being stimulated (Marasco et al., 2009). More details about TMR and TSR can be found in Chapter 15 ("Myoelectric Prostheses and Targeted Reinnervation") of this book.

FUTURE DIRECTIONS

Technological advances in power sources, actuators, and transmission systems, as well as sensors and control systems build the mechatronic foundation for more functional and sophisticated prosthetic devices. The improved research funding situation for artificial limbs over the past decades, in particular in North America and Europe, has considerably increased the research interest in this field with a focus on mechatronics and neural–machine interfaces. Current anthropomorphic artificial hands in research such as those hands mentioned previously utilize these new technological advances. Despite this progress in research, these new dexterous devices are not yet clinically viable, mostly because they cannot be controlled satisfactorily by amputees. There are basically two ways to overcome this problem. One way is to increase the reliability and information transfer rate of the neural–machine interface. The other way is to develop new control strategies that employ more automatic control embedded in the prosthesis to reduce the attention required by the user. Such low-level control mechanisms are already available to a limited extent in some of the commercial systems. For instance, the SensorHand Speed can adjust the grip strength automatically to prevent objects from slipping. As another example, the i-LIMB automatically performs a gripping pattern till sufficient grip is achieved and then locks the fingers. In both cases, no intervention from the amputee is necessary during these low-level control processes. Of course, the amputee has to initiate these control mechanisms by sending appropriate (high-level) commands over the human–machine interface, but then the action is performed automatically by the prosthesis. Future commercial prostheses may have more sensors and more low-level control algorithms and allow sophisticated combinations of high-level control by the amputee and low-level control by the embedded system. This "shared control" between amputee and prosthesis could be individually adapted to the needs of amputees (Micera et al., 2010; Cipriani et al., 2008). Amputees with limited high-level control capabilities (e.g., due to a difficult stump condition) could benefit greatly from the embedded low-level control. In contrast, amputees able to achieve sufficient information transfer rates with the available human–machine interfaces may feel more comfortable taking over most or even all of the low-level control by themselves. In any case, new mechatronic developments in prosthetics will have to satisfy the requirements mentioned earlier, and these new shared-control strategies will have to be intuitive or at least easy to learn. Although such approaches may depend a little less on efficient communication channels between amputee and prosthesis, there is no doubt that improved human–machine interfaces are of great importance for more reliable and functional upper limb prostheses. In fact, the lack of sufficiently powerful human–machine interfaces—or, more specifically, neural–machine interfaces—is the major limiting factor at the moment in upper limb prosthetics. Considerable research has been devoted to improving this situation by developing advanced signal processing methods and by designing new interface technologies.

Advanced Signal Processing and Pattern Recognition

Conventional direct control uses simple signal processing to calculate the amplitude or the amplitude change rate of the signals recorded. To work sufficiently well in terms of reliability and response time, this approach requires largely independent muscle activity. With surface EMG electrodes, only a small number of isolated EMG signals are available. Usually, only one or two electrodes are used with this approach in commercial systems. Even with TMR the number of independent signals may remain still too limited for controlling prostheses with a large number of DOFs by direct control. Advanced signal processing has the potential to extract more information from the recorded signals by decomposing these signals into components with increased independence, improved signal-to-noise ratio, and better spatial resolution. The goal is to transform the recorded signals into independent control signals that allow simultaneous and proportional control of multiple DOFs with a performance that is sufficient to carry out ADL tasks in an intuitive or at least easy-to-learn way (Parker et al., 2006).

Many different signal processing methods have been suggested since the early 1990s. Most methods require multivariate input recorded from a larger number of electrode sites. There are methods that transform the recorded signals into continuous control signals by employing linear or nonlinear transformations that are based on physiologically inspired models or derived from the recorded EMG signals by machine learning procedures (Englehart et al., 2000; Jiang et al., 2010; Ajiboye and Weir, 2009; Antfolk et al., 2010; Jiang et al., 2009; Muceli et al., 2010; Nielsen et al., 2011). Another group of methods categorizes the recorded signals into certain patterns that are generated by muscle activities. These pattern recognition methods transform the recorded signals not in continuous but rather discrete control signals (Oskoei and Hu, 2007). More information about these signal processing approaches can be found in Chapter 15 ("Myoelectric Prostheses and Targeted Reinnervation") in this book.

There are a number of methods that have achieved classification rates of more than 90% in classification tasks with up to 12 or even more classes. This might seem impressive at first sight; however, the clinical impact of pattern recognition approaches in prosthetic control is very limited. In fact, so far there is no commercial system available that makes use of these approaches. One reason for this unsatisfactory transfer from academic research to clinical practice can be explained by the fact that classification accuracy is the most often applied performance measure. Classification accuracy alone, however, is not sufficient for assessing the usability and clinical viability of a prosthetic system. More specifically, even high classification rates are not very meaningful if the controller is evaluated offline, no strict separation between training and testing data is performed, and signals are not recorded from amputees but rather from able-bodied subjects. Furthermore, many experiments reported in the literature were carried out under ideal lab conditions. These ideal environmental and experimental parameters are not available in practical applications. Signals are often susceptible to artifacts and their statistical properties are nonstationary; that is, they change their statistical properties over time. In fact, there are a number of ways signal parameters can change. The reasons include electrode impedance changes, mechanical displacement of electrodes, user fatigue, and user adaptation. New research efforts have been trying to increase the robustness and reliability of pattern recognition methods (Huang et al., 2010; Kaufmann et al., 2010; Hargrove et al., 2008; Scheme et al., 2010; Simon et al., 2009). One direction of research involves attempts to consider signal parameter changes in the *setup* of the classifier by including parameter variations such as electrode displacements in the training data (Hargrove et al., 2008). Another research direction investigates adaptive methods to alleviate the performance degeneration due to signal nonstationarities (Sensinger et al., 2009). The lack of adequate adaptive signal processing and classification is one of the limiting factors in the clinical applicability of pattern recognition approaches (Zecca et al., 2002). To be clinically useful, the algorithmic complexity of these methods has to be small enough so that they can run on embedded systems with modest computational power and memory resources. In any case more research is necessary before these methods will become available in commercial systems.

Advanced Neural–Machine Interfaces

Improved signal processing methods have the ability to increase the information transfer rate of human–machine interfaces. On the other hand, improved signal acquisition methods could simplify signal processing and increase the robustness of the overall human–machine interface. When the available information is limited by the signal acquisition method, then even advanced signal processing cannot improve much. This underpins the importance of effective and robust signal acquisition. With surface electrodes, for instance, the myoelectric signal of individual muscles can be difficult to record. Due to volume conduction and the simultaneous activation of muscles, a mixture of signals generated by different muscles is usually recorded from one electrode site. Although advanced multivariate signal processing can effectively reduce this cross talk, signal acquisition approaches that record closer to the signal sources may eliminate this problem completely. Moreover, the difficulties with establishing sufficient and consistent electrode contact, correct electrode placement, and the influence of other environmental factors could be eliminated or at least greatly reduced with invasive interfaces. In other words, invasive signal acquisition could solve many problems inherent in noninvasive approaches.

A number of interfaces with muscles or with the peripheral nervous system using different levels of invasiveness have been proposed (see, e.g., Micera et al., 2010; Yoshida et al., 2010; Navarro et al., 2005). Invasive myoelectric interfaces use either epimysial or intramuscular electrodes. **Epimysial electrodes** are placed on the muscle surfaces. **Intramuscular electrodes** are usually placed into the belly of the muscle. In contrast to surface electrodes, both approaches can record focal myoelectric activity from deep and small muscles. Fully implantable systems designed for prosthetic control using epimysial electrodes (Ruff et al., 2010) and intramuscular electrodes (Weir et al., 2009; Farnsworth et al., 2009; Light et al., 2002) are currently being developed.

To work, muscle interfaces require muscles or at least remnants of muscles that can still be activated to produce myoelectric activity. Depending on the amputation level and the condition of the residual limb, the number of suitable muscles can be insufficient. Targeted muscle reinnervation may be a solution for many of these cases, but interfaces to the peripheral nervous system (PNS) provide direct access to motor and sensory nerves. Since both afferent and efferent residual nerves remain intact long after amputation (Dhillon et al., 2004), interfaces to the PNS offer a bidirectional interface for control and feedback.

The design of electrodes is still an active area of research. Reliable electrodes are the key for the success of invasive human–machine interfaces. To be clinically useful, invasive electrodes have to be integrated into a fully implantable recording and stimulation system with wireless data transfer and power supply. Percutaneous systems for transmitting signals from the nerve to an external device are not an option, because of the risk of infections and wire breakage. In fact, the system has to be suitable for chronic, long-term, stable, maintenance-free, and safe implantation. It has to be fully biocompatible and mechanically robust, and the implantation procedure should be as simple as possible—ideally in a single surgical intervention. Before such invasive muscle or PNS interfaces will become common clinical practice more research and long-term *in vivo* tests are necessary.

Details about invasive interface technologies can be found in Chapter 4 ("Progress in Peripheral Neural Interfaces"), Chapter 5 "Multimodal, Multisite Neuronal Recordings for Brain Research"), and Chapter 16 ("Controlling Prostheses Using PNS-Invasive Interfaces for Amputees") in this book.

Sensory Feedback

As already mentioned earlier, visual information is the primary form of feedback in upper limb prosthetics. Visual feedback, however, often fails to reliably provide the information needed. Or at least, getting the required information by visual feedback alone may be cumbersome. For instance, the prostheses can block the view and compensatory movements are necessary to get the required visual information. As another example, a nontransparent container could contain a heavy load or could be empty, but visual information does not help to decide that. Applying sufficient grip for such an object can be difficult. Although local control mechanisms such as the one in the Otto Bock Sensorhand (see previous discussion) can help to prevent the object from slipping, sufficiently graded somatosensory feedback would be very important for intuitive prosthetic control. Sensory substitution, which replaces a lost sensory modality with a still existing modality, is one possible solution to close the feedback loop more explicitly than in current prostheses (Lundborg and Rosén, 2001). Noninvasive techniques for applying vibration, pressure, temperature, and electrical stimulation have been suggested to induce feedback (Micera et al., 2010; Antfolk et al., 2010; Rosén et al., 2009). These techniques try to map information such as forces and position captured by sensors in the prosthesis onto skin areas of amputees. Most often tactile feedback is induced. Proprioceptive feedback is more difficult to realize, but it would be important for more intuitive prosthetic control (Wheeler et al., 2010). In any case, unnatural feedback modalities and sensory substitution requires training. The amputees need to learn to interpret the unnatural stimuli and correctly map this feedback to the information it represents. Targeted sensory reinnervation (TSR) has the potential to simplify this learning process. In TSR, afferent nerves that once innervated the lost limb are redirected to previously deinnervated skin areas in or near the residual limb (Kuiken et al., 2007; Marasco et al., 2009). After the reinnervation process, sensory information from these areas is linked to the somatosensory areas in the brain that once were connected to the receptors in the missing limb. As a consequence, when these areas are mechanically stimulated, the amputee perceives the sensation not on these reinnverated skin areas but in the missing limb (phantom limb). TSR may, therefore, be used as a noninvasive portal to the sensory system to provide intuitive sensory feedback to the users of prostheses. Invasive interfaces that can directly connect to sensory pathways have the potential to provide a more selective feedback for both exteroceptive and proprioceptive perception. But as explained earlier there are many challenges to overcome before such invasive interfaces and the required electrodes will become clinical practice.

CONCLUSION

Over the last decades, advances in commercial prosthetics for upper limbs have been modest. The designs of body-powered or externally powered commercial systems have changed very little for a long time. New materials, rechargeable batteries with enhanced energy-to-weight density, and increasingly powerful actuators have led to redesigns of the established prosthetic components. These incremental design improvements have resulted in stronger and faster prosthetic components with improved durability, but they have not significantly changed the functionality, or the degrees of freedom, or the way these prostheses are controlled. Only recently have new commercial prosthetic developments such as the Touch Bionics i-LIMB or the Otto Bock Michelangelo hand become available. Although these new prosthetic hands offer more DOFs, increased functionality, and compliance components, their control mechanisms based on surface myoelectric signals are still very similar to conventional systems. That is, only one DOF at a time can be controlled, and controlling more than one DOF is cumbersome and difficult to learn.

New research efforts supported by increased funding opportunities over the last years have suggested a number of ways to improve the current state of the art in commercial systems. The possible solutions include shared-control approaches, advanced signals processing, selective nerve transfer, new signal acquisition methods, and methods for providing somatosensory feedback. A key component of advancing the state of the art is the human–machine interface for controlling the prosthetic device. Different invasive and noninvasive approaches have been investigated, with promising results for all levels of amputation. However, many of these new technologies still require considerably more research to evaluate and improve reliability and long-term stability. Most of the results have been achieved in lab environments under ideal conditions. To determine the advantages of a new approach and, most important, its clinical relevance and benefit for the amputee, extensive testing in realistic environments with practical evaluation protocols and comparable performance measures is required. Additional research is needed to determine the best methods for validation. Standardized performance measures for prosthetic devices are required to objectively compare the efficacy and clinical success of new approaches. Such performance measures are also needed to justify the increased costs of new developments. Given the rather small number of upper limb amputees and the complexity of the advanced solutions discussed in this chapter as well as the excessive costs for clinical evaluation, there is no doubt that prosthetic products based on these technologies will have to be expensive. Insurance companies will very carefully assess these products, and they will only cover the expenses when the ratio between costs and clinical benefit is appropriate. Often the commercial success—and thus the viability and availability of a prosthetic device—depends on the willingness of insurance companies to pay for it. Consequently, new prosthetic products have to be both clinically and commercially viable.

ACKNOWLEDGMENTS

The authors thank Dr. Lüder Mosler for his helpful comments and for proofreading this chapter. This work has been supported by the German Ministry for Education and Research (BMBF) via the Bernstein Focus Neurotechnology (BFNT) Göttingen and the BMBF project MYOPLANT.

REFERENCES

Ajiboye A.B. and Weir R.F. (2009). Muscle synergies as a predictive framework for the EMG patterns of new hand postures. *J Neural Eng* 6(3). Epub 036004.

Antfolk C., Cipriani C., Controzzi M., Carrozza M.C., Lundborg G., Rosén B., and Sebelius F. (2010). Using EMG for real-time prediction of joint angles to control a prosthetic hand equipped with a sensory feedback system. *J Med Biol Eng* 30(6):399–406.

Atkins D.J., Heard D.C.Y., and Donovan W.H. (1996). Epidemiologic overview of individuals with upper-limb loss and their reported research priorities. *J Prosthetics Orthotics* 8(1):2–11.

Bear M.F., Connors B.W., and Paradiso M.A. (2006). *Neuroscience: Exploring the Brain*. Philadelphia: Lippincott Williams & Wilkins.

Biddiss E. (2009). Need-directed design of prostheses and enabling resources. In: *Amputation, Prosthesis Use, and Phantom Limb Pain: An Interdisciplinary Perspective*, pp. 7–21. New York: Springer.

Biddiss E.A. and Chau T.T. (2007). Upper limb prosthesis use and abandonment: a survey of the last 25 years. *Prosthetics Orthotics Int* 31(3):236–257.

Butterfass J., Grebenstein M., Liu H., and Hirzinger G. (2001). DLR-Hand II: next generation of a dextrous robot hand. In: *Proceedings 2001 ICRA, IEEE International Conference on Robotics and Automation*, 2001, 1:109–114.

Carrozza M.C., Cappiello G., Micera S., Edin B.B., Beccai L., and Cipriani C. (2006). Design of a cybernetic hand for perception and action. *Biol Cybernetics* 95:629–644.

Changmok C., Mihye S., Suncheol K., Wonil P., and Jung K. (2008). Understanding of hands and task characteristics for development of biomimetic robot hands. In: *8th IEEE-RAS International Conference on Humanoid Robots*, pp. 413–417.

Chappell P.H., Cranny A., Cotton D.P.J., White N.M., and Beeby S.P. (2007). Sensory motor systems of artificial and natural hands. *Int J Surgery* 5(6):436–440.

Cipriani C., Zaccone F., Micera S., and Carrozza M.C. (2008). On the shared control of an EMG-controlled prosthetic hand: analysis of user–prosthesis interaction. *IEEE Trans Robotics* 24(1):170–184.

Cipriani C., Controzzi M., and Carrozza M.C. (2009). Progress towards the development of the SmartHand transradial prosthesis. In: *IEEE International Conference on Rehabilitation Robotics, ICORR 2009*, pp. 682–687.

Dhillon G.S., Lawrence S.M., Hutchinson D.T., and Horch K.W. (2004). Residual function in peripheral nerve stumps of amputees: implications for neural control of artificial limbs. *J Hand Surgery* 29(4):605–615.

Dietl S. (2008). Der Weg zur Gedanken gesteuerten Prothese—Steps towards mind-controlled prosthetics. *Orthopädie Technik* 5(6):340–344.

Dillingham T.R., Pezzin L.E., and MacKenzie E.J. (1998). Incidence, acute care length of stay, and discharge to rehabilitation of traumatic amputee patients: an epidemiologic study. *Arch Physical Med Rehabil* 79(3):279–287.

Englehart K., Hudgins B., and Parker R. (2000). Multifunction control of prostheses using the myoelectric signal. In: *Intelligent Systems and Technologies in Rehabilitation Engineering* (ed. Teodorescu H.-N.L. and Jain L.C.). CRC Press, pp. 153–208.

Farnsworth B.D., Talyor D.M., Triolo R.J., and Young D.J. (2009). Wireless in vivo EMG sensor for intelligent prosthetic control. In: *TRANSDUCERS 2009—15th International Conference on Solid-State Sensors, Actuators and Microsystems*, pp. 358–361.

Gaiser I., Pylatiuk C., Schulz S., Kargov A., Oberle R., and Werner T. (2009). The FLUIDHAND III: a multifunctional prosthetic hand. *J Prosthetics Orthotics* 21(2): 91–96.

Hargrove L., Englehart K., and Hudgins B. (2008). A training strategy to reduce classification degradation due to electrode displacements in pattern recognition based myoelectric control. *Biomed Signal Process Control* 3(2):175–180.

Huang H., Zhang F., Sun Y.L., and He H. (2010). Design of a robust EMG sensing interface for pattern classification. *J Neural Eng* 7(5). Epub 056005.

Jiang N., Englehart K.B., and Parker P.A. (2009). Extracting simultaneous and proportional neural control information for multiple-dof prostheses from the surface electromyographic signal. *IEEE Trans Biomed Eng* 56(4):1070–1080.

Jiang N., Falla D., D'Avella A., Graimann B., and Farina D. (2010). Myoelectric control in neurorehabilitation. *Crit Rev Biomed Eng* 38(4):381–391.

Kaufmann P., Englehart K., and Platzner M. (2010). Fluctuating EMG signals: investigating long-term effects of pattern matching algorithms. In: *2010 Annual International Conference of the IEEE Engineering in Medicine and Biology Society, EMBC'10*, pp. 6357–6360.

Kuiken T.A., Miller L.A., Lipschutz R.D., Lock B.A., Stubblefield K.A., Marasco P.D., Zhou P., and Dumanian G.A. (2007). Targeted reinnervation for enhanced prosthetic arm function in a woman with a proximal amputation: a case study. *Lancet* 369(9559):371–380.

Kuiken T.A. , Li G., Lock B.A., Lipschutz R.D., Miller L.A., Stubblefield K.A., and Englehart K.B. (2009). Targeted muscle reinnervation for real-time myoelectric control of multifunction artificial arms. *JAMA—J Am Med Assoc* 301(6):619–628.

Light C.M. and Chappell P.H. (2000). Development of a lightweight and adaptable multiple-axis hand prosthesis. *Med Eng Physics* 22(10):679–684.

Light C.M., Chappell P.H., Hudgins B., and Engelhart K. (2002). Intelligent multifunction myoelectric control of hand prostheses. *J Med Eng Technol* 26(4):139–146.

Liu H., Meusel P., Hirzinger G., Jin M.H., Liu Y.W., and Xie Z.W. (2007). The modular multisensory DLR-HIT-Hand. *Mechanism Machine Theory* 42(5):612–625.

Lundborg G. and Rosén B. (2001). Sensory substitution in prosthetics. *Hand Clin* 17(3):481–488.

Marasco P.D., Schultz A.E., and Kuiken T.A. (2009). Sensory capacity of reinnervated skin after redirection of amputated upper limb nerves to the chest. *Brain* 132(6):1441–1448.

Massa B., Roccella S., Carrozza M.C., and Dario P. (2002). Design and development of an underactuated prosthetic hand. In: *Proceedings ICRA '02, IEEE International Conference on Robotics and Automation*, 4:3374–3379.

Micera S., Carpaneto J., and Raspopovic S. (2010). Control of hand prostheses using peripheral information. *IEEE Rev Biomed Eng* 3:48–68.

Miller L.A., Lipschutz R.D., Stubblefield K.A., Lock B.A., Huang H., Williams T.W., Weir R.F., and Kuiken T.A. (2008a). Control of a six degree of freedom prosthetic arm after targeted muscle reinnervation surgery. *Arch Phys Med Rehabil* 89(11):2057–2065.

Miller L.A., Stubblefield K.A., Lipschutz R.D., Lock B.A., and Kuiken T.A. (2008b). Improved myoelectric prosthesis control using targeted reinnervation surgery: a case series. *IEEE Trans Neural Syst Rehabil Eng* 16(1):46–50.

Muceli S., Boye A.T., D'Avella A., and Farina D. (2010). Identifying representative synergy matrices for describing muscular activation patterns during multidirectional reaching in the horizontal plane. *J Neurophysiol* 103(3):1532–1542.

Muzumdar A. (2004). *Powered Upper Limb Prostheses. Control, Implementation and Clinical Application*. New York: Springer,

Näder H.G., Ed. (2011). *Otto Bock Prosthetic Compendium: Upper Extremity Prostheses*, 2nd ed. Duderstadt: Hans Georg Näder.

Navarro X., Krueger T.B., Lago N., Micera S., Stieglitz T., and Dario P. (2005). A critical review of interfaces with the peripheral nervous system for the control of neuroprostheses and hybrid bionic systems. *J Peripheral Nervous Syst* 10(3):229–258.

Nielsen J., Holmgaard S., Jiang N., Englehart K.B., Farina D., and Parker P.A. (2011). Simultaneous and proportional force estimation for multifunction myoelectric prostheses using mirrored bilateral training. *IEEE Trans Biomed Eng* 58(3):681–688.

Ohnishi K., Weir R., and Kuiken T.A. (2007). Neural machine interfaces for controlling multifunctional powered upper-limb prostheses. *Expert Rev Med Devices* 4(1):43–53.

Oskoei M. and Hu H. (2007). Myoelectric control systems—a survey. *Biomed Signal Process Control* 2(4):275–294.

Parker P., Englehart K.B., and Hudgins B. (2006). Myoelectric signal processing for control of powered limb prostheses. *J Electromyogr Kinesiol* 16(6):541–548.

Pons J.L., Rocon E., Ceres R., Reynaerts D., Saro B., Levin S., and Van Moorleghem W. (2004). The MANUS-HAND Dextrous Robotics upper limb prosthesis: mechanical and manipulation aspects. *Autonomous Robots* 16(2):143–163.

Pons J.L., Ceres R., and Calderon L. (2008). Introduction to wearable robots. In: *Wearable Robots*, pp. 2–16. Hoboken, NJ: Wiley.

Puchhammer G. (2008). Michelangelo 03—a versatile hand prosthesis, featuring superb controllability and sophisticated bio mimicry. In: *Proceedings of the Symposium on Measuring Success in Upper Limb Prosthetics*, Fredericton, NB, p. 162.

Pylatiuk C., Schulz S., and Döderlein L. (2007). Results of an Internet survey of myoelectric prosthetic hand users. *Prosthetics Orthotics Int* 31(4):362–370.

Rosén B., Ehrsson H.H., Antfolk C., Cipriani C., Sebelius F., and Lundborg G. (2009). Referral of sensation to an advanced humanoid robotic hand prosthesis. *Scand J Plastic Reconstructive Surgery Hand Surgery* 43(5):260–266.

Ruff R., Poppendieck W., Gail A., Westendorff S., Russold M., Lewis S., Meiners T., and Hoffmann K.P. (2010). Acquisition of myoelectric signals to control a hand prosthesis with implantable epimysial electrodes. In: *2010 Annual International Conference of the IEEE Engineering in Medicine and Biology Society*, pp. 5070–5073.

Scheme E., Fougner A., Stavdahl Ø., Chan A.D.C., and Englehart K. (2010). Examining the adverse effects of limb position on pattern recognition based myoelectric control. In: *2010 Annual International Conference of the IEEE Engineering in Medicine and Biology Society*, pp. 6337–6340.

Schultz A.E. and Kuiken T.A. (2011). Neural interfaces for control of upper limb prostheses: the state of the art and future possibilities. *PM & R: J Injury, Function, Rehabil* 3(1):55–67.

Schultz A.E., Baade S.P., and Kuiken T.A. (2007). Expert opinions on success factors for upper-limb prostheses. *J Rehabil Res Dev* 44(4):483–489.

Schulz S., Pylatiuk C., Reischl M., Martin J., Mikut R., and Bretthauer G. (2005). A hydraulically driven multifunctional prosthetic hand. *Robotica* 23(3):293–299.

Sensinger J.W., Lock B.A., and Kuiken T.A. (2009). Adaptive pattern recognition of myoelectric signals: exploration of conceptual framework and practical algorithms. *IEEE Trans Neural Syst Rehabil Eng* 17(3):270–278.

Simon A.M., Hargrove L.J., Lock B.A., and Kuiken T.A. (2009). A strategy for minimizing the effect of misclassifications during real time pattern recognition myoelectric control. In: *2009 Annual International Conference of the IEEE Engineering in Medicine and Biology Society*, pp. 1327–1330.

U.S.G.P.O. (2010). *Care of the Combat Amputee*, 2009 ed. Washington, D.C.: U.S. Government Printing Office.

U.K. Prosthetics Services (2008). Amputee Statistical Database for the UK: 2006/07 Report. UK Prosthetics Services.

Weir R., Mitchell M., Clark S., Puchhammer G., Haslinger M., Grausenburger R., Kumar N., Hofbauer R., Kushnigg P., Cornelius V., Eder M., Eaton H., and Wenstrand D. (2008). The intrinsic hand—a 22 degree-of-freedom artificial hand–wrist replacement. In: *Proceedings of the Symposium on Measuring Success in Upper Limb Prosthetics*, Fredericton, NB, pp. 233–237.

Weir R.F., Troyk P.R., DeMichele G.A., Kerns D.A., Schorsch J.F., and Maas H. (2009). Implantable myoelectric sensors (IMESs) for intramuscular electromyogram recording. *IEEE Trans Biomed Eng* 56(1):159–171.

Wheeler J., Bark K., Savall J., and Cutkosky M. (2010). Investigation of rotational skin stretch for proprioceptive feedback with application to myoelectric systems. *IEEE Trans Neural Syst Rehabil Eng* 18(1):58–66.

Yoshida K., Farina D., Akay M., and Jensen W. (2010). Multichannel intraneural and intramuscular techniques for multiunit recording and use in active prostheses. *Proc IEEE* 98(3):432–449.

Zecca M., Micera S., Carrozza M.C., and Dario P. (2002). Control of multifunctional prosthetic hands by processing the electromyographic signal. *Crit Rev Biomed Eng* 30(4–6):459–485.

15

REVIEW CHAPTER

MYOELECTRIC PROSTHESES AND TARGETED REINNERVATION

LEVI HARGROVE,[1,2] ERIK SCHEME,[3] AND KEVIN ENGLEHART[3]

[1]*Center for Bionic Medicine, Rehabilitation Institute of Chicago, Chicago, Illinois, USA*

[2]*Department of Physical Medicine and Rehabilitation, Northwestern University Feinberg School of Medicine, Chicago, Illinois, USA*

[3]*Institute of Biomedical Engineering, University of New Brunswick, Fredericton, New Brunswick, Canada*

SUMMARY

Amputation is a major cause of disability and is often treated most effectively with a prosthetic device. Prostheses may be passive, body powered, or externally powered by means of a battery. Notwithstanding the challenges associated with mechanical design, the natural and intuitive control of prostheses has been the major factor limiting the functionality restored to the patient.

Neural information extracted through the surface electromyographic (EMG) signal has been used successfully as a control source for electrically powered prostheses since the 1940s. A number of clinically viable myoelectric control strategies have been incorporated into a variety of commercially available devices. These strategies, referred to as conventional control, are intuitive when EMG signals are measured from physiologically appropriate muscles, but the number of degrees of freedom restored to the patient is limited by the number

Introduction to Neural Engineering for Motor Rehabilitation, First Edition. Edited by Dario Farina, Winnie Jensen, and Metin Akay.

of independent EMG control sites on the residual limb. There are two developments, however, that promise to considerably advance the state of the art in EMG-controlled prosthetics: pattern recognition and targeted muscle reinnervation.

Pattern recognition based myoelectric control has been an active area of research since the 1970s and has the potential to restore more degrees of freedom than conventional techniques. Pattern recognition systems operate on the assumption that a set of features describing the myoelectric signal at a given electrode location will be repeatable for a given state of muscle activation and distinct from one state of activation to another. Like conventional control strategies, pattern recognition systems are limited by the number of control sites from which EMG may be measured, but the control sites need not be independent.

Targeted muscle reinnervation (Basmajian and De Luca, 1985) is a new surgical technique that restores additional EMG control sites that can be combined with existing conventional control strategies to restore a great deal of functionality to amputees, especially to those with high functional deficits. The reinnervated muscle signals are often the result of complex motor patterns and, consequently, have also been shown to perform exceptionally well with pattern recognition techniques.

This chapter provides an overview of existing myoelectric control strategies, highlighting the benefits and limitations of both conventional and pattern recognition techniques, and targeted muscle reinnervation (TMR). The chapter concludes with a summary of exciting emerging technologies that have the potential to further enhance the field of myoelectric control.

INTRODUCTION

Problem Statement

Amputees suffer severe functional deficits as a result of amputation. Passive and body-powered prostheses have long been used to restore some degree of functionality to patients but they fall drastically short of what can be achieved by an intact limb. Strong and lightweight robotic arms and hands have been designed. The limiting factor in their suitability to act as effective functional replacements is a lack of natural and intuitive control. This is despite the fact that patients often have volitional control over significant musculature remaining on their residual limbs. If decoded properly, electromyographic (EMG) signals can provide pertinent neural information that may be used to help control an electrically powered prosthesis.

The concept of myoelectric control was first introduced in the late 1940s at the Export Mess in Hanover, but little immediate follow-up work was completed (McLean and Scott, 2004). It was not until 1960 that a clinically useful myoelectrically controlled prosthesis appeared and attracted considerable attention (Kobrinski, 1960). This new prosthesis, along with the development of semiconductor devices, led to a dramatic increase in the research area of myoelectric control. An excellent historical perspective of the early development of myoelectric control and prosthetic limbs development was given by McLean and Scott (2004).

A number of factors have led to a resurgence of myoelectric control research since the early 2000s. First, low-power electronics have developed to the point where multichannel pattern recognition algorithms can readily be implemented on an embedded system. Second, due to a large number of high-level amputees resulting from recent military conflicts, governments have initiated well-funded programs to improve neural interfaces for prosthetics. Finally, new and innovative neural–machine interfaces, such as targeted muscle reinnervation (TMR), have been developed to provide a rich source of neural information from which control signals can be derived.

Terminology

Prior to proceeding, it is useful to provide some definitions that are often used in the field of myoelectric control for powered prostheses and are used in the remainder of the chapter. An EMG **control site** is the area on the surface of the skin above a muscle from which the EMG signal is measured. A **channel** of EMG data refers to a bipolar measurement from electrodes located over a control site. A **feature** of the data refers to some descriptor of the measured signal over a given window of time. A **degree of freedom** (DOF) is a rotation about a joint or coordinate axis. A **motion class** is any actuation that the prosthesis is capable of making in one direction along one or multiple DOFs. For example, elbow flexion and elbow extension are motion classes and, together, form the DOF that is rotation about the elbow joint. A motion class also includes preprogrammed, coordinated movements about multiple DOFs such as multiarticulate hand grasps. **Proportional control** maps the intensity of a muscular contraction to a velocity for the actuated class. **Physiologically appropriate** muscles are muscles that act as the prime movers for a given DOF. The biceps is the physiologically appropriate muscle for elbow flexion. A **physiologically appropriate control site** is a control site that is located over a physiologically appropriate muscle. **Independent simultaneous control** refers to the separate control of multiple motion classes at the same time. **Sequential control** refers to the use of multiple motion classes in series.

STATE OF THE ART

The availability of one or more EMG control sites on the residual limb is a *fundamental requirement* of myoelectric control. The level of amputation and the residual limb musculature often determine the number of available EMG control sites and, consequently, the functionality that may be restored to the patient. As discussed in the next two subsections ("Conventional control strategies" and "Pattern recognition"), if sufficient control sites remain, conventional control or pattern recognition control may be used to extract prosthesis commands from the neural information contained in the EMG. As discussed in the third subsection ("Targeted muscle reinnervation"), if there are few remaining control sites, then TMR may be used to restore physiologically appropriate control sites to some amputees. Regardless of the number of control sites or the method of control, a patient may require additional therapy, as described in the fourth subsection ("Physical and occupational therapy"), to reap maximum benefits from the prosthesis.

Conventional Control Strategies

The simplest form of myoelectric control can be termed **direct control**. Direct control measures the EMG from a control site and uses basic signal processing to extract an amplitude feature from the signal, usually a low-pass filtered mean absolute value (MAV) or root mean square (RMS) value of the signal. The amplitude is then compared with a preset threshold to determine if the user intends to make a motion. A simple extension of this type of control, termed **direct proportional control**, uses the value of the amplitude feature to control the speed of actuation of the prosthesis, as it is well known that the amplitude features are proportional to the intensity of a muscular contraction (Parker et al., 1977; Basmajian and De Luca, 1985; Hogan, 1976). Both direct control and direct proportional control strategies have gained widespread clinical use and are intuitive provided that physiologically appropriate control sites are available from which to measure the EMG.

Amplitude-based multistate myoelectric control is closely related to direct control, described previously; however, in this case the dynamic range of the amplitude feature is divided into discrete levels. Each of the levels is mapped to an appropriate motion class of the prosthesis. In real-time operation, the extracted amplitude feature is compared with the level thresholds, and the appropriate motion class is actuated. The three-state controller is the most popular amplitude multistate controller (Dorcas and Scott, 1969). Although it is possible to partition the amplitude range into more than three states, it becomes difficult for the user to control due to the increased cognitive load and the random nature of the EMG (Parker et al., 1977). The benefit of using multistate controllers is that more motion classes can be controlled with a limited number of control sites. Unfortunately, the control sites for those motion classes may no longer be physiologically appropriate, and proportional control is significantly more difficult than in the case of direct proportional control.

Instead of comparing the amplitude feature with a predetermined threshold, the rate of change of the amplitude can also be used for control purposes (Childress, 1966). In this type of control, slowly changing amplitude corresponds to one motion class and quickly changing amplitude corresponds to another. After the motion is rate-activated, proportional control is derived as in the direct control case. Thus, three-state proportional control can be achieved using this type of rate-sensitive controller; however, it too is limited by a lack of physiologically appropriate control.

Mode switching is another important aspect of conventional control (Williams, 1990). The conventional control systems described previously are frequently limited to a single DOF, and a mode switch may be used to support additional DOFs. Mode switching, which is typically activated using muscle cocontraction or a mechanical input, is used to toggle the DOF being controlled. This, for example, can allow a transhumeral amputee to alternate control between elbow and hand DOFs using a single set of control sites over their residual biceps and triceps muscles. While this can require greater user focus, it is often the only way to provide sequential control of multiple DOFs using limited control sites.

Pattern Recognition

Pattern recognition based myoelectric control systems operate on the assumption that, at a given electrode location, the set of features describing the myoelectric signal will be repeatable for a given state of muscle activation and will be different from one state of activation to another (Graupe and Cline, 1975). These systems require that training data be collected from each user so that patterns can be learned by the system. The state of the art in pattern recognition systems has evolved to contain some or all of the components shown in Figure 15.1.

Unlike conventional control, pattern recognition does not require independent EMG signal control sites (Hargrove et al., 2007a). Muscle cross talk can actually add discriminatory information that improves classification performance. Thus, it is possible to record from many more channels than there are independent control sites. Hargrove (Hargrove et al., 2007b) and Li (Li et al., 2009) have shown that there is a dramatic improvement in classification performance for transradial amputees when increasing the number of EMG

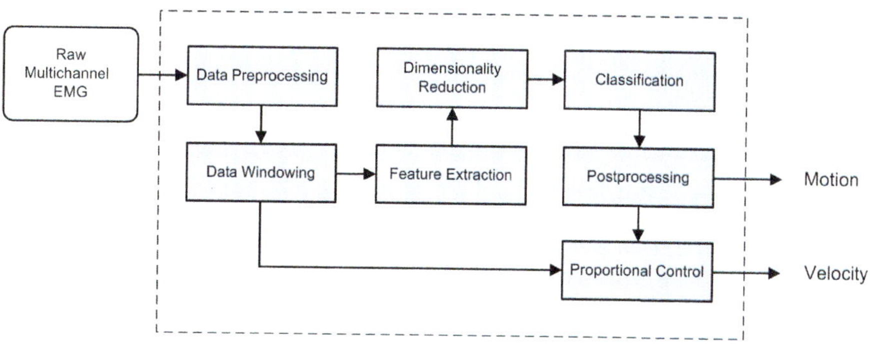

Figure 15.1. Flowchart of state-of-the-art pattern recognition based control.

measurement channels up to approximately 4–6 channels, after which there are diminishing returns. Huang (H. Huang et al., 2008) found that having up to 10 channels was beneficial for decoding neural information from high-level TMR amputees, who tend to have rich and complex neural information sources. The practicality of physically embedding the electrodes in a socket is now the major limitation to the number of channels that can be used as input to the pattern recognition system. Based on the works just cited, it is recommended that *at least* 4–6 EMG channels be used as input to pattern recognition systems to control multifunction hands, wrists, and arms.

Signal preprocessing typically consists of amplification to a convenient value and band-pass filtering between approximately 20 and 500 Hz to make use of the available surface EMG spectrum while minimizing motion artifacts and high-frequency interference. Optionally, notch filters are placed at 50 or 60 Hz and associated harmonics to reduce power-line interference.

The stochastic characteristics of the EMG signal make the instantaneous value unsuitable for control purposes. Consequently, an analysis window must be formed from which features are extracted. The window must be short enough that an excessive delay is not introduced to the controller (Farrell and Weir, 2007), yet long enough that temporally stable features can be extracted so a reliable decision can be formed (Englehart and Hudgins, 2003). Smith (Smith et al., 2011) performed a controllability study in a virtual environment and determined that window lengths in the range of 150–250 ms yielded the most controllable pattern recognition system, supporting a previous claim for pattern recognition systems made by Farrell and Weir (2008).

A feature extraction stage is used to increase the information density of the EMG signals. Ideally, information that discriminates motion classes should be retained while irrelevant information is discarded. Many different feature sets have been proposed, including time-domain (TD) coefficients (Hudgins et al., 1993), autoregressive (AR) coefficients (Graupe et al., 1985), mel-frequency cepstral coefficients (MFCC; Kang et al., 1997), short-time frequency transform, wavelet transform, wavelet packet transform (Englehart et al., 2001; Lucas et al., 2008), and concatenated TD and AR (TDAR) features (Y.H. Huang et al., 2005). A comparison of feature sets has shown that, for slowly varying EMG patterns, a concatenated TDAR feature set outperforms all others (Hargrove et al., 2007a; Y.H. Huang et al., 2005), but the slight improvement in performance over simple time-domain features incurs considerable processing overhead.

If the resultant feature set occupies a high-dimensional space (as is the case with most time–frequency transforms), it may be beneficial to reduce the dimensionality of the feature space by selection of a subset of features (Zhou et al., 2007), or some combination of the features such as averaging (Lucas et al., 2008), principal components analysis (Hargrove et al., 2007b; Englehart et al., 1999; Hargrove et al., 2009a), uncorrelated linear discriminant analysis (Hargrove et al., 2010), independent components analysis (Naik et al., 2008), or nonlinear projection (Chu et al., 2006).

Just as researchers have pursued an extensive range of feature sets, many forms of statistical and learning classifier have been investigated and compared for use in myoelectric control. An extensive examination of classifiers is given in Scheme et al. (2011) and the results are summarized in Figure 15.2. A common observation is that, if an effective representation of the EMG signals is in place (an appropriate feature set and a sufficient number of channels), most modern classifiers will have similar performance. The effect of the feature set is shown in Figure 15.3, which shows the mean and standard deviation of the classification accuracy over 10 able-bodied subjects completing 11 classes of motion, and 5 amputee subjects completing 7 classes of motion. The

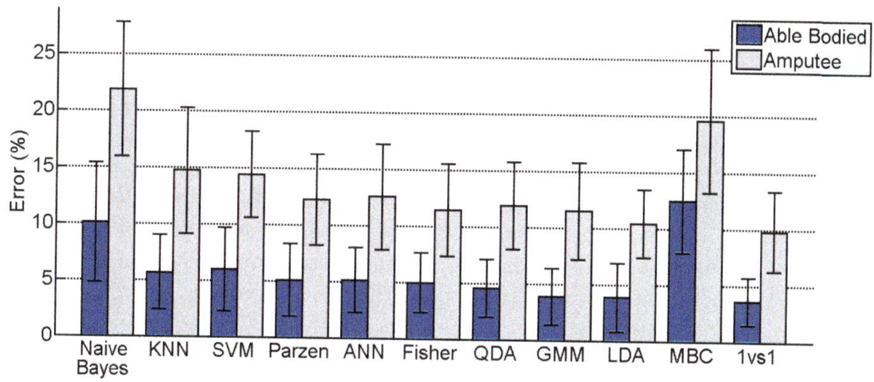

Figure 15.2. Comparison of classifier types for 10 able-bodied users performing 11 classes of motion (left, blue bar), and 5 amputee users performing 7 classes of motion (right, gray bar). Time-domain (TD) features were used for feature extraction.

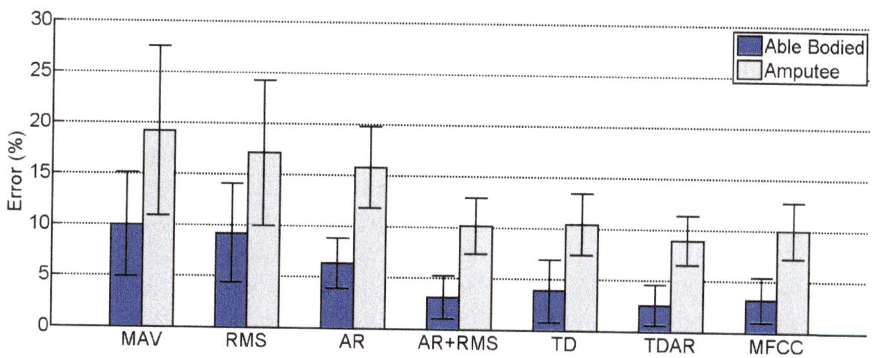

Figure 15.3. Comparison of feature types for 10 able-bodied users performing 11 classes of motion (left, blue bar), and 5 amputee users performing 7 classes of motion (right, gray bar). Classification was performed using an LDA classifier.

implication is that information other than the amplitude of the EMG signal can be used to add valuable class discrimination information and, with an appropriate feature representation, the classification task is essentially a linear problem. The most popular choices of classifiers in recent research are therefore based on only marginal advantages in classification performance. Scheme and colleagues (Scheme et al., 2011) compared the performance of 10 commonly used classification techniques. *K*-Nearest neighbors (KNN) is a simple machine learning algorithm that classifies based on proximity to training exemplars (Cover and Hart, 1967). Support vector machines (SVMs) are increasingly popular machine learning tools that use kernels to map data into separable hyperplanes (Lucas et al., 2008). A Parzen classifier estimates the class probability density functions nonparametrically using kernel density estimation (Parzen, 1962). A neural network (NN) is an iterative learning algorithm that minimizes the error between system outputs and desired value (Hudgins et al., 1993). Fisher discriminant analysis uses data projection to maximize the distance between class means while minimizing the within-class variance (Duda et al., 2000). Quadratic discriminant analysis (QDA) and linear discriminant analysis (LDA), much like Fisher, are statistically based classifiers that seek to find combinations of features that optimally discriminate between classes. LDA is a simplified version of QDA, assuming homoscedasticity (common class covariance). Gaussian mixture models (GMM) use an iterative expectation-maximization algorithm to maximize the likelihood output from each model given the class input data (Y.H. Huang et al., 2005). Multiple binary classifiers (MBCs) employ heuristics to interpret a one-versus-all type classification scheme preprocessed with uncorrelated linear discriminant analysis (ULDA) data feature reduction of the underlying two-class problems (Hargrove et al., 2010). A variation of the technique that yields slightly higher classification accuracy may be used by employing heuristics to interpret a one-versus-one classification scheme preprocessed with ULDA feature reduction of a set of two-class problems (Scheme et al., 2011).

In clinically implemented systems, the LDA is a common choice because of its relative ease of implementation (especially as an embedded process) and ease of training. Consequently, pattern classifications can be made very quickly and postprocessing can be used to refine the decision stream. Majority voting is useful for removing spurious misclassifications in the decision stream and to provide smooth transitioning between motions (Englehart and Hudgins, 2003; Chan and Englehart, 2005). A sliding buffer of preceding decisions is retained, and the control output is overridden to be its most frequently appearing motion class. This acts as a form of low-pass filter, eliminating abrupt changes due to misclassification. Another particularly effective form of postprocessing is a velocity ramp function (Hudgins et al., 1993; Simon et al., 2010). In this approach a buffer is again retained, but instead, the proportion of decisions that coincide with the current decision is computed. This result is then passed through an activation function and used to attenuate the speed of the prosthesis. As with majority vote, the ramp function smoothes spurious mis-

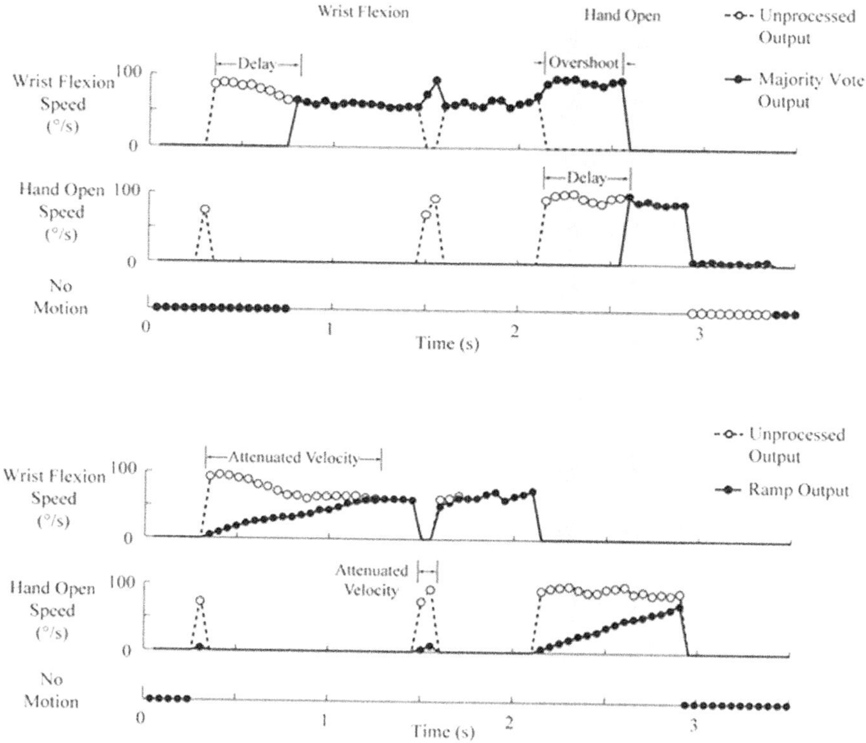

Figure 15.4. Comparison of majority vote (top) and velocity ramp (bottom) postprocessing methods.

classification and transitions between motion classes, but it also eliminates some of the problems associated with it, such as controller delay and overshoot (as shown in Figure 15.4).

When both majority vote and the velocity ramp were tested in a functional block-stacking task, it was found that both methods reduced misclassifications (there were fewer dropped blocks than when no postprocessing method was used) but the velocity ramp postprocessing method yielded more stacked blocks.

Proportional control is more challenging in pattern recognition systems compared with conventional control systems because there is a many-to-many mapping between channels and the DOFs under control. A naïve method of deriving a proportional control signal is to take the average RMS of all channels used in pattern recognition. A more physiologically correct method is to compute a linear combination of the RMS values across these channels (Lock et al., 2005). This intensity calculation can then be normalized by a motion class specific factor which is found using training data (Hargrove et al., 2009a).

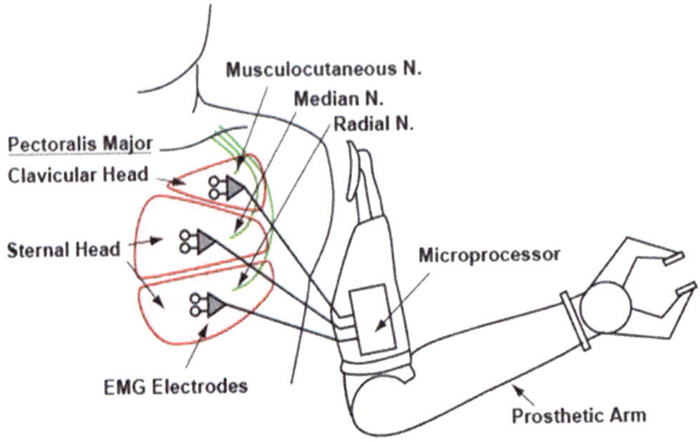

Figure 15.5. Illustration of targeted muscle reinnervation. (Figure courtesy of the Rehabilitation Institute of Chicago.)

Targeted Muscle Reinnervation

Conventional and pattern recognition control techniques both require control sites from which to measure EMG. Furthermore, these methods are most effective when physiologically appropriate control sites are used. Unfortunately, those with high-level limb deficiency, such as shoulder disarticulation or transhumeral amputees, have the most functional deficit but the fewest remaining control sites. To overcome this paradox, TMR surgery was developed and refined by Dr. Todd Kuiken and colleagues at the Rehabilitation Institute of Chicago (Kuiken et al., 2009, 2007; Kuiken, 2003). TMR, as illustrated in Figure 15.5, uses the residual nerves from an amputated limb and transfers them onto alternative muscles that are not biomechanically functional since they are no longer attached to the missing arm.

During the nerve transfer procedure, target muscles are denervated from their original nerve supply, and the remnant nerves from the amputated limb are anastomosed onto these muscles. Reinnervation typically occurs within 3–6 months of the TMR procedure. Once reinnervated, the muscles then serve as natural biological amplifiers of the nerve signals and the neural information can readily be detected using surface EMG electrodes.

The most compelling use for TMR is for shoulder disarticulation or transhumeral amputees. In such patients the functional deficit is severe and the control sites required to restore elbow, wrist, and hand functions are limited. For shoulder disarticulation amputees, the four major brachial nerves (median, radial, musculocutaneous, ulnar) can be transferred to distinct regions on the pectoralis and/or serratus muscles (Kuiken et al., 2007, 2009). For transhumeral amputees, the neural information pertaining to hand and wrist function can be restored by transferring the median nerve to the medial biceps, and the distal radial nerve to the brachialis or lateral triceps. The remaining heads of the biceps and triceps may still be used to provide control sites for elbow function. TMR can also be performed for transradial amputees. In these patients, there is usually sufficient remaining musculature to control the degrees of freedom at the wrist and a simple hand grasp. TMR may be used to regain neural information from intrinsic hand muscles to allow additional and more complex hand grasps to be restored.

Targeted muscle reinnervation control sites can be used with either conventional or pattern recognition based control strategies. Direct proportional control may be used with four TMR control sites to provide simultaneous, proportional control of the elbow and hand. Furthermore, mode switching can be used so that the control sites for the hand can alternatively be used to operate a wrist. In a recent study, this scheme was found to provide a 250% increase in a "box-and-blocks" functional test and a 26% increase in speed to complete a "clothespin placement" function test for a bilateral shoulder disarticulation amputee in comparison to the performance before TMR surgery. The improvements for a transhumeral amputee were even more pronounced, yielding a 300% increase in box-and-blocks performance and 86% increase in clothespin placement speed (Kuiken et al., 2005).

By employing pattern recognition using the TMR sites, the neural information can be decoded into a large number of motion classes. Zhou (Zhou et al., 2007) demonstrated that a shoulder disarticulation patient was able to discriminate 27 classes of upper limb function with greater than 97% accuracy. This finding was very exciting and can be explained by the fact that the remnant nerves are usually transferred from a rather proximal location along their course, resulting in a mixture of various branches of the nerve within the fascicles. The consequence is that the new EMG sites, in general, contain a mixture of functions corresponding to the various branches of each nerve. TMR in combination with pattern recognition has been found to perform well in real time with both shoulder disarticulation and transhumeral amputees, and clinical experience suggests that pattern recognition will provide further performance increases over conventional techniques.

Physical and Occupational Therapy

Proper therapy can dramatically impact a patient's ability to successfully use an EMG-controlled prosthesis to complete activities of daily living. Residual limb muscles often atrophy because of disuse after amputation. Consequently, the EMG signal amplitudes are often much lower than those measured from able-bodied individuals. A therapist should work the patient to establish muscle strengthening exercises to develop as much muscle as possible. When using conventional control techniques, a therapist must also train the patient to independently activate agonist and antagonist muscles. Alternatively, if only one control site is available, the therapist must train the patient to contract at distinct and repeatable levels so that three-state control can be reliably achieved. This phenomenon was illustrated by Scott and colleagues (Scott et al., 1978) over a six-month experiment in which users controlled a three- and five-state myoelectric control system for an hour each day. The results indicated that the subjects significantly improved performance over the first 10 days of the experiment, after which time they plateaued at a reasonable performance level. When tested four months later during a follow-up experiment, the subjects were found to have retained their training and saw only modest further increases. This evidence strongly argues for intense therapy over the course of several days during the initial prosthesis fitting.

There have been no long-duration studies that have investigated the role of therapy in the performance of pattern recognition control systems, but improvements in performance have been noted during short-duration studies (Hargrove et al., 2010). Clinical observations suggest that therapy is very important to help patients make repeatable contractions. For unilateral amputees, the therapist often instructs the patient to move the unaffected side at the same time they are attempting to move their phantom limb. In the case of bilateral amputees, EMG feedback is often the only mechanism available to determine if distinct patterns can be generated.

Finally, the therapist must teach the patient alternative strategies to complete activities of daily living. In the case of unilateral amputees, the prosthesis always acts as the nondominant hand and is therefore used exclusively in a supporting manner. Additionally, current pattern recognition systems only provide sequential control, requiring that the patient make a conscious effort not to generate EMG patterns corresponding to combined motions. As pattern recognition systems are deployed in clinical situations, it is expected that more therapeutic techniques will become available to improve control in real-life situations.

CURRENT CHALLENGES AND FUTURE DIRECTIONS

As demonstrated by many studies, the accuracy of current EMG pattern recognition systems is sufficient for reliable control under ideal conditions. Classification accuracies of 90% or greater have been shown to be sufficient for reliable real-time control in controlled laboratory settings (Hargrove et al., 2010; Kuiken et al., 2009; Hargrove et al., 2007a). Clearly, this is conditioned by the capabilities of the user and an appropriate selection of motion classes, but, when coupled with appropriate postprocessing and proportional control, this is a reasonable expectation for many users. The reality, however, is that ideal conditions do not exist in practical use. The most significant issues to be overcome to improve the previously described control schemes arise from the act of wearing a prosthesis and using it in a functional manner: electrode shift, variation in force, variation in position of the residual limb, and transient changes in EMG.

Advances in electronics packaging, telemetry methods, and biocompatible materials have led to a renewed interest in developing implantable EMG measurement systems. A robust implantable measurement system would alleviate at least some of the problems listed previously. Furthermore, implantable EMG systems should allow for more independent control sites to be localized. This may allow different signal processing techniques to be used to achieve simultaneous, proportional control of multiple DOFs, which is currently not possible using pattern recognition methods and is very difficult to achieve even with TMR surgery and conventional myoelectric control.

Clinical Challenges Affecting Pattern Recognition Based EMG Control

Each time a user dons a prosthesis, it is likely that the electrodes will settle in a slightly different position, relative to the underlying musculature. The electrodes may also shift during use due to loading and positioning of the limb. It has been shown (Hargrove et al., 2006, 2008) that a shift of 1 cm of four electrodes placed circumferentially about the forearm will increase classification error in a 10-class experiment from roughly 5% to 20% (if shifted distally) and to 40% (if rotated about the forearm). It was also shown that, by including shifted versions of the data in the training session, degradation due to shift in the test set could be virtually eliminated.

It is possible that the variability in electrode position from day to day may necessitate a daily training session when donning a prosthesis. A daily calibration is probably also warranted to accommodate other influences that may affect the EMG, such as electrode impedance (due to skin dryness, humidity), muscle hypertrophy or hypotrophy, and learning effects as the amputee becomes a more experienced user.

As previously discussed, pattern recognition control relies on clustering repeatable patterns of EMG activity into discernable classes. While proportional control is an integral part of usability, contractions performed at different intensities may be very different from one another—and therefore present a challenge to a pattern classifier. The effects of force variation are described in Scheme and Englehart (2011), in which users performed contractions at 20% to 80% (in 10% increments) of their strongest voluntary class-specific contraction levels. As would be expected, training and testing a classifier with data collected from the same force level yielded low classification errors on the order of 10% across all motions except at the lowest level, which yielded classification errors on the order of 20%. When the classifier was trained with data from a single contraction level and tested with data from all other contraction levels, the error rates tripled to approximately 30% classification error. In an attempt to train a classifier that better generalizes to new force levels, a restricted protocol was investigated using only the lowest (20%) and highest (80%) force levels. In such a system, the classification error decreased to 19% when tested with data from all levels. While this is certainly not ideal it indicates that thoughtful training can improve the usability of a pattern recognition based system.

Similarly, the unrestricted functional use of an upper limb prosthesis necessitates using the limb in a variety of positions. For a transradial or transhumeral amputee, this will impose a loading of the muscles inside the socket from which EMG is being recorded. This will alter the nature of the EMG due to compression of the muscle and possibly elicitation of eccentric contractions or mechanical stimulation of the muscle. As well, various limb positions will impose different gravitational influences, causing relative displacement of the muscles. Direct skeletal attachment (osseointegration) may help alleviate some of the socket loading issues but it is likely that there will still be changes in muscle activation patterns as the limb is moved in space.

This effect is illustrated in Scheme et al. (2010), which characterizes the influence of limb position on pattern recognition accuracy. EMG data corresponding to eight classes of motion (wrist flexion–extension, wrist pronation–supination, hand open, power grip, pinch grip, and a no-motion–rest class) were performed in each of eight limb positions that comprised the work space. In a manner similar to the contraction intensity experiment, classifiers were trained in each individual position and tested in each individual position. As expected, the intraposition error was low, averaging 6.9% over all positions, compared with the average interposition error (35%). As with electrode shift and contraction intensities, training was completed with exemplars from multiple positions in order to reduce the interposition error. This strategy proved to be an effective approach, yielding an average error of 7.4%, which was only slightly higher than the average intraposition error of 6.9%.

Currently, the only strategy to effectively reduce the degradation caused by electrode shift and by force and position variation is to build these data into a training protocol. This places considerable burden on the amputee and prohibits a quick and easy daily recalibration.

There are additional factors that confound the use of the EMG that are a result of short- and long-term variations in the recording environment during use. External interference, electrode impedance changes, electrode shift and lift, and muscle fatigue may alter the EMG and present challenges to clinical robustness. Adaptive EMG pattern recognition systems have been investigated by several groups (Fukuda et al., 2003; Nishikawa et al., 1999; Sensinger et al., 2009; Biron and Englehart, 2010) to accommodate these changes. Adaptation of a pattern classifier is challenging because the system must know not only *how* to adapt, but *when* to adapt. For a classifier to know *how* to adapt properly, it must have knowledge (or at least a reliable estimate) of the intended motion corresponding to the data to which it must adapt. If the user is directed (supervised) during the data acquisition, the classes will be known, and the task is straightforward. However, the greatest use of adaptation is during normal, unsupervised use, and the system must therefore guess the intended class. Knowing *when* to adapt is equally critical. When parsing a data stream in real time, the system must be confident that the data are representative of the estimated class. Some success has been reported through the use of entropy measures to characterize the uncertainty in the data, and therefore attempt to select appropriate data for retraining. Improvement as a result of adaptation has been shown during extended use (Fukuda et al., 2003) and for specifically induced effects (electrode shift, fatigue; Biron and Englehart, 2010) but the dynamics of adaptation and the interaction with the user are still not well understood. A completely stable, unsupervised solution has yet to be realized but is of great clinical interest.

Implantable EMG Systems

Surface and intramuscular EMG both contain sufficient neural information to train accurate pattern recognition control systems, but intramuscular EMG may provide additional benefits not available to surface EMG systems. Ideally, an implantable EMG system would be inserted and anchored into targeted muscles through a minor procedure under local anaesthesia (Weir et al., 2009). After implantation, telemetry would be used to acquire the detected EMG signals through the skin by a receiver mounted on the prosthesis or embedded in the socket. This would ostensibly eliminate many of the problems associated with electrode shift, electrode lift-off, and skin–electrode impedance changes.

The required power to operate the implant and transmit the data through the residual limb tissue, however, is a major design challenge. The system should be able to operate continuously for approximately 8–12 hours and must be powered from the battery already contained in the prosthesis. Encapsulation and migration of the implants are also important secondary design challenges. Long-term studies need to be completed to ensure that high-quality EMG measurements can be made over a number of years. Finally, implantable EMG measurement systems will be much more expensive compared with surface EMG electrodes. Thus, even if excellent implantable systems become commercially available, there will still be a need for control systems that rely only on surface EMG as inputs.

Independent Simultaneous Multifunction Control

Intuitive independent and simultaneous multifunction control is the ultimate goal for a myoelectric control system but current pattern recognition systems are unable to achieve this. An increase in the number of available independent control sites through implantable EMG systems may allow for multiple pattern classifiers to be trained in parallel to achieve this goal. In the simplest case, a mutually exclusive set of EMG channels would be input to each classification system, but this is difficult or impossible to achieve with most amputees. Another possibility would be to use a blind source separation algorithm as preprocessing and use the outputs to train parallel classifiers. A similar method has been used with surface EMG to improve classification accuracies but was not able to achieve independent parallel classification (Hargrove et al., 2009b).

Other approaches have attempted to restore independent and simultaneous control (Yatsenko et al., 2007; Jiang et al., 2009; Nielsen et al., 2011; Ajiboye and Weir, 2009) but have not been implemented clinically. One of the more promising approaches has been the use of muscle synergies (Tresch et al., 2006): muscles activated by the central nervous system in coordinated patterns. Jiang (Jiang et al., 2009) used blind source separation techniques to determine the driving force functions at the wrist joint. This has also been shown to work with a unilateral amputee using a mirrored training protocol with the intact limb (Nielsen et al., 2011). These force functions could eventually be mapped directly as control signals for a powered wrist. The method performed reasonably well but muscle cross talk was cited as a confounding factor. Implantable EMG may overcome this limitation and yield a more robust system but a stable muscle synergy matrix would still need to be computed for each amputee. Considerable research remains before independent simultaneous multifunction control can be implemented robustly in a clinical setting.

CONCLUSIONS

Current commercially available myoelectric prostheses use control algorithms that were largely developed prior to the 1980s. Improvements in the design of prosthetic devices and in clinically viable control algorithms have been incremental, at best. More recently, however, advances in electronics, computing power, and surgical procedures have opened the door to new promises of dexterous control. These advances, driven by renewed international interest, have pushed the field to a point where pattern recognition based myoelectric control systems are finally being tested in clinics with new wearable prototypes of multi-DOF prostheses.

While clinical testing has reinforced the suitability of myoelectric control strategies, it has also highlighted challenges not faced in classical off-line testing. Factors such as electrode shift, variations in force and limb position, and temporal changes in the EMG are being investigated but currently remain obstacles to truly robust clinical outcomes. Nonetheless, the push toward clinically viable pattern recognition based EMG control has elevated the expectations of myoelectric control research to new levels. New and exciting neural–machine interfaces such as targeted muscle reinnervation and implantable EMG measurement systems provide more independent EMG control sites to be localized, auguring greater dexterity of control and opening new promising avenues of research.

REFERENCES

Ajiboye A.B. and Weir R.F. (2009). Muscle synergies as a predictive framework for the EMG patterns of new hand postures. *J Neural Eng* 6:036004.

Basmajian J. and De Luca C. (1985). *Muscles Alive: Their Functions Revealed by Electromyography*, 5th ed. Baltimore, MD: Williams and Wilkins.

Biron K. and Englehart K. (2010). EMG pattern recognition adaptation. In: *18th Congress of the International Society of Electrophysiology and Kinesiology*, Aalborg, Denmark.

Chan A.D.C. and Englehart K.B. (2005). Continuous myoelectric control for powered prostheses using hidden Markov models. *IEEE Trans Biomed Eng* 52:121–124.

Childress D. (1966). A myoelectric three state controller using rate sensitvity. Paper presented at the ICMBE, Chicago.

Chu J.U., Moon I., and Mun M.S. (2006). A real-time EMG pattern recognition system based on linear–nonlinear feature projection for a multifunction myoelectric hand. *IEEE Trans Biomed Eng* 53:2232–2239.

Cover T.M. and Hart P.E. (1967). Nearest neighbor pattern classification. *IEEE Trans Inform Theory* 13:21–27.

Dorcas D.S. and Scott R. N. (1969). A three-state myoelectric controller. *Med Biol Eng* 4:367–372.

Duda R., Hart P., and Stork D. (2000). *Pattern Classification*, 2nd ed. New York: Wiley-Interscience.

Englehart K. and Hudgins B. (2003). A robust, real-time control scheme for multifunction myoelectric control. *IEEE Trans Biomed Eng* 50:848–854.

Englehart K., Hudgins B., Parker P., and Stevenson M. (1999). Classification of the myoelectric signal using time-frequency based representations. *Med Eng Physics* 21:431–438.

Englehart K., Hudgins B., and Parker P.A. (2001). A wavelet-based continuous classification scheme for multifunction myoelectric control. *IEEE Trans Biomed Eng* 48:302–311.

Farrell T.R. and Weir R.F. (2007). The optimal controller delay for myoelectric prostheses. *IEEE Trans Neural Syst Rehabil Eng* 15:111–118.

Farrell T.R. and Weir R. (2008). Analysis window induced controller delay for multifunctional prostheses. In: *Myoelectric Controls Symposium*, Fredericton, NB, pp. 225–228.

Fukuda O., Tsuji T., Kaneko M., and Otsuka A. (2003). A human-assisting manipulator teleoperated by EMG signals and arm motions. *IEEE Trans Robotics Automation* 19:210–222.

Graupe D. and Cline W.K. (1975). Functional separation of EMG signals via ARMA identification methods for prosthesis control purposes. *IEEE Trans Systems Man Cybernetics* 2:252–258.

Graupe D., Salahi J., and Zhang D.S. (1985). Stochastic-analysis of myoelectric temporal signatures for multifunctional single-site activation of prostheses and orthoses. *J Biomed Eng* 7:18–29.

Hargrove L., Englehart K., and Hudgins B. (2006). The effect of electrode displacements on pattern recognition based myoelectric control. In: *Proceedings of the 28th IEEE EMBS Annual International Conference*, New York City, pp. 2203–2206.

Hargrove L., Englehart K., and Hudgins B. (2007a). A comparison of surface and intramuscular myoelectric signal classification. *IEEE Trans Biomed Eng* 54:847–853.

Hargrove L., Losier Y., Lock B.A., Englehart K., and Hudgins B. (2007b). A real-time pattern recognition based myoelectric control usability study implemented in a virtual environment. In: Proceedings of the 29th Annual International Conference of the IEEE EMBS, Lyons, France, pp. 4842–4845.

Hargrove L., Englehart K., and Hudgins B. (2008). A training strategy to reduce classification degradation due to electrode displacements in pattern recognition based myoelectric control. *Biomed Signal Process Control* 3:175–180.

Hargrove L., Li G., Englehart K., and Hudgins B. (2009a). Principal components analysis preprocessing for improved classification accuracies in pattern-recognition-based myoelectric control. *IEEE Trans Biomed Eng* 56:1407–1414.

Hargrove L., Zhou P., Englehart K., and Kuiken T. (2009b). The effect of ECG interference on pattern recognition based myoelectric control for targeted muscle reinnervated patients. *IEEE Trans Biomed Eng* 56:2197–2201.

Hargrove L., Scheme E., Englehart K., and Hudgins B. (2010). Multiple binary classifications via linear discriminant analysis for improved controllability of a powered prosthesis. *IEEE Trans Neural Systems Rehabil Eng* 18:49–57.

Hogan N. (1976). Review of methods of processing EMG for use as a proportional control signal. *Biomed Eng* 11:81–86.

Huang H., Zhou P., Li G., and Kuiken T.A. (2008). An analysis of EMG electrode configuration for targeted muscle reinnervation based neural machine interface. *IEEE Trans Neural Systems Rehabil Eng* 16:37–45.

Huang Y.H., Englehart K.B., Hudgins B., and Chan A.D.C. (2005). A Gaussian mixture model based classification scheme for myoelectric control of powered upper limb prostheses. *IEEE Trans Biomed Eng* 52:1801–1811.

Hudgins B., Parker P., and Scott R.N. (1993). A new strategy for multifunction myoelectric control. *IEEE Trans Biomed Eng* 40:82–94.

Jiang N., Englehart K., and Parker P. (2009). Extracting simultaneous and proportional neural control information of multiple degree of freedom from the surface electromyographic signal. *IEEE Trans Biomed Eng* 56:1070–1080.

Kang W.J., Shiu J.R., Cheng C.K., Lai J.S., Tsao H.W., and Kuo T.-S. (1997). The effect of electrode arrangement on spectral distance measures for discrimination of EMG signals. *IEEE Trans Biomed Eng* 44:1020–1023.

Kobrinski A.E. (1960). Problems of bioelectric control. In: 1st IFAC International Congress, London, England, pp. 619–623.

Kuiken T., Miller L., Lipschutz R., Stubblefield K., and Dumanian G. (2005). Prosthetic command signals following targeted hyper-reinnervation nerve transfer surgery. *Conf Proc IEEE Eng Med Biol Soc* 7:7652–7655.

Kuiken T.A. (2003). Consideration of nerve–muscle grafts to improve the control of artificial arms. *J Technol Disability* 15:105–111.

Kuiken T.A., Miller L.A., Lipschutz R.D., Lock B.A., Stubblefield K., Marasco P.D., Zhou P., and Dumanian G.A. (2007). Targeted reinnervation for enhanced prosthetic arm function in a woman with a proximal amputation: a case study. *Lancet* 369:371–380.

Kuiken T.A., Li G., Lock B.A., Lipschutz R.D., Miller L.A., Stubblefield K.A., and Englehart K.B. (2009). Targeted muscle reinnervation for real-time myoelectric control of multifunction artificial arms. *JAMA* 301:619–628.

Li G., Schultz A.E., and Kuiken T.A. (2009). Quantifying pattern recognition-based myoelectric control of multifunctional transradial prostheses. *IEEE Trans Neural Systems Rehabil Eng* 18:185–192.

Lock B.A., Englehart K., and Hudgins B. (2005). Real-time myoelectric control in a virtual environment to relate usability vs. accuracy. In: *MyoElectric Controls Symposium*, Fredericton, NB, pp. 122–127.

Lucas M., Gaufriau A., Pascual S., Doncarli C., and D. Farina D. (2008). Multi-channel surface EMG classification using support vector machines and signal-based wavelet optimization. *Biomed Signal Process Control* 3:169–174.

McLean L. and Scott R.N. (2004). The early history of myoelectric control of prosthetic limbs. In: *Powered Upper Limb Prosthetics: Control, Implementation, and Clinical Application* (Muzumdar A., Ed.), pp. 1–15. New York: Springer.

Naik G.R., Kumar D.K., and Palaniswami M. (2008). Surface EMG based hand gesture identification using semi blind ICA: validation of ICA matrix analysis. *Electromyogr Clin Neurophysiol* 48:169–180.

Nielsen J., Holmgaard S., Jiang N., Englehart K., Farina D., and Parker P. (2011). Simultaneous and proportional force estimation for multifunction myoelectric prostheses using mirrored bilateral training. *IEEE Trans Biomed Eng* 58:681–688.

Nishikawa D., Yu W., Yokoi H., and Kakazu Y. (1999). EMG prosthetic hand controller discriminating ten motions using real-time learning method. In: *IEEE IROS*, Kyongju, South Korea, pp. 1592–1597.

Parker P., Stuller J., and Scott R.N. (1977). Signal processing for the multistate myoelectric channel. *Proc IEEE* 65:662–674.

Parzen E. (1962). An estimation of probability density function and mode. *Ann Math Statist* 33:1065–1076.

Scheme E. and Englehart K. (2011). EMG pattern recognition for the control of powered upper limb prostheses: state-of-the-art and challenges for clinical use. *J Rehabil Res Development* 48(6):643–660.

Scheme E., Fougner A., Stavdahl O., Chan A., and Englehart K. (2010). Examining the adverse effects of limb position on pattern recognition based myoelectric control. In: 32nd Annual International Conference of the IEEE Engineering in Medicine and Biology Society, Buenos Aires, Argentina.

Scheme E., Englehart K., and Hudgins B. (2011). Selective classification for improved robustness of myoelectric control under non-ideal conditions. *IEEE Trans Biomed Eng* 58:1698–1705.

Scott R.N., Paciga J.E., and Parker P.A. (1978). Operator error in multistate myoelectric control systems. *Med Biol Eng Computing* 16:296–301.

Sensinger J.W., Lock B.A., and Kuiken T.A. (2009). Adaptive pattern recognition of myoelectric signals: exploration of conceptual framework and practical algorithms. *IEEE Trans Neural Systems Rehabil Eng* 1:270–278.

Simon A., Hargrove L., Lock B.A., and Kuiken T. (2010) A decision-based velocity ramp for minimizing the effect of misclassifications during real-time pattern recognition control. *IEEE Trans Neural Systems Rehabil Eng*. 58(8):2360–2368.

Smith L.H., Hargrove L., Lock B.A., and Kuiken T. (2011). Determining the optimal window length for pattern recognition-based myoelectric control: balancing the competing effects of classification error and controller delay. *IEEE Trans Neural Systems Rehabil Eng* 19:186–192.

Tresch M., Cheung V., and d'Avella A. (2006). Matrix factorization algorithms for the identification of muscle synergies: evaluation on simulated and experimental data sets. *J Neurophysiol* 95:2199–2212.

Weir R.E.F., Troyk P.R., DeMichele G.A., Kerns D.A., Schorsch J.F., and Maas H. (2009). Implantable myoelectric sensors (IMESs) for intramuscular electromyogram recording. *IEEE Trans Biomed Eng* 56:159–171.

Williams T.W. 3rd (1990). Practical methods for controlling powered upper-extremity prostheses. *Assist Technol* 2:3–18.

Yatsenko D., McDonnall D., and Guillory K.S. (2007). Simultaneous, proportional, multi-axis prosthesis control using multichannel surface EMG. In: *29th Annual International Conference of the IEEE Engineering and Medicine in Biology Society*, Lyons, France, pp. 6133–6136.

Zhou P., Lowery M.M., Englehart K.B., Huang H., Li G., Hargrove L., Dewald J.P., and Kuiken T.A. (2007). Decoding a new neural–machine interface for control of artificial limbs. *J Neurophysiol* 98:2974–2982.

16

REVIEW CHAPTER

CONTROLLING PROSTHESES USING PNS INVASIVE INTERFACES FOR AMPUTEES

JACOPO CARPANETO,[1] LUCA CITI,[1] STANISA RASPOPOVIC,[1,2]
JACOPO RIGOSA,[1] AND SILVESTRO MICERA[1,2]

[1]*Biorobotics Institute, Scuola Superiore Sant'Anna, Pisa, Italy*
[2]*Center for Neuroprosthetics and Institute of Bioengineering, School of Engineering, Ecole Polytechnique Federale de Lausanne (EPFL), Lausanne, Switzerland*

SUMMARY

Several attempts have been made to enhance the control of highly sensorized prostheses. One of the main goals is to create a bidirectional link between the prosthesis and the nervous system. Several approaches based on noninvasive and invasive interfaces have been developed and tested. Among these, interfaces with the peripheral nervous system (PNS), especially longitudinal intrafascicular electrodes, seem to be a promising solution. The potential and the limits of these interfaces for the control of prosthetic devices are presented in this chapter.

INTRODUCTION

The human hand is a very complex system, with a large number of degrees of freedom (DOFs), sensors embedded in its structure, actuators and tendons,

Introduction to Neural Engineering for Motor Rehabilitation, First Edition. Edited by Dario Farina, Winnie Jensen, and Metin Akay.

and a complex hierarchical control (Kapandji, 1982). Its loss due to disease, traumatic injury, and amputation causes severe physical as well as mental illness (e.g., inability to grasp and manipulate objects, to sense and explore the surrounding world, and to use gestures to support speech and express emotions). The incidence rate of major upper limb amputees (e.g., above elbow, below elbow and multiple upper limb extremity) in a part of the United States is 0.14 per 100,000 persons per year (Dillingham et al., 1998). Hand prostheses represent a current technological aid for these persons. These devices are of three types: (i) passive (cosmetic), (ii) body powered, and (iii) myoelectric. Yet surveys (Atkins et al., 1996) on the use of these artificial hands reveal that 30–50% of amputees do not use their prosthetic hand regularly. This is basically due to its low functionality, poor cosmetic and unnatural appearance, lack of sensory feedback, and low controllability. This situation calls for the development of a versatile prosthetic limb with intuitive motor control and realistic sensory feedback that will allow amputees to perform tasks that are necessary for activities of daily living (ADLs).

Intuitive prosthetic control may be developed by extracting the amputee's intention from signals recorded in noninvasive or invasive ways from the peripheral nervous system (PNS). Electromyographic (EMG) signals, collected at the skin surface, have been used for the control of upper limb prosthetic devices since 1948 (Reiter, 1948), because they provide an easy and noninvasive access to physiological processes that cause contraction of the muscles. At present, EMG signal processing is the most common approach used for controlling active prosthetic hands (Micera et al., 2010a; Oskoei and Hu, 2007). On the other hand, invasive interfaces (Grill et al., 2009; Navarro et al., 2005) with the PNS are another interesting way to create a bidirectional link between the user's nervous system and an artificial device. These interfaces can be used to induce activity in nerve fibers through electrical stimulation and to deliver information into the nervous system. Conversely, information from the nervous system could be retrieved by recording the electrical activity of the nerve, the electroneurographic (ENG) signal. Given a chronically stable device acting on an appropriate set of nerve fibers, such an interface could be used as part of a functional electrical stimulation (FES) system to restore function to paralyzed limbs (Popovic et al., 1993) or in a brain-controlled robotic limb application (Micera et al., 2006; Ohnishi et al., 2007).

In all these applications, recorded raw EMG or ENG signals are amplified, filtered, and fed into a signal processing unit for the control of artificial limbs (e.g., hand prosthesis). These signals differ on the basis of which neural or muscular structures have been recorded (e.g., muscle signals for EMG, nerve signals for ENG) and of the level of invasiveness used (e.g., surface or intramuscular EMG, epineural or intraneural ENG). The main approaches used for controlling neuroprostheses and hybrid bionic systems are briefly summarized in Table 16.1.

The aim of this chapter is to review the state of the art of artificial hand control based on the use of invasive interfaces with the PNS.

TABLE 16.1. Comparison of Various Approaches for the Control of Artificial Limbs Using Peripheral Information (from Micera et al., 2010b)

Main Approach	Main Advantages	Disadvantages
Surface EMG (Micera et al., 2010a; Oskoei and Hu, 2007; Englehart and Hudgins, 2003; Zecca et al., 2002)	Noninvasive	Unnatural control strategies need to be learned by the subject especially in case of more proximal amputations
Implantable EMG (Farina et al., 2008; Weir et al., 2009)	Improved quality of EMG signals	Same as surface EMG
Targeted muscle reinnervation (Huang et al., 2008; Kuiken et al., 2004)	More natural control strategies, effective sensory feedback	Requires a surgical implantation but works with noninvasive signals; more suitable for amputations at the shoulder level
Implantable peripheral interfaces (Micera et al., 2010a; Navarro et al., 2005; Dhillon and Horch, 2005; Dhillon et al., 2005; Jia et al., 2007; Micera et al., 2008)	Potentially selective and versatile for natural sensory feedback	Limitations in terms of controllable DOFs, invasiveness, acquisition of noisy signals

Source: Micera et al., 2010b.

STATE OF THE ART

Interfaces with the Peripheral Nervous System

The use of invasive neural interfaces directly connected to the PNS is potentially appealing because, in most cases, it is able to provide an "almost physiological" condition in which efferent and afferent fibers, previously connected with the natural hand, return to their role in controlling prosthetic limbs or hands (Micera et al., 2008). This is theoretically possible since a significant number of peripheral nerve fibers, as well as spinal cord and brain connections and relays dedicated to controlling the amputated limb, survive in time and remain available to restore a physiological condition. A bidirectional interface for hand prosthesis control in amputees should allow, on the one hand, the recording of neural efferent signals that can be used for controlling mechanical prosthesis motion (Edell, 1986), and on the other, the provision of sensory feedback from tactile and force sensors to the user through stimulation of afferent nerve fibers within the residual limb (Riso et al., 1999). Several invasive PNS interfaces have been developed in the past (Grill et al., 2009; Navarro et al., 2005; Stieglitz et al., 2005) and can be grouped according to their level of selectivity and invasiveness (see Fig. 16.1).

Although most interfaces were originally devoted to FES in spinal cord injured persons (Fisher et al., 2009; Guiraud et al., 2006; Hoffer and Kallesøe,

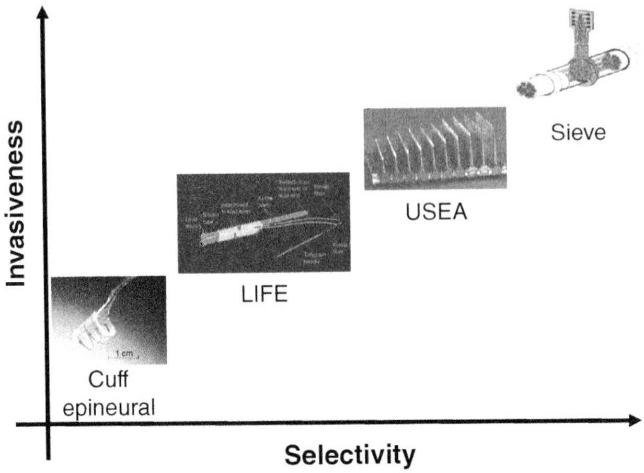

Figure 16.1. Schematic plot of selectivity and invasiveness of different implantable electrodes (from Micera et al., 2010b).

2001), they can also be the key component of neurocontrolled hand prostheses. In this case, they are used to record efferent motor signals and to stimulate afferent nerves (i.e., in a complementary manner with FES systems; Dhillon and Horch, 2005; Jia et al., 2007; Micera et al., 2010b; Dhillon et al., 2004; Rossini et al., 2010).

Thanks to their low invasiveness, cuff electrodes are the most highly used PNS interface not only in research with animal models but also in human trials. They are composed of an insulating tubular sheath that completely encircles the nerve and contains two or more electrode contacts exposed at their inner surface that are connected to insulated lead wires (Stieglitz et al., 2005). Although innovative cuff structures (e.g., flat interface nerve electrode (FINE) or multipolar spiral electrodes; Fisher et al., 2009; Polasek et al., 2009; Tyler and Durand, 2002) and advanced processing techniques (Tesfayesus and Durand, 2007; Micera et al., 2001) have been developed, the use of this electrode for the control of cybernetic prosthesis is still limited by its reduced selectivity during recording (e.g., signals recorded are a mixed activity of the enclosed fibers) and stimulation.

Regeneration-type (or sieve) electrodes are designed to interface a high number of nerve fibers by using an array of "via holes," with electrodes built around them, implanted between the severed stumps of a peripheral nerve. Regenerating axons eventually grow through the via holes, making it possible to record action potentials (spikes) and to stimulate individual axons or small groups. The most logical and challenging application of regenerative electrodes is implantation in the severed nerves of an amputee limb in order to provide a bidirectional interface. In fact, regenerative electrodes allow interfacing with the injured axons that originally innervated the lost limb, thereby increasing selectivity. Although promising results on the use of regenerative electrodes have been achieved in experiments with animal models (Dario et al., 1998; Navarro et al., 1998; Ceballos et al., 2002; Lago et al., 2005; Lago et al., 2007b), their clinical usability is still limited by various challenges (e.g., higher regeneration speed for small nerve fibers with respect to large nerve fibers, underrepresentation of sensory and motor axons within the holes of the electrode, risk of compressive axonopathy at the sieve electrode level, and loss of normal topographical architecture in regenerated nerves; Lago et al., 2005).

Intraneural electrodes inserted into the PNS (Nannini and Horch, 1991; Yoshida and Horch, 1993) represent another conceptual approach for the development of interfaces characterized by enhanced selectivity (with respect to extraneural electrodes) and increased signal-to-noise ratio (SNR) of recordings. An example of intraneural electrodes is the multielectrode array (MEA; Nordhausen et al., 1994), composed of tens of needles inserted transversely into the nervous system and made of different materials such as silicon, glass, or polyimide. A modified version named Utah slanted electrode array (USEA) has also been used within the PNS, allowing the selective recording of single-unit responses and low-current highly selective stimulation of motor fibers (McDonnall et al., 2004). MEAs have been used in several times in experimental work with animal models (Branner et al., 2004, 2001; Branner and Normann, 2000) and also in a human volunteer (Warwick et al., 2003). They offer the important advantage of having a large number of electrical contacts that increase selectivity. However, they also appear to have some limitations, such as the rigid structure of the electrodes and the tethering forces by lead wires that represent a concern in terms of damage to the limb nerves during long-term implantation, as well as the lack of stability of the recorded sensory signals over time during chronic studies (Branner et al., 2004; Warwick et al., 2003).

Longitudinal intrafascicular electrodes (LIFEs; see Fig. 16.2) are another kind of intraneural electrodes, constructed from thin insulated conducting wires (such as Pt–Ir or metalized Kevlar fibers). They are inserted longitudinally into the nerve tissue (Lawrence et al., 2004; Yoshida and Stein, 1999) and are designed to run in between and parallel to the nerve fibers.

Recently, a novel electrode ("transverse intrafascicular multichannel electrode," TIME; see Fig. 16.2) that transversely penetrates the peripheral nerve has been developed in order to selectively activate subsets of axons in different fascicles within the nerve (Boretius et al., 2010). Gathering signals from only a small number of axons will allow more selective recordings than cuff electrodes, for example. On the other hand, LIFE electrodes are less invasive than USEAs, which are inserted transversely into the nerve—leading to a higher risk of nerve damage. LIFEs have been under development as a neuroprosthetic device since the late 1980s (Nannini et al., 1991; Yoshida and Horch, 1993) and more recently have been implanted semichronically in amputees, showing great promise as a selective neural human–machine interface for

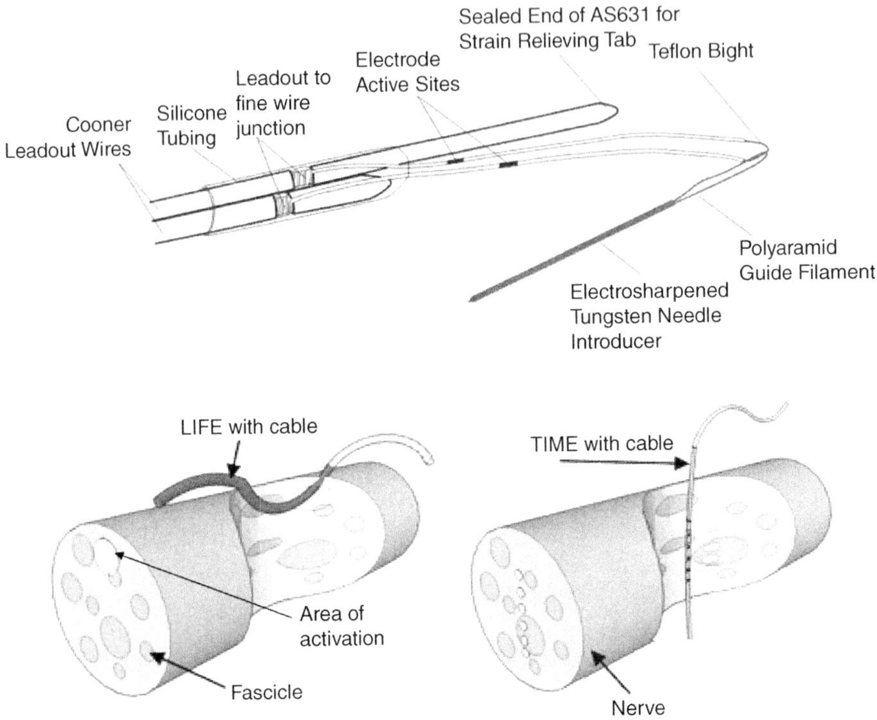

Figure 16.2. (top) Scheme of chronic dual-channel wire LIFE (from Yoshida et al., 2010). (bottom) Scheme of LIFE and TIME electrodes with predicted areas of activation during electrical stimulation (white areas) (from Boretius et al., 2010).

advanced prosthetics (Dhillon and Horch, 2005; Dhillon et al., 2005; Micera et al., 2010b; Dhillon et al., 2004; Rossini et al., 2010).

ENG-Based Control of Hand Prosthesis

Decoding algorithms represent a critical step for ENG-based hand prosthesis control. Algorithms have been mainly developed for the closed-loop control of FES systems. Nonetheless, these findings can provide an insight into the information that can be decoded from these signals, giving hints about which algorithms to use. The main problem and the reason for the development of ENG algorithms with afferent signals is the difficulty of obtaining voluntary efferent ENG activity (most of the experiments have been carried out in anesthetized animals). For cuff electrodes, higher order statistics algorithms have been used in order to switch the electrical stimulation off-and-on (Upshaw and Sinkjaer, 1998), whereas a continuous decoding of afferent signals has been obtained by means of power-based features and support vector machine

(SVM) classifiers (Raspopovic et al., 2010). In the case of FINE electrodes, two interesting approaches have been proposed: blind source separation (Tesfayesus and Durand, 2007) and beam-forming algorithms (Wodlinger and Durand, 2009) for signal separation. These algorithms could also be used in closed-loop prosthetic devices.

After appropriate experiments to test their biocompatibility (Lago et al., 2007a) and efficacy in animals (Citi et al., 2008; Navarro et al., 2007), LIFE electrodes have been used to develop and test algorithms for the neurocontrol of hand prostheses in amputees (Dhillon et al., 2004; Dhillon and Horch, 2005; Dhillon et al., 2005; Jia et al., 2007; Rossini et al., 2010). In these experiments, electrodes have been implanted for some weeks in severed nerves of human volunteers. From signals recorded with LIFEs, it is possible to detect spikes with peak-to-peak amplitudes of about 20 μV and power spectrum lying between 500 Hz and 7 kHz (peaking at about 2 kHz). As reported in these studies, the signal obtained is very weak and prone to different noise sources (mainly due to the EMG). In an intraoperative study (Jia et al., 2007), six intrafascicular electrodes were inserted into the ulnar, radial, and median nerves in the stump of an amputee. The signal amplitude from the radial nerve was 5.5 μV ± 0.8, which was greater than the amplitudes from the ulnar (2.5 μV ± 0.4) and median (2.2 μV ± 0.3) nerves.

In Dhillon et al. (2004), subjects were able to control a cursor on a PC screen making missing limb movements (see Fig. 16.3, left). During these movements, LIFE signals were compared with a threshold in order to detect volitional motor nerve activity. Then, the firing rate of action potentials was used to control the cursor position (see Fig. 16.4, left). In another experiment (Micera et al., 2010b; Rossini et al., 2010), a new version of LIFE electrodes, thin-film LIFEs (tfLIFEs) were implanted in the nerves of an amputee for three weeks. Nerve signals were used to control three different grips of a cybernetic hand prosthesis. The decoding algorithm was based on (i) wavelet denoising, (ii) spike sorting, and (iii) classification of the amputee's intention using sorted spike templates (see Fig. 16.3, right; Citi et al., 2008).

A wavelet-denoising technique, previously exploited in the field of microneurography (Diedrich et al., 2003), has been used to eliminate the noise from ENG signals and leave the informative spiking activity only. The algorithm presented in Diedrich et al. (2003) was modified and tested during the decoding of afferent rabbit signals (Citi et al., 2008). Wavelet denoising was followed by a spike detection and sorting phase. The shape of the spike, as recorded by an extracellular electrode, is determined by several factors such as the distance between the fiber and the electrode (the tissue in between acts as a low-pass filter: the longer the distance, the smoother, smaller, and longer the spike shape), the relative orientation of the nodes of Ranvier in the nerve fiber, and the inhomogeneity of the intrafascicular space. Thus, the signals related to different nerve fibers can be identified and extracted on the basis of the shape recorded from a multiunit recording. Since different nerve fibers carry different information (e.g., which motor units to activate to perform a given grasp),

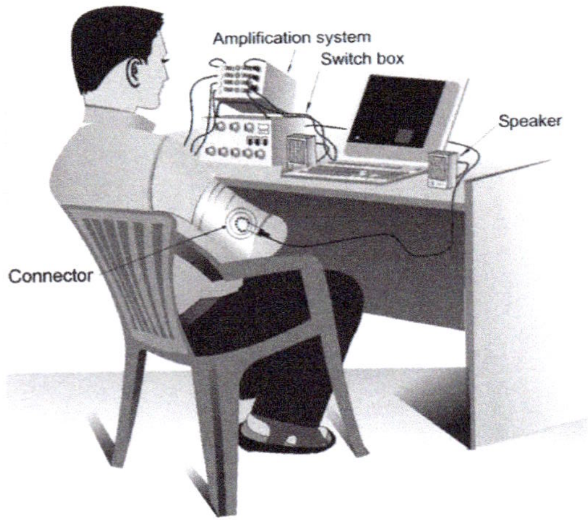

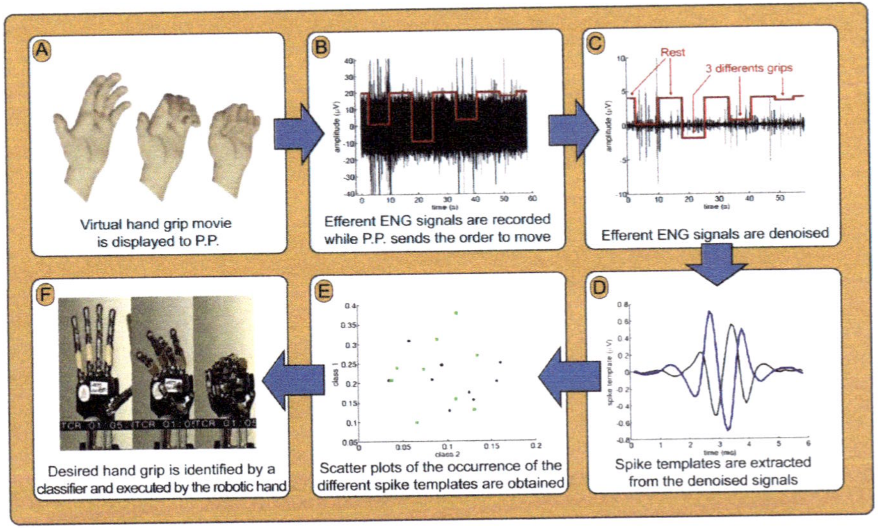

Figure 16.3. Examples of external device control by means of PNS activity recorded with LIFE electrodes (left, from Dhillon et al., 2004; right, from Rossini et al., 2010).

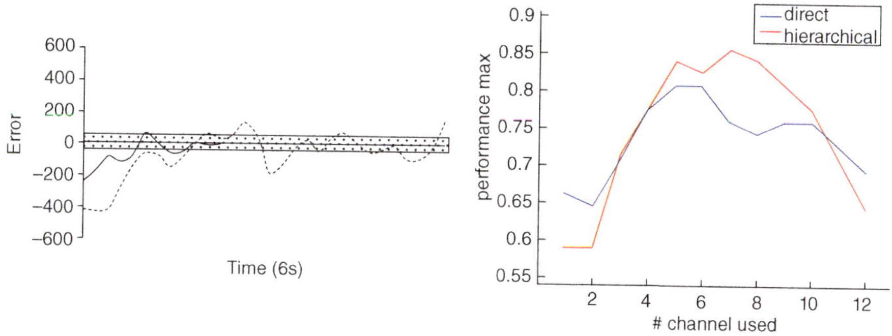

Figure 16.4. Examples of accuracy in the control of external devices using PNS activity recorded with LIFE electrodes. (left) Error during the control of a cursor on a screen (from Dhillon et al., 2004). (right) Control of a multi-DOF hand prosthesis (from Micera et al., 2010b). In this case two different classification methods are shown: direct discrimination of four classes simultaneously (rest and three grip types) and hierarchical discrimination (rest versus activities and then selection of grip type).

spike sorting can be used to infer the grasp selected by the user. In fact, SVMs (Cortes and Vapnik, 1995) were trained to infer the stimulus applied to the rabbit from the sorted spike waveforms (Citi et al., 2008).

The same algorithm, with small modifications, was exploited in a case study with one amputee (Micera et al., 2010b; Rossini et al., 2010), showing for the first time the possibility to decode specific grip types and also finger movements. Figure 16.4, right panel, shows an example of the accuracy obtained during the classification of three different grasps.

In a study regarding afferent neural activity decoding (Djilas et al., 2010), the multiscale continuous wavelet transform was exploited; it outperformed the simple amplitude thresholding method, especially in low-SNR cases, typical in real ENG recording systems. Interestingly, all of the previously mentioned decoding studies showed that spike denoising and recognition algorithms outperform classical amplitude thresholding and rectified integral bin (RIB) studies, and enhancement in performance when using several independent channels on LIFEs.

These experiments indicate that ENG signals can be used to achieve more than a simple on–off control and that neural signals could solve some of the problems of EMG-based control of hand prostheses for amputees. The basic drawback of this denoising and decoding solution is the high computational burden, which becomes a critical factor when aiming to develop a real-time controller. Other algorithms for spike sorting can also be applied as in the case of central nervous system intracortical signals (Brown et al., 2004; Lewicki, 1998; Quiroga et al., 2004). However, it is important to point out that intracortical signals have much higher absolute values, whereas in PNS signals the spikes are more difficult to observe and generally have a lower SNR.

Sensory Feedback

Although there is no dedicated device available for delivering sensory feedback in current commercial prostheses, various studies have shown the great importance of delivering feedback to the user to achieve effective prosthesis control (Lundborg and Rosen, 2001; Carrozza et al., 2006). Several techniques have been developed to provide noninvasive tactile or proprioceptive information to the amputees (Kaczmarek et al., 1991). Electrical (Beeker et al., 1967; Mulvey et al., 2009; Shannon, 1979) and mechanical (Cipriani et al., 2008; Schultz et al., 2009; Rosén et al., 2009) skin stimulation are two of the most commonly used techniques for tactile feedback.

Neural interfaces with the PNS have been shown to be an effective and natural approach to elicit a sensory feedback to prosthesis users by stimulating the afferent nerves through LIFEs (Dhillon and Horch, 2005; Dhillon et al., 2004; Rossini et al., 2010). Information about the protocols used for afferent stimulation have been designed starting from previous studies on neural stimulation (Cogan, 2008; Merrill et al., 2005).

Usually, in order to investigate the possibility of eliciting sensations associated with the lost limb, each active site of the electrodes is separately tested by electrically stimulating with monophasic or biphasic (e.g., rectangular) charge balanced pulses of different duration (e.g., from 10 to 300 μs), pulse amplitude (e.g., from 1 to 200 μA), and frequency (e.g., from 10 to 500 Hz). Threshold (the lowest average pulse current at which the subject could feel a sensation reliably) and upper-limit (the current at which the nature or the location of the sensation changed or when the sensation became uncomfortable) pulse amplitudes are evaluated together with a psychometric staircase method to quantify sensation (from the minimal perceived threshold to discomfort).

In Dhillon et al. (2005), the authors enrolled eight amputees and all of them reported either tactile or proprioceptive sensations from one or more LIFE electrodes (see Fig. 16.5, right upper panel). In half of the subjects, both types of sensation were elicited (through different electrodes). Usually, sensations were discrete, unimodal (i.e., touch, finger flexion, or static joint position), and repeatable and could be painlessly elicited over the duration of the study. These sensations were stable in time from a topographic point of view, suggesting a stable electrode position. The same finding, regarding stable positioning versus fascicular territories, has been confirmed in a single-case study (Rossini et al., 2010), performed with multipolar tfLIFEs (see Fig. 16.5, left upper panel).

Sensations of touch or pressure usually spread from distal (digit tip) to proximal locations as a function of the stimulus current. Moreover, a systematic change of referred sensations of touch–pressure or proprioception was obtained by modulating stimulus frequency without modification of the modality or of the topography (Dhillon et al., 2005). A logarithmic regression gave the best fit for the ratings of sensation magnitudes (psychometric sensory

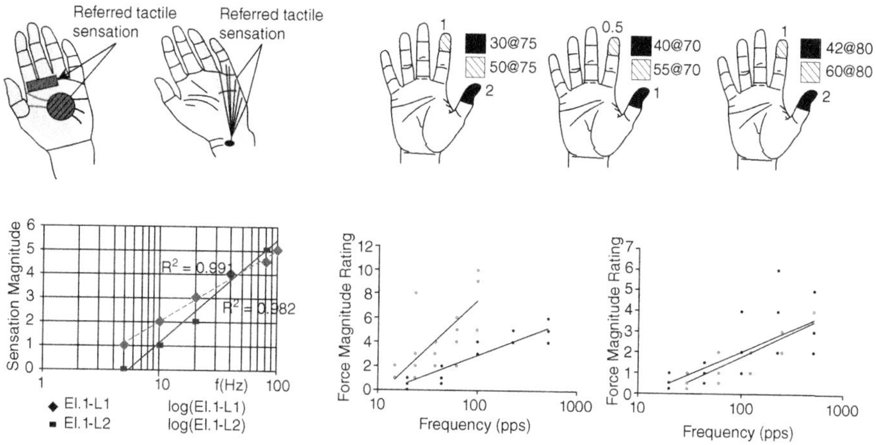

Figure 16.5. (left) The positions of referred sensation on the phantom limb and the modulation of the referred sensation (from Rossini et al., 2010); (right) touch and proprioceptive sensation were elicited in a multisubject study (Dhillon et al., 2005). In both studies, the touch is modulated linearly with respect to the log of frequency, when using psychometric sensory magnitude rating (from Dhillon et al., 2005).

magnitude rating) versus frequency of nerve stimulation (see Fig. 16.5, right bottom panel).

Similar results are also reported in Rossini et al. (2010), where a psychometric sensory magnitude rating was used (see Fig. 16.5, left, bottom panel). However, by indicating perceived finger position and matching it with the contralateral finger, a more linear relationship between both stimulus amplitude and frequency and the reported phantom finger position was produced (see Fig. 16.5, right, bottom panel). The resting position of a given phantom digit was consistently reported in full extension by the amputees. A complete description of the different behaviors obtained as a function of the different stimulation parameters (e.g., pulse duration and amplitude, stimulation frequency) can be found in Dhillon et al. (2005) and Rossini et al. (2010). The presence of a statistically significant increase in both the threshold and the upper limit for eliciting painless and unitary sensations of touch–pressure or joint movement is a critical aspect. Very similar results have been observed in Rossini et al. (2010), where, after 10 days, it was impossible (owing to the limited charge injectable through the multipolar LIFE electrodes used) to elicit any sensations. This could be due to the encapsulation answer from tissue, which increases electrode impedance over time. The two studies at issue, however, were relatively short; instead, in chronic stimulation studies with cuff electrodes (Polasek et al., 2009) the changes in stimulation threshold currents were reported for the first 20 weeks after implantation, with values subsequently returning close to those found in acute conditions.

Finally, an interesting demonstration of direct neural feedback from an artificial arm in amputees has been reported by Dhillon and Horch (2005). In this study, three amputees were able to judge and set grip force and joint position in an artificial arm, in the absence of visual input.

CURRENT CHALLENGES AND FUTURE DIRECTIONS

Electromyographic signals and more recently ENG signals have been used to develop more effective hand prosthesis control (Ohnishi et al., 2007; Micera et al., 2010a). Processing EMG signals to control hand prostheses is a potentially interesting approach (achieving over 90% classification performance for several DOFs as shown, e.g., by Englehart and co-workers); however, clinical applicability is currently still limited. To address this issue, several research groups are working on implantable solutions (invasive EMG or ENG signals; Navarro et al., 2005; Farina et al., 2008; Weir et al., 2009) or on "hybrid" solutions (targeted muscle reinnervation where surface EMG signals are recorded and processed but after a surgical intervention; Kuiken et al., 2009).

With regard to implantable solutions based on PNS neural interfaces, ENG signals can probably provide interesting performance in decoding motor commands (e.g., proportional control and different grip state control; Dhillon and Horch, 2005; Rossini et al., 2010). Moreover, the advantage of providing natural (i.e., by using existing neural "pathways") sensory feedback can increase the robustness of prosthetic device control. Their greatest drawbacks are the invasiveness of the implantation procedure, the relatively blind approach (e.g., repositioning procedures can be attempted during the surgical insertion), and the current lack of long-term chronic studies. If the results from these studies confirm the good performance of this approach, these implants could enter clinical practice in the future, as in the case of cochlear implants, pacemakers, and bladder stimulators.

REFERENCES

Atkins D.J., Heard D.C.Y., and Donovan W.H. (1996). Epidemiologic overview of individuals with upper-limb loss and their reported research priorities. *J Prosthet Orthot* 8(1):2–11.

Beeker T.W., During J., and Hertog A.D. (1967). Artificial touch in a hand-prosthesis. *Med Biol Eng* 5(1):47–49.

Boretius T. et al. (2010). A transverse intrafascicular multichannel electrode (TIME) to interface with the peripheral nerve. *Biosens Bioelectron* 26(1):62–69.

Branner A. and Normann R.A. (2000). A multielectrode array for intrafascicular recording and stimulation in sciatic nerve of cats. *Brain Res Bull* 51(4):293–306.

Branner A., Stein R.B., and Normann R.A. (2001). Selective stimulation of cat sciatic nerve using an array of varying-length microelectrodes. *J Neurophysiol* 85(4):1585–1594.

Branner A. et al. (2004). Long-term stimulation and recording with a penetrating microelectrode array in cat sciatic nerve. *IEEE Trans Biomed Eng* 51(1):146–157.

Brown E.N., Kass R.E., and Mitra P.P. (2004). Multiple neural spike train data analysis: state-of-the-art and future challenges. *Nat Neurosci* 7(5):456–461.

Carrozza M.C. et al. (2006). Design of a cybernetic hand for perception and action. *Biol Cybernet* 95(6):629–644.

Ceballos D. et al. (2002). Morphologic and functional evaluation of peripheral nerve fibers regenerated through polyimide sieve electrodes over long-term implantation. *J Biomed Mater Res* 60(4):517–528.

Cipriani C. et al. (2008). On the shared control of an EMG-controlled prosthetic hand: analysis of user-prosthesis interaction. *IEEE Trans Robot* 24(1):170–184.

Citi L. et al. (2008). On the use of wavelet denoising and spike sorting techniques to process electroneurographic signals recorded using intraneural electrodes. *J Neurosci Methods* 172(2):294–302.

Cogan S.F. (2008). Neural stimulation and recording electrodes. *Annu Rev Biomed Eng* 10:275–309.

Cortes C. and Vapnik V. (1995). Support-vector networks. *Mach Learn* 20:273–297.

Dario P. et al. (1998). Neural interfaces for regenerated nerve stimulation and recording. *IEEE Trans Rehabil Eng* 6(4):353–363.

Dhillon G.S. and Horch K.W. (2005). Direct neural sensory feedback and control of a prosthetic arm. *IEEE Trans Neural Syst Rehabil Eng* 13(4):468–472.

Dhillon G.S. et al. (2004). Residual function in peripheral nerve stumps of amputees: implications for neural control of artificial limbs. *J Hand Surg Am* 29(4):605–615; discussion 616–618.

Dhillon G.S. et al. (2005). Effects of short-term training on sensory and motor function in severed nerves of long-term human amputees. *J Neurophysiol* 93(5):2625–2633.

Diedrich A. et al. (2003). Analysis of raw microneurographic recordings based on wavelet de-noising technique and classification algorithm: wavelet analysis in microneurography. *IEEE Trans Biomed Eng* 50(1):41–50.

Dillingham T.R., Pezzin E.L., and MacKenzie E.J. (1998). Incidence, acute care length of stay, and discharge to rehabilitation of traumatic amputee patients: an epidemiologic study. *Arch Phys Med Rehabil* 79(3):279–287.

Djilas M. et al. (2010). Spike sorting of muscle spindle afferent nerve activity recorded with thin-film intrafascicular electrodes. *Comput Intell Neurosci* p. 836346.

Edell D.J. (1986). A peripheral nerve information transducer for amputees: long-term multichannel recordings from rabbit peripheral nerves. *IEEE Trans Biomed Eng* 33(2):203–214.

Englehart K. and Hudgins B. (2003). A robust, real-time control scheme for multifunction myoelectric control. *IEEE Trans Biomed Eng* 50(7):848–854.

Farina D. et al. (2008). Multichannel thin-film electrode for intramuscular electromyographic recordings. *J Appl Physiol* 104(3):821–827.

Fisher L.E. et al. (2009). Chronic stability and selectivity of four-contact spiral nerve-cuff electrodes in stimulating the human femoral nerve. *J Neural Eng* 6(4):046010.

Grill W.M., Norman S.E., and Bellamkonda R.V. (2009). Implanted neural interfaces: biochallenges and engineered solutions. *Annu Rev Biomed Eng* 11:1–24.

Guiraud D. et al. (2006). Original electronic design to perform epimysial and neural stimulation in paraplegia. *J Neural Eng* 3(4):276–286.

Hoffer J.A. and Kallesøe K. (2001). How to use nerve cuffs to stimulate, record or modulate neural activity. In: *Neural Prostheses for Restoration of Sensory and Motor Function* (Moxon E.K.A. and Chapin J.K., Eds.). Boca Raton, FL: CRC Press.

Huang H. et al. (2008). An analysis of EMG electrode configuration for targeted muscle reinnervation based neural machine interface. *IEEE Trans Neural Syst Rehabil Eng* 16(1):37–45.

Jia X. et al. (2007). Residual motor signal in long-term human severed peripheral nerves and feasibility of neural signal-controlled artificial limb. *J Hand Surg Am* 32(5):657–666.

Kaczmarek K.A. et al. (1991). Electrotactile and vibrotactile displays for sensory substitution systems. *IEEE Trans Biomed Eng* 38(1):1–16.

Kapandji I.A. (1982). *The Physiology of the Joints. The Upper Limb*, vol. 1. New York: Churchill Livingstone.

Kuiken T.A. et al. (2004). The use of targeted muscle reinnervation for improved myoelectric prosthesis control in a bilateral shoulder disarticulation amputee. *Prosthet Orthot Int* 28(3):245–253.

Kuiken T.A. et al. (2009). Targeted muscle reinnervation for real-time myoelectric control of multifunction artificial arms. *JAMA* 301(6):619–628.

Lago N. et al. (2005). Long term assessment of axonal regeneration through polyimide regenerative electrodes to interface the peripheral nerve. *Biomaterials* 26(14): 2021–2031.

Lago N. et al. (2007a). Assessment of biocompatibility of chronically implanted polyimide and platinum intrafascicular electrodes. *IEEE Trans Biomed Eng* 54(2):281–290.

Lago N. et al. (2007b). Neurobiological assessment of regenerative electrodes for bidirectional interfacing injured peripheral nerves. *IEEE Trans Biomed Eng* 54(6): 1129–1137.

Lawrence S.M. et al. (2004). Acute peripheral nerve recording characteristics of polymer-based longitudinal intrafascicular electrodes. *IEEE Trans Neural Syst Rehabil Eng* 12(3):345–348.

Lewicki M.S. (1998). A review of methods for spike sorting: the detection and classification of neural action potentials. *Network* 9(4):R53–R78.

Lundborg G. and Rosen B. (2001). Sensory substitution in prosthetics. *Hand Clin* 17(3):481–488, ix–x.

McDonnall D., Clark G.A., and Normann R.A. (2004). Interleaved, multisite electrical stimulation of cat sciatic nerve produces fatigue-resistant, ripple-free motor responses. *IEEE Trans Neural Syst Rehabil Eng* 12(2):208–215.

Merrill D.R., Bikson M., and Jefferys J.G.R. (2005). Electrical stimulation of excitable tissue: design of efficacious and safe protocols. *J Neurosci Methods* 141:171–198.

Micera S. et al. (2001). Neuro-fuzzy extraction of angular information from muscle afferents for ankle control during standing in paraplegic subjects: an animal model. *IEEE Trans Biomed Eng* 48(7):787–794.

Micera S. et al. (2006). Hybrid bionic systems for the replacement of hand function. *Proc IEEE* 94:1752–1762.

Micera S. et al. (2008). On the use of longitudinal intrafascicular peripheral interfaces for the control of cybernetic hand prostheses in amputees. *IEEE Trans Neural Syst Rehabil Eng* 16(5):453–472.

Micera S., Carpaneto J., and Raspopovic S. (2010a). Control of hand prostheses using peripheral information. *IEEE Rev Biomed Eng* 3:48–68.

Micera S. et al. (2010b). Decoding information from neural signals recorded using intraneural electrodes: toward the development of a neurocontrolled hand prosthesis. *Proc IEEE* 98:407–417.

Mulvey M.R. et al. (2009). The use of transcutaneous electrical nerve stimulation (TENS) to aid perceptual embodiment of prosthetic limbs. *Med Hypotheses* 72(2):140–142.

Nannini N. and Horch K. (1991). Muscle recruitment with intrafascicular electrodes. *IEEE Trans Biomed Eng* 38(8):769–776.

Navarro X. et al. (1998). Stimulation and recording from regenerated peripheral nerves through polyimide sieve electrodes. *J Periph Nerv Syst* 3:91–101.

Navarro X. et al. (2005). A critical review of interfaces with the peripheral nervous system for the control of neuroprostheses and hybrid bionic systems. *J Peripher Nerv Syst* 10(3):229–258.

Navarro X. et al. (2007). Neurobiological evaluation of thin-film longitudinal intrafascicular electrodes as a peripheral nerve interface. In: Proc. IEEE 10th Int. Conf. Rehabilitation Robotics ICORR, pp. 643–649.

Nordhausen C.T., Rousche P.J., and Normann R.A. (1994). Optimizing recording capabilities of the Utah Intracortical Electrode Array. *Brain Res* 637(1–2):27–36.

Ohnishi K., Weir R.F., and Kuiken T.A. (2007). Neural machine interfaces for controlling multifunctional powered upper-limb prostheses. *Expert Rev Med Devices* 4(1):43–53.

Oskoei M.A. and Hu H. (2007). Myoelectric control systems—a survey. *Biomed Signal Proc Contr J* 2:275–294.

Polasek K.H. et al. (2009). Stimulation stability and selectivity of chronically implanted multicontact nerve cuff electrodes in the human upper extremity. *IEEE Trans Neural Syst Rehabil Eng* 17(5):428–437.

Popovic D.B. et al. (1993). Sensory nerve recording for closed-loop control to restore motor functions. *IEEE Trans Biomed Eng* 40(10):1024–1031.

Quiroga R.Q., Nadasdy Z., and Ben-Shaul Y. (2004). Unsupervised spike detection and sorting with wavelets and superparamagnetic clustering. *Neural Computation* 16(8):1661–1687.

Raspopovic S. et al. (2010). On the identification of sensory information from mixed nerves by using single channel cuff electrodes. *J Neuroeng Rehabil* 7(1):17.

Reiter R. (1948). Eine neue Elecktrokunsthand. *Grenzgebiete der Medizin* 4:133–135.

Riso R.R. (1999). Strategies for providing upper extremity amputees with tactile and hand position feedback—moving closer to the bionic arm. *Technol Health Care* 7(6):401–409.

Rosén B. et al. (2009). Referral of sensation to an advanced humanoid robotic hand prosthesis. *Scand J Plast Reconstr Surg Hand Surg* 43(5):260–266.

Rossini P.M. et al. (2010). Double nerve intraneural interface implant on a human amputee for robotic hand control. *Clin Neurophysiol* 121(5):777–783.

Schultz A.E., Marasco P.D., and Kuiken T.A. (2009). Vibrotactile detection thresholds for chest skin of amputees following targeted reinnervation surgery. *Brain Res* 1251:121–129.

Shannon G.F. (1979). Sensory feedback for artificial limbs. *Med Prog Technol* 6(2):73–79.

Stieglitz T., Schuetter M., and Koch K.P. (2005). Implantable biomedical microsystems for neural prostheses. *IEEE Eng Med Biol Mag* 24(5):58–65.

Tesfayesus W. and Durand D.M. (2007). Blind source separation of peripheral nerve recordings. *J Neural Eng* 4(3):S157.

Tyler D.J. and Durand D.M. (2002). Functionally selective peripheral nerve stimulation with a flat interface nerve electrode. *IEEE Trans Neural Syst Rehabil Eng* 10(4):294–303.

Upshaw B. and Sinkjaer T. (1998). Digital signal processing algorithms for the detection of afferent nerve activity recorded from cuff electrodes. *IEEE Trans Rehabil Eng* 6(2):172–181.

Warwick K. et al. (2003). The application of implant technology for cybernetic systems. *Arch Neurol* 60(10):1369–1373.

Weir R.F. et al. (2009). Implantable myoelectric sensors (IMESs) for intramuscular electromyogram recording. *IEEE Trans Biomed Eng* 56(1):159–171.

Wodlinger B. and Durand D.M. (2009). Localization and recovery of peripheral neural sources with beamforming algorithms. *IEEE Trans Neural Syst Rehabil Eng* 17(5):461–468.

Yoshida K. and Horch K. (1993). Selective stimulation of peripheral nerve fibers using dual intrafascicular electrodes. *IEEE Trans Biomed Eng* 40(5):492–494.

Yoshida K. and Stein R.B. (1999). Characterization of signals and noise rejection with bipolar longitudinal intrafascicular electrodes. *IEEE Trans Biomed Eng* 46(2):226–234.

Yoshida K. et al. (2010). Multichannel intraneural and intramuscular techniques for multiunit recording and use in active prostheses. *Proc IEEE* 98(3):432–449.

Zecca M. et al. (2002). Control of multifunctional prosthetic hands by processing the electromyographic signal. *Crit Rev Biomed Eng* 30(4–6):459–485.

17

REVIEW CHAPTER

EXOSKELETAL ROBOTICS FOR FUNCTIONAL SUBSTITUTION

JOSÉ LUIS PONS, JUAN C. MORENO, AND EDUARDO ROCON
Bioengineering Group, Spanish Research Council, CSIC, Arganda del Rey, Spain

SUMMARY

Technologies for functional substitution in motor disorders have been applied in chronic phases of neurological conditions, whenever no functional recovery is possible. Among these technologies, exoskeletal robots (ERs) might find an application niche. Exoskeletal robots are worn by patients, and immediately thereafter issues related to wearability, safety, and dependability emerge. Because most of the time they are applied to the functional substitution of walking, energy consumption is at stake and there is evidence that most of the times when use is discontinued it is related to the high metabolic cost of transport when using the ER.

This chapter advocates for considering functional substitution ERs for rehabilitation during chronic phases of motor disorders. As a consequence many of the technologies (control, mechanisms, actuation, sensors) being applied in the broad field of robotic-based physical therapy for rehabilitation can be imported to functional substitution robots. Therefore, this chapter reviews how these technologies have been applied in the past to control the flow of mechanical power and the flow of information between human and robot. The former constitutes

Introduction to Neural Engineering for Motor Rehabilitation, First Edition. Edited by Dario Farina, Winnie Jensen, and Metin Akay.

the physical interface between both actors; it is orchestrated by control technologies and mediated by exoskeletal mechanisms and actuator technologies. The latter constitutes the cognitive interaction between human and robot and is used to relay command information between the patient and the technology and feedback information from the robot to the user.

The conclusion is that, for truly wearable solutions (especially in the case of substitution during walking), the passive dynamics of humans during walking must be exploited. This implies that a better understanding of how humans orchestrate different neuromotor mechanisms is required. This new knowledge must then fully inform the design of the functional substitution technologies. If this is achieved, then the concept of "assist as needed" can be revisited and imported from rehabilitation scenarios to functional substitution scenarios.

The chapter identifies technological bottlenecks in the development of functional substitution ERs. One is backdrivable actuator technologies. We propose that a way to circumvent this technological limitation might be based on hybrid functional substitution technologies comprising an ER and a motor neuroprosthesis (MNP). In this way, latent motor capabilities of patients can be preserved or even augmented by orchestrating the action of MNPs and exoskeletal robots in a scheme that seeks to exploit the advantages of each technology.

INTRODUCTION

Exoskeletal robots (ERs) are person-oriented robots. Human users wear them, whether to supplement the function of a limb or to replace it completely. Exoskeletal robots operate alongside human limbs (Pons, 2008) and can be divided into two main categories: (1) human-empowering technology (HET) and (2) robotic rehabilitation devices. The latter can be used to assist weakened muscles, to preserve and protect human joints, or to rehabilitate people (Herr, 2009). The former are typically used for functional substitution.

Target populations for both rehabilitation and functional substitution ERs are patients suffering from neurological conditions leading to severe motor disorders—in particular, cerebrovascular accident (CVA), spinal cord injury (SCI), cerebral palsy (CP), and pathological tremor. All these conditions lead to altered sensorimotor capabilities, and one of the most disabling sequelae is diminished locomotion capability.

A common method used to improve motor control is repetitive physical therapy and rehabilitation training over several months. Both neuroplasticity—in particular, brain plasticity—and feedback are at the root of rehabilitation. When acting passively (without active intervention from the patient), muscle activity is reduced during rehabilitation or functional substitution and thus neuroplasticity is not triggered. Feedback has been shown to play a crucial role in eliciting both new and permanent motor patterns (Kim et al., 2010).

Although physical therapy is very labor intensive, it could be proven that task-specific repetitive training in conventional and robotics rehabilitation therapies is beneficial (Hesse et al., 1995). However, permanent improvements in gait patterns could not be shown. The main problems are attributed to the passive role the patient was given during training, the lack of task-specific training in ambulatory conditions or even at home, and the user's noncompliance with the therapy due to low motivation.

Rehabilitation ERs were originally designed to allow for repetitive training over longer therapy sessions. Eventually, the ultimate goal is to allow for the extension of physical therapy programs out of the clinical environment. In some neurological conditions (e.g., cerebral palsy, especially in children), exoskeletal technologies are focused on making the child prone to lifelong use of orthotic devices (Piazza et al., 2010).

In studies involving both SCI patients and professionals (Ditunno et al., 2008), motor rehabilitation or motor substitution of walking has been established as a high-priority objective in patients with SCI. The study concludes that walking is a high priority for SCI patients irrespective of severity of injury, time of injury, and age at time of injury. In this chapter we mostly focus on ERs for motor substitution during walking.

There is consensus in that user-perceived value of the rehabilitation or functional substitution technology is at the core of acceptance. The United States requires rehabilitation facilities to document patient satisfaction for certification. In the United Kingdom, a clinical checklist—needs assessment checklist—is used as an outcome measure to ensure that rehabilitation services address the needs of the patient (Kennedy and Hamilton, 1999).

Active patient involvement in the rehabilitation process enhances patient compliance and is a good predictor of a positive outcome (Duff et al., 2004). The inadequacy of providing effective results in rehabilitation could be explained by the difference between the user's intention and the robot's actions, and the occasional absence of the user's focus on lower limb motion during gait.

As for evaluating the user's intention, different methods have been tried (e.g., online force measurement during gait), but in general they have not led to significant improvements. A new way of providing an effective rehabilitation or functional substitution ER would be to gain insight into the user's neural activity during therapy to assess the intention; then applying an adequate action at the robot's end-effectors that might not only allow the patient to walk more efficiently but also result in increased activation of the cortical areas engaged in neural recovery.

The design of ambulatory exoskeletons requires systems that are lightweight but at the same time powerful. These contradictory requirements might be the reason no commercial systems are available yet. Still, open questions remain as to robotics-based physical therapy and motor substitution: Can chronic patients (stroke, SCI, CP) improve gait (or motor control capabilities in general)? Can robots alter motor patterns of healthy users and physically impaired patients? Can functional substitution be attained in an energetically efficient way? What is the role of feedback? What is the best feedback modality? Positive answers to some of these questions would help extend rehabilitation robotics to chronic phases and thus allow us to treat functional substitution ERs as the last stage in a continuous and permanent physical therapeutic program.

STATE OF THE ART

An obvious growing interest in exoskeletal robots both for rehabilitation and for functional substitution can be observed by simply looking at the number of articles on the subject over the past decades (Mohammed and Amirat, 2009). Despite the several decades of work in the field a number of open issues can be highlighted: dressing and undressing, that is, putting the systems on and taking them off; safety, and as a consequence dependability; and energy consumption (Dollar and Herr, 2008), among others.

These open issues can be traced back to issues related to the fundamentals of the cognitive and physical interaction between the user and the ER for rehabilitation or functional substitution.

In any instance of a functional substitution ER, a **cognitive human–robot interface** (cHRI) supports the bidirectional flow of information between the human and the robot. Information is any type of variable (e.g., electrical signals, proprioception, visual perception) required for control and feedback of the exoskeletal robot (Pons, 2008). Likewise, a **physical human–robot interface** (pHRI) supports bidirectional flow of mechanical power between the human and robot in order to interact with the external world. Any type of force interaction or motion must be possible without impeding or endangering each other (Pons, 2008). Ultimately, the previously mentioned open issues could be directly related to how these cHRIs and pHRIs are implemented.

Therefore, this section approaches the analysis of the state of the art in functional substitution ERs in relation to their cHRI and pHRI. Since we advocate that functional substitution ERs be considered as true rehabilitation robots in chronic phases, in a comprehensive approach, we also include rehabilitation ERs in this discussion.

Physical Human–Robot Interface

The flow of mechanical power between the human and robot is orchestrated by control strategies and mediated by both actuators and exoskeletal mechanisms. When controlling rehabilitation or functional substitution ERs, trajectory-based stiff controllers have often led to reduced kinematic error, which is a key signal for driving motor rehabilitation (Wolbrecht et al., 2007). In addition, a stiff controller would complete the task without active participation from the patient, thus leading to a reduced motivation during the therapy.

To overcome these problems a number of control algorithms have been conceived to provoke neuroplasticity. Most of these controllers can be classified in two categories: (1) assistive controllers and (2) challenge-based controllers. They are briefly analyzed in the following subsection.

Control. Assistance during physical therapy is usually manually implemented in clinical practice. A nonexhaustive list of rationales for active robot assisted physical therapy is (1) to make physical therapy safer to allow more intense practice, (2) to enhance somatosensory input that is correlated with motor output, (3) to reinforce normative sensory–motor pathways, (4) to increase task success to motivate practice, and (5) to allow more repetitions because movements are completed more efficiently.

Two main drawbacks can be attributed to physically assisting movement: the so-called guidance hypothesis (Schmidt and Bjork, 1992) and slacking hypothesis (Wolbrecht et al., 2007). The former is due to the fact that assisting a movement changes the dynamics of the task and therefore what is learnt is not the target task. The latter refers to the phenomenon by which the robotic device reduces the patient's recovery if it leads to a slacking by decreasing patient effort or attention.

Strategies to promote effort and self-initiated movements include (1) allowing some error variability around the target trajectory in which no assistance is provided, (2) triggering assistance only when the participant achieves a force or velocity threshold, (3) making the robot compliant, and (4) including a forgetting factor in the robotic assistance (Marchal-Crespo and Rienkensmeyer, 2009).

A way of avoiding too much assistance, in line with strategy (4), is to follow the "assist-as-needed" (AAN) paradigm. The AAN strategy is intended to lead to simultaneous activation of efferent motor pathways and afferent sensory pathways, which are prerequisites for cortical reorganization and thus are the cornerstone of successful learning during gait rehabilitation (Emken et al., 2005). The assumption is made that the human resists applied forces by internally modelling the force and then counteracting it appropriately. It is hypothesized that the human thereby uses an error-based learning controller. In robotic control the attempt is made to copy this adaptive strategy. The robotic controller takes the form of an error-based learning controller with a "forgetting factor" comparable to the human neuromotor control according to the following expression:

$$G_{i+1} = fG_i = g\varepsilon_i$$

where G is the control parameter that is adapted and ε is a performance indicator related to the ability of the patient to initiate or reach the target motor task. In AAN gait training, the "forgetting factor" f, for instance, reduces assistance as a function of the measured trajectory error. In this particular example, G would represent a force gain and ε would be the error in the trajectory of the motor task.

The robotic controller is able to adapt to varying gait profiles and levels of assistance. One method of implementation is to use impedance control (Hogan,

1985). The stiffness and the position of the actuator's output are chosen as functions of the current position error or can be manipulated manually by the therapist (Veneman et al., 2007). If the patient wants to move freely, the output impedance is set to zero—referred to as zero-impedance control mode. This type of control has been implemented in several devices (Duschau-Wicke et al., 2010).

Similar patient-cooperative control strategies have been developed: "force-field control" (Banala et al., 2010) and "path control" (Duschau-Wicke et al., 2004). Both methods use a virtual environment to influence the timing and impedance of the limb movement.

Challenge-based robotic strategies are, in a way, opposite to assistive strategies. While the latter help reaching task targets, the former is aimed at rendering the task more challenging. Among the various challenge-based strategies, resistive strategies and error-amplification strategies are worth mentioning. The latter are based on evidence that kinematic errors are key neural signals driving motor adaptation (Emken et al., 2007). The former relies on well-established evidence that resistive exercise requiring higher effort can help improve motor function (Weiss et al., 2000).

When one is confronted with exoskeletal robotics for functional substitution, wearability and ambulation are musts. All control approaches previously described for motor rehabilitation require complex mechanisms and actuation principles (see the next subsection, "Mechanisms and actuation"), which in turn result in bulky exoskeletal robots that cannot be taken out of the clinical environment. More important, these actuator technologies and the bulky exoskeletal structures can render ambulation not viable by impeding the passive dynamics of human limbs in efficient walking. Energy cost of these control strategies is thus another factor limiting the development of ambulatory functional substitution ERs.

However, human walking is an extremely efficient process. This suggests the design of controllers must be founded on insights into human motion control: (1) exploitation of passive dynamics by robot-limb systems and (2) stabilizing robot–human walking by means of reflexes, ranging from local reflexes to complex supraspinal postural reflexes (Luksch, 2010). For this to be done, mechanism and actuation systems for exoskeletal robots must exhibit natural muscle-like dynamic characteristics, for example, adaptable backdrivability. These topics are further addressed in the next subsection.

Mechanisms and Actuation. When designing an exoskeletal robot, one of the first questions needing clarification is which joints and how many degrees of freedom (DOF) need to be covered by the ER. An exoskeleton is defined as an active mechanical device that is essentially anthropomorphic, is worn by a user and fits closely to his or her body, and works in concert with the operator's movements. Therefore, the robot's kinematics must be compatible with anatomical kinematics; that is, instantaneous rotation axes of robot and anatomy must be aligned throughout joint excursion.

As to the alignment of anatomical and robot joints, increasing the number of DOFs, that is, designing redundant kinematics, is a way to solve this problem (Schiele and van der Helm, 2006). A second alternative is designing robotics joints to exactly follow anatomical joints. While the latter requires a customization process, the former is a general solution across users. However, quick and accurate alignment of joints remains a problem in future exoskeleton designs, even though elastic supports can dynamically compensate for some misalignment.

Another point to clarify is which of these DOFs need to be actively controlled. Here we consider only devices that provide means of augmenting power at one or more joints of the lower extremities. This includes both adding and dissipating power, as well as the controlled release of energy stored in springs during various phases of gait.

In general, the human leg can be thought of as a seven-DOF structure, with three rotational DOFs at the hip, one at the knee, and three at the ankle. By considering the energy balance at ankle, knee, and hip during normal human walking (Dollar et al., 2008), powered ERs often apply the following actions: (1) adding power at the hip, (2) dissipating power at the knee, and (3) storing and releasing energy at the ankle using primarily passive elastic structures.

Rehabilitation ERs inherit features previously explored in exoskeletons for power augmentation such as those developed in the framework of the U.S. DARPA's Exoskeletons for Human Performance Augmentation (EHPA) program: BLEEX exoskeleton, SARCOS exoskeleton, and MIT exoskeleton.

One of the most visible of the DARPA program's exoskeletons has been the Berkeley Lower Extremity Exoskeleton (BLEEX). This exoskeletal robot comprises three DOFs at the hip, one DOF at the knee, and three DOFs at the ankle. Of these DOFs, four are actuated: hip flexion–extension, hip abduction–adduction, knee flexion–extension, and ankle flexion–extension.

Also in the framework of EHPA, a quasi-passive exoskeleton concept has been proposed at the Massachusetts Institute of Technology (MIT) (Walsh, 2006). This concept seeks to exploit the passive dynamics of human walking in order to create lighter and more efficient exoskeletal devices. The MIT exoskeleton employs a quasi-passive design that does not use any actuators for adding power at the joints. The design relies completely on the controlled release of energy stored in springs during the (negative power) phases of the walking gait.

While the MIT exoskeleton was proposed in the context of power augmentation, a similar concept, the GAIT exoskeletal robot, was recently proposed in the context of functional substitution in patients suffering from unilateral post-poliomyelitis syndrome (Moreno et al., 2009). The exoskeleton spanned ankle and knee joints. This time the concept seeks to exploit the passive dynamics of the robot-limb system through a limit cycle walking approach as in passive bipedal walkers. Spring-based passive actuators were implemented in a zero-energy-consumption solution to assist knee stability during stance

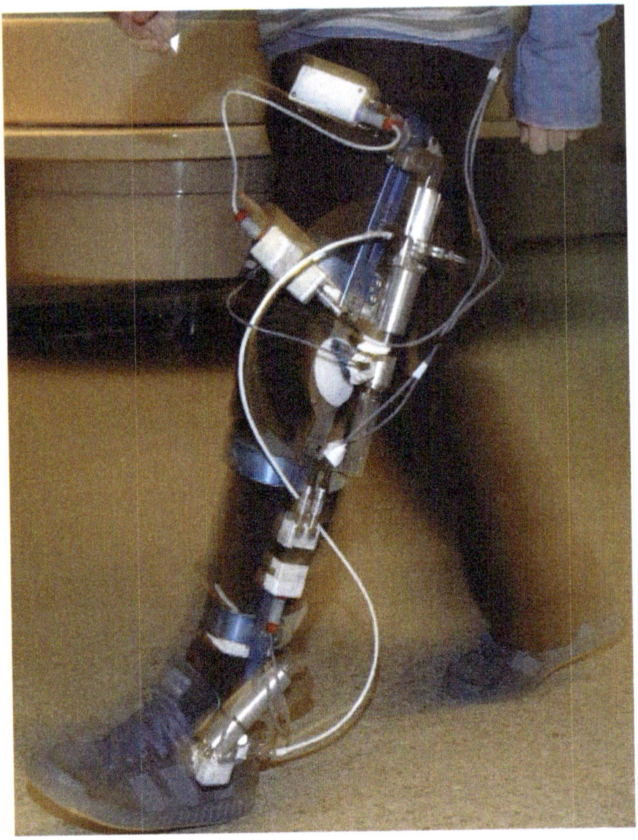

Figure 17.1. The GAIT exoskeletal robot, a zero-energy-consumption functional substitution device for poststroke and post-polio syndrome patients.

and knee extension at final swing phase (Cullell et al., 2009); see Figure 17.1. To do this, normal-gait average data was used to model springlike behavior for stance and swing. A rigid and a soft version of linear springs were then used to substitute for quadriceps weakness; see Figure 17.2. The ankle was also assisted to avoid drop foot during swing and uncontrolled plantar flexion (plantar fall) upon heel contact as well as push-off. Results have demonstrated that, upon application of functional substitution in the lower limb, significant improvement with respect to the current gait of the subjects is achieved (Moreno et al., 2008).

A comparison of metabolic power with and without the ER is a good determinant of whether there is any energetic advantage to using the device versus other assistive technologies. Metabolic studies with quasi-passive ERs showed a 10% increase in walking metabolic cost of transport, or the metabolic energy required to transport unit weight per unit distance (Dollar and Herr, 2008). It

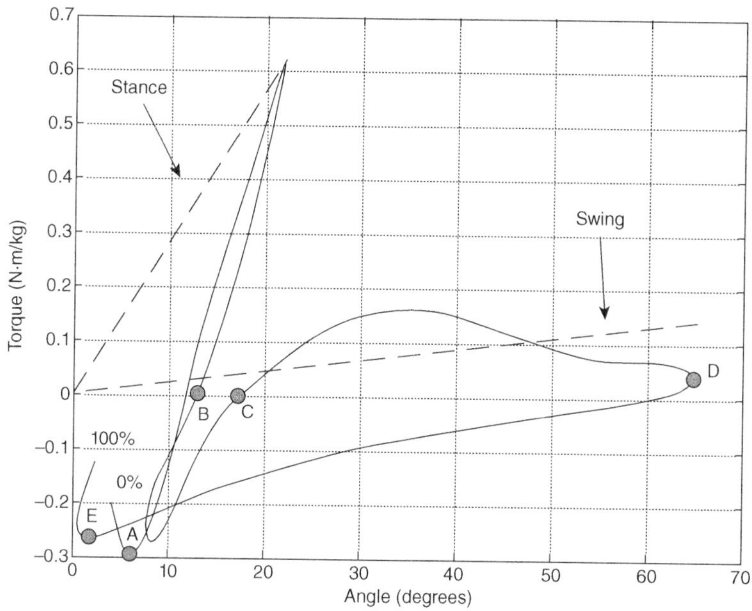

Figure 17.2. Normal-gait average biomechanical data. Knee angle versus torque in the sagittal plane (complete gait cycle) and optimal elastic adjustment.

is thought that a dominant cause for the observed increase in cost of transport is the additional kinematic constraints inadvertently imposed on the wearer, upsetting the efficient dynamics normally seen during human walking (Walsh, 2006).

Upright posture and walking can benefit people with lower limb paralysis. Benefits include reduced incidence of joint contractures, bone fractures, and pressure sores; prevention of osteoporosis; improved bowel function, urinary drainage, and peripheral circulation; and stimulation of leg growth for growing children. In addition, upright ambulation has psychosocial advantages, such as improved interaction with peers and the environment, more positive perception of the body, and greater self-respect (Johnson et al., 2009).

Mechanisms to obtain erect gait and postural balance have been proposed (Johnson et al., 2009), for example, reciprocating mechanisms (see the next paragraphs) where extension and flexion patterns at hip level are mechanically coupled to the contralateral limb,.

Many assistive devices have been developed to help people with lower limb paralysis stand and walk. These devices range from the relatively simple, conventional hip–knee–ankle–foot orthosis (HKAFO; Fatone, 2006) to the more complex reciprocating gait orthosis (RGO; Douglas et al., 1983); see Figure 17.3. Both orthoses immobilize the knees and ankles in an appropriate standing alignment and allow ambulation with crutches or a walker. The

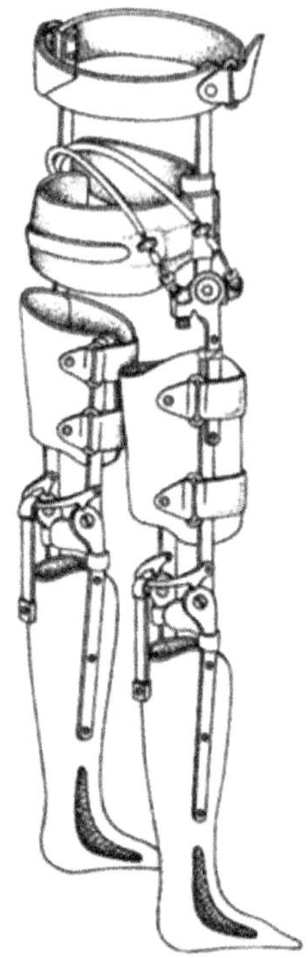

Figure 17.3. Schematic representation of a reciprocating gait orthosis (RGO): flexion–extension motor patterns are mechanically linked in contralateral limbs as a mechanism to promote upright walking.

conventional HKAFO also immobilizes the hip joint; hence, users usually adopt a swing-to or swing-through gait pattern.

Despite the advantages of upright ambulation, discontinuation of these reciprocating mechanisms is high, ranging from roughly 60% to 90% for children (myelomeningocele) and from roughly 45% to 55% in adults (spinal cord injury; Johnson et al., 2009). The effort needed to ambulate with the orthosis is one of the main reasons for limiting RGO use.

Drawbacks of reciprocating mechanisms are chiefly related to limitations of the stride length. A possible solution is the use of the so-called variable hip constraint mechanism (VHCM; To et al., 2006), in which it is proposed to use hybrid solutions that combine an ER and a motor neuroprosthesis (MNP) based on functional electrical stimulation (FES).

However, a hybrid ER combining robotics structures and FES is not a new concept. Two trends could be identified: (1) FES is used to actively drive the limbs while the ERs are mostly energy dissipation devices (Kobetic et al., 2002); and (2) FES and ERs are both torque-generating devices (Obinata et al., 2007; Popovic et al., 1989).

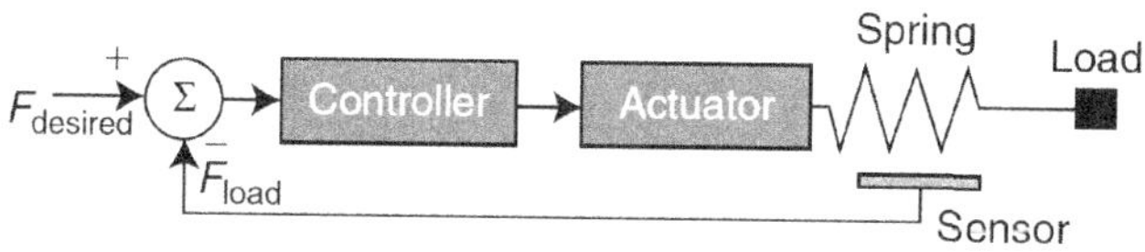

Figure 17.4. Concept of a series elastic actuator (SEA): precise and compliant force control can be obtained with good position actuators by means of a spring mechanically in series.

In most existing exoskeleton designs (e.g., Lokomat [Hocoma, Switzerland] and Reoambulator [Motorika, USA]; see Chapter 18 and Chapter 26) electric direct drives and brushless DC-motor–gearbox combinations are mainly used. Pneumatic actuators, for example, pneumatic artificial muscles (PAMs), are being employed in many ER projects (Ferris et al., 2005) as they offer a relatively large power-to-weight ratio of several kilowatts per kilogram and are compliant. However, noise and energy source are issues.

Compliant actuator designs and concepts have been structured in an excellent review article (Van Ham et al., 2009). Series elastic actuation (SEA) has been experiencing high popularity among researchers in prosthetics and orthotics; see Figure 17.4 for a schematic representation of this concept. The advantage of such a configuration lies in its ability to act as quasi-perfect torque or force source; for example, electric motors are excellent in positioning but inaccurate for low-force display. When the drive is turned off the SEA remains backdrivable to a certain extent and can absorb shocks, which improves safety for both the human and the robot. The development of SEA has been expanded to hydroelastic actuators (Pratt et al., 2004) and rotational hydroelastic actuators (rHEA; Stienen et al., 2010). The principle of SEA has also been expanded to an active series elastic clutch (SEC) system (Martinez-Villalpando and Herr, 2009). Basically an SEC system is a combination of two antagonistic SEAs acting on the same joint. By properly selecting the impedance delivered in each actuation direction the joint can be effectively blocked or actuated.

Backdrivability and altered passive dynamics of limbs (by wearing exoskeletal robots) are clear limitations in efficiently exploiting natural gait dynamics. Both problems are closely related to actuator technologies and robot mechanisms. Hybrid concepts combining mechanical actions of motor neuroprostheses and ER for motor substitution might be a possible way of circumventing technological bottlenecks; this is addressed in more detail in the third section ("Current challenges, future directions").

Cognitive Human–Robot Interface

Cognitive human–robot interfaces (cHRIs) support the flow of information between the exoskeletal robot functional substitution robot and the patient.

Whenever information is relayed from the user to the ER it is basically used for commanding the assisting technology. Whenever the information is relayed from the robot to the user it constitutes a means for implementing feedback strategies.

Since, in motor substitution applications, the command information being relayed to the robot in terms of motor intentions and executions, there is a current trend in sourcing the information from all stages in the motor planning–execution process in humans. This means that, in general, a combination of information on motor planning, for example, electroencephalography (EEG), functional near-infrared (fNIR) activity, peripheral motor execution (e.g., electromyographic, EMG, activity), and motion analysis (e.g., limb motion), can be used to robustly command the exoskeletal robot.

Brain-to-machine interfaces (BMIs), or the recently coined term, brain and neural to machine interfaces (BNMIs), which constitutes a multimodal interface to the humans' neural system, is gaining momentum as a method to command exoskeletal rehabilitation and functional substitution robots. BNMIs might constitute a new means to improve current AAN strategies for robotic-based gait training and substitution. BNMIs have the potential to improve AAN controllers for gait training by demanding neural control within the involved cortical network. BNMIs also include systems that rely on the following: (1) passive monitoring (which might assess motor intention without providing real-time feedback); (2) information derived from the peripheral nervous system (such as peripheral action potentials that might directly trigger muscle flexion); and (3) indirect measures of neural activity (such as EMG).

The analysis of information of motor planning in slow cortical potentials (motor-related cortical potentials, MRCPs) might be related to movement preparation, initiation, and finalization; therefore they can constitute means for volitionally driving exoskeletal robots. Correlates between EMG–EEG activities (e.g., analysis of corticomuscular coherence) can be regarded as a potential methodology to distinguish higher control mechanisms in bipedal gait, which can be useful in closing AAN loops to correct involuntary or weak activation during training.

Also, the sensory fusion at different levels of the neuromotor control system in humans can be used to ascertain whether motor information is consistent with voluntary movement activity. For instance, spasms do not produce the same EEG activity pattern as voluntary, planned movement intentions. Hence, if EMG activity indicates movement but EEG activity shows that this movement was unintentional, the patient may be having a spasm or any sort of involuntary movement. Then the robot can be commanded accordingly.

Due to research with event-related desynchronization (ERD) in EEG activity, scientists now have a fairly good understanding of how the brain manages actual, intended, and perceived movements (Pfurtscheller and Lopes da Silva, 1999). This will help integrate such knowledge into robust ways of relaying motor commands to the functional substitution ER.

Electromyographic signals are most often used to investigate which muscle groups are active during rehabilitation or to detect latent muscular activity in an empowering approach. This is the case with HAL 5, which acts as an EMG amplifier by empowering the users (Hayashi et al., 2005). HAL 5 is not for functional substitution but it is a clear example of how EMG amplification can also be adopted for functional substitution or for rehabilitation. In addition to HAL 5, only a few rehabilitation devices have been reported to be using EMG to control the exoskeleton.

More recently EMG control has been proposed in the context of controlling ankle exoskeletons (Kinnaird and Ferris, 2009). The authors showed that when soleus electromyography was used to control the amount of plantar flexion assistance from a robotic ankle exoskeleton, subjects significantly reduced their soleus activity to quickly return to normal gait kinematics.

CURRENT CHALLENGES AND FUTURE DIRECTIONS

The following subsections present a nonexhaustive list of potential future research directions and challenges in ER for motor rehabilitation and substitution.

Understanding Motor Function in Humans for Analogous Design of Exoskeletal Robots

In healthy humans, gait emerges naturally from a hierarchical organization and combination of motor control mechanisms. These neuromotor mechanisms include central nervous system (CNS) modulation of spinal rhythmic motor patterns and reflex-driven mechanisms. The result of this process is a highly efficient, stable, and robust gait.

Functional substitution ERs are devices specifically designed to substitute for human motor control mechanisms involved in gait whenever they are impaired as a consequence of neurological conditions. As such, the design and successful application of these devices will face challenges in terms of portability and usability, subjective perception or user-perceived value of the system.

If a system is not subjectively perceived as helpful, a discontinuation in its use will most likely follow. Perceived value is closely related to dependability and safety. Studies on the safety of the human operator, who is strapped into the exoskeletal functional substitution device, have yet to be performed.

Exoskeletal functional substitution robots are devices worn by users in their daily life activities to support mobility. Therefore, wearability and ambulation are key characteristics of these systems. The issue of efficiency and low energy consumption emerges immediately. The aspects of assisted gait dynamics that contribute to mechanical inefficiencies (high metabolic cost) remain unclear. Further understanding of the gait dynamics of healthy and assisted ambulation is needed for reduction of the high metabolic energy expenditure required of users. This is a prerequisite for truly ambulatory functional substitution ERs.

An improved understanding of neuromotor mechanisms and how they are orchestrated and modulated by the CNS, as well as an understanding of musculoskeletal and tendon function in walking and other movement tasks, may shed light on more effective exoskeletal architectures. This involves understanding and exploiting the following: (1) the passive system (the hybrid human–robot system) dynamics as a means for reducing control effort and energy; (2) the self-stabilizing effects of low latency and supraspinal reflexes in the robot controllers; and (3) the way the CNS modulates reflex and spinal rhythmic patterns depending on the current task or gait phase.

Gait models based on actual machine elements that capture the major features of human locomotion may enhance the understanding of human leg morphology and control—and lead to analogous improvements in the design of efficient and low-mass exoskeletons.

A Change of Paradigm: Functional Substitution as a Continuous Rehabilitation Process in Chronic Stages

There is evidence of continuous improvement in motor function, even in chronic phases of motor disorders. This allows us to advocate for long-term rehabilitation of motor impairments instead of, or at least in combination with, motor substitution. This implies that concepts from rehabilitation scenarios, as "assist-as-needed" and "challenge-based" paradigms can be imported to motor substitution ERs.

The concept of the assist-as-needed control approach to promoting motor improvement needs to be revisited. Currently, it is conceived as an approach in which, upon active participation of the user (e.g., to start a task), the device helps attain the goal while still preserving the task as a challenging one. In this concept, any insight into the cause of the motor disorder is not fully exploited.

If insight is gathered on how motor mechanisms are orchestrated (see the preceding discussion), this insight can inform the design of exoskeletal controllers that mimic how control is organized in the biological counterpart. If this is combined with specific knowledge of motor status gathered during the diagnosis phase of a particular patient, then ER controllers can be specifically customized to "assist" exactly the motor mechanisms that are impaired. This can be illustrated with the example of an SCI patient in whom an exacerbated reflex is observed. Then the motor rehabilitation ER can be set to assist this specific reflex mechanism (e.g., by using facilitating therapies to exploit a cutaneous reflex), thus triggering and exploiting it at the right time in the gait cycle. We advocate extending the concept of AAN to the very conception and customization of the motor control scheme in combination with traditional AAN approaches.

Clinical Challenges

Following the concept of considering functional substitution as the last stage in motor rehabilitation, the role of feedback becomes of paramount importance. There is currently, and there will be in the future, a vivid discussion on the nature of feedback mechanisms to better improve motor function.

Research on ERs still has to demonstrate that when they are integrated into motor rehabilitation therapies they attain a sustained improvement in motor recovery. Clinical evidence must still be gathered to support rehabilitation and functional substitution exoskeletal robotics.

The Technical Challenges

Reduced metabolic cost while using functional substitution ERs is one of the key aspects to improve. For this to happen the passive dynamics of the human–robot system must be exploited and a technical requirement is backdrivable motors. Therefore, how to backdrive motors is expected to be a very active area of research in the years to come, including the different alternatives being proposed as potential technical solutions to the backdrivability problem. These solutions include series elastic actuators (SEAs), series damping actuators (SDAs), and series elastic clutch (SEC) concepts.

In the short term, none of the emerging actuator technologies based on active or smart materials (shape memory actuators, electroactive polymer actuators, etc.) is foreseen as a true alternative in exoskeletal functional substitution robotics. The only exception would be magnetorheological fluid actuators (MRFs), which already have been integrated into commercial prosthetic devices. The main drawback of this technology is that it can only be used to dissipate energy; therefore it might find an application niche in hybrid FES and ER applications.

The investigation of nonanthropomorphic kinematic architectures of mechanisms for ERs may provide solutions to some of the problems associated with closely matching the structure of the exoskeleton to the wearer, such as the need for close alignment between the joints of the robot and those of the wearer. It is expected that research will continue in this area as a way of improving mechanical human–robot interaction. The target will most likely be reducing undesired interaction forces due to noncompliant kinematics of robot and human anatomy. Additionally, effective strategies for interfacing an exoskeleton with the human body, both mechanically and neurally, are important areas for future research.

Which DOFs are critical for rehabilitation and how many of these DOFs are to be actuated are still valid questions. These questions have implications for neuroscience (see section "Physical human–robot interface"), but also for ER mechanisms and actuator technologies. Lightweight ERs need to be developed that preserve the natural passive dynamics of human gait. Given the current limitations in mechanisms and actuator technologies, the concept of hybrid MNPs in combination with ERs will likely gain momentum (see the next subsection).

Orchestration of Motor Neuroprostheses, Exoskeletal Robots, and Latent Human Motor Capabilities

Both ERs and MNPs are technologies that seek to restore or substitute motor function. MNPs constitute an approach to restoring function using artificial control of human muscles or muscle nerves with FES. Exoskeletal robots use volitional commands for controlling the application of controlled forces to drive paralyzed or weak limbs. ERs are more versatile in the implementation of motor actions: (1) precise kinematics and impedance control are readily available; (2) biomimetic control architectures, for example, spinal pattern generators (SPGs), internal models (model-based feedforward control), and reflexes, can be readily applied; and (3) built-in artificial proprioception (limb awareness) and vestibular sensors can be used to drive nonvolitional motor actions, for example, control of balance. As a serious drawback, the transmission of motor actions (forces) to the human musculoskeletal system takes place through soft tissues. MNPs are better at preserving the natural passive dynamics of limbs as they do not impose artificial nonbackdrivable actuator technologies and do not significantly modify dynamic parameters of limbs.

Both MNPs and ERs can share the same cHRI to decipher volitional commands from the user as input for the MNP and ER controllers.

In a 2009 three-day scientific meeting on advances in neural rehabilitation engineering[1] (in which some of the authors of this book actively participated) the international context of the field of neural rehabilitation research was thoroughly analyzed. The consensus indicated that the previously stated limitations and shortcomings of MNPs and ERs for neural rehabilitation still hold.

Investigating MNPs and ERs is a very active scientific area involving numerous disciplines. The administrations of the United States, Japan, and European Union actively support the field. Just to mention the case of the European administration, MNP and ER research is supported by the EU 7th Framework Programme (FP7) in a number of programs: for example, the Virtual Physiological Human Programme of eHealth; ICT, Cognitive Systems and Robotics as well as eInclusion; and FET Proactive, Brain-Inspired ICT and Human–Computer Confluence.

In this context, future research activities will likely aim at the following:

1. Restoring motor function in patients with neurological conditions (tremor, SCI, CVA, CP, and others) through an integrated application of ERs and MNPs in a functional compensation approach

[1]Symposium on Advances in Neural Rehabilitation Engineering, August 24–26, 2009. Center for Sensory–Motor Interaction (SMI), Aalborg University, Aalborg, Denmark.

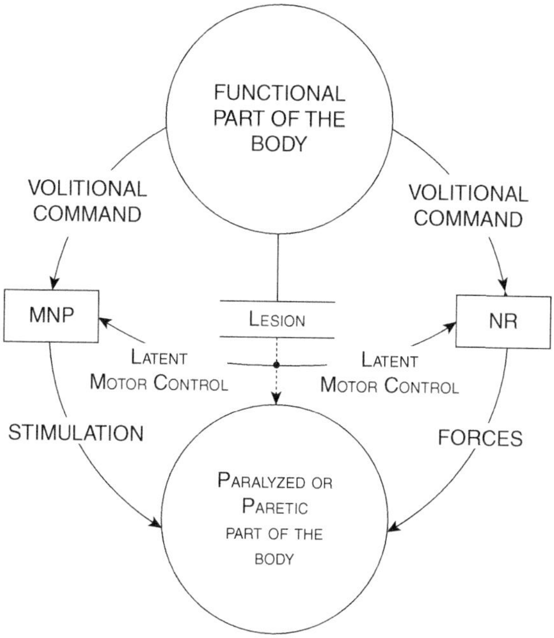

Figure 17.5. Functional scheme of a hybrid motor rehabilitation or substitution approach combining latent motor capabilities of patients, an MNP, and an ER.

2. Promoting motor control relearning in patients suffering all these conditions, via an integrated use of ERs and MNPs

A new and unconventional approach is required. Motor control of human limbs during manipulation and locomotion will be the result of three systems acting mechanically in parallel (see Fig. 17.5):

1. The biological system through the latent motor capabilities of the patients
2. Motor control of human limbs using MNPs
3. Motor control of human limbs using a lightweight and wearable ER

Meeting this ambitious goal will require coordinated research in at least the following topics:

- Development of (1) an accurate (engineering) model of the human musculoskeletal system (both upper and lower limbs) for a safe and efficient application of FES by MNPs and (2) an accurate (engineering) model of the human musculoskeletal system (joints both of upper and of lower limbs) and soft tissues for a safe and efficient transmission of forces from the ER to the user

- Development of engineering models of the human neuromotor mechanisms being implemented in control of walking, standing, reaching, and grasping in healthy humans
- Development of control structures and algorithms mimicking the biological model for an efficient and safe orchestration of the three actors: (1) human latent characteristics, (2) MNP, and (3) ER (and to particularize them for each application scenario)
- Development of actuator, sensor, and energy-management technologies so that the ER–MNP can be embodied in a portable, ambulatory, usable, and acceptable solution for the patient and the therapist
- Development of a multimodal cHRI capable of deciphering volitional commands in a robust manner (which also involves handling undesired motion, e.g., spasm or tremors, and assessing human latent capability)

CONCLUSIONS

Exoskeletal robot technologies are not yet ready for full deployment as a solution for functional substitution in motor disorders. They are not truly wearable technologies. Still, limitations in terms of control, mechanisms, and actuator technologies can be identified.

Functional substitution technologies have been mostly applied to walking impairments. This is supported by the subjective opinion of patients and health professionals who independently identified walking as a high priority across different impairments and severities.

At the core of user acceptance of these technologies are subjective issues such as usability, dependability and safety, comfort, and aesthetics. Therefore, user opinion should drive future development, and methods to accurately grasp key design parameters need to be fully exploited.

This chapter has identified future research directions as well as scientific, technological, and clinical challenges. Also, it has proposed a new concept of a hybrid MNP-and-ER solution for rehabilitation and functional substitution technologies. A better understanding of how humans orchestrate neuromotor mechanisms should back this concept. Then technologies can be analogously designed and the concept of "assist-as-needed" revisited.

REFERENCES

Banala S.K., Agrawal S.K., Kim S.H., and Scholz J.O. (2010). Novel gait adaptation and neuromotor training results using an active leg exoskeleton. *IEEE/ASME Trans Mechatronics* (99):1–10.

Cullell A., Moreno J.C., Rocon E., Forner-Cordero A., and Pons J.L. (2009). Biologically based design of an actuator system for a knee–ankle–foot orthosis. *Mechanism Machine Theory* 44:860–872.

Ditunno P.L., Patrick M., and Ditunno J.F. (2008). Who wants to walk? Preferences for recovery after SCI: a longitudinal and cross-sectional study. *Spinal Cord* 46:500–506.

Dollar A.M. and Herr H. (2008). Lower extremity exoskeletons and active orthoses: challenges and state-of-the-art. *IEEE Trans Robotics* 24(1):1–15.

Douglas R., Larson P.F., D'Ambrosia R., and McCall R. (1983). The LSU reciprocating gait orthosis. *Orthopedics* 6:834–839.

Duff J., Evans M.J., and Kennedy P. (2004). Goal planning: a retrospective audit of rehabilitation processes and outcome. *Clin Rehabil* 18:275–286.

Duschau-Wicke A., von Zitzewitz J., Caprez A., Lunenburger L., and Riener R. (2010). Path control: a method for patient-cooperative robot-aided gait rehabilitation. *IEEE Trans Neural Systems Rehabil Eng* 18(1):38–48.

Emken J.L., Bobrow J.E., and Reinkensmeyer D.J. (2005). Robotic movement training as an optimization problem: designing a controller that assists only as needed. In: *ICORR 2005 9th International Conference on Rehabilitation Robotics*, pp. 307–312.

Emken J.L., Benitez R., and Reinkensmeyer D.J. (2007). Human–robot cooperative movement training: learning a novel sensory motor transformation during walking with robotic assistance-as-needed. *J Neuroeng Rehabil* 4:8.

Fatone S. (2006). A review of the literature pertaining to KAFOs and HKAFOs for ambulation. *J Prosthet Orthot* 18(7):137–168.

Ferris D.P., Czerniecki J.M., Hannaford B., and Sound V.A.P. (2005). An ankle–foot orthosis powered by artificial pneumatic muscles. *J Appl Biomech* 21(2):189–197.

Hayashi T., Kawamoto H., and Sankai Y. (2005). Control method of robot suit HAL working as operator's muscle using biological and dynamical information. In: *Proceedings of IEEE/RSJ International Conference on Intelligent Robots and Systems (IROS 2005)*, pp. 3455–3460.

Herr H.M. (2009). Exoskeletons and orthoses: classification, design challenges and future directions. *J Neuroeng Rehabil* 6(21):1–9.

Hesse S., Bertelt C., Jahnke M.T., Schaffrin A., Baake P., Malezic M., and Mauritz K.H. (1995). Treadmill training with partial body weight support compared with physiotherapy in nonambulatory hemiparetic patients. *Stroke* 26(6):976–981.

Hogan N. (1985). Impedance control: an approach to manipulation. Part i—theory. Part ii—implementation. Part iii—applications. *J Dynam Systems Measurement Control* 107:1–24.

Johnson W.B., Fatone S., and Gard S.A. (2009). Walking mechanics of persons who use reciprocating gait orthoses. *J Rehabil Res Dev* 46(3):435–446.

Kennedy P. and Hamilton L.R. (1999). The needs assessment checklist: a clinical approach to measuring outcome. *Spinal Cord* 37:136–139.

Kim S.H., Banala S.K., Brackbill E.A., Agrawal S.K., Krishnamoorthy V., and Scholz J.P. (2010). Robot-assisted modifications of gait in healthy individuals. *Exp Brain Res* 202:809–824.

Kinnaird C.R. and Ferris D.P. (2009). Medial gastrocnemius myoelectric control of a robotic ankle exoskeleton for human walking. *IEEE Trans Neural Systems Rehabil Eng* 17:31–37.

Kobetic R., Marsolais E.B., Triolo R.J., Davy D.T., Gaudio R., and Tashman S. (2002). Development of a hybrid gait orthosis: a case report. *J Spinal Cord Med* 26:254–258.

Luksch T. (2010). Human-like control of dynamically walking bipedal robots. Ph.D. dissertation, University of Kaiserslautern.

Marchal-Crespo L. and Rienkensmeyer D. (2009). Review of control strategies for robotic movement training after neurologic injury. *J Neuroeng Rehabil* 6:20.

Martinez-Villalpando E.C. and Herr H.M.. (2009). Agonist–antagonist active knee prosthesis: a preliminary study in level-ground walking. *J Rehabil Res Dev* 46(3):361–373.

Mohammed S. and Amirat Y. (2009). Towards intelligent lower limb wearable robots: challenges and perspectives—state of the art. In: *Proceedings of the International Conference on Robotics and Biomimetics*, pp. 312–317.

Moreno J.C., Brunetti F., Rocon E., and Pons J.L. (2008). Immediate effects of a controllable knee ankle foot orthosis for functional compensation of gait in patients with proximal leg weakness. *Med Biol Eng Comput* 46(1):43–53.

Moreno J.C., Brunetti F., Navarro E., Forner-Cordero A., and Pons J.L. (2009). Analysis of the human interaction in the design of a wearable lower-limb exoskeleton. *Appl Bionics Biomech* 6(2):245.

Obinata G., Fukada S., Matsunaga T., Iwami T., Shimada Y., Miyawaki K., Hase K., and Nakayama A. (2007). Hybrid control of powered orthosis and functional neuromuscular stimulation for restoring gait. In: *Conference Proceedings IEEE Engineering in Medicine and Biology Society*, pp. 4879–4882.

Pfurtscheller G. and Lopes da Silva F.H. (1999). Event-related EEG/MEG synchronization and desynchronization: basic principles. *Clin Neurophysiol* 110(11):1842–1857.

Piazza S., Calanca A., and Fiorini P. (2010). Force control system for pneumatic actuators of an active gait orthosis. *IEEE International Conference on Biomedical Robotics and Biomechatronics*.

Pons J.L., Ed. (2008). *Wearable Robots: Biomechatronic Exoskeletons*. Chichester, UK: Wiley.

Popovic D., Tomovic R., and Schwirtlich L. (1989). Hybrid assistive system—the motor neuroprosthesis. *IEEE Trans Biomed Eng* 36(7):729–737.

Pratt J.E., Krupp B.T., Morse C.J., and Collins S.H. (2004). The roboknee: an exoskeleton for enhancing strength and endurance during walking. *Proceedings of the IEEE International Conference on Robotics and Automation* 3:2430–2435.

Schiele A. and van der Helm F.C.T. (2006). Kinematic design to improve ergonomics in human machine interaction. *IEEE Trans Neural Systems Rehabil Eng* 14(4):456–469.

Schmidt R.A. and Bjork R.A. (1992). New conceptualizations of practice: common principles in three paradigms suggest new concepts for training. *Psychol Sci* 3(4):207–217.

Stienen A.H.A., Hekman E.E.G., ter Braak H., Aalsma A.M.M., van der Helm F.C.T., and van der Kooij H. (2010). Design of a rotational hydroelastic actuator for a powered exoskeleton for upper limb rehabilitation. *IEEE Trans Biomed Eng* 57(3):728–735.

To C.S., Kobetic R., and Triolo R.J. (2006). Hybrid orthosis system with a variable hip coupling mechamism. In: *Proceedings of the 28th IEEE EMBS Annual International Conference*, pp. 2928–2931.

Van Ham R., Sugar T., Vanderborght B., Hollander K., and Lefeber D. (2009). Compliant actuator designs: review of actuators with passive adjustable compliance/controllable stiffness for robotic applications. *IEEE Robotics Automation Magazine* 16(3):81–94.

Veneman J.F., Kruidhof R., Hekman E.E.G., Ekkelenkamp R., van Asseldonk E.H.F., and van der Kooij H. (2007). Design and evaluation of the lopes exoskeleton robot for interactive gait rehabilitation. *IEEE Trans Neural Systems Rehabil Eng* 15(3):379–386.

Walsh C.J. (2006). Biomimetic design of an underactuated leg exoskeleton for load-carrying augmentation. Master's thesis, Department of Mechanical Engineering, Massachusetts Institute of Technology, Cambridge, MA.

Weiss A., Suzuki T., Bean J., and Fielding R.A. (2000). High intensity strength training improves strength and functional performance after stroke. *Am J Phys Med Rehabil* 79(4):369–376.

Wolbrecht E.T., Chan V., Le V., Cramer S.C., Reinkensmeyer D.J., and Bobrow J.E. (2007). Real-time computer modeling of weakness following stroke optimizes robotic assistance for movement therapy. In: *3rd International IEEE/EMBS Conference on Neural Engineering, CNE*, pp. 152–158.

PART IV

FUNCTION RESTORATION

As discussed in the General Introduction, the three main approaches of neural engineering to rehabilitation of impaired motor functions are restoration, replacement, and neuromodulation. The focus of this section is on function restoration, where the aim is to assist a residual, but impaired, part of the motor apparatus. A review of methods available for movement restoration is first provided in Chapter 18 by Popovic and Popovic. Methods that are accepted as favorable for rehabilitation include robot assistance for balance and body weight in the training of walking, reaching assistance for training shoulder and elbow functions, and assistive systems based on electrical or magnetic stimulation.

The next four chapters all demonstrate specific examples of the use of electrical stimulation in function restoration. In Chapter 19, by Corbett and collaborators, the focus is on the use of invasive BCIs as an interface for improving the control of reaching and hand grasp for neuroprostheses that use functional electrical stimulation (FES). The authors argue that there is a need to develop more "effortless" and low-cognitive-burden interfaces that allow users to control FES systems in a natural manner, which may only be offered through invasive BCIs. In Chapter 20, Došen and Popović suggest a novel approach to the control of multichannel FES for walking; it relies on musculoskeletal modeling and simulations and allows the design of subject-specific, sensor-driven control of FES. In Chapter 21, Larsen and Patriciu

Introduction to Neural Engineering for Motor Rehabilitation, First Edition. Edited by Dario Farina, Winnie Jensen, and Metin Akay.

report on the progress of the clinical testing of the commercial system ActiGait, which is a four-channel partly implantable neuroprosthesis to control ankle dorsiflexion during the swing phase of gait. Finally, in Chapter 22 Jensen and Harreby report on the stimulation selectivity of a specific implantable neural interface. Many neural prosthesis applications based on stimulation rely on our ability to interface with individual or specific populations of nerve fibers, typically motor or sensory. It is therefore important to consider and assess how well different types of neural interfaces selectively access specific nerve fibers, and the importance of design and placement of the electrode must be carefully taken into consideration before choosing a neural interface for a specific neural prosthesis application.

18

REVIEW CHAPTER

METHODS FOR MOVEMENT RESTORATION

DEJAN B. POPOVIĆ[1,2] AND MIRJANA B. POPOVIĆ[1,2,3]

[1]*University of Belgrade, Faculty of Electrical Engineering, Belgrade, Serbia*
[2]*Aalborg University, Department of Health Science and Engineering, Aalborg, Denmark*
[3]*University of Belgrade, Institute for Multidisciplinary Research, Belgrade, Serbia*

SUMMARY

Restoration of movement in humans with motor disability is an important element of rehabilitation that allows reintegration into healthy life. The rehabilitation comprises the use of various technologies in addition to person-to-person interaction. The desired scenario of rehabilitation is to augment the natural recovery leading to normal-like functions (e.g., hands-free standing, walking, reach, and grasp) with no or minimum external assistance. However, in many cases the level of impairment requires lifelong use of assistive systems (e.g., orthosis, neural prosthesis) to assist the impaired function. This reveals two tasks to be addressed: (1) to design technologies that are used for therapy for a limited time, and (2) to design neural prostheses that are an integral part of the body for the rest of the life of the person with the disability. In this chapter we consider both tasks. We intentionally concentrate only on methods and systems for assisting the function, and we leave out artificial limbs.

The basis of movement restoration is the external assistance that provides the compromised function for the person with the disability. This assistance can come

Introduction to Neural Engineering for Motor Rehabilitation, First Edition. Edited by Dario Farina, Winnie Jensen, and Metin Akay.

from the therapist, a robot, or the external activation of neural tissues (e.g., electrical or magnetic stimulation). The assistance should in all cases be linked to voluntary effort of the user to perform the movement or, even better, a functional task. The assistance for restoring standing and walking is very different from the assistance required for arm–hand movement. Assistance for standing and walking deals with a relatively small repertoire of cyclic-like activities and high power requirements ("antigravity" function). Reach-and-grasp tasks deal with complex three-dimensional control of a highly redundant system (more than 20 degrees of freedom for one arm–hand) that has never been replicated by an artificial system so far; but the power requirements are low. The other element that makes a major difference is the type of disability: persons with bilateral paralysis benefit from assistance and definitely need it; persons with unilateral disability could replace the missing function by a variety of compensatory strategies, especially the reach-and-grasp tasks.

Here we present methods of assistance and illustrate them with some systems that are currently being accepted as favorable for rehabilitation: balance and body weight robot assistants for training of walking (e.g., Lokomat, Advanced Gait Trainer, Walkaround); reaching assistants for training shoulder and elbow functions (e.g., Armeo, InMotion, Braccio di Ferro); assistive systems based on electrical stimulation (e.g., Bioness L300 and Bioness H200, Parastep, Stiwell 4med, Actigait, Freehand). We also briefly present the possible use of magnetic or direct electrical stimulation of brain structures and spinal cord to enhance motor function and have therapeutic effect.

INTRODUCTION

Injury or disease of the spinal cord or brain greatly modifies both afferent and efferent communication between the upper motor neuron and peripheral sensory–motor systems: limbs become paralyzed, spasticity develops, sensations are depreciated or missing, muscle atrophy starts developing, and so on. This impairment leads to disability that compromises normal life activities and greatly decreases the quality of life. A spinal cord injury (SCI) at the cervical level may result in complete or partial paralysis of arms and legs bilaterally (tetraplegia) and insufficient control of trunk posture. A spinal cord injury at the thoracic level leads to bilateral paralysis of legs and partial loss of trunk control (paraplegia). A spinal cord injury at the low thoracic level and lower results in flaccid paralysis and denervation of leg muscles (paraplegia). Traumatic brain injury or stroke leads to unilateral paralysis or paresis (hemiplegia) that is expressed in upper and lower extremities. Due to improved immediate treatment most persons suffering from an injury or disease have an incomplete lesion. The plasticity of the upper motor neuron allows incomplete lesions to recover function to various degrees, and this is the basis for therapeutic effects. For more details see Chapter 1 in this book.

The lost natural communication between the upper motor neuron and the periphery caused by a central nervous system (CNS) lesion can be partly

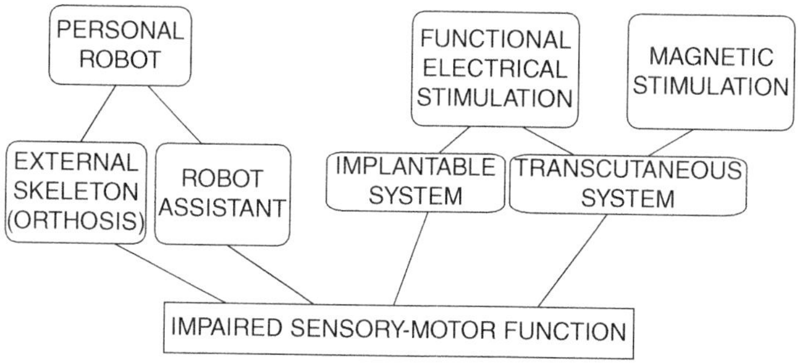

Figure 18.1. Possible methods for assisting motor function in a person with sensory–motor deficit leading to paralysis.

replaced with an externally controlled assistive system called **neural prosthesis.** This neural prosthesis can be a mechanical system (robot) where externally powered actuators interface with the paralyzed limb and drive the movement, or it could be an electronic activator of neural systems that uses natural actuators (paralyzed muscle) in order to generate desired movement. In the case of denervated muscles robotic assistance is currently the only alternative (Fig. 18.1).

ROBOT-BASED ASSISTANCE OF MOVEMENT

The use of mechanical systems as movement assistants can be divided into two categories: (1) permanent assistance to one or more joints (orthosis); (2) facilitation of movement during exercise using a therapeutic rehabilitation robot that is a self-standing machine used for a period of time.

Orthoses need to be lightweight, easy to don and doff, and cosmetically acceptable, and they need to fit body contours and natural axes of joint rotations. A major initiative in designing orthoses for walking was related to control of the ankle joint (correction of drop foot), stabilization of the knee joint with systems frequently called knee–ankle–foot orthoses (KAFOs), and to a great extent hip control. The systems called knee–ankle–foot–hip orthoses (hip-KAFOs, or HKAFOs) typically include a reciprocating mechanism at the pelvic level; thereby, reciprocating left–right hip flexion–extension provided. The KAFO and HKAFO are efficient methods of assisting paraplegics with flaccid paralysis stand and walk. There are only a very few orthoses for upper extremities (Popović and Sinkjær, 2000).

Therapeutic rehabilitation walking robots provide necessary antigravity assistance during postural tasks such as standing or walking (e.g., Lokomat, Hocoma, Switzerland; Advanced Gait Trainer, Rehastim, Germany). The alternative to stationary machines is walkerlike assistive systems that provide

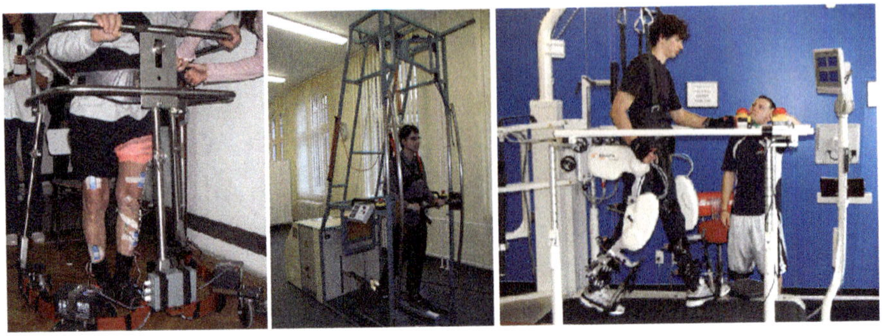

Figure 18.2. Walkaround (left panel), Advanced Gait Trainer (middle panel), and Lokomat (right panel) for exercising by walking with reduced body weight, over ground and on a powered treadmill.

partial body-weight support and allow long training sessions in the free environment (e.g., Walkaround, Tecnalia, Spain).

Locomotion therapy supported by the Lokomat has been shown, in many cases, to be an effective intervention for improving over-ground walking function impaired by neurological diseases and injuries. Lokomat relieves therapists from strain and allows a single-operator mode (Colombo et al., 2000). The current versions of the Lokomat provide the following: (1) increased motivation to exercise through visualized performance feedback; (2) individual adjustments to the patient's needs; (3) assessments of stiffness, forces, and range of movements.

The LokomatPro system[1] consists of the robotic gait orthosis and Lokobasis (a body-weight support system) combined with a treadmill (Fig. 18.2, right panel). The patient's legs are guided on the treadmill according to a preprogrammed physiological gait pattern. The LokomatPro utilizes computer-controlled motors (drives) which are integrated in the gait orthosis at each hip and knee joint to achieve a physiologically meaningful gait pattern. Force transducers at the joints accurately measure the interaction between the patient and the LokomatPro. The LokomatPro robotic gait orthosis is adjustable to the patient's anatomy. Body-weight support provides two functions: lifting of the patient from the wheelchair and partial unloading of the body weight. A specially designed pediatric orthosis allows adapting the Lokomat for the training of children.

The Advanced Gait Trainer (Fig. 18.2, middle panel) simulates the phases of gait, supports the subjects according to their abilities, and controls the center of mass (CoM) in the vertical and horizontal directions (Hesse et al., 1995). The patterns of sagittal lower limb joint kinematics resemble the healthy-like

[1]http://www.hocoma.com/en/products/lokomat/lokomatRpro/features-functions/

pattern. Gait movements on the trainer are highly symmetrical and impact free. The vertical displacement of the CoM is biphasic during each gait cycle. The CoM oscillates sinusoidally in the vertical and horizontal directions. The amplitude of the double-frequent vertical movement is approximately 2 cm and of the monofrequent horizontal movement approximately 4 cm. The gait trainer is based on a doubled crank and rocker gear system. It consists of two footplates positioned on two bars (couplers), two rockers, and two cranks that provide the propulsion. The low backward movement of the footplates simulates the stance phase while the forward movement simulates the swing phase. The system generates a different movement of the tip and of the rear of the footplate during the swing. The tip of the plate follows an arclike movement corresponding to the length of the rocker. The rear end is lifted during swing so that the footplate itself is inclined. Furthermore, the crank propulsion is modified by a planetary gear system to provide a ratio of 60% to 40% between stance and swing phases. The system allows the selection of the gait speed from 0 to 1.1 m/s.

The third system we describe is the Walkaround, a hands-free powered walker with wheels for assisting posture, providing controlled body-weight support, and preventing falls (Fig. 18.2, left panel). The basis for postural control is a special lumbar belt connected to a rigid frame of the Walkaround with adjustable suspensors. The suspension system and the belt provide the required stability and orientation of the body with respect to environment and—which is instrumental—provide safety against falls and subsequent injury. The lumbar belt integrates an inflatable chamber that provides ideal fit to the body contours. The suspension system is instrumented with powered suspensors (artificial muscles), which allow mimicking the motion of the CoM while walking (Veg and Popović, 2008). The joystick, controlled by the therapist or the user, is implemented for selection of speed (v = 0.1 and 1.5 m/s) and direction of motion of the Walkaround. The Walkaround supports forward and backward walking, rotation in place, and curvilinear walking.

The best-known haptic robot for exercise of upper extremities is the MIT-MANUS (Krebs et al., 1998, 1999), followed by a series of other robots (Lum et al., 1997; Scott, 1999; Gomi and Kawato, 1997; Micera et al., 2003; Casadio et al., 2006). These devices can be used to guide or assist the movements of a patient and in close interaction with him or her, just like a human physical therapist. Another possibility is to use the robot to generate perturbations: in patients with spasticity, elbow torque perturbations were shown to increase the elbow range of motion (Cozens, 1990). However, a haptic robotic workstation can do much more than emulate the behavior of a human therapist. By combining haptic with virtual reality technology, it is capable of emulating an extremely rich and adaptable environment that includes visual, auditory, kinesthetic, and/or proprioceptive stimulations. As a consequence, it is possible to study quantitatively the mechanisms of motor learning and sensorimotor adaptation (Shadmehr and Mussa-Ivaldi, 1994) in order to design interacting protocols that are optimal for the individual patient.

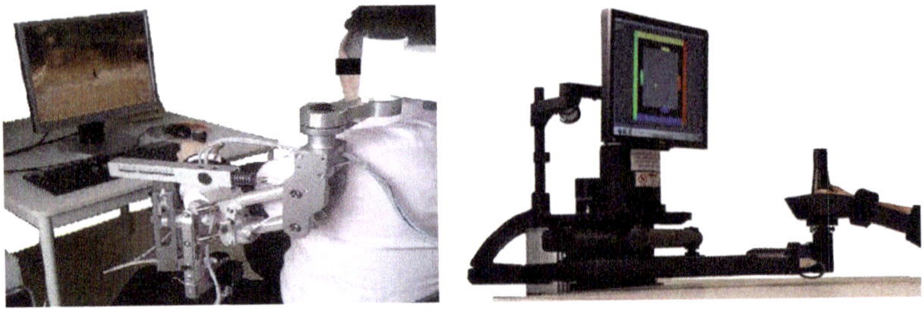

Figure 18.3. InMotion (right panel) and ArmeoSpring (left panel) haptic robots for exercising the upper extremities. See text for details.

InMotion robots (http://www.interactive-motion.com) for upper extremities follow the design and characteristics of the MIT-MANUS robot (Fig. 18.3, right panel). The InMotion robots have the capacity for measurement and immediate interactive response. The InMotion robot assesses the patient's movement and responds to a patient's continually changing ability, and it guides the exercise treatment according to the patient's status: (1) if the patient is unable to move, the robot gently assists the patient to initiate movement toward the target; (2) if coordination is a problem, the robot "guides" the movement, allowing the patient to move toward the target and making certain that the patient is practicing the movement correctly; (3) as the patient gains movement control, the robot provides less assistance and continually challenges the patient; and (4) the robot provides quantifiable feedback on progress and performance. The InMotion robots are made to assist reaching, wrist motion, and grasp: the InMotion 2.0 Shoulder/Arm Robot augments elbow and shoulder control; the InMotion 5.0 Hand Robot provides active (motor-driven) grasp, release, and pinch features; and the InMotion 3.0 Wrist Robot allows three degrees of movement (pronation–supination, flexion–extension, and radial–ulnar deviation).

The Hocoma company developed a series of robots for assisting upper extremities. This series comprises three Armeo products (http://www.hocoma.com/en/products/armeo/), all driven from a single software platform. The Armeo product line (ArmeoSpring, shown in Fig. 18.3, left panel; ArmeoBoom; and Armin) has the following features: arm weight support, augmented feedback, and assessments tools. Together, these features enable the patient to achieve a higher intensity of self-directed movement exercises, while remaining continuously motivated and capable of assessing and tracking the progress toward recovery. The ArmeoSpring is specifically suited for patients who are beginning to regain active movement of the arm and hand, and it has already proven to be successful in many clinics worldwide. Mounted on a trolley for quick and easy positioning, the ArmeoSpring offers various self-initiated repetitive therapies to increase the patient's range of motion and selective control. The self-directed exercises motivate the patient to exert intense levels of both concentration and coordination. The ergonomic and adjustable arm support is an exoskeleton with integrated springs. It embraces the whole arm, from shoulder to hand, and counterbalances the weight of the patient's arm, enhancing any residual function and neuromuscular control, and assisting active movement across a large three-dimensional workspace. The pressure-sensitive handgrip is not only an input device for exercises but also a computer interface for the software and computer games, and it can be removed for functional training of real-life tasks.

NEURAL PROSTHESES TO ASSIST LOST MOTOR FUNCTION

A neural prosthesis based on electrical stimulation generates or augments movement. Functional electrical stimulation (FES) elicits neural activity by delivering a controlled amount of electrical charge, patterned as bursts of electrical charge pulses (Popović and Sinkjær, 2000). FES activates motoneurons or reflex pathways by stimulating efferent or afferent nerve fibers, respectively. FES can be delivered using monopolar or bipolar configuration. In the bipolar configuration two electrodes are positioned in the vicinity of the neural pathways to the muscle that should be activated. In the monopolar configuration an active electrode (cathode) is positioned in the vicinity of the neural pathway to be stimulated, while a common electrode (anode) is positioned distant from the stimulated structure, yet somewhere along the neural pathway to the CNS. FES delivers trains of the electrical charge pulses, mimicking to an extent the natural flow of excitation signals generated by the CNS in non-impaired structures. FES operation can be modeled with a relatively simple electric circuit: stimulator, electrodes, and tissue. The tissue is an ionic conductor with an impedance of about 10–100 Ω, and electrodes are capacitive conductors whose electrical properties depend on many variables, but their impedance is typically between 1.5 and 5 kΩ with resistive and capacitive properties. Stimulators are referred to as constant-current or constant-voltage devices. High-output-impedance devices will deliver the desired current to the tissue, regardless of the electrode's properties. These constant-current stimulators should be correctly termed **current-regulated stimulators**. Low-output-impedance devices, termed **voltage-regulated stimulators**, provide constant voltage, and the charge depends on the electrode's properties. The current-regulated stimulators are preferred since they allow better control of the amount of charge. The amplitude and duration of stimulus pulses determine the electrical charge that will be delivered to neural tissues within one pulse, and the frequency of stimulation determines the total charge within a given interval. The stimulus waveform selected for the excitation process must take into consideration the desired physiological effect (action-potential generation), potential damage to the tissue, and potential degradation of the electrode (Mortimer, 1981; Scheiner et al., 1990; McCreery et al., 1995). The waveform selected is in most cases rectangular. A nonrectangular pulse could be used, but the rise time must be sufficiently fast so that the nerve membrane does not accommodate and fails to open its channels. The stimulus waveform may be unidirectional (monophasic) or bidirectional (biphasic). Biphasic, charge-compensated stimulus is recommended for two reasons: minimization of tissue damage and minimization of electrode damage. Tissue damage is significantly related to the pH change at the electrode–tissue interface. At the cathode, the pH may increase due to the production of OH^-, whereas at the anode the environment will become more acidic (Scheiner et al., 1990). While some buffering capacity for pH changes exists in the tissue, the changes with monophasic stimulation are greater than those with biphasic stimulation. Although reactions at the electrodes are not completely reversed with the biphasic stimulation, this stimulus allows significantly greater charge injection before tissue damage is encountered (Mortimer, 1981). The shape of the secondary (charge compensation) pulse is also important. Conceptually, one would like the electrode reactions to be totally reversible, suggesting a rapid current reversal in the secondary phase. With the metallic conductors presently used, the electrode reactions are not completely reversed. The biphasic pulse limits the extent of irreversibility of the reactions.

During biphasic stimulation both electrodes receive a pulse of each polarity. By convention, the electrode that receives the negative stimulus pulse first is called the cathode. During the cathodic phase (assuming a monopolar configuration) the charge is delivered first to the active (working) electrode, which is near the depolarization site. This is important in order to minimize the risk of possible tissue damage. The anodic pulse is then applied after a short interphase delay following the cathodic pulse. This second pulse will allow the action potential to fully develop on the nerve since immediate current reversal may stop the firing of a nerve fiber, which is just at its threshold. This prevents the electrode potential developed on the working electrode from corroding due to the application of excessive anodic potentials. The cathodic pulse does not contribute to the corrosion of the electrode. For example, using nonnoble intramuscular electrodes, the cathodic amplitude may be 20–40 times greater than the anodic phase.

The surface area of the electrode is important in defining the safety of the stimulation. The geometric surface area may be many times less than the real surface area, depending upon the surface structure of the stimulating electrodes. Large surface areas will diffuse the current and may not effect the excitation desired. A small surface area may result in a high charge density and current density. When selecting an electrode one must know these values in order to assess the safety of the stimulation. With intramuscular electrodes, safe stimulation is demonstrated at a charge density of 2.0 $\mu C/mm^2$ in the cathodic phase using balanced biphasic monopolar stimulation with a current-regulated stimulus and a capacitively coupled recharge phase.

Either amplitude modulation (AM) or pulse width (duration) modulation (PWM) may govern the level of recruitment. A comparison of the method of recruitment modulation to be selected should include consideration of recruitment linearity, charge injected (for safety reasons), and ease of implementation and control of the circuitry that generates the stimulus pulses. This comparison has been made for intramuscular electrodes. The results were that, in consideration of recruitment, the differences between AM and PWM were small. PWM used a slightly lower charge density than AM to evoke a response of equal magnitude, a result that has been theoretically confirmed. Since timing circuits (i.e., regulating pulse width) can be easily constructed and controlled with a resolution of 1 μs or less, many stimulator designers use this technique. In general the pulse duration should be relatively short (200–500 μs), resulting in the activation of the nerve.

The threshold for excitation of the fibers of a peripheral nerve is proportional to the diameter of the fiber. Since the nerve is composed of a mixture of afferent and efferent fibers with a spectrum of fiber diameters, short pulses of constant amplitude will excite large afferent and efferent fibers. Longer pulses may also excite smaller fibers, including afferents normally carrying information of noxious stimuli, and therefore may be painful to the subject. For this reason and in order to minimize the electrical charge injection, short pulse duration is preferred.

In a physiological contraction, the recruitment order is fixed; slow, fatigue-resistant motor units are active at a lower voluntary effort than larger, fast, fatigable units. In an electrically induced recruitment, the recruitment order is not known a priori, but depends upon the variables of position and geometry as well as fiber size. An inverse order of electrically induced recruitment is typical when applying FES: the largest fibers are easily excited, compared with small fibers. This implies that recruitment has to be considered at all times in order to provide controlled and graded externally induced activation. The recruitment of nerve fibers with increasing stimulus pulse amplitude or duration is sigmoidal with saturation. For this reason, a linear increase of muscle output force cannot be achieved by a linear change in the input. The selection of the most effective parameter for regulation of recruitment has been studied by many (Bajzek and Jaeger, 1987; Baratta and Solomonov, 1990; Crago et al., 1974, 1980; Popović et al., 1991).

The second mechanism affecting the overall force developed by the muscle is temporal summation. Stimulus pulses applied to the nerve in rapid succession will produce a mechanically additive effect of the twitch response. At low frequencies the response is unfused, and variations of the muscle force are noticeable. As the mechanical responses sum with increasing frequency, the force variability ceases and the force increases. The frequency at which the mechanical responses produced are sufficiently smooth is known as the **fusion frequency**. At this frequency sufficient smoothness of muscle force is expected. The point at which fusion is achieved depends upon the speed of contraction of the activated muscle fibers, and therefore ultimately upon the level of recruitment. In most human upper extremity muscles, the fusion occurs at less then 20 pulses per second (pps). Very often people use the unit hertz (Hz) instead of "pps." Up to 40% or 50% of the maximum muscle force may be regulated by temporal summation from fusion to tetanus. Temporal summation leads to temporal modulation being inversely proportional to the frequency (f) of stimulation. To grade the force, muscle temporal modulation of the interpulse interval (IPI = $1/f$), or a combination of recruitment and temporal modulation, can be selected.

Electrodes

Surface electrodes (Bowman and Baker, 1985) are placed on the skin surface over the area where the stimulus is to be delivered. McNeal and Baker (1988) and McNeal and Reswick (1976) defined the criteria for surface electrodes: low impedance and even distribution of current; flexibility, to maintain good skin contact; ease of application and removal; and suitable to be mounted for days without irritating the skin. A surface electrode has three elements: the conductor, the interfacial layer, and the adhesive. The earliest electrodes used metal plates for the conductor, gel- or saline-saturated fabric for the interfacial layer, and tape or circumferential bands for maintaining position. Substantial improvements of these electrodes have been directed toward resolving some of the problems, such as drying of the conductor, difficulties in application, and difficulties in maintaining position and good electrical contact of the entire electrode surface, especially on skin surfaces with a large radius of curvature (back and abdomen). Conductive polymers and conductive adhesives have proven to be effective for clinical and home use. Surface electrodes for most applications have a rather large surface area of 5 cm^2 or more. The stimulus parameters required for activation using surface electrodes depend on the stimulus waveform, the surface area, electrode materials, placement, skin impedance, and other factors. Typically, for the rectangular pulsatile waveform frequently used, threshold values are 30 mA or greater for a pulse width of 100–300 μs. Stimulus pulses shorter than 50 μs cause stronger unpleasant sensation (pain); thus, they are not used. The frequency of stimulation depends on the application but it is typically between 16 and 50 pps.

The primary limitation encountered with surface electrodes is that small muscles generally cannot be selectively activated, and deep muscle cannot be activated without first exciting muscles that are more superficial. Furthermore, fine gradation of force can be difficult because relative movement between the electrode and muscle will alter the stimulation–force relationship. Physical movement of the electrode can cause such movement, from length changes in the muscle induced by the contraction process, or from internal change in the nerve–electrode geometry during isometric contraction. Pain is definitely a limiting factor in applying surface electrodes in subjects with preserved sensory and diminished motor functions.

Subcutaneous electrodes (Fig. 18.4) can be divided into those in which the electrode is secured to the muscle exciting the motoneurons and those that contact the nerve containing the motoneurons. The advantages of subcutane-

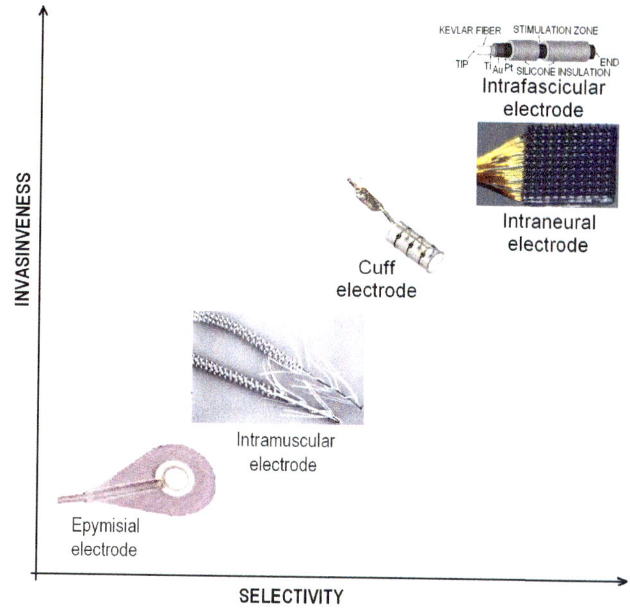

Figure 18.4. Implantable electrodes for interfacing the neural tissues.

ous electrodes versus surface electrodes are better selectivity, repeatable excitation, and permanent positioning. The sensation is much more comfortable to the users since the electrodes are placed away from the pain receptors, and the current amplitude is much lower. The potential disadvantage of implanted electrodes is the damage that can result from improper design and implantation (e.g., irreversible deleterious effects to the neural tissue; physical failure of an electrode, requiring an invasive revision procedure).

Subcutaneous electrodes, which are secured to the muscle, include two types: the **intramuscular electrode** (Bowman and Erickson, 1985; Handa et al., 1989; Smith et al., 1994), which can be injected using a hypodermic needle either nonsurgically through the closed skin or surgically through an open incision; and the **epimysial electrode** (Grandjean and Mortimer, 1986), which is fixed to the muscle surface and must be placed surgically. The intramuscular electrode is a helical coil fabricated from a multiple-strand wire. Such a configuration provides a structure that is able to sustain multiple flexions without fracture. Generally nonnoble alloys are employed (e.g., type 316L stainless steel) and the wire insulation is Teflon. A hook formed at the end of the coil keeps the electrode from being pressed out of the needle during insertion and assists in securing the electrode when the needle is withdrawn. The intramuscular electrode is used as the cathode electrode in a monopolar configuration with a surface electrode as the indifferent (anode) electrode. The intramuscular electrode is implanted, using the hypodermic needle as the carrier, into a site near a motor point of the target muscle. Placement of multiple intramuscular electrodes within a single muscle enables one to employ sequential stimulation techniques. Intramuscular electrodes generally elicit a maximal muscular contraction with a 20-mA, 200-μs stimulus. This is on the order of 10% of the stimulus charge required by surface electrodes. The impedance of the intramuscular electrode is typically below 1 kΩ, but the entire load impedance of tissue and surface anode may be as high as 2.5 kΩ. The probability of functional operation of intramuscular electrodes in the upper extremity is 80% after one year. The primary disadvantage of percutaneous electrodes is the maintenance of the skin interface; yet reports show that only a few infections occurred in implantation of over 2000 electrodes, some implanted for more than five years (Peckham, 1988).

The epimysial electrode (Grandjean and Mortimer, 1986) is a disk-shaped piece of metal with reinforced polymer for shielding the surface away from the muscle and for suturing to the muscle. The electrode is surgically placed on the muscle near the motor point. The conductive surface of the disk is about 3–5 mm in diameter. The impedance and physiologic characteristics of stimulation over this electrode are also similar to the intramuscular electrode. The recruitment is nonlinear with either pulse-amplitude or pulse-width modulation and may be approximated by piecewise linear segments.

Nerve electrodes have the potential for producing the most desired physiological response (Sweeney and Mortimer, 1986; Naples et al., 1988; Rutten et al., 1991; Sweeney et al., 1990; Loeb and Peck, 1996). The electrode must be designed with an appreciation for the sensitivity of the nerve to mechanical trauma (manifested by swelling), its longitudinal mobility during muscle movement, and the necessity of maintaining a constant orientation between the nerve fibers and the electrode. The nerve electrodes are characterized by their placement relative to the nerve: encircling or intraneural. **Cuff electrodes** encircle the nerve; they have either tube or spiral configuration. The latter is a loose, open helix, which is wrapped around the nerve in a monopolar configuration. Cuff electrodes come in a variety of configurations; they all have a longitudinal slit to allow installation on the nerve without damaging it. The cuff is formed of a polymer (usually silicone rubber or polyamide), sometime reinforced with Dacron. The electrodes within the cuff are made out of metal, being circumferential around the inner surface of the cuff or as small patches. A self-wrapping cuff (Naples et al., 1988) is self-sizing; thus it eliminates the problem of selecting the appropriate size. Both cuff and spiral electrode configurations may be used in various monopolar, bipolar, or multipolar configurations (Sweeney and Mortimer, 1986; Naples et al., 1988).

With the choice of the proper geometric relationship between the nerve and electrode and the configuration, it is possible to restrict the direction of current flow and generate action potentials that propagate unidirectionally along the nerve. Encircling cuff electrodes typically require one-tenth the charge required with muscle (intramuscular or epimysial) electrodes, with maximal responses elicited with stimulus amplitudes on the order of 2 mA and pulse widths shorter than 300 μs. The recruitment gain is often higher compared with the intramuscular electrodes, and similar nonlinearities are encountered. The new multipolar cuff electrodes are oriented toward selective stimulation of specific fascicles within the same nerve (Sweeney et al., 1995, Veraart et al., 1993).

Intraneural electrodes use a conductor that invades the epineurium. There is sufficient evidence that maximal contraction is elicited at stimulation levels an order of magnitude lower than with nerve cuff electrodes (200 μA, pulse duration shorter than 300 μs). The variations of such a design would be a valuable clinical tool, yet the connectors, fixation, and neural damage are still not resolved to allow clinical use.

Electronic Stimulators

An electronic stimulator for FES application has to be a self-contained device with low power consumption, and it must be small and lightweight with the simplest possible user interface (e.g., Bijak et al., 1999; Ilić et al., 1994). The stimulator should be programmable, and the programming should be done wirelessly, although using wires is acceptable. The stimulator needs a setup mode—a mode of programming when communicating with the host computer. Once the programming is finished, the stimulator should be turned to autonomous mode.

An electronic stimulator consists of the following elements: power source or communication link to a remote unit that delivers energy for operation; DC/DC converter; output stage that generates current-regulated or voltage-regulated pulses; and controller that defines the shape, intensity, and frequency of pulses (Ilić et al., 1994). The DC/DC is a system that generates the needed high voltage for stimulation. In some cases the required voltage must be as high as 300 V, yet in others it is less than 10 V. The output stages are parts that ensure that the pulse applied will be effective, yet not harmful.

Implantable stimulators for FES may be separated into single- or multiple-channel devices. Single-channel implants, which have been fabricated, are all radio-frequency (RF) powered and controlled devices. They use relatively few discrete components and have a receiving antenna that is integrated into the circuitry. The packaging materials are epoxy or glass–ceramic. The most common single-channel configuration is one in which lead wires are used to place the electrode away from the site of the receiver unit.

Alternatively, the electrodes may be an integral part of the packaging of the circuitry, allowing the entire device to be placed adjacent to the nerve, as it is in the Ljubljana-designed implantable foot-drop system (Strojnik et al., 1987).

Two alternative schemes have been considered for multimuscle excitation. It is possible, in principle, to use several one-channel units that are controlled from one controller (Cameron et al., 1997; Strojnik et al., 1990), or to use a single implantable stimulator that will connect with multiple electrodes (Holle et al., 1984; Rushton, 1990; Strojnik et al., 1990; Smith et al., 1996; Davis et al., 1997).

The single-channel devices were developed originally for foot-drop correction and have been implanted in many subjects with positive experience, yet the development was continued in a different direction, that is, toward multichannel stimulators (Strojnik et al., 1990). The wireless single-channel stimulator has been developed for extensive use in restoring motor functions, and animal experiments have shown great promise (Cameron et al., 1997; Loeb et al., 2001). The BION single-channel wireless stimulator is sealed with glass beads and uses an anodized tantalum and surface-activated iridium electrodes to minimize tissue damage. The diameter of the glass tube (ceramic package) is about 2–3 mm, and the length of the whole device is between 13 and 20 mm. The BION is powered by inductive coupling from an external coil at 2 MHz. Many units can be driven from a single controller. The problem with all implantable devices without batteries, which require relatively high power to drive sensory and motor systems, is the low efficiency of RF transmission. To transmit energy the emitting and receiving antenna must be close and aligned; this is very difficult if a stimulator is injected into a deep muscle.

The alternative solution, accepted by most other research teams developing electronic stimulators, is a miniature, implantable, multichannel device that will excite as many muscles as needed. The difficulty is that such a stimulator will be remote from the stimulation points, and therefore connectors and leads have to be used between the stimulator and stimulation points. The use of long leads should be eliminated after the technique of selective nerve stimulation, including potential stimulation of the spinal cord directly, of spinal roots is perfected. The experiments of Haugland et al. (1999) with the two-channel fully implanted RF-driven stimulators integrated in the cuff electrode led to the development of the ActiGait assistive system for hemiplegic patients.

Sensors

Sensors for neuroprosthetic applications should provide information to both the system and the user regarding the conditions of the neural prosthesis. In some cases, it is not obvious that the user does need instant information (e.g., if automatic execution follows the desired trajectory), yet if anything unexpected is happening, the sensory warning may prevent catastrophic consequences. Sensors are needed in FES systems for the command interface (e.g., activating the neural prosthesis, changing the mode of operation). The optimal sensor system will use the natural information from the biological sensors (Hoffer et al., 1995).

The sensor system to be used should provide information of various kinds, such as the contact force or pressure over the area of contact (grasping, standing, and walking), the position of the joints (prehension, reaching, standing, and walking), and perhaps the activity of the muscle. The dynamic range, resolution, and frequency response of sensors must be determined by the application. For example, force sensors for walking and standing must withstand several times body weight under dynamic loading, and joint position sensors must allow unrestrained movement over the entire range of motion of the joint.

The constraints imposed on the sensors for FES systems are significant; they must be cosmetically acceptable and easy to mount, they should be self-contained, should have low power consumption, and must provide adequate information. In most available FES systems sensors are placed externally. The sensor positioned on the surface of the body is not a suitable solution for many situations (e.g., an external-force sensor on the digits of the hand requires donning and needs a cable to communicate with the control box, and it should work in variable temperature conditions and hazardous environments). The alternative is to use implanted sensors. They have to meet the same performance specifications while functioning in a more hostile environment. These sensors should communicate with the remote control box, and the device must be powered via RF link. The ultimate solution is to use available sensors in the organism: to record from nerves and muscles and process the information in a real-time useful signal. This solution requires the ability to interface without the nerves and interpret the signals they are supplying to the CNS.

A controller for FES is a system that has to be programmable to provide the following functions: (1) use data from analog and analog sensors in real time for sensory-driven feedforward or closed-loop control, (2) use switches for finite state control, and (3) generate signals that define the shape, amplitude, duration, and repetition rate of impulses to be delivered. Today many stimulation patterns are needed that require sophisticated programming. A controller must produce a sufficient number of output signals that accommodates the stimulator's number of channels. Today most microcontrollers can do simple real-time control, yet most are not capable of executing complex model-based control algorithms. Digital signal processors can be used, yet their power consumption is at this stage still questionable for portable, self-contained systems, especially implantable systems.

The controller should communicate with the host computer. This communication today can be easily accomplished by wireless communication. The host computer needs an interface that allows the user to efficiently program the stimulation protocol with respect to number of channels to be used, number and type of sensors and switches that will be applied, timing sequences, special events, and so on.

FUNCTIONAL ELECTRICAL STIMULATION FOR RESTORING UPPER AND LOWER EXTREMITY FUNCTIONS

Functional electrical stimulation systems for upper limbs are being developed to establish and augment the independence to the user. The target population for many years was tetraplegics with diminished, yet preserved, shoulder and elbow functions, lacking the wrist control and grasping ability (Nathan, 1989; Scott et al., 1996; Triolo et al., 1996; Prochazka et al., 1997; Peckham and Knutson, 2005). Controlling the movement of the whole arm also received more attention (Crago et al., 1998; Klose et al., 1997; Popović and Popović, 1998a,b, 2000, 2001; Smith et al., 1996; Wieler et al., 1999).

Functional electrical stimulation systems for lower limbs addressed two problems: assistance of hemiplegic drop foot (e.g., Bogataj et al., 1989; Burridge et al., 1997) and assisting standing and walking in SCI patients (Kralj and Bajd, 1989).

Examples of FES-Based Neural Prostheses for Therapy

The **Ness H200** (http://www.bioness.com/NESS_H200_for_Hand_Rehab.php) is a portable neural prosthesis comprising a spiral forearm–wrist–hand splint that allows fast and simple placing of surface electrodes to the dorsal and volar aspects of the forearm and thennar muscle group (Fig. 18.5, left panel). The stimulator delivers constant-voltage biphasic symmetrical pulses (stimulation frequency 36 Hz, duty cycle 40%) and is preprogrammed to sequentially stimulate muscles to open the hand after the trigger signal, then close the hand, providing safe and stable grasp. The user can select between palmar and lateral grasps. The splint limits the movement at the wrist joint. The stimulation intensity is set by the user at a level that provides prehension, grasp, and comfort. Therapy with the Ness H200 can be prolonged, which is necessary in chronic stroke patients. The original fitting of the system and teaching of the user need to be done by a trained therapist. In the following period, patients work out an intensive stimulation program at home. The stimulation starts with three sessions of 20 minutes per day, gradually increasing to 60 minutes or more.

The **Ness L300** (http://www.bioness.com/NESS_L300_for_Foot_Drop.php) is a system that uses more or less the same stimulation technology as the Ness H200, but it is applied for drop-foot correction in stroke patients (Fig. 18.5,

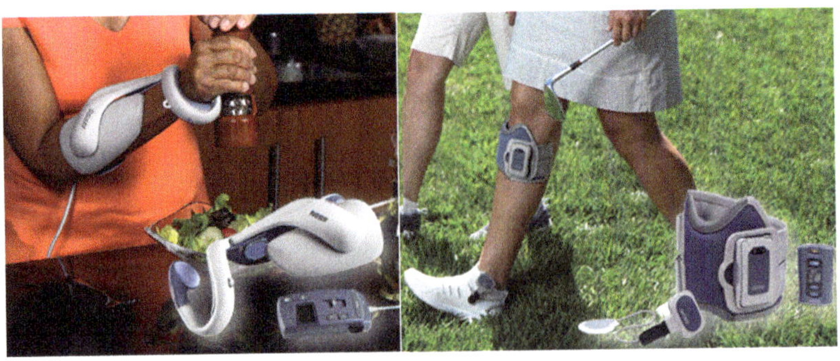

Figure 18.5. Bioness H200 (right) and Bioness L300 that can be used for therapy or as orthoses.

right panel). The stimulator and the electrodes are held with a cuff that needs to be carefully positioned on the shank of the user. The device is controlled by a switch that detects heel contact and heel-off and triggers the stimulation of dorsiflexion.

The **UNAFET-4** (http://www.unasistemi.com/) stimulators have four channels of stimulation via surface electrodes. The operation of the stimulator is based on preprogrammed sequences of stimulation that are automatically triggered by sensors. The UNAFET-4 was applied for functional electrical therapy (FET) of upper extremities (see Chapter 24, on functional electrical therapy, in this book), but also for training of walking (Kojović et al., 2009, 2011). The basic difference in the operation of UNAFET-4 compared with most other systems is the cloning of near-normal muscle activations, with cocontractions for stabilization of the joints, and simplicity for selecting the type of grasp or type of other activity. The UNAFET-8 device (eight channels) has been used for the extended control of elbow, forearm, and hand in experimental setups (Popović and Popović, 2009).

The **WalkAide** (http://www.walkaide.com/en-US/Pages/default.aspx) is designed to assist with the ability to lift the foot for those individuals who have suffered an injury to their CNS. The WalkAide is not designed to work with people who have damage to the lower motor neurons or peripheral nerves. The WalkAide uses functional electrical stimulation to restore typical nerve-to-muscle signals in the leg and foot, effectively lifting the foot at the appropriate time. WalkAide uses advanced sensor technology (Dai et al., 1996) to analyze the movement of the leg and foot. The system then sends electrical signals to the peroneal nerve, which controls movement of the ankle and foot. WalkAide consists of a battery-operated, single-channel electrical stimulator, two electrodes, and electrode leads. WalkAide is applied directly to the leg. A shank cuff holds the system comfortably in place, and it can be worn discreetly under most clothing. WalkAide needs a specially trained medical professional for initial fitting and customization. Using WalkAnalyst, a multifaceted computer software program, the clinician tailors WalkAide to the individual walking pattern for optimal effectiveness. WalkAide also allows the use of a preprogrammable exercise mode for a set period of time as prescribed by the therapist.

Examples of Implantable FES-Based Neural Prostheses

The **STIMuSTEP**,[2] formally known as the Finetech Implanted Dropped Foot Stimulator, makes walking easier, safer, and faster for people who have a dropped foot as a result of an incomplete spinal cord injury. Implantation of the device requires an operation that takes about 1 hour and is performed using a spinal anesthetic. A single incision of approximately 5 cm on the

[2]http://www.salisburyfes.com/STIMuSTEP%20for%20web%20page.htm

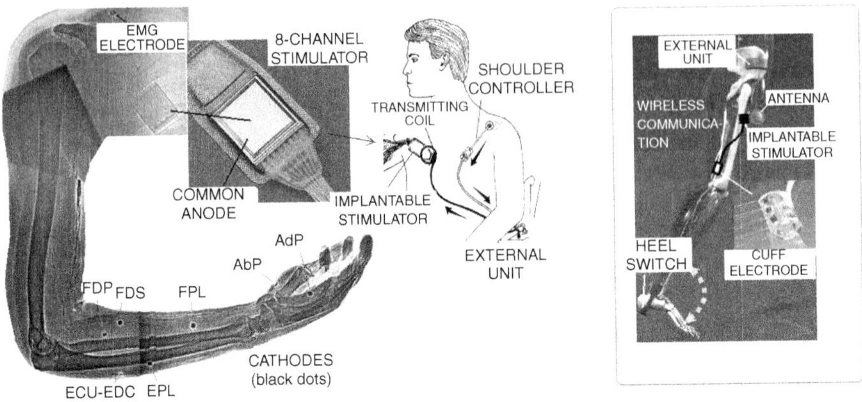

Figure 18.6. The implantable systems for control of movement of paralyzed extremities: Freehand system for control of grasping in tetraplegic individuals (left), and ActiGait for assisting the dorsiflexion in hemiplegic individuals (right).

outside of the lower leg just below the knee is required to site the device. The electrode is positioned at the branches of the common peroneal nerve. At a point just below the knee, this nerve splits into two branches, the deep branch and the superficial branch. The deep branch goes to the muscles that lift (dorsiflex) and turn inward (inversion) the foot while the superficial branch supplies the muscles that turn the foot outward (eversion). In normal walking, a combination of these movements is required. Therefore an electrode will be surgically inserted in both nerves, enabling the movements to be controlled separately. Figure 18.6 shows the electrodes, which are attached to the implanted receiver–stimulator using flexible silicon leads. To stimulate the nerve, very short pulses of electricity are passed through each electrode.

ActiGait[3] uses a multipolar cuff that is implanted around the common peroneal nerve (Fig. 18.6, right panel). The cuff is connected to the implantable unit that is positioned at the convenient position above the knee. The multichannel stimulator allows the setting of the electrode configuration that forms the cuff and the stimulation parameters after the surgery. The stimulator is triggered to activate muscles responsible for ankle flexion during the swing, and stop stimulating at the stance. A wirelessly communicating and cosmetically acceptable robust switch is positioned in the shoe of the user.

The **NeuroControl Freehand System** (http://www.markfelling.com/id450.htm) is a surgically implanted device that restores hand function in people with tetraplegia by neuroelectric stimulation of forearm and hand muscles (Kilgore et al., 1997). The Freehand System (Fig. 18.6, left panel) is designed for use by individuals who have some use of an upper arm or shoul-

[3]http://www.ottobock.com/cps/rde/xchg/ob_com_en/hs.xsl/4762.html

der but do not have control of the forearm, wrist, and hand. The system includes electrodes surgically attached to the muscles of the hand and forearm. The electrode wires are tunneled up the arm to a control box located under the skin in the pectoral region. A pacemaker-style stimulator is implanted in the chest, and a transmitting coil is worn externally on the skin over the location of the implanted stimulator. Shoulder movements, monitored by a shoulder position sensor, control the Freehand System. There are other control interfaces (e.g., wrist controller, electromyographic controller). A separate external controller provides the power supply and the "intelligence" for the system. The tendon transfer surgery may also be necessary (Johnson and Whitworth, 2002; Keith et al., 1996). With the system, small shoulder movements (wrist, electromyogram) control the opening and closing of the hand. Depending upon the individual, the system may enable independent feeding or grooming, use of the telephone, writing, and so on. The NeuroControl Freehand System allows for two different hand grasps: lateral (key pinch) and palmar prehension grasp.

FUTURE DIRECTIONS

To maximize the use of the preserved neural networks after the injury of CNS it could be effective to stimulate higher CNS structures, rather than to directly activate the last-order motoneurons. There is evidence in humans that the neural circuitry of the spinal cord is capable of generating complex behaviors with coordinated muscle activity (Dimitrijević, 1998). Epidural cortical stimulation (ECS) depends greatly on various technical parameters, including stimulation frequency and intensity, pulse width, duty cycle, montage (monopolar vs. bipolar), electrode polarity (anode vs. cathode), and the distance between the electrodes and neural elements (Ranck Jr., 1975). The stimulator is implanted epidurally near the regions of interest to be activated. It was suggested that the thickness of the cerebrospinal fluid layer between the dura mater and the underlying cortex affects the strength and the distribution of the electrical field induced in the brain (Manola et al., 2005, 2007).

The stimulation of spinal cord is currently receiving more attention (Mushahwar and Horch, 1997, 2000). The behaviors that can be mediated by the isolated spinal cord include reaching-like limb movements (Lemay and Grill, 1999), standing, and walking. Spinal circuits can be activated by epidural and intraspinal electrical stimulation (Barbeau et al., 1999). Two methods are equally appealing: intraspinal microstimulation enables direct activation of spinal neurons, and afferent input makes synaptic contacts on spinal neurons; thus, they also could be used for stimulation of CNS.

Brain stimulation can upregulate or downregulate the excitability of lesioned and intact hemispheres, which could be used to facilitate relearning and might ultimately lead to the recovery of function in stroke patients. Possible mechanisms mediating these effects may include the correction of abnormally persistent interhemispheric inhibitory drive from the intact hemisphere to the lesioned hemisphere in the process of generation of voluntary movements by the paretic hand, a disorder which is correlated with the magnitude of impairment. We describe here three techniques for brain stimulation: (1) transcranial magnetic stimulation (TMS), (2) direct transcranial current stimulation (dTCS), and (3) epidural cortical stimulation (ECS). These techniques are intended to increase the cortical excitability of the affected hemisphere of the brain and decrease the excitability of the nonaffected side (Harvey and Nudo, 2007; Lefaucheur, 2008; Ogura, 2008).

Transcranial magnetic stimulation (TMS) uses a rapidly changing magnetic field generated by a coil positioned in closely proximal skull bone to painlessly induce controlled electrical currents in well-targeted regions in the brain. The geometry of the activated structures depends on the waveform of the magnetic pulse and on the type and orientation of the coil. Repetitive TMS (rTMS) produces sustained changes in cortical excitability and regional brain activities. Low-frequency rTMS (0.2–1 Hz) consists of a single continuous train of pulses, whereas high-frequency rTMS (5–20 Hz) employs intermittent bursts. High-frequency rTMS can also be applied in continuous mode (>100 pulses). Long-term synaptic changes explain why rTMS effects persist for a few hours after treatment. Repetition of sessions can reinforce and prolong rTMS aftereffects. In addition, methods that can increase the duration and size of rTMS aftereffects and produce substantial changes in cortical excitability have been introduced (Fitzgerald et al., 2006; Hallett, 2007; Sheffler and Chae, 2007). These methods include:

1. Paired-pulse rTMS, consisting of a subthreshold pulse followed by a suprathreshold pulse (Sommer et al., 2001), two subthreshold pulses (Khedr et al., 2004), and two suprathreshold pulses (Thickbroom et al., 2006)
2. Theta burst stimulation (TBS), consisting of short bursts of three low-intensity pulses with inner high frequency (50 Hz, within the gamma range) delivered at 5 Hz (within the theta range) (Huang et al., 2005)
3. Paired associative stimulation (PAS), consisting of the association between a single electrical stimulus delivered at peripheral level and a single TMS pulse delivered over the motor cortex (Stefan et al., 2000)

Direct TCS is a technique that involves generating weak constant direct currents by placing electrodes directly on the scalp (Fregni and Pascual-Leone, 2007). In general, cortical excitability is reduced by cathodal dTCS and increased by anodal dTCS, likely due to neuronal hyperpolarization–depolarization (Bindman et al., 1964). However, the direction of the polarizing effects depends strictly on the orientation of dendrites and axons in the induced electrical field. In addition, either inhibitory or excitatory networks can be affected by DC stimulation. Therefore, functional and clinical changes induced by dTCS are influenced by technical parameters, such as the placement of the reference electrode or the intensity of stimulation (Nitsche and Paulus, 2000; Priori, 2003).

REFERENCES

Bajzek T.J. and Jaeger R.J. (1987). Characterization and control of muscle response to electrical stimulation. *Ann Biomed Eng* 15:485–501.

Baratta R. and Solomonow M. (1990). The dynamic response model of nine different skeletal muscles. *IEEE Trans Biomed Eng* BME-37:243–251.

Barbeau H., McCrea D.A., O'Donovan M.J., Rossignol S., Grill W.M., and Lemay M.A. (1999). Tapping into spinal circuits to restore motor function. *Brain Res Rev* 30:7–51.

Bijak M., Sauerman S., Schmutterer C., Lanmueller H., Unger E., and Mayr W. (1999). A modular PC-based system for easy setup of complex stimulation patterns. In: *Proceedings of the 4th International Conference IFESS*, Sendai, Japan.

Bindman L.J., Lippold O.C., and Redfearn J.W. (1964). The action of brief polarizing currents on the cerebral cortex of the rat (1) during current flow and in the production of long-lasting after-effects. *J Physiol (London)* 17:369–382.

Bogataj U., Gros N., Maležič M., Kelih B., Kljajić M., and Aćimović R. (1989). Restoration of gait during two to three weeks of therapy with multichannel electrical stimulation. *Phys Ther* 69(5):319–327.

Bowman B. and Baker L. (1985). Effects of waveform parameters on comfort during transcutaneous neuromuscular electrical stimulation. *Ann Biomed Eng* 13:59–74.

Bowman R.B. and Erickson R.C. (1985). Acute and chronic implantation of coiled wire intraneural electrodes during cyclical electrical stimulation. *Ann Biomed Eng* 13:75–93.

Burridge J.H., Taylor P.N., Hagan S.A., and Swain I.D. (1997). Experience of clinical use of Odstock dropped foot stimulator. *Artif Organs* 21:254–260.

Cameron T., Loeb G.E., Peck R.A., Schulman J.H., Strojnik P., and Troyk P.R. (1997). Micromodular implants to provide electrical stimulation of paralyzed muscles and limbs. *IEEE Trans Biomed Eng* BME-44(9):781–790.

Casadio M., Sanguineti V., Morasso P.G., and Arrichiello V. (2006). Braccio di Ferro: a new haptic workstation for neuromotor rehabilitation. *Technol Health Care* 14:123–142.

Colombo G., Joerg M., Schreier R., and Dietz V. (2000). Treadmill training of paraplegic patients using a robotic orthosis. *J Rehabil Res Develop* 37:693–700.

Cozens J.A. (1990). Robotic assistance of an active upper limb exercise in neurologically impaired patients. *IEEE Trans Rehabil Eng* 7:254–256.

Crago P.E., Peckham P.H., Mortimer J.T., and Van der Meulen J. (1974). The choice of pulse duration for chronic electrical stimulation via surface, nerve and intramuscular electrodes. *Ann Biomed Eng* 2:252–264.

Crago P.E., Peckham P.H., and Thrope G.B. (1980). Modulation of muscle force by recruitment during intramuscular stimulation. *IEEE Trans Biomed Eng* BME-27 (12):679–684.

Crago P.E., Memberg W.D., Usey M.K., Keith M.W., Kirsch R.F., Chapman G.J., et al. (1998). An elbow extension neuroprosthesis for individuals with tetraplegia. *IEEE Trans Rehabil Eng* TRE-6(1):1–6.

Dai R., Stein R.B., Andrews B.J., James K.B., and Wieler M. (1996). Application of tilt sensors in functional electrical stimulation. *IEEE Trans Rehabil Eng* TRE-4(2):63–72.

Davis R., Houdayer T., Andrews B.J., Emmons S., and Patrick J. (1997). Paraplegia: prolonged closed-loop standing with implanted nucleus FES-22 stimulator and Andrews' foot–ankle orthosis. *Stereotact Funct Neurosurg* 69:281–287.

Dimitrijević M.R. (1998). Head injuries and restorative neurology. *Scand J Rehab Med* 17(Suppl):9–13.

Fitzgerald P.B., Fountain S., and Daskalakis Z.J. (2006). A comprehensive review of the effects of rTMS on motor cortical excitability and inhibition. *Clin Neurophysiol* 117:2584–2596.

Fregni F. and Pascual-Leone A. (2007). Technology insight: noninvasive brain stimulation in neurology—perspectives on the therapeutic potential of rTMS and tDCS. *Nat Clin Pract Neurol* 3:383–393.

Gomi H. and Kawato M. (1997). Human arm stiffness and equilibrium point trajectory during multijoint movement. *Biol Cybernet* 76:163–171.

Grandjean P.A. and Mortimer J.T. (1986). Recruitment properties of monopolar and bipolar epimysial electrodes. *Ann Biomed Eng* 14:53–66.

Hallett M. (2007). Transcranial magnetic stimulation: a primer. *Neuron* 55:187–199.

Handa Y., Hoshimiya H., Iguchi Y., and Oda T. (1989). Development of percutaneous intramuscular electrode for multichannel FES system. *IEEE Trans Biomed Eng* BME-36(7):705–710.

Harvey R.L. and Nudo R.J. (2007). Cortical brain stimulation: a potential therapeutic agent for upper limb motor recovery following stroke. *Top Stroke Rehabil* 14:54–67.

Haugland M., Lickel A., Haase J., and Sinkjær T. (1999). Control of FES thumb force using slip information obtained from the cutaneous electroneurogram in quadriplegic man. *IEEE Trans Rehab Eng* TRE-7(2):215–227.

Hesse S., Bertelt C., Jahnke M.T., et al. (1995). Treadmill training with partial body weight support compared with physiotherapy in nonambulatory hemiparetic patients. *Stroke* 26:976–981.

Hoffer J.A., Stein R.B., Haugland K.K., Sinkjær T., Durfee W.K., Schwartz A.B., Loeb G.E., and Kantor C. (1995). Neural signals for command control and feedback in functional electrical stimulation. *J Rehabil* 33(2):145–157.

Holle J., Frey M., Gruber H., Kern H., Stohr H., and Thoma H. (1984). Functional electrostimulation of paraplegics: experimental investigations and first clinical experience with an implantable stimulation device. *Orthopaedics* 7:1145–1160.

Huang Y.Z., Edwards M.J., Rounis E., Bhatia K.P., and Rothwell J.C. (2005). Theta burst stimulation of the human motor cortex. *Neuron* 45:201–206.

Ilić M., Vasiljević D., and Popović D.B. (1994). A programmable electronic stimulator for FES systems. *IEEE Trans Rehabil Eng* TRE-2:234–239.

Johnson D. and Whitworth H. (2002). Recent developments in plastic surgery. *Br Med J* 325:319–322.

Keith M.W., Kilgore K.L., Peckham P.H., et al. (1996). Tendon transfers and functional electrical stimulation for restoration of hand function in spinal cord injury. *J Hand Surg (Am)* 21(1):89–99.

Khedr E.M., Gilio F., and Rothwell J.C. (2004). Effects of low frequency and low intensity repetitive paired pulse stimulation of the primary motor cortex. *Clin Neurophysiol* 115:1259–1263.

Kilgore K.L., Peckham P.H., Keith M.W., et al. (1997). An implanted upper-extremity neuroprosthesis. *J Bone Joint Surg Am* 79:533–541.

Klose K.J., Jacobs P.L., Broton J.G., Guest R.S., Needham-Shropshire B.M., Lebwohl N., et al. (1997). Evaluation of a training program for persons with SCI paraplegia using the Parastep 1 ambulation system: part 1. Ambulation performance and anthropometric measures. *Arch Phys Med Rehabil* 78:789–793.

Kojović J., Djurić-Jovičić M., Došen S., Popović M.B., and Popović D.B. (2009). Sensor-driven four-channel stimulation of paretic leg: functional electrical walking therapy. *J Neurosci Methods* 181:101–105.

Kojović J., Miljković N., Janković M., and Popović D.B. (2011). Recovery of motor function after stroke: a polymyography-based analysis. *J Neurosci Methods* 194(2):321–328.

Kralj A. and Bajd T. (1989). *Functional Electrical Stimulation: Standing and Walking after Spinal Cord Injury*. Boca Raton, FL: CRC Press.

Krebs H.I., Hogan N., Aisen N.M., and Volpe B. (1998). Robot-aided neurorehabilitation. *IEEE Trans Rehabil Eng* 6:75–87.

Krebs H.I., Hogan N., Volpe N.T., Aisen M.L., Edelstein L., and Diels C. (1999). Overview of clinical trials with MIT-MANUS: a robot-aided neuro-rehabilitation facility. *Technol Health Care* 7:419–423.

Lefaucheur J.P. (2008). Principles of therapeutic use of transcranial and epidural cortical stimulation. *Clin Neurophysiol* 119:2179–2184.

Lemay M.A. and Grill W.M. (1999). Spinal force fields in the cat spinal cord (abstract). *Soc Neurosci* 25:1396.

Loeb G.E. and Peck R.A. (1996). Cuff electrodes for chronic stimulation and recording of peripheral nerve activity. *J Neurosci Methods* 64:95–103.

Loeb G.E., Peck R.A., Moore W.H., and Hood K. (2001). BION™ system for distributed neural prosthetic interfaces. *Med Eng Phys* 23:9–18.

Lum P., Lehman S., and Reikensmeyer D. (1997). The use of a robotic device for post-stroke movement therapy. In: *Proceedings of the International Conference on Rehabilitation Robotics, ICORR'97*. Available at http://people.bath.ac.uk/mpsmrh/icorr97/lum.pdf (accessed April 8, 2013).

Manola L., Roelofsen B.H., Holsheimer J., Marani J., and Geelen J. (2005). Modelling motor cortex stimulation for chronic pain control: electrical potential field, activating functions and responses of simple nerve fibre models. *Med Biol Eng Comput* 43:335–343.

Manola L., Holsheimer J., Veltink P.H., and Buitenweg J.R. (2007). Anodal vs cathodal stimulation of motor cortex: a modeling study. *Clin Neurophysiol* 118:464–474.

McCreery D.B., Agnew W.F., Yuen T.G., and Bullara L.A. (1995). Relationship between stimulus amplitude, stimulus frequency and neural damage during electrical stimulation of sciatic nerve of cat. *Med Biol Eng Comput* 33:426–429.

McNeal D.R. and Baker L. (1988). Effects of joint angle, electrodes and waveform on electrical stimulation of the quadriceps and hamstrings. *Ann Biomed Eng* 16:299–310.

McNeal D.R. and Reswick J.B. (1976). Control of skeletal muscle by electrical stimulation. *Adv Biomed Eng* 6:209–256.

Micera S., Carrozza M.C., Guglielmelli E., Barboni L., Freschi C., and Dario P. (2003). On the use of robotic devices for neurorehabilitation after stroke. In: *Proceedings of 11th International Conference on Advanced Robotics*, Coimbra, pp. 185–190.

Mortimer T. (1981). Motor prosthesis. In: *Handbook of Physiology* (Brooks V.B., Ed.), Sect. 1, Vol. 11, Part 1(5), pp. 155–187. Bethesda, MD: American Physiological Society.

Mushahwar V.K. and Horch K.W. (1997). Proposed specifications for a lumbar spinal cord electrode array for control of lower extremities in paraplegia. *IEEE Trans Rehabil Eng* TRE-5:237–243.

Mushahwar V.K. and Horch K.W. (2000). Muscle recruitment through electrical stimulation of the lumbo-sacral spinal cord. *IEEE Trans Rehab Eng* TRE-8:22–28.

Naples G.G., Mortimer J.T., Scheiner A., and Sweeney J.D. (1988). A spiral nerve cuff electrode for peripheral nerve stimulation. *IEEE Trans Biomed Eng* BME-35:905–916.

Nathan R.H. (1989). An FNS-based system for generating upper limb function in the C4 quadriplegic. *Med Biol Eng Comp* 27:549–556.

Nitsche M.A. and Paulus W. (2000). Excitability changes induced in the human motor cortex by weak transcranial direct current stimulation. *J Physiol (London)* 527:633–639.

Ogura K. (2008). Epidural motor cortex stimulation might be a novel neurosurgical modality for the recovery of motor impairment following stroke: a review and perspective. *Neurol Surg* 36:667–675.

Peckham P.H. (1988). Functional electrical stimulation. In: *Encyclopedia of Medical Devices and Instrumentation* (Webster J.G., Ed.), pp. 1341–1358. New York: Wiley.

Peckham P.H. and Knutson J.S. (2005). Functional electrical stimulation for neurosmuscular application. *Annu Rev Biomed Eng* 7:327–360.

Popović D.B. and Popović M.B. (1998a). Tuning of a nonanalytical hierarchical control system for reaching with FES. *IEEE Trans Biomed Eng* BME-45:203–212.

Popović D.B. and Popović M.B. (1998b). Belgrade grasping system. *J Electronics* 2:21–28.

Popović D.B. and Popović M.B. (2000). Nonanalytical control for assisting reaching in humans with disability. In: *Control of Posture and Movement: Neuro-musculo-skeletal Interaction and Organization Principles* (Winters J.M. and Crago P.E., Eds.), Chapter 39, pp. 535–548. New York: Springer Verlag.

Popović D.B. and Popović M.B. (2001). Control for an elbow neuroprosthesis: cloning biological synergies. *IEEE Med Biol Eng Mag* 20:74–81.

Popović D.B. and Popović M.B. (2009). Automatic determination of the optimal shape of the surface electrode: selective stimulation. *J Neurosci Methods* 178:174–181.

Popović D.B. and Sinkjær T. (2000). *Control of Movement for the Physically Disabled: Control for Rehabilitation Technology*. London: Springer.

Popović D.B., Gordon T., Rafuse V., and Prochazka A. (1991). Properties of implanted electrodes for functional electrical stimulation. *Ann Biomed Eng* 19:303–316.

Priori A. (2003). Brain polarization in humans: a reappraisal of an old tool for prolonged non-invasive modulation of brain excitability. *Clin Neurophysiol* 114:589–595.

Prochazka A., Gauthier M., Wieler M., and Kenwell Z. (1997). The Bionic glove: an electrical stimulator garment that provides controlled grasp and hand opening in quadriplegia. *Arch Phys Med Rehabil* 78:608–614.

Ranck J.B. Jr. (1975). Which elements are excited in electrical stimulation of mammalian central nervous system: a review. *Brain Res* 98:417–440.

Rushton D.N. (1990). Choice of nerves roots for multichannel leg controller implant. In: *Advances in External Control X* (Popović D.B., Ed.), pp. 99–108. Belgrade: Nauka.

Rutten W.L., van Wier H.J., and Put J.H. (1991). Sensitivity and selectivity of intraneural stimulation using a silicon electrode array. *IEEE Trans Biomed Eng* 38:192–198.

Scheiner A., Mortimer J.T., and Roessmann U. (1990). Imbalanced biphasic electrical stimulation: muscle tissue damage. *Ann Biomed Eng* 18:407–425.

Scott S.H. (1999). Apparatus for measuring and perturbing shoulder and elbow joint positions and torques during reaching. *J Neurosci Methods* 89:119–127.

Scott T.R.D., Peckham P.H., and Kilgore K.L. (1996). Tri-state myoelectric control of bilateral upper extremity neuroprosthesis for tetraplegic individuals. *IEEE Trans Rehab Eng* TRE-4:251–263.

Shadmehr R. and Mussa-Ivaldi F.A. (1994). Adaptive representation of dynamics during learning of a motor task. *J Neurosci* 14:3208–3224.

Sheffler L.R. and Chae J. (2007). Neuromuscular electrical stimulation in neurorehabilitation. *Muscle Nerve* 35:562–590.

Smith B.T., Betz R.R., Mulcahey M.J., and Triolo R.J. (1994). Reliability of percutaneous intramuscular electrodes for upper extremity functional neuromuscular stimulation in adolescents with C5 tetraplegia. *Arch Phys Med Rehabil* 75:939–945.

Smith B.T., Mulcahey B.J., and Betz R.P. (1996). Development of an upper extremity FES system for individuals with C4 tetraplegia. *IEEE Trans Rehab Eng* TRE-4:264–270.

Sommer M., Tergau F., Wischer S., and Paulus W. (2001). Paired-pulse repetitive transcranial magnetic stimulation of the human motor cortex. *Exp Brain Res* 139:465–472.

Stefan K., Kunesch E., Cohen L.G., Benecke R., and Classen J. (2000). Induction of plasticity in the human motor cortex by paired associative stimulation. *Brain* 125:572–584.

Strojnik P., Aćimović-Janežič R., Vavken E., Simić V., and Stanič U. (1987). Treatment of drop foot using an implantable peroneal under knee stimulator. *Scand J Rehabil Med* 19:37–43.

Strojnik P., Whitmoyer D., and Schulman J. (1990). An implantable stimulator for all seasons. In: *Advances in External Control of Human Extremities X* (Popović D.B., Ed.), pp. 335–344. Belgrade: Nauka.

Sweeney J.D. and Mortimer J.T. (1986). An asymmetric two electrode cuff for generation of unidirectionally propagated action potentials. *IEEE Trans Biomed Eng* BME-33:541–549.

Sweeney J.D., Ksienski D.A., and Mortimer J.T. (1990). A nerve cuff technique for selective excitation of peripheral nerve trunk regions. *IEEE Trans Biomed Eng* BME-37(7):706–715.

Sweeney J.D., Crawford N.R., and Brandon T.A. (1995). Neuromuscular stimulation selectivity of multiple-contact nerve cuff electrode arrays. *Med Biol Eng Comp* 33:418–425.

Thickbroom G.W., Byrnes M.L., Edwards D.J., and Mastaglia F.L. (2006). Repetitive paired-pulse TMS at I-wave periodicity markedly increases corticospinal excitability: a new technique for modulating synaptic plasticity. *Clin Neurophysiol* 117:61–66.

Triolo R., Nathan R., Handa Y., Keith M., Betz R.R., Carroll S., and Kantor C. (1996). Challenges to clinical deployment of upper limb neuroprostheses. *J Rehabil Res Dev* 33:111–122.

Veg A. and Popović D.B. (2008). Walkaround®: mobile balance support for therapy of walking. *IEEE Trans Neural Syst Rehab Eng* TNSRE-16:264–269.

Veraart C., Grill W.M., and Mortimer J.T. (1993). Selective control of muscle activation with a multipolar nerve cuff electrode. *IEEE Trans Biomed Eng* BME-40:640–653.

Wieler M., Stein R.B., Ladouceur M., et al. (1999). Multicenter evaluation of electrical stimulation systems for walking. *Arch Phys Med Rehabil* 80:495–500.

19

FOCUSED CHAPTER

ADVANCED USER INTERFACES FOR UPPER LIMB FUNCTIONAL ELECTRICAL STIMULATION

ELAINE A. CORBETT,[3] CHRISTIAN ETHIER,[2] EMILY R. OBY,[2] KONRAD KORDING,[1,2] ERIC J. PERREAULT,[1,3] AND LEE E. MILLER[1,2,3]

[1]*Department of Physical Medicine and Rehabilitation, Feinberg School of Medicine, Northwestern University, Chicago, Illinois, USA*

[2]*Department of Physiology, Feinberg School of Medicine, Northwestern University, Chicago, Illinois, USA*

[3]*Department of Biomedical Engineering, Northwestern University, Evanston, Illinois, USA*

SUMMARY

Spinal cord injury (SCI) paralyzes approximately 12,000 people each year in the United States. Individuals with an injury at and above the sixth cervical vertebra (C6) lose function in the upper and lower limbs. To provide greater independence to this population, the restoration of reaching and grasping movements is critically important. Functional electrical stimulation (FES) is currently the only clinical approach for reanimating paralyzed muscles. While it has been used with great success in this population, especially for the restoration of hand grasp, the user interfaces currently available are not sufficient to control high-dimensional, dexterous movements of the hand and arm. Some neuroprostheses use purely

Introduction to Neural Engineering for Motor Rehabilitation, First Edition. Edited by Dario Farina, Winnie Jensen, and Metin Akay.

logic-based command signals to switch between preprogrammed grasp patterns, while others use logic to select features such as grasp type in combination with proportional control of hand opening and closing. Furthermore, many of these approaches rely on unnatural actions not directly related to the desired movement and place considerable cognitive burden on the user. There is a need to develop "effortless" or low-cognitive-burden interfaces that allow users to control FES systems in a natural manner. Recent advances in physiological recording technologies and signal processing may provide solutions to this problem.

The cortical areas that are involved in the normal generation of movement commands are still intact in SCI patients and may provide a more natural user interface. By recording directly from the cortex it may be possible for a user to control many degrees of freedom of the hand and arm using methods similar to those that are already enabling paralyzed patients to use computer interfaces and perform robot control tasks.

A further way to improve control may to combine information from multiple signal sources depending on the needs of the individual patient: the brain, residual movements, electromyograms (EMGs), and gaze direction. In the context of Bayesian statistics, an optimal combination of the user's control signals with prior knowledge of the probabilistic nature of the desired movement can be formulated. For example, if we want to predict a reaching movement, we might have some knowledge of the likely reach target. Since the reach trajectory is generally stereotypical, the reach target informs us about the likely trajectory of the arm. By using this information to improve our estimate of the intended movement, less input may be required from the user.

This chapter reviews currently available interfaces and addresses aspects of these promising new technologies with a view to improving control of reaching and hand grasp for FES neuroprostheses. We focus on the development of more natural user interfaces that may enable more effortless control, while recognizing that the needs and abilities vary widely between individual patients.

INTRODUCTION

Patients with spinal cord injury (SCI) at the C6 level or above have extensive paralysis affecting both the upper and lower limbs. These patients often consider arm function, including hand grasp, to be the most desirable function to restore (Anderson, 2004). Functional electrical stimulation (FES), the electrical stimulation of motor nerves or muscles, is a particularly promising means of realizing this goal. Achieving robust control of upper limb movement using FES requires both the means to activate individual muscles with adequate force and an appropriate control signal. There are many challenges associated with the muscle-activation part of this problem that are being addressed through innovative electrode design and stimulation protocols. Here, however, we focus on the need for—and the approaches available to obtain—an intuitive interface to allow a paralyzed user to control the stimulation as effortlessly as possible. In addition to sensors of residual motion or activity of individual muscles, this interface may now include a "decoder" component that computes a low-dimensional control signal from a high-dimensional set of signals.

The chapter starts by reviewing existing technologies for obtaining a control signal that is usable for an FES neuroprosthesis. This is followed by a discussion of the promise that recent advances in brain–machine interfaces (BMIs) hold for more natural user interfaces. Differences in the information content of potential signal sources suggest that enhanced control signals may be generated through an efficient combination of the available sources from each individual. Techniques combining signal sources in ways that are statistically optimal are reviewed in the following section. The final section discusses the important topic of the relation between off-line decoder accuracy and online user performance. Throughout the chapter, we try to highlight the relative benefits and drawbacks of the available technologies as well as the specific applications in which they appear to be most applicable. Two important examples are in the challenges associated with restoring hand grasp and reach, problems that may be best solved by decoders with distinct characteristics.

EXISTING METHODS FOR FUNCTIONAL ELECTRICAL STIMULATION AND PROSTHETIC CONTROL

The arm has a total of 28 degrees of freedom, 21 of them in the hand. There are roughly 30 muscles that normally control grasping movement of the hand. Identifying user intent for each of these degrees of freedom is a daunting task. Fortunately, the kinematic trajectories for a fairly large range of simple grasping movements can be described reasonably well with only one or two degrees of freedom (Santello et al., 1998; Ingram et al., 2008). Consequently, it has proven possible to implement useful hand grasp neuroprostheses using preprogrammed patterns of stimulation controlled by a single degree of freedom user interface.

Although the proximal arm has many fewer degrees of freedom, achieving useful control of reaching through FES has proven to be an even greater challenge than grasping. In part, this may be because the need to position and orient the hand for even simple reaches requires a relatively large proportion of the limb's total degrees of freedom. There also are problems associated with generating appropriate muscle force that can impact the development of appropriate user interfaces. Accelerating the mass of the limb and continuously opposing gravity pose substantial problems given the rapid onset of fatigue through FES-activated muscles (Shields, 1995). Finally, the unfortunate irony is that those individuals with high-level injuries, who have the greatest need for replaced function, also have the most limited means of available control.

In this section we discuss the control options currently available to individuals with SCI. While the emphasis is on FES applications, there are many parallels to the control of artificial prosthetic limbs, in which a mechanical device is used to replace arm function. For more detailed information on particular FES neuroprostheses that have been tested, or are in clinical use, see Lauer and Hunter Peckham (2002), Popovic et al. (2002) Peckham and Knutson (2005), and Ragnarsson (2008).

Residual Movement

Movements of the body that remain under voluntary control constitute an important component of the user interface. Movements of the residual arm have long been used to control prosthetic limbs. Cable-actuated mechanical prostheses for upper limb amputees have been in use since World War II. Simpler versions with leather straps were used as early as World War I. The approach has been so successful that, even with the introduction of more complex, electrically powered prostheses, many patients prefer the simpler version because of their lesser weight, greater speed, lower cost, and the intrinsic feedback provided through the cable tension (Pylatiuk et al., 2007).

Residual movements have also been widely used for the control of FES hand grasp systems. Among the FES neuroprostheses that have been made clinically available, most have been controlled through mechanical actions, including the NESS H200 (formerly the Handmaster; Fig. 19.1A) currently used by several thousand patients worldwide (Snoek et al., 2000; Alon and McBride, 2003). Continuous control of stimulation for hand grasp has been

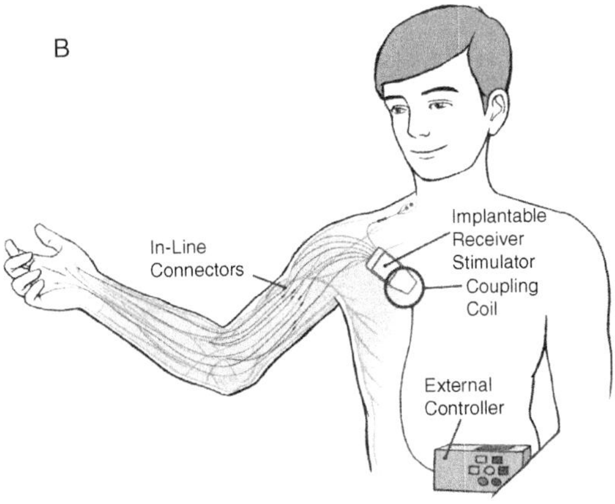

Figure 19.1. Existing upper limb FES devices in clinical use. (A) The NESS H200 Hand Rehabilitation System marketed by BioNess. (B) The Freehand Neuromuscular Stimulator. Although effective, and still in use by several hundred patients, the Freehand was withdrawn from the market in 2002 by its supplier, Neurocontrol Corp., for financial reasons. Figure used with the permission of the Cleveland FES Center.

achieved through the use of a position sensor mounted on the contralateral shoulder (Smith et al., 1987; Keith et al., 1989). More natural control has been achieved by taking advantage of the synergies that often occur between wrist flexion–extension and hand closing–opening linking remaining wrist extension with appropriate grasp stimulation patterns (Prochazka et al., 1997; Bhadra et al., 2002; Peckham and Knutson, 2005).

Residual movements have also been used to control reaching. The NEC FESMate (Handa et al., 1992) used a variety of methods, including head switches, voice, sip-and-puff, and shoulder motions, to control reach and grasp. A synergistic restoration of elbow extension has been integrated with a grasping neuroprosthesis by detecting the position of the arm in space and automatically compensating for gravity (Grill and Peckham, 1998). While residual movements are useful sources of control, they are restrictive by their nature as they occupy the user's remaining functional movements. Furthermore, the small set of movements that can be used limits the potential for high-dimensional control.

Electromyograms

Use of residual motion to control an artificial prosthetic limb has the distinct advantage that the voluntary motion can be coupled mechanically to the end effector, providing a sense of effort as well as effecting movement. This coupling is indirect for an FES neuroprosthesis, which requires a transducer to sense and convert the motion to an electrical signal. Given this added complexity, it is reasonable to consider alternative FES control methods, including electromyograms (EMG) sampled from muscles that remain under voluntary control. The first EMG-controlled FES neuroprosthesis used surface recordings of the contralateral trapezius muscle to drive stimulation of extensor digitorum communis (Long, 1963; Vodovnik et al., 1965). EMG control has been further developed in more recent FES neuroprostheses. The Compex Motion is a programmable stimulator intended to provide palmar and lateral grasp through transcutaneous stimulation (Popovic and Keller, 2005).

Implanted EMG sensors have also been used for FES control, providing significant improvements in cosmesis, convenience, and selectivity, compared with surface electrodes. The Freehand FES neuroprosthesis (Fig. 19.1B) uses fully implanted stimulating electrodes, as well as sensors implanted in the ipsilateral limb that include both mechanical sensors of shoulder and wrist positions and EMG electrodes implanted in a variety of muscles (Hart et al., 1998; Johnson et al., 1999). In addition to its other advantages, this allows the option of bilateral implants. This neuroprosthesis is currently used by approximately 250 patients.

Whether controlled by residual movement or muscle activity, all current FES neuroprostheses have a relatively simple control system that triggers and perhaps controls the amplitude of a preprogrammed stimulator output pattern. The advantage of this approach, and arguably the main reason for its commercial and clinical success, is its ease of use. Once properly calibrated, the patient can rely on these systems for their rehabilitation training or activities of daily living without continuous, professional assistance. The limitation of the simple control system is the restricted set of motor tasks that is achievable.

USE OF SIGNALS RECORDED FROM THE CEREBRAL CORTEX

Unfortunately, patients with SCI above C5 have even more limited control options, and much more demanding control needs. Combining more highly selective intramuscular recordings of EMG with higher level indications of the patient's intent such as gaze direction (described later) is one promising approach that may be used to implement natural interfaces with a low cognitive burden. Ultimately, however, signals recorded directly from the brain may be the most effective means to achieve this goal. Although significant technical challenges remain, it is possible to record on the order of 100 neurons with current technology. Monkey subjects have achieved information rates of about 5 bits per second in a target selection task using single- and multiple-unit neuronal signals recorded from the premotor cortex (Santhanam et al., 2006). It is likely that it will be possible to record larger numbers of neurons with correspondingly greater information rates in the near future.

Single-Neuron Recording

Nearly 20 years ago, Georgopoulos and colleagues hypothesized that it should be possible to predict the direction of hand movement simply by combining the information from a large population of neurons using the activity of each neuron to weight a vector pointing in that neuron's "preferred direction" (Georgopoulos et al., 1983). This "population vector analysis" was finally applied in real time in 2002, allowing a monkey to control the movement of a computer cursor in three dimensions (Taylor et al., 2002). More recently, similar methods allowed a monkey to feed itself with a robotic arm, both positioning and opening and closing the gripper to grasp pieces of food (Velliste et al. 2008).

More typical than the population vector method is the use of multiple input, optimal linear filters computed between the neural firing rate and the kinematics of normal movement (Humphrey et al., 1970; Chapin et al., 1999; Wessberg et al., 2000; Carmena et al., 2003; Serruya et al., 2003; Hochberg et al., 2006). These "decoders" can then be used to transform subsequent neural recordings into kinematic control signals.

Recursive Bayesian decoding schemes have also been developed, where a trajectory model represents the limb kinematics as a time-evolving state. An observation model then specifies the probability distribution for the neural activity given the state. These approaches are flexible and allow for both linear

Gaussian (Wu et al., 2006) and nonlinear Poisson-like (Brockwell et al., 2004; Truccolo et al., 2005) representations of the neural activity. However, recent comparisons have shown that the simple linear Gaussian model is often as good as and sometimes better than the more complicated nonlinear observation models (Santhanam et al., 2009; Lawhern et al., 2010). The Bayesian framework also facilitates nonlinear switching (Wu et al. 2004) and mixture models (Yu et al., 2007) that have worked well. In general, however, increases in performance by nonlinear methods have been modest (Kim et al., 2006).

In addition to the more common kinematic applications, primary motor cortex M1 discharge has been used to predict various kinetic aspects of movement, including joint torque or impedance during reaching (Kim et al., 2007; Fagg et al., 2009), grasp or endpoint force (Carmena et al., 2003; Gupta and Ashe, 2009), and the activity of individual muscles (Santucci et al., 2005; Westwick et al., 2006; Pohlmeyer et al., 2007; Oby et al., 2010). It is notable that many of these kinetic predictions are equally accurate or more accurate than the corresponding kinematic predictions despite their greater bandwidth and, in the case of EMG signals, their stochastic nature. Figure 19.2 shows the prediction of EMG activity in four different wrist muscles during the generation of isometric flexion and extension torque.

The ability to predict muscle activity continuously provides an attractive option for FES control that would circumvent the limitation imposed by the need to use only a small set of preprogrammed stimulus sequences. Our group has now demonstrated this ability in several monkey subjects, by temporarily blocking the peripheral nerves at the elbow to mimic some of the effects of C5 spinal cord injury (Pohlmeyer et al., 2009a). Prior to the block (typically the preceding day) we record neural and EMG signals from the monkey and use the data to train a linear or Wiener cascade decoder. On subsequent sessions, we use real-time predictions to control the stimulation of four or five muscles. Several monkeys have achieved voluntary control of different levels of isometric wrist flexion and extension force with these methods (Pohlmeyer et al., 2009b).

We have also trained a monkey to pick up and manipulate balls of different sizes and weights in a paradigm modeled after the box-and-block test used by occupational therapists working with stroke and SCI patients (Mathiowetz et al., 1985). Using the cortically controlled FES methods described previously, we restored voluntary grasp function to two monkeys, despite a complete paralysis of the hand and wrist flexor muscles induced by median and ulnar nerve block (Oby et al., 2010; Ethier et al., 2012) (Fig. 19.3). The neuroprosthesis improved the monkeys' performance at least sevenfold in terms of the number of successfully completed trials in the course of 20 experimental sessions. We have thus shown that chronic recordings of fewer than 100 neurons can provide a reliable, high-information-bandwidth signal appropriate to controlling a range of FES applications.

The decoding of intended muscle activation is especially useful for hand grasp control, with its many degrees of freedom and strong descending cortical

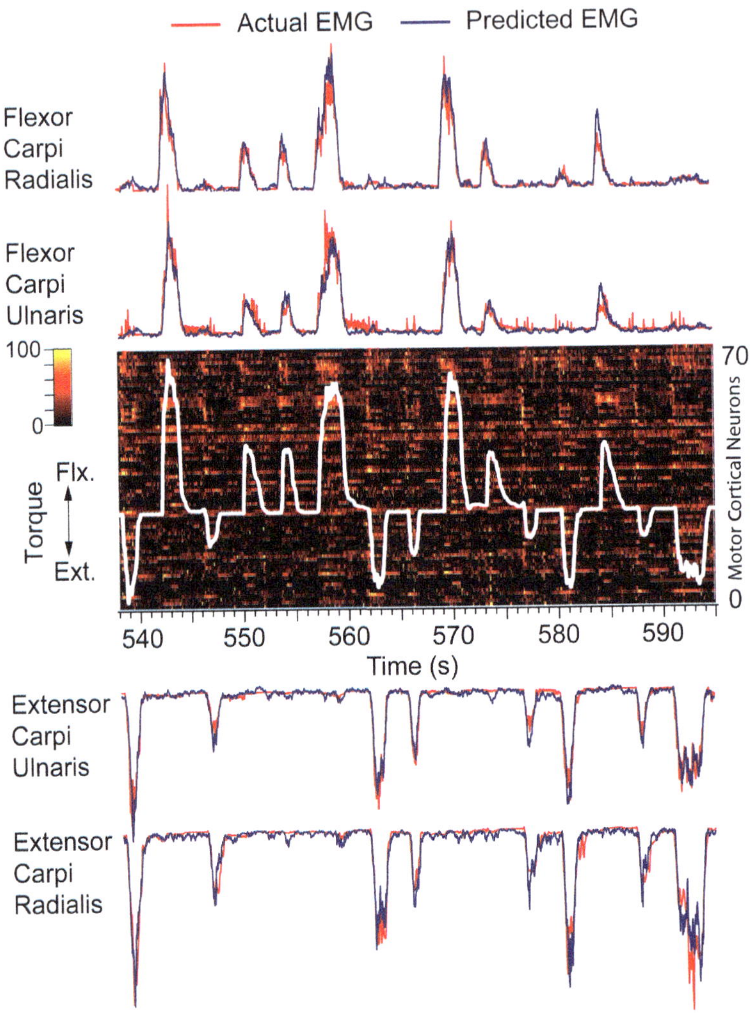

Figure 19.2. EMG predictions for four wrist muscles that generate isometric wrist torque. Predictions were based on the activity of 74 neurons. The discharge rate of each neuron has been normalized to its maximum firing rate within this period. The corresponding flexion–extension torque signal is overlaid.

control. It remains an open question if this is also the most appropriate strategy for restoration of more proximal arm function. Movement of the proximal arm and the associated muscle activations are more influenced by gravity and its impact on muscle fatigue. Small errors in proximal arm control could have a large influence on hand location, requiring constant monitoring and correction from the user. Such errors could be easily corrected using a kinematic controller with feedback compensation (Lan et al., 1991), a task that is more

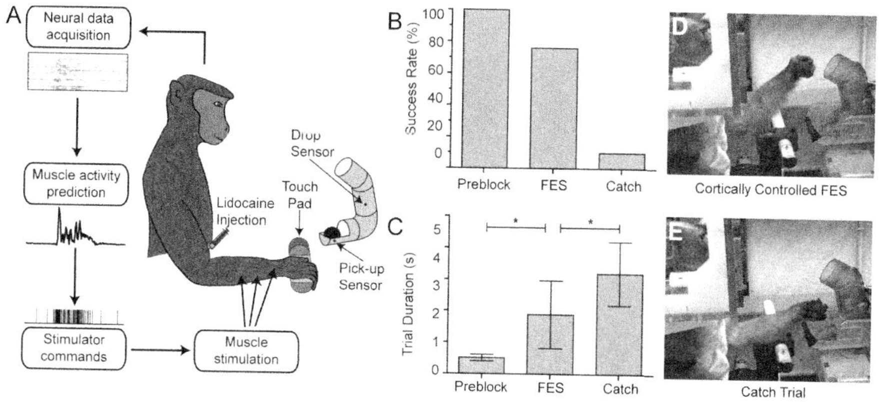

Figure 19.3. Cortically controlled FES neuroprosthesis for grasp. (A) The monkey's forearm flexor muscles were temporarily paralyzed by injections of lidocaine to the median and ulnar nerves. EMG predictions based on M1 recordings were used to modulate the intensity of stimulation, much like the Freehand neuroprosthesis in Figure 19.1B. (B,C) Success rates for normal, FES, and catch-trial conditions comprising over 6500 trials for each of two monkeys. Both monkeys completed nearly 80% of the trials with FES but only 11% (B) or essentially 0% (C) when the stimulators were turned off. (D) Example of FES-controlled grasp, executed much like the monkey's normal grasp. (E) Example of the alternative strategy adopted by one of the monkeys during catch-trials, which involved scooping, rather than grasping, the ball. Figure adapted from Ethier et al. (2012) with permission.

feasible for proximal control due to the reduced degrees of freedom and increased space for sensor location. Finally, there is evidence that the supraspinal control of the proximal arm musculature is more distributed than that of the hand, which could make the development of a muscle-based decoder more challenging (Palmer and Ashby, 1992; Turton and Lemon, 1999; Krutky and Perreault, 2007). As a result, we are actively pursuing both kinematic and muscle-based decoders of proximal limb function to evaluate the relative merits of each.

Field Potential Recordings

Despite its potential for high information rates, single-neuron recording technology currently suffers from the relatively limited lifespan of the implants. Most chronically implanted electrode arrays appear to be limited to 1–2 years, after which it is no longer possible to record single neurons. Even during the early period there is a constant turnover of neurons, estimated to be on the order of 50% of neurons per week (Dickey et al., 2009). This turnover means that the BMI user must be constantly adapting to a changing map from neural state to decoded output, a condition that may ultimately limit performance.

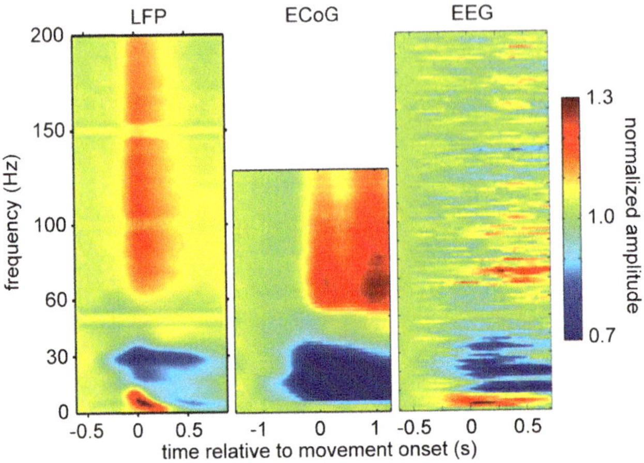

Figure 19.4. Grand-average time-resolved amplitude spectrograms during center-out movements for the different recording techniques (LFP, ECoG, EEG). Spectrograms depict trial-average across all movement directions averaged across multiple subjects. Recording sites were in or above the motor cortex contralateral to center-out hand–arm movements. LFP recordings from motor cortex of monkeys (taken from Rickert et al., 2005), ECoG recordings from primary motor cortex of epilepsy patients (taken from Ball et al., 2009), and EEG (reanalyzed from Waldert et al., 2008). (Figure adapted from Waldert et al., 2009.)

An alternative to using neural discharge is to record low-frequency field potentials that are the result of the summed activity of hundreds to many thousands of neurons. The nature of these signals suggests that they may be more stable and ultimately have a longer useful lifespan despite much lower spatial resolution. The signals are divided into several bands, often designated Delta (<4 Hz), Theta (4–8 Hz), Alpha (8–12 Hz), Beta (12–30 Hz), and Gamma (30–100 Hz). Differential power in these bands is thought to reflect different cortical states and has been used to discriminate the kinematics of both reaching and grasping (Fig. 19.4). For a review of field potential recordings, see Waldert et al. (2009).

The local field potential (LFP) is typically defined as the component of the intracortical signal below 200 Hz. It has been studied primarily in animals, since it requires the same invasive methods needed to record single neurons (Rickert et al., 2005; Scherberger et al., 2005; Spinks et al., 2008; Baumann et al., 2009; Zhuang et al., 2010). On the other hand, patients with intractable epilepsy who are undergoing weeklong cortical monitoring prior to surgery provide the opportunity to study the characteristics of signals recorded with a large grid of electrodes placed beneath the dura on the surface of the cortex (Freeman et al., 2000; Leuthardt et al., 2004; Mehring et al., 2005; Pistohl et al., 2008). These electrocorticographic (ECoG) signals are analyzed in much the same manner as LFPs. One study described a small number of ECoG channels whose voltage varied in proportion with the kinematics of movement, much like the single-unit firing rate (Schalk et al., 2007).

There may be somewhat less risk involved in the placement of these grids compared with penetrating, intracortical electrodes (Loeb et al., 1977; Bullara et al., 1979). There is almost certainly less risk in recording from grid electrodes placed on the dural surface, but the information content of these signals is likely to be reduced somewhat from that of ECoG. For applications not requiring a large bandwidth, these epidural recordings may represent an appropriate compromise between safety and efficacy (Slutzky et al., 2010).

Electroencephalography (EEG) provides a completely noninvasive approach for recording from the cortex. It has been used as a communication interface, achieving information rates of 10–20 bits/min (Blankertz et al., 2003; Serby et al., 2005; Nijboer et al., 2008) and even for off-line predictions of limb movement, although with much less accuracy than that typical of single-neuron-based predictions (Bradberry et al., 2010). With sufficient practice, paralyzed individuals can guide a wheelchair, cursor, or robot using EEG (Wolpaw et al., 1991; Bell et al., 2008; Galan et al., 2008; McFarland et al., 2010). However, in addition to its low bandwidth, EEG is particularly prone to artifacts and requires that the user regularly don and doff a set of recording electrodes, an operation typically requiring the help of an assistant. Hence, these systems are less likely to be useful in FES control.

The different techniques that can be used to record neural signals appear to lie on a continuum, from single-cell recordings that are highly invasive and somewhat fragile over time, to EEG recordings that are noninvasive, but could be applied literally every day for the life of the patient. Depending on the specific application these different techniques may be used to solve different classes of problems.

COMBINATIONS OF SOURCES

As we have described, a number of different signals can be recorded from each patient, potentially including movement, muscle, and neural activity. A combination of these inputs may prove to be the best option for many patients. Clearly, there are advantages and limitations to each of the available signal sources representing different levels of invasiveness, information content, extent of required adaptation, and potential for multidimensional control. Ultimately, it makes sense to consider all of the potentially useful signals available. In addition to enhanced performance, the redundancy of combined sources would help avoid a loss of control in the case of a sensor failure in one of the signals.

Bayesian Approach to the Optimal Combination of Input Sources

Many state-of-the-art decoders use a Bayesian approach, which is particularly useful in the context of combined, noisy information sources. Bayesian statistics provides tools to formulate a combination of sources. State estimation is performed by optimally combining the predictions of the trajectory model of the time-varying limb kinematics and the observation model, which maps the kinematic states to the control signals. If both models are linear and Gaussian, the Kalman filter (KF) performs optimal inference. While linear Gaussian models have in some cases been found to be as effective (Santhanam et al. 2009; Lawhern et al. 2010), nonlinear Poisson-like models are frequently used to model the spiking statistics of neurons (Pillow et al., 2008). This difference may be accommodated in the generalized linear model framework (Truccolo et al., 2005).

Combining Trajectory and Target Information

Depending on the source, we might get information about the reach target location or instantaneous information about the trajectory; promising research has looked at combining such discrete and continuous information in the context of Bayesian decoding. With perfect knowledge of the intended reach target, a preprogrammed trajectory could be implemented, requiring minimal effort from the user. However, this approach would provide little flexibility in the face of obstacles or different grasp objects. On the other hand, trajectory-related signals may give the user full control of the reach dynamics, but at the cost of cognitive effort. However, we know that reaches typically have a straight hand path and bell-shaped, minimum jerk speed profiles (Hogan, 1984; Flash and Hogan, 1985). Knowledge of the reach target is thus a potentially rich source of information about the reach trajectory, which may be used to enhance the trajectory model. With a Bayesian approach, an optimal balance can be found between using the signals under direct user control and using prior information, which may serve to reduce the cognitive burden.

There are many potential sources of target information. While the M1 encodes information about the reach trajectory, signals from the dorsal premotor cortex (PMd) may provide more information about the reach target (Hatsopoulos et al., 2004). Similarly, it has been suggested that the parietal reach region (PRR) of the posterior parietal cortex (PPC) represents goal information in visual coordinates (Batista et al., 1999). Both regions have been used for classification among a fixed set of targets (Hatsopoulos et al., 2004; Musallam et al., 2004; Santhanam et al., 2006).

Ideally, an FES controller would allow reaches anywhere in the workspace without advance knowledge of the distribution of possible targets. One way of obtaining such information is by tracking eye movements. In the period just before a reach, a subject is likely saccade to a small number of locations, with a high probability that one of those locations corresponds to the intended reach target. While there is uncertainty associated with such a continuous source of target information, it may nonetheless be useful in a Bayesian framework.

A trajectory model trained toward a single target can have parameters tuned to capture reaches to that target. One recent effort to deal with target-specific reach kinematics was the mixture of trajectory models (MTM) presented by Yu and co-workers (Yu et al., 2007). They trained a different trajectory model for each of several targets and performed a mixture over the models, producing a final estimate that was a weighted sum of the predictions based on the likelihood of the observations for each model. This algorithm produced impressive results, particularly for a known, fixed set of targets. However, it would not perform as well if asked to generalize to novel targets.

Other groups have explored ways to incorporate information about the reach target into a single trajectory model, making the assumption that the target has a linear effect on the arm state (Srinivasan et al., 2006; Mulliken et al., 2008b). Srinivasan et al. (2006) augmented a single trajectory model with a time-varying, target-related command term by assuming knowledge of the time at which the target should be reached. Alternatively, adding the target to the state space vector is a natural way of incorporating target information (Kemere and Meng, 2005; Mulliken et al., 2008b; Corbett et al., 2012). Using this approach, Mulliken and co-workers continuously decoded both the likely endpoint of the reach and its trajectory from neural activity in the PPC (Mulliken et al., 2008b).

We will refer to the use of a Kalman filter with the target added to the state space as a KFT model. This linear approach can also be extended to account for multiple potential targets in which an instance of the model is performed for each possible target in order to obtain a probabilistically weighted sum of the predictions (Kulkarni and Paninski, 2008; Corbett et al., 2012).

Comparison of the Discrete and Continuous Approaches to Target Information

The MTM may be considered to be a discrete method of incorporating target information, as it is designed to be used for the fixed set of targets on which it is trained. On the other hand, the KFT is an example of a continuous approach, which should generalize to any target location in the workspace. We compared predictions from these algorithms for reaches by unimpaired subjects within a large, three-dimensional workspace, using eye-tracking (Eye-Trac H6; ASL, Bedford, MA) to provide estimates of the target location (Corbett et al., 2010). In addition to the target, the state vector consisted of the hand and joint angle positions, velocities, and accelerations, which we measured with an active marker optical tracking system (Optotrak 3020; NDI, Waterloo, ON). We also measured the activity of pectoralis major and the three heads of the deltoid muscle to use as observations (Fig. 19.5A). These muscles approximately mimic those available to subjects with C5-level SCI.

Initially we assumed perfect knowledge of the reach target and constructed a single trajectory model (STM) for each target as the ideal case of the MTM.

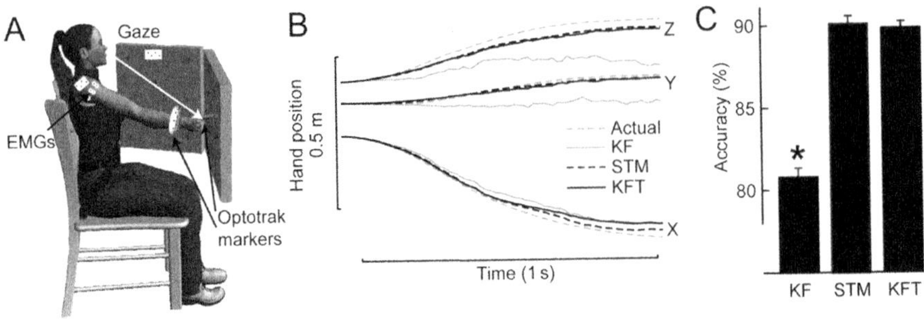

Figure 19.5. (A) Experimental setup. (B) Three-dimensional finger position from a sample reach with predictions by KF, STM, and KFT. (C) Mean and standard error of prediction accuracies from two subjects (approximately 200 reaches). (Reproduced from Corbett et al., 2010.)

Both STM and KFT produced significant improvements over the generic Kalman filter, whose trajectory model is similar to a random walk (Fig. 19.5B,C). The KFT performed as well as the STM, suggesting that a continuous trajectory model can be very effective if target information is available.

From the recordings of gaze direction in the one-second period preceding each reach we obtained an estimate for a prior probability distribution of possible reach targets. This estimate was solely dependent on the subject's gaze. We performed a mixture of KFTs (mKFT) over each of the possible target estimates, without assuming any knowledge of the actual target locations. We also used this prior distribution to initialize the MTM weights. In both cases, the weights usually converged quickly to the most appropriate model; the probabilistic mixture model approach was effective in dealing with uncertain target information. The accuracy was nearly as good as when the algorithms had perfect target information (Fig. 19.6). However, the advantage of the mKFT approach became evident when we tested how well the algorithms generalized to novel targets. We used a leave-one-out cross-validation test of each reach direction, training the algorithms on all but the tested direction. As anticipated, the MTM, which relies on having training data for all possible targets, did not generalize well. Thus the mKFT approach allows for more flexibility by incorporating linear effects of the target in a single, comprehensive trajectory model, thereby enabling reaches to anywhere in the workspace.

CLOSED-LOOP CONTROL

Open-Loop versus Closed-Loop Performance

In many approaches to decoder development, especially those that focus on interfaces that attempt to mimic the user's normal motor behaviors, control

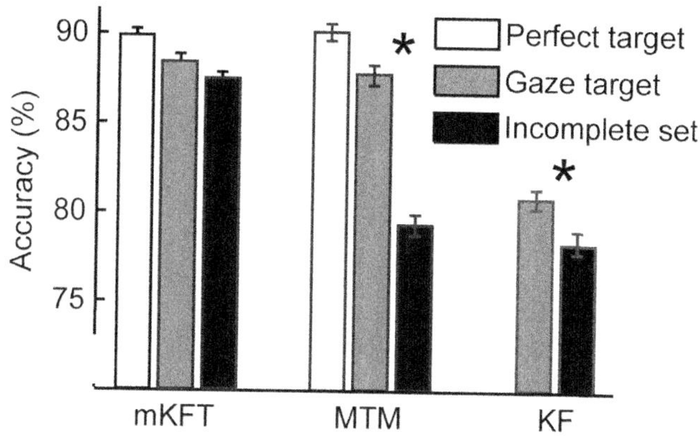

Figure 19.6. Mean and standard errors of prediction accuracy for algorithms with full knowledge of the reach target, and uncertain target information obtained from eye-tracking with full and partial training set for leave-one-out cross-validation. The asterisk indicates a statistically significant effect between the two training types.

signals are initially recorded during normal limb movements. These signals are then used to reconstruct the movement off-line. As the subject is unaware of the decoder's output and therefore unable to correct for errors, the decoder predictions are referred to as **open-loop** performance. Decoder performance is evaluated based on how similar the predictions are to the actual movement. This is a useful approach, as it allows the rapid comparison of multiple decoding algorithms on a single data set. However, it is not clear to what extent the open-loop performance of an algorithm corresponds to its utility for real-time control.

In actual use, the prediction of the decoding algorithm is used to control the device of interest in real time; the user has visual feedback of the decoder's output through the device, allowing for correction of errors and adaptation in the presence of persistent, systematic errors. It is therefore important that decoding methods for FES be evaluated with the user in the loop, which we refer to as **closed-loop** performance. An important limitation of these methods for application to paralyzed patients is the impossibility of recording both neural inputs and kinematic outputs during normal movement to compute an initial off-line decoder. As an alternative method of developing kinematic decoders, several groups have recorded the neural discharge rate while the user simply watches the motion of a cursor (Hochberg et al., 2006; Tkach et al., 2007). The resulting decoders are sufficiently accurate to allow them to be used for real-time control.

The question of whether open-loop decoder prediction errors are indicative of problems in closed-loop performance was considered by Kass and colleagues. They found that subjects' closed-loop performance using an optimal linear estimator was similar to that using the population vector algorithm, even though the latter performed far worse in open loop as a result of bias induced by a nonuniform distribution of neurons' preferred directions (Chase et al., 2009). In contrast, the same group found that the level of smoothing, which would alter the prediction error variance, had a large influence on decoder performance in the closed loop (Koyama et al., 2010). These conclusions may be related to the results of Kim et al. (2008), who found that tetraplegic patients with implanted cortical arrays could control cursor velocity better than position, signals that differ by an integration operator that acts as a smoothing function. These studies are among the relatively rare examples of closed-loop comparisons of decoding approaches. Much work remains before we can completely understand all of the factors that influence closed-loop performance. It will be interesting to see whether improvements achieved in open-loop performance with more complicated state models in the Bayesian framework will also translate to improved real-time control (Srinivasan et al., 2006; Yu et al. 2007; Corbett et al., 2012).

User Learning and Adaptation

Closed-loop operation provides the opportunity for the user to learn and adapt to the interface. Performance improvements with practice have been reported in the context of EMG control of both actual (Bouwsema et al., 2008; Kuiken et al., 2009) and virtual (Corbett et al., 2011) prosthetic arms. Similar changes occur for users of cortical interfaces. Gradual performance changes typically continue for tens of sessions as subjects learn to modulate the sensorimotor rhythms recorded by EEG (McFarland et al., 2010). A similar effect has been shown for control of two-dimensional cursor movements using the power in the 65–100 Hz band of ECoG signals. Over the course of a week, monkeys were able to learn to modulate the power in two electrodes that were randomly assigned to control movement along the X and Y axes (Rouse and Moran, 2009).

Likewise, several studies of BMIs controlled by single neurons found changes in performance within and across sessions, in some cases under conditions in which the decoders were intentionally made suboptimal (Fetz, 1969; Chapin et al., 1999; Taylor et al., 2002; Carmena et al., 2003; Musallam et al., 2004; Hochberg et al., 2006; Jarosiewicz et al., 2008; Moritz et al., 2008; Mulliken et al., 2008a; Ganguly and Carmena, 2009). A critical observation in this context was that simply recomputing a decoder, even if it and the former provided equivalent off-line prediction, forced the monkey to go through a readaptation process that caused decreased closed-loop performance extending even into the subsequent one or two sessions (Ganguly and Carmena, 2009).

This observation has important implications for a number of studies that have used some form of adaptive algorithm in combination with the user's own adaptation (Taylor et al., 2002; Wolpaw and McFarland, 2004; Wahnoun et al., 2006; DiGiovanna et al., 2009). Algorithm adaptation potentially has a number of useful applications, including the optimization of initial decoder performance and compensation for gradual changes in the information encoded by the inputs. However, it is critical that the combination of the two adapting systems (the user and algorithm) remain stable. This problem was studied in human subjects by Danziger et al. (2009), who used finger movements measured by a high-dimensional data glove to control a two degree of freedom virtual arm. These subjects were unable to learn if the mapping from the glove space to the two-dimensional limb space was adapted too rapidly. The development of adaptive algorithms that enhance, rather than limit, the user's own performance adaptation is an appropriate challenge to consider. Such coadaptation may be particularly important for a BMI implemented in a paralyzed user for whom an optimal decoder based on normal kinematics cannot be calculated.

CONCLUSIONS

Since the early 1980s, FES neuroprostheses have made great strides in providing increased function and independence to individuals paralyzed by SCI. The restoration of the ability to reach for and grasp objects is essential in enabling activities of daily living. There are currently a number of promising systems for this purpose, all of which generate preprogrammed grasp patterns using a simple user interface. One of the most difficult problems in the development of systems with more degrees of freedom is finding a viable means of control for the patient to use. Various signal sources have been used to interface with FES neuroprostheses, most of which have limited potential for high-dimensional control due to the large cognitive burden they impose. This is particularly true for patients with high-level injuries who need to control whole arm movements in addition to hand grasp.

To address these problems, research has endeavored to create user interfaces that enable nearly effortless control of more degrees of freedom. Brain–machine interfaces, using direct cortical recordings, provide one promising approach for developing effortless user interfaces. Also promising is the possibility to combine information from multiple sources, including the brain. Such combinations can take advantage of the unique information content available in each source and can be tailored to the unique needs and abilities of each individual with SCI. Together, these developments have the potential to provide greater control of FES neuroprostheses, resulting in greater independence and an improved quality of life for these individuals.

REFERENCES

Alon G. and McBride K. (2003). Persons with C5 or C6 tetraplegia achieve selected functional gains using a neuroprosthesis. *Arch Phys Med Rehabil* 84:119–124.

Anderson K.D. (2004). Targeting recovery: priorities of the spinal cord–injured population. *J Neurotrauma* 21:1371–1383.

Ball T., Schulze-Bonhage A., Aertsen A., Mehring C. (2009). Differential representation of arm movement direction in relation to cortical anatomy and function. *J Neural Eng* 6(1):016006.

Batista A.P., Buneo C.A., et al. (1999). Reach plans in eye-centered coordinates. *Science* 285(5425):257–260.

Baumann M.A., Fluet M.C., et al. (2009). Context-specific grasp movement representation in the macaque anterior intraparietal area. *J Neurosci* 29:6436–6448.

Bell C., Shenoy P., et al. (2008). Control of a humanoid robot by a noninvasive brain–computer interface in humans. *J Neural Eng* 5:214.

Bhadra N., Peckham P.H., et al. (2002). Implementation of an implantable joint-angle transducer. *J Rehabil Res Development* 39:411–422.

Blankertz B., Dornhege G., et al. (2003). Boosting bit rates and error detection for the classification of fast-paced motor commands based on single-trial EEG analysis. *IEEE Trans Neural Syst Rehabil Eng* 11:127–131.

Bouwsema H., van der Sluis C.K., et al. (2008). The role of order of practice in learning to handle an upper-limb prosthesis. *Arch Phys Med Rehabil* 89:1759–1764.

Bradberry T.J., Gentili R.J., et al. (2010). Reconstructing three-dimensional hand movements from noninvasive electroencephalographic signals. *J Neurosci* 30:3432–3437.

Brockwell A.E., Rojas A.L., et al. (2004). Recursive bayesian decoding of motor cortical signals by particle filtering. *J Neurophysiol* 91:1899–1907.

Bullara L.A., Agnew W.F., et al. (1979). Evaluation of electrode array material for neural prostheses. *Neurosurgery* 5:681–686.

Carmena, J.M., Lebedev M.A., et al. (2003). Learning to control a brain–machine interface for reaching and grasping by primates. *PLoS Biol* 1:193–208.

Chapin J.K., Moxon K.A., et al. (1999). Real-time control of a robot arm using simultaneously recorded neurons in the motor cortex. *Nat Neurosci* 2:664–670.

Chase S.M., Schwartz A.B., et al. (2009). Bias, optimal linear estimation, and the differences between open-loop simulation and closed-loop performance of spiking-based brain–computer interface algorithms. *Neural Networks* 22:1203–1213.

Corbett E.A., Perreault E.J., et al. (2010). Mixture of time-warped trajectory models for movement decoding. *Adv Neural Inform Process Syst* 23.

Corbett E.A., Perreault E.J., et al. (2011). Comparison of electromyography and force as interfaces for prosthetic control. *J Rehabil Res Dev* 48:629–641.

Corbett E.A., Perreault E.J., et al. (2012). Decoding with limited neural data: a mixture of time-warped trajectory models for directional reaches. *J Neural Eng* 9:036002.

Danziger Z., Fishbach A., et al. (2009). Learning algorithms for human–machine interfaces. *IEEE Trans Biomed Eng* 56:1502–1511.

Dickey A.S., Suminski A., et al. (2009). Single-unit stability using chronically implanted multielectrode arrays. *J Neurophysiol* 102:1331–1339.

DiGiovanna J., Mahmoudi B., et al. (2009). Coadaptive brain–machine interface via reinforcement learning. *IEEE Trans Biomed Eng* 56:54–64.

Ethier C., Oby E.R., et al. (2012). Restoration of grasp following paralysis through brain-controlled stimulation of muscles. *Nature* 485:368–371.

Fagg A.H., Ojakangas G.W., et al. (2009). Kinetic trajectory decoding using motor cortical ensembles. *IEEE Trans Neural Syst Rehabil Eng* 17:487–496.

Fetz E.E. (1969). Operant conditioning of cortical unit activity. *Science* 163:955–958.

Flash T. and Hogan N. (1985). The coordination of arm movements: an experimentally confirmed mathematical model. *J Neurosci* 7:1688–1703.

Freeman W.J., Rogers L.J., et al. (2000). Spatial spectral analysis of human electrocorticograms including the alpha and gamma bands. *J Neurosci Methods* 95:111–121.

Galan F., Nuttin M., et al. (2008). A brain-actuated wheelchair: asynchronous and non-invasive brain–computer interfaces for continuous control of robots. *Clin Neurophysiol* 119:2159–2169.

Ganguly K. and Carmena J.M. (2009). Emergence of a stable cortical map for neuroprosthetic control. *PLoS Biol* 7:e1000153.

Georgopoulos A.P., Caminiti R., et al. (1983). Spatial coding of movement: a hypothesis concerning the coding of movement direction by motor cortical populations. *Exp Brain Res* 7(Suppl):327–336.

Grill J.H. and Peckham P.H. (1998). Functional neuromuscular stimulation for combined control of elbow extension and hand grasp in C5 and C6 quadriplegics. *IEEE Trans Rehabil Eng* 6:190–199.

Gupta R. and Ashe J. (2009). Offline decoding of end-point forces using neural ensembles: application to a brain–machine interface. *IEEE Trans Neural Syst Rehabil Eng* 17:254–262.

Handa Y., Handa T., et al. (1992). Functional electrical stimulation (FES) systems for restoration of motor function of paralyzed muscles—versatile systems and a portable system. *Frontiers Med Biol Eng Int J Japan Soc Med Electronics Biol Eng* 4:241.

Hart R.L., Kilgore K.L., et al. (1998). A comparison between control methods for implanted FES hand-grasp systems. *IEEE Trans Rehabil Eng* 6:208–218.

Hatsopoulos N., Joshi J., et al. (2004). Decoding continuous and discrete motor behaviors using motor and premotor cortical ensembles. *J Neurophysiol* 92:1165–1174.

Hochberg L.R., Serruya M.D., et al. (2006). Neuronal ensemble control of prosthetic devices by a human with tetraplegia. *Nature* 442:164–171.

Hogan N. (1984). An organizing principle for a class of voluntary movements. *J Neurosci* 4:2745–2754.

Humphrey D.R., Schmidt E.M., et al. (1970). Predicting measures of motor performance from multiple cortical spike trains. *Science* 170:758–761.

Ingram J.N., Kording K.P., et al. (2008). The statistics of natural hand movements. *Exp Brain Res* 188:223–236.

Jarosiewicz B., Chase S.M., et al. (2008). Functional network reorganization during learning in a brain–computer interface paradigm. *Proc Natl Acad Sci U S A* 105:19486–19491.

Johnson M.W., Peckham P.H., et al. (1999). Implantable transducer for two–degree of freedom joint angle sensing. *IEEE Trans Rehabil Eng* 7:349–359.

Keith M.W., Peckham P.H., et al. (1989). Implantable functional neuromuscular stimulation in the tetraplegic hand. *J Hand Surg [Am]* 14:524–530.

Kemere C. and Meng T. (2005). Optimal estimation of feed-forward-controlled linear systems. In: Proceedings (ICASSP'05), IEEE International Conference on Acoustics, Speech, and Signal Processing, p. 5.

Kim H.K., Carmena J.M., et al. (2007). The muscle activation method: an approach to impedance control of brain–machine interfaces through a musculoskeletal model of the arm. *IEEE Trans Biomed Eng* 54:1520–1529.

Kim S.P., Sanchez J.C., et al. (2006). A comparison of optimal MIMO linear and nonlinear models for brain–machine interfaces. *J Neural Eng* 3:145–161.

Kim S.P., Simeral J.D., et al. (2008). Neural control of computer cursor velocity by decoding motor cortical spiking activity in humans with tetraplegia. *J Neural Eng* 5:455–476.

Koyama S., Chase S., et al. (2010). Comparison of brain–computer interface decoding algorithms in open-loop and closed-loop control. *J Comput Neurosci* 29:73–87.

Krutky M.A. and Perreault E.J. (2007). Motor cortical measures of use-dependent plasticity are graded from distal to proximal in the human upper limb. *J Neurophysiol* 98:3230–3241.

Kuiken T.A., Li G., et al. (2009). Targeted muscle reinnervation for real-time myoelectric control of multifunction artificial arms. *JAMA* 301:619–628.

Kulkarni J.E. and Paninski L. (2008). State-space decoding of goal-directed movements. *IEEE Signal Processing Magazine* 25:78.

Lan N., Crago P.E., et al. (1991). Control of end-point forces of a multijoint limb by functional neuromuscular stimulation. *IEEE Trans Biomed Eng* 38:953–965.

Lauer R. and Hunter Peckham P. (2002). Motor prostheses. In: *Handbook of Neuroprosthetic Methods*, Boca Raton, FL: CRC Press.

Lawhern V., Wu W., et al. (2010). Population decoding of motor cortical activity using a generalized linear model with hidden states. *J Neurosci Methods* 189:267–280.

Leuthardt E.C., Schalk G., et al. (2004). A brain–computer interface using electrocorticographic signals in humans. *J Neural Eng* 1:63–71.

Loeb G.E., Walker A.E., et al. (1977). Histological reaction to various conductive and dielectric films chronically implanted in the subdural space. *J Biomed Mater Res* 11:195–210.

Long C. 2nd (1963). An electrophysiologic splint for the hand. *Arch Phys Med Rehabil* 44:499–503.

Mathiowetz V., Volland G., et al. (1985). Adult norms for the Box and Block Test of manual dexterity. *Am J Occupational Therapy* 39:386.

McFarland D., Sarnacki W., et al. (2010). Electroencephalographic (EEG) control of three-dimensional movement. *J Neural Eng* 7:036007.

Mehring C., Nawrot M.P., et al. (2005). Comparing information about arm movement direction in single channels of local and epicortical field potentials from monkey and human motor cortex. *J Physiol Paris* 98:498–506.

Moritz C.T., Perlmutter S.I., et al. (2008). Direct control of paralysed muscles by cortical neurons. *Nature* 456:639–642.

Mulliken G.H., Musallam S., et al. (2008a). Decoding trajectories from posterior parietal cortex ensembles. *J Neurosci* 28:12913–12926.

Mulliken G.H., Musallam S., et al. (2008b). Forward estimation of movement state in posterior parietal cortex. *Proc Natl Acad Sci U S A* 105:8170–8177.

Musallam S., Corneil B.D., et al. (2004). Cognitive control signals for neural prosthetics. *Science* 305:258–262.

Nijboer F., Sellers E.W., et al. (2008). A P300-based brain–computer interface for people with amyotrophic lateral sclerosis. *Clin Neurophysiol* 119:1909–1916.

Oby E.R., Ethier C., et al. (2010). Getting a grip on spinal cord injury: a novel application of a brain machine interface. In: *Statistical Signal Processing for Neuroscience and Neurotechnology* (O'Weiss K.G., Ed.), pp. 369–406. Academic Press, Elsevier.

Palmer E. and Ashby P. (1992). Corticospinal projections to upper limb motoneurones in humans. *J Physiol* 448:397–412.

Peckham P.H. and Knutson J.S. (2005). Functional electrical stimulation for neuromuscular applications. *Annu Rev Biomed Eng* 7:327–360.

Pillow J.W., Shlens J., et al. (2008). Spatio-temporal correlations and visual signalling in a complete neuronal population. *Nature* 454:995–999.

Pistohl T., Ball T., et al. (2008). Prediction of arm movement trajectories from ECoG-recordings in humans. *J Neurosci Methods* 167:105–114.

Pohlmeyer E.A., Solla S.A., et al. (2007). Prediction of upper limb muscle activity from motor cortical discharge during reaching. *J Neural Eng* 4:369–379.

Pohlmeyer E.A., Jordon L.R., et al. (2009a). A fully implanted drug delivery system for peripheral nerve blocks in behaving animals. *J Neurosci Methods* 182:165–172.

Pohlmeyer E.A., Oby E.R., et al. (2009b). Toward the restoration of hand use to a paralyzed monkey: brain-controlled functional electrical stimulation of forearm muscles. *PLoS ONE* 4:e5924.

Popovic M.R. and Keller T. (2005). Modular transcutaneous functional electrical stimulation system. *Med Eng Physics* 27:81–92.

Popovic M.R., Popovic D.B., et al. (2002). Neuroprostheses for grasping. *Neurol Res* 24:443–452.

Prochazka A., Gauthier M., et al. (1997). The bionic glove: an electrical stimulator garment that provides controlled grasp and hand opening in quadriplegia. *Arch Phys Med Rehabil* 78:608–614.

Pylatiuk C., Schulz S., et al. (2007). Results of an Internet survey of myoelectric prosthetic hand users. *Prosthet Orthot Int* 31:362–370.

Ragnarsson K.T. (2008). Functional electrical stimulation after spinal cord injury: current use, therapeutic effects and future directions. *Spinal Cord* 46:255–274.

Rickert J., Oliveira S.C., et al. (2005). Encoding of movement direction in different frequency ranges of motor cortical local field potentials. *J Neurosci* 25:8815–8824.

Rouse A.G. and Moran D.W. (2009). Neural adaptation of epidural electrocorticographic (EECoG) signals during closed-loop brain computer interface (BCI) tasks. In: *Engineering in Medicine and Biology Society (EMBC 2009), Annual International Conference of the IEEE*, pp. 5514–5517.

Santello M., Flanders M., et al. (1998). Postural hand synergies for tool use. *J Neurosci* 18:10105–10115.

Santhanam G., Ryu S.I., et al. (2006). A high-performance brain–computer interface. *Nature* 442:195–198.

Santhanam G., Yu B.M., et al. (2009). Factor-analysis methods for higher-performance neural prostheses. *J Neurophysiol* 102:1315–1330.

Santucci D.M., Kralik J.D., et al. (2005). Frontal and parietal cortical ensembles predict single-trial muscle activity during reaching movements in primates. *Eur J Neurosci* 22:1529–1540.

Schalk G., Kubanek J., et al. (2007). Decoding two-dimensional movement trajectories using electrocorticographic signals in humans. *J Neural Eng* 4:264.

Scherberger H., Jarvis M.R., et al. (2005). Cortical local field potential encodes movement intentions in the posterior parietal cortex. *Neuron* 46:347–354.

Serby H., Yom-Tov E., et al. (2005). An improved P300-based brain–computer interface. *IEEE Trans Neural Systems Rehabil Eng* 13:89–98.

Serruya M., Hatsopoulos N., et al. (2003). Robustness of neuroprosthetic decoding algorithms. *Biol Cybern* 88:219–228.

Shields R. (1995). Fatigability, relaxation properties, and electromyographic responses of the human paralyzed soleus muscle. *J Neurophysiol* 73:2195.

Slutzky M.W., Jordan L.R., et al. (2010). Optimal spacing of surface electrode arrays for brain–machine interface applications. *J Neural Eng* 7:26004.

Smith B., Peckham P.H., et al. (1987). An externally powered, multichannel, implantable stimulator for versatile control of paralyzed muscle. *IEEE Trans Biomed Eng* 34:499–508.

Snoek G.J., Mj I.J., et al. (2000). Use of the NESS handmaster to restore handfunction in tetraplegia: clinical experiences in ten patients. *Spinal Cord* 38:244–249.

Spinks R.L., Kraskov A., et al. (2008). Selectivity for grasp in local field potential and single neuron activity recorded simultaneously from M1 and F5 in the awake macaque monkey. *J Neurosci* 28:10961–10971.

Srinivasan L., Eden U.T., et al. (2006). A state-space analysis for reconstruction of goal-directed movements using neural signals. *Neural Computation* 18:2465–2494.

Taylor D.M., Tillery S.I., et al. (2002). Direct cortical control of 3D neuroprosthetic devices. *Science* 296:1829–1832.

Tkach D., Reimer J., et al. (2007). Congruent activity during action and action observation in motor cortex. *J Neurosci* 27:13241–13250.

Truccolo W., Eden U.T., et al. (2005). A point process framework for relating neural spiking activity to spiking history, neural ensemble, and extrinsic covariate effects. *J Neurophysiol* 93:1074–1089.

Turton A. and Lemon R.N. (1999). The contribution of fast corticospinal input to the voluntary activation of proximal muscles in normal subjects and in stroke patients. *Exp Brain Res* 129:559–572.

Velliste M., Perel S., et al. (2008). Cortical control of a prosthetic arm for self-feeding. *Nature* 453:1098–1101.

Vodovnik L., Long C. 2nd, et al. (1965). Myo-electric control of paralyzed muscles. *IEEE Trans Biomed Eng* 12:169–172.

Wahnoun R., He J., et al. (2006). Selection and parameterization of cortical neurons for neuroprosthetic control. *J Neural Eng* 3:162–171.

Waldert S., Preissl H., Demandt E., Braun C., Birbaumer N., Aertsen A., Mehring C. (2008). Hand movement direction decoded from MEG and EEG. *J Neurosci* 28(4):1000–1008.

Waldert S., Pistohl T., et al. (2009). A review on directional information in neural signals for brain–machine interfaces. *J Physiol Paris* 103:244–254.

Wessberg J., Stambaugh C.R., et al. (2000). Real-time prediction of hand trajectory by ensembles of cortical neurons in primates. *Nature* 408:361–365.

Westwick D.T., Pohlmeyer E.A., et al. (2006). Identification of multiple-input systems with highly coupled inputs: application to EMG prediction from multiple intracortical electrodes. *Neural Comput* 18:329–355.

Wolpaw J., McFarland D., et al. (1991). An EEG-based brain–computer interface for cursor control. *Electroencephalogr Clin Neurophysiol Suppl* 78:252–259.

Wolpaw J.R. and McFarland D.J. (2004). Control of a two-dimensional movement signal by a noninvasive brain–computer interface in humans. *Proc Natl Acad Sci U S A* 101:17849–17854.

Wu W., Black M.J., et al. (2004). Modeling and decoding motor cortical activity using a switching Kalman filter. *IEEE Trans Biomed Eng* 51:933–942.

Wu W., Gao Y., et al. (2006). Bayesian population decoding of motor cortical activity using a Kalman filter. *Neural Comput* 18:80–118.

Yu B.M., Kemere C., et al. (2007). Mixture of trajectory models for neural decoding of goal-directed movements. *J Neurophysiol* 97:3763–3780.

Zhuang J., Truccolo W., et al. (2010). Decoding 3-D reach and grasp kinematics from high-frequency local field potentials in primate primary motor cortex. *IEEE Trans Biomed Eng* 57:1774–1784.

20

FOCUSED CHAPTER

CUSTOMIZED MODELING AND SIMULATIONS FOR THE CONTROL OF FES-ASSISTED WALKING OF INDIVIDUALS WITH HEMIPLEGIA

STRAHINJA DOŠEN[1] AND DEJAN B. POPOVIĆ[2,3]

[1]*Department of Neurorehabilitation Engineering, University Medical Center, Georg-August University, Göttingen, Germany*
[2]*School of Electrical Engineering, University of Belgrade, Belgrade, Serbia*
[3]*Department of Health Science and Technology, Center for Sensory–Motor Interaction, Aalborg University, Aalborg, Denmark*

SUMMARY

Clinical studies have demonstrated that hemiplegic patients benefit from gait training assisted by surface functional electrical stimulation (FES). However, the effective control of multichannel FES for walking is a difficult task. The human musculoskeletal system is very complex, with highly nonlinear and time-varying dynamics, and redundant actuators. To cope with this complexity, we advocate an approach termed sensor-driven control that is inspired by the biological mechanisms for the control of walking. To create an online controller, we rely on musculoskeletal modeling and simulations, and we developed methods specifically tailored for our intended application. Since different hemiplegic patients

Introduction to Neural Engineering for Motor Rehabilitation, First Edition. Edited by Dario Farina, Winnie Jensen, and Metin Akay.

require different levels of assistance, we use a modular approach. The basic building block is a single joint–single segment system actuated by two antagonistic muscle groups (flexors and extensors). This system is used to create the models of the ankle–foot complex (single block), lower leg (two blocks), and the entire affected leg (three blocks). Inputs for the simulation are the selected model customized with the parameters determined from the prospective user of FES and a desired walking pattern recorded from a healthy individual. The simulation based on optimal control outputs the pulse trains (stimulation profiles) that should be delivered to the user's muscles to generate the selected gait. A user-friendly software tool (OptiWalk) implements the simulations, data management, and visualization. OptiWalk is used to test the feasibility of different walking patterns and/or to generate and fine tune stimulation profiles, and this comprises the foundation for the design of a subject-specific sensor-driven control of FES.

INTRODUCTION AND BACKGROUND

Clinical studies have demonstrated that hemiplegic patients in both acute (Kojović et al., 2009; Yan et al., 2005) and chronic (Mesci et al., 2009; Stevens and Hunsacker, 2010) phase benefit from gait training assisted by surface functional electrical stimulation (FES; see Box 20.1). The simplest form of assistance is to use a single channel of stimulation to activate ankle dorsiflexor muscles, preventing drop foot during the swing phase of gait (Lyons et al., 2002). Multiple stimulation channels can be employed to provide a more comprehensive support during both stance and swing (Malezic et al., 1994, 1987; Strojnik et al., 1979). For example, one more channel can activate ankle plantar flexors to augment the push-off (Embrey et al., 2010). Knee extensors can be stimulated to assist weight transfer and weight bearing during stance, and also to extend the knee at the end of swing, making the step longer. The stimulation of knee flexors facilitates proper foot clearance during swing and may prevent knee hyperextension at the beginning of stance. Finally, gluteal muscles can be externally activated to support the pelvis and assist hip extension. The precise number of channels is selected by assessing the level of disability of the prospective user of FES (Bogataj et al., 1997).

However, the effective control of multichannel FES for assisting walking is a difficult task. The system to be controlled is very complex, is highly nonlinear (body and muscle dynamics) and time varying (muscle fatigue), is mechanically unstable (inverted pendulum), and has multiple actuators that are partially functionally redundant. To cope with this complexity, we advocate a hierarchically organized sensor-driven control of FES, a method that is inspired by the biological mechanisms for the control of walking. It is well known that the motor systems are organized hierarchically, with many pathways and centers operating in parallel. Furthermore, the assumption is that the biologi-

BOX 20.1. Functional Electrical Stimulation and Therapy

Functional electrical stimulation (FES) is a technique for restoring movements in paralyzed persons. Electrical current pulses are delivered to the tissue to stimulate motor neurons and thereby elicit muscle contractions. The goal of FES is to activate the muscles in a coordinated and controlled manner so that a functional movement is generated (e.g., reaching, grasping, standing, and walking). In surface FES, the electrodes are placed on the surface of the skin above the motor points of the targeted muscles.

Functional electrical therapy (FET) refers to the application of FES to facilitate a program of intensive functional exercises in clinical settings (e.g., gait training, training of reaching and grasping). The goal is to provide a substrate for "retraining the brain," that is, extensive practice to relearn the movements that have been lost as a consequence of nervous system injury or disease. Clinical studies have shown that FET promotes functional recovery. For more information see Chapter 18, Chapter 24, and Chapter 25.

cal controller simplifies the task of motor control by constructing complex control strategies from a set of basic functional modules (Neptune et al., 2009). Namely, during development and through practice, the controller builds a repertoire of useful muscle synergies (Ivanenko et al., 2006; see also Chapter 3). A synergy can be defined as a functionally relevant, spatially organized (across muscles), and temporally coordinated pattern of activation of a group of muscles during the motor task. Essentially, these are learned feedforward commands (i.e., motor programs) that are triggered by sensory events (discrete action) and/or modulated by sensory feedback (continuous action). An artificial sensor-driven controller uses a similar concept. It comprises three levels as shown in Figure 20.1. The function of the highest level is to detect the intention of the user (i.e., brain–machine interface; see Chapter 12) in order to activate relevant lower-level components. The coordination level organizes the activity of electrically stimulated muscles spatially and temporally; in other words, it encodes the relevant muscle synergies in the form of IF–THEN rules and releases them in response to detected gait events (i.e., "IF event 1 THEN release synergy 1"). The task of the lowest level (execution stage) is to specify the actual stimulation profiles for each muscle by taking into account the individual properties of each actuator. This chapter presents the methods for the design of the lowest level of a sensor-driven controller.

To determine muscle stimulation profiles for the execution level of a sensor-driven controller, one can use musculoskeletal modeling and simulations. Many musculoskeletal models have been developed (for review see Erdemir et al., 2007), and there are many software packages for the design, analysis, and simulation of these models (e.g., SIMM, Musculographics, Santa Rosa, CA, USA; AnyBody, Damsgaard et al., 2006), some of which are even distributed

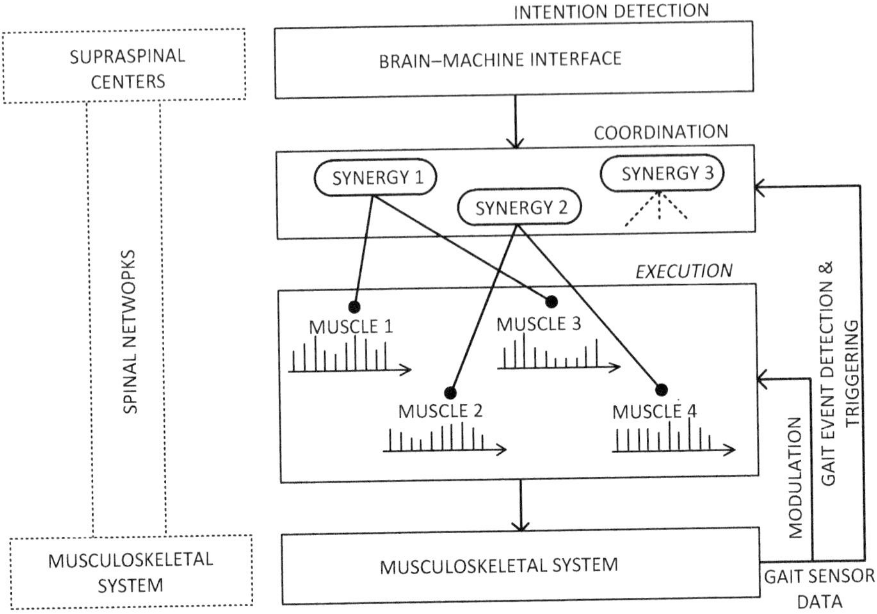

Figure 20.1. Sensor-driven control of functional electrical stimulation. The hierarchically organized controller comprises intention detection, coordination, and execution stages. Intention detection level is implemented as a brain–machine interface. Coordination level encodes the patterns of coordinated activations of groups of muscles (i.e., muscle synergies) that are triggered by detected gait events. Execution level stores predefined muscle stimulation profiles that are determined by taking into account individual muscle properties. This control scheme is inspired by the biology, and the tentative mapping between the biological structures and the stages of the artificial controller is shown on the left.

under public license and available as open source (e.g., OpenSim, Delp et al., 2007). Within these frameworks, the human body is modeled as a highly sophisticated, three-dimensional system comprising many segments connected by different types of joints and actuated by many, individually represented, muscles. The simulations of walking employing these models are instrumental for better understanding of gait kinematics and dynamics, as well as for gaining important insights into the biological mechanisms for the control of walking.

Musculoskeletal models of lower extremities have been applied also in the context of FES-assisted standing and walking (for review see Riener, 1999). Yamaguchi and Zajac (1990) used an 8-degree-of-freedom (DOF) model driven by 10 musculotendon actuators to construct a set of suboptimal, discretely varying stimulation patterns needed for the model to sustain a step. Riener and Fuhr (1998) tested two strategies for FES-supported standing up by applying them to a generic planar, 3-segmental model with 9 mono- and bi-articular muscle groups. Gartman and co-workers (Gartman et al., 2008) used a model with 15 DOFs and 40 muscles to determine an optimal set of muscles to be stimulated by a 16-channel neuroprosthesis for the restoration of standing. Zhang and Zhu (2007) presented a comprehensive, biologically inspired control system for FES of walking and tested it by using a whole-body model comprising 7 segments and 18 muscles. Park and Durand (2008) proposed a control scheme separating the steady-state properties from the dynamic properties and tested the method by controlling a 2-DOF ankle–subtalar joint model with 8 muscles. Audu and co-workers (Audu et al., 2011) investigated the control of posture shifting using FES in individuals with spinal cord injury. They used a comprehensive musculoskeletal model comprising 21 bone segments connected by 21 joints and actuated by 32 musculotendon units. Evidently, the aforementioned musculoskeletal models are very sophisticated and useful for answering many important questions, such as assessing the feasibility of FES-induced motions, selecting target muscles, or developing and testing different control strategies. However, they are also quite complex, including many actuators and model parameters, some of which are very difficult or even impossible to identify experimentally (e.g., maximum force, optimal fiber length, and tendon slack length for a specific muscle). Therefore, it is difficult to customize the simulation so that it reflects the characteristics of an individual impaired subject and captures the specifics of using surface FES to activate the muscles (e.g., low selectivity). Furthermore, the simulations based on complex models can be extremely time-consuming (Pandy, 2001).

Here, we present our modeling and simulation approach for the control of surface FES for assisting walking in individuals with hemiplegia. The model is carefully reduced into a form that is tractable and therefore convenient for practical application, while still capturing the most important features of the real plant. The parameters describing the model are experimentally identifiable, and the model can be customized to represent the affected limb of a specific patient. We also developed a practical simulation method based on optimal control that takes a desired walking pattern as input and outputs the stimulation profiles that are necessary to generate the pattern. The resulting stimulation profiles can be directly used for the design of a sensor-driven FES controller; specifically, they comprise its execution stage (see Fig. 20.1).

METHODOLOGY AND RESULTS

Musculoskeletal Model

Our modeling approach was developed by carefully considering the context of the intended application. As explained in the introduction, different stroke patients need different levels of assistance: from one to five stimulation channels can be applied to activate ankle and/or knee and/or hip muscles, and the precise number used depends on the patient's individual level of disability. To accommodate these different configurations, we adopted a modular approach.

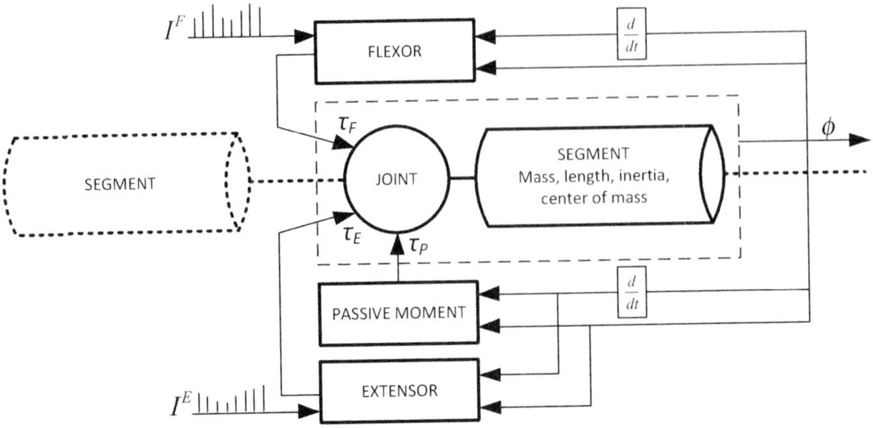

Figure 20.2. A basic building block (single joint–single segment system). A pair of antagonistic muscles, modeling the flexor and extensor muscle groups, actuates the joint by producing flexion (τ_F) and extension (τ_E) moments. The generated moment depends on the stimulation profile $I = [I_n]$, where I_n is the intensity of the nth pulse within the profile, joint angle (ϕ), and angular velocity ($d\phi/dt$). A passive moment (τ_P) represents viscoelastic joint properties. The inputs for the basic modeling block are the stimulation profiles (pulse trains) for the flexor (I^F) and extensor (I^E), and the output is the generated joint angle.

The basic building block for our models is a single joint–single segment system (see Fig. 20.2). To actuate the joint, we use one channel of stimulation to activate the flexor muscles and one channel to activate the extensors. Only one DOF (joint flexion–extension) is actively assisted by FES. The model is therefore reduced to one plane, and the anatomical joint is modeled as a simple pin joint with the rotation axis perpendicular to the sagittal plane. The leg segment is modeled as a rigid body described by the body mass, length, moment of inertia about its center of mass, and distance of the center of mass from the proximal joint.

The actuators for the model are the groups of synergistic muscles. This reflects the fact that surface FES is not selective enough to activate the muscles individually. As described previously, joint flexors are activated simultaneously by using a single channel of stimulation. Hence, all flexor muscles are lumped together and treated as a single entity, the so-called **equivalent muscle**, and the same is assumed for the extensors. The action of this **equivalent actuator** is given by a total moment that its component muscles generate around the joint. Therefore, the joint is driven by a pair of muscles generating opposing moments (antagonistic muscles). The control inputs for the equivalent muscles are the trains of electrical pulses, and the model output is the trajectory of the segment (i.e., simulation-generated joint angle).

As the model of the muscle, we use a well-known three-component multiplicative model (see Fig. 20.3a) formulated in terms of the joint-level variables

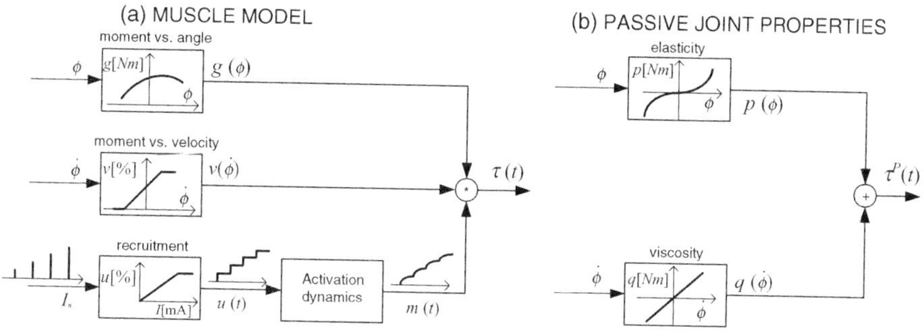

Figure 20.3. (a) Muscle model. Active moment $\tau(t)$ generated by the muscle is given as the product of three terms: (1) angle-dependent maximal isometric moment $g(\phi)$, (2) velocity-dependent scaling factor $v(\dot{\phi})$, and (3) muscle active state $m(t)$, which is controlled by the electrical pulses I_n delivered to the muscle. (b) Passive moment at the joint $\tau^P(t)$ is the sum of linear viscosity $q(\dot{\phi})$ and a nonlinear (double-exponential) elasticity $p(\phi)$. Notation: ϕ = joint angle; $\dot{\phi}$ = joint angular velocity; $u(t)$ = muscle excitation.

(Riener and Fuhr, 1998; Anderson et al., 2007; Shue et al., 1995). To model the physiological relationship between the maximal isometric moment and joint angle, we use a quadratic function. The fact that the maximal moment a muscle can produce scales up or down with the velocity of contraction is approximated by a piecewise linear function. Muscle response to stimulation is modeled by a static nonlinearity (recruitment curve) followed by activation dynamics (Riener and Fuhr, 1998; Khang and Zajac, 1989). Within the recruitment block, we use a correction factor for the muscle excitation levels as suggested in Riener and Fuhr (1998) in order to accommodate different frequencies of stimulation. The passive properties of the joints are also taken into account (see Fig. 20.3b). Again, the common approach is used; that is, the viscosity is modeled as a linear function (Stein et al., 1996), and the elastic moment as a double exponential curve (Anderson et al., 2007; Riener and Edrich, 1999). Finally, the total moment at the joint is the sum of active and passive contributions (see Fig. 20.2). More details about the modeling approach can be found in Popović et al. (1999, 2003).

The described single joint–single segment structure can be used to represent the ankle–foot complex, and this model can be exploited to design the control for a single- or dual-channel drop-foot stimulator. It can be also regarded as a basic building block to create more complex models (see Fig. 20.4). If two of these blocks are connected, the model becomes a shank–foot system, which can be used to design assistance for the lower leg, employing four channels of stimulation to activate ankle and knee flexors and extensors. Finally, if three blocks are connected, one obtains the model of the entire paretic leg. In this case, the assistance can also include stimulation of the hip extensors (five stimulation channels).

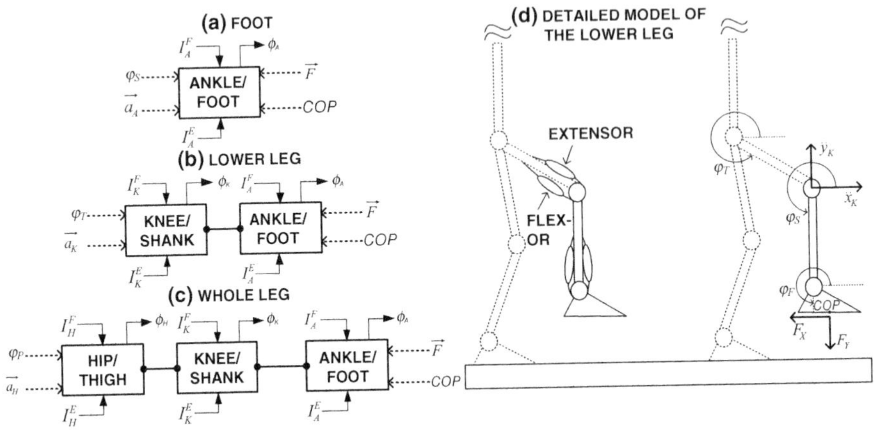

Figure 20.4. Models of the (a) foot, (b) lower leg, and (c) the entire affected leg shown in the form of block diagrams with inputs and outputs, and (d) the model of the lower leg depicted in detail (continuous lines, modeled subsystem; dotted lines, the rest of the body). The inputs for the simulation are as follows: (1) the interaction of the modeled subsystem with the rest of the body, that is, the acceleration of the most proximal joint, and the angle of the segment to which the subsystem is attached; (2) ground contact, that is, ground-reaction forces and the trajectory of the center of pressure; and (3) control inputs, that is, the stimulation profiles for the equivalent flexors and extensors. The outputs are the joint angles generated as the response to stimulation. Notation: φ_P, φ_T, φ_S, φ_F = pelvis, thigh, shank, and foot segment angles from the horizontal axis; a_H, a_K, a_A = hip, knee, and ankle accelerations; $\boldsymbol{F} = [F_X\ F_Y]^T$ = vector of horizontal and vertical ground-reaction forces; ϕ_H, ϕ_K, ϕ_A = hip, knee, and ankle joint angles; $I = [I_n]$ = stimulation profile, that is, the array of pulse intensities I_n, for flexors (F) and extensors (E) acting at the knee (K), hip (H), and ankle (A); $\boldsymbol{COP} = [x_{COP}\ y_{COP}]^T$ = the trajectory of the center of pressure; $\ddot{x}_K$, $\ddot{y}_K$ = horizontal and vertical acceleration of the knee.

All of the models shown in Figure 20.4 can be regarded as planar pendulums comprising one or more segments, with the most proximal joint being a moving hanging point. The pendulum is connected with the rest of the body and contacts the ground during the stance phase of walking. These interactions are taken into account as follows: (1) the rest of the body controls the acceleration of the hanging point; (2) the orientation of the rest of the body with respect to the modeled subsystem is given by the absolute angle of the body segment to which the subsystem is attached; and (3) the ground contact is modeled by a ground-reaction force acting at the center of pressure (COP), which moves along the sole of the foot during the stance phase of gait. The aforementioned signals are given as inputs for the simulation together with muscle control signals (stimulation profiles). The mathematical model can be derived by applying Newton–Euler equations and expressed in the following matrix form:

$$\boldsymbol{M}_{n\times n}(\boldsymbol{\varphi})\ddot{\boldsymbol{\varphi}}+\boldsymbol{C}_{n\times n}(\boldsymbol{\varphi})\dot{\boldsymbol{\varphi}}^2+\boldsymbol{G}_{n\times 1}(\boldsymbol{\varphi})+\boldsymbol{J}_{n\times 2}(\boldsymbol{\varphi})\boldsymbol{f}+\boldsymbol{B}_{n\times 2}(\boldsymbol{\varphi})\boldsymbol{a}_P=\boldsymbol{T}_{n\times n}(\boldsymbol{\tau}^F-\boldsymbol{\tau}^E-\boldsymbol{\tau}^P) \quad (20.1)$$

Notation

$\boldsymbol{\varphi}_{n\times 1}$—Segment angles

$\boldsymbol{f}=[F_X\ F_Y]^T$—Horizontal and vertical ground-reaction forces

$\boldsymbol{a}_P=[\ddot{x}\ \ddot{y}]^T$—Horizontal and vertical accelerations of the moving hanging point

$\boldsymbol{\tau}^F_{n\times 1}$—Joint moments produced by the equivalent flexor

$\boldsymbol{\tau}^E_{n\times 1}$—Joint moments produced by the equivalent extensors

$\boldsymbol{\tau}^P_{n\times 1}$—Passive joint moments

$\boldsymbol{M}_{n\times n}(\boldsymbol{\varphi})$—Inertia matrix

$\boldsymbol{J}_{n\times 2}(\boldsymbol{\varphi})$—Jacobian matrix

$\boldsymbol{T}_{n\times n}$—Constant matrix specifying how the joint moments contribute to the segmental moments

The parameter n is the number of segments comprising the model ($n = 1,2,3$ for the foot, lower leg, and entire leg, respectively). The individual terms on the left-hand side of Eq. 20.1 from left to right are the moments due to inertial, centrifugal, gravitational, and ground-reaction forces, and the moments due to the interaction with the rest of the body. These are balanced by the segmental moments, which appear on the right-hand side of the equation.

Simulation

The goal of the simulation is to calculate the stimulation profiles (pulse trains) that should be delivered to the muscles in order to generate a selected desired walking pattern. Mathematically, this can be formulated as an optimization problem. Let $\boldsymbol{\varphi}^*(t)$ and $\boldsymbol{\varphi}(t)$ represent the vectors of desired and simulation-generated joint angles, respectively. Since we assume that the pulse width and frequency of stimulation are predefined and constant, the only unknown parameters are the pulse intensities, and they should be determined so that the following cost function is minimized:

$$\min J[\boldsymbol{I}(n)]=\min\int_{t=0}^{T_D}(\boldsymbol{e}^T(t)\boldsymbol{Q}\boldsymbol{e}(t)+\boldsymbol{m}^T(t)\boldsymbol{R}\boldsymbol{m}(t)]dt \quad (20.2)$$

subject to

$$\frac{d}{dt}\begin{bmatrix}\boldsymbol{\varphi}\\ \dot{\boldsymbol{\varphi}}\end{bmatrix}=\begin{bmatrix}\dot{\boldsymbol{\varphi}}\\ M^{-1}(\boldsymbol{\varphi})(\boldsymbol{\tau}[\boldsymbol{m}(t)]-\boldsymbol{N}(\boldsymbol{\varphi},\dot{\boldsymbol{\varphi}}))\end{bmatrix} \quad (20.3)$$

where $\boldsymbol{e}(t)=\boldsymbol{\varphi}^*(t)-\boldsymbol{\varphi}(t)$ is the vector of tracking errors from the desired trajectories, $\boldsymbol{m}(t)$ is the vector of muscle activations, and $\boldsymbol{I}(n)$ is the vector of pulse intensities (one intensity per muscle) for the nth pulse of the muscle pulse trains. The parameter T_D is the simulation duration, and $\boldsymbol{Q}$ and $\boldsymbol{R}$ are the diagonal matrices comprising weighting (normalization) factors. The first term in Eq. 20.2 minimizes the sum of the squares of the tracking errors, while the second term is the penalty on the sum of the squares of the muscle activation levels. Equation 20.3 is the mathematical model given in Eq. 20.1 expressed in the canonical form that is suitable for the simulation. The parameter optimization problem defined by Eq. 20.2 and Eq. 20.3 can be solved by using nonlinear programming (Bryson, 1999).

We developed a method for an efficient calculation of pulse intensities termed **moving-window dynamic optimization** (Došen and Popović, 2009). Instead of solving a large-scale parameter optimization problem defined by Eq. 20.2 and Eq. 20.3 to determine the values of all the pulses at once, we minimize the cost function J within the time window spanning only K pulses, where K is an adjustable parameter. In other words, only the segment of the desired trajectory that is within the current time window is tracked by the model. The output of this smaller scale optimization is the K pulses. We keep only the first pulse, and then shift the time window for one stimulation period toward the end of the desired trajectory. This procedure is repeated until all of the pulses are determined. The algorithm is illustrated in Figure 20.5. The parameter K can be used to control the trade-off between the quality of tracking and optimization duration, as demonstrated in Došen and Popović (2009). A longer optimization window decreases the tracking error but increases the overall processing time.

OptiWalk Software Tool

For the practical application, we have developed a user-friendly software tool named OptiWalk implementing our customized modeling and simulation approach. The graphical user interface was developed in C# by using Visual Studio 2008; the optimization engine was implemented in MATLAB 2010 by using object-oriented programming and then compiled and exported as a .NET component to be included in the main application.

The software integrates a database of experimentally identified model parameters and a database of desired walking patterns recorded from healthy subjects. Mechanical parameters of the leg segments are computed by using standard regression equations (Zatsiorsky and Seluyanov, 1983; Winter, 1990) or experimentally measured by using methods explained in Stein et al. (1996; e.g., pendulum tests). Passive joint properties are determined using a pull test (Stein et al., 1996). The profiles of muscle moment versus angle and of muscle moment scaling factor versus angular velocity are obtained by conducting isometric and isokinetic tests with a stimulated muscle (Anderson et al., 2007). Desired walking patterns are recorded using an optoelectronic motion capture system (eight ProReflex MCU240 cameras, Qualisys, Sweden) and a force plate (OR6-5, Advanced Mechanical Technology, United States) in the Gait

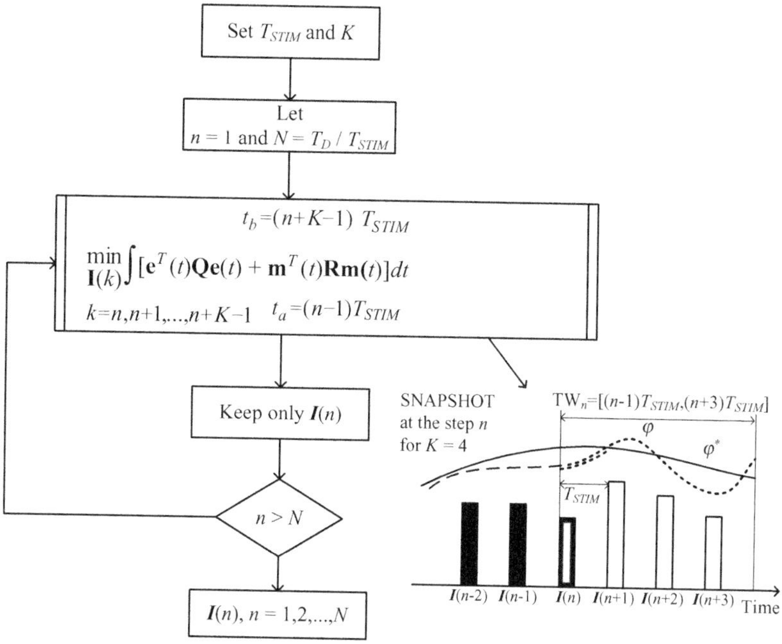

Figure 20.5. Moving-window dynamic optimization. The overall algorithm is depicted (see text for explanation) together with the snapshot of the algorithm for an arbitrary step n. The stimulation pulses are calculated one by one, by tracking a segment of the desired trajectory within the time window TW_n spanning a predefined number of pulses (K), and then moving the time window pulse-by-pulse toward the end of the desired trajectory. The continuous line is the desired trajectory (φ^*), the dashed line is the part of the trajectory (φ) generated in the simulation up to the current step (n), the dotted line is the trajectory generated in the current step, and the double dotted line is the segment of the trajectory that will be kept as the final solution in this step. The black pulses are previously determined, the white pulses are obtained in the current step, and only the white pulse with a thick border is kept as the final solution in this step. Notation: T_{STIM} = stimulation period; K = number of pulses; T_D = duration of simulation (i.e., the length of the desired trajectory); $\boldsymbol{e}(t)$ = the vector of tracking errors; $\boldsymbol{m}(t)$ = the vector of muscle activations; $\boldsymbol{I}(n)$ = the vector of pulse intensities (one intensity per muscle) for the nth pulse of the muscle pulse trains; $\boldsymbol{Q}$, $\boldsymbol{R}$ = weighting factors.

laboratory at the Center for Sensory–Motor Interaction. For out-of-lab recordings, we developed a mobile sensor system comprising accelerometers (ADXL203, Analog Devices, United States) and force-sensing resistors (Interlink Electronics, United States), and a method based on machine learning for estimating the simulation inputs from the data recorded by the practical sensors (Došen and Popović, 2008).

A typical scenario for use of OptiWalk proceeds as follows (see Fig. 20.6 and Fig. 20.7). Model type (foot only, lower leg, or the entire affected leg) is

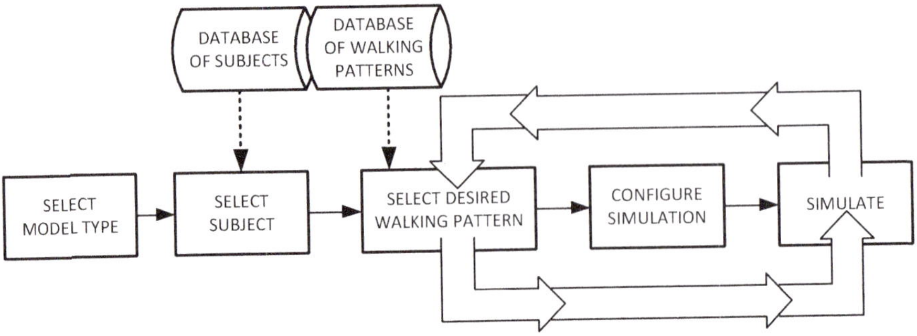

Figure 20.6. OptiWalk software tool implements our modeling and simulation method, and integrates data management and visualization functions. Once the type of assistance (foot, lower leg, or the entire affected leg) and model parameters are selected, multiple simulations with different walking patterns and/or optimization parameters can be conducted until satisfactory results are obtained.

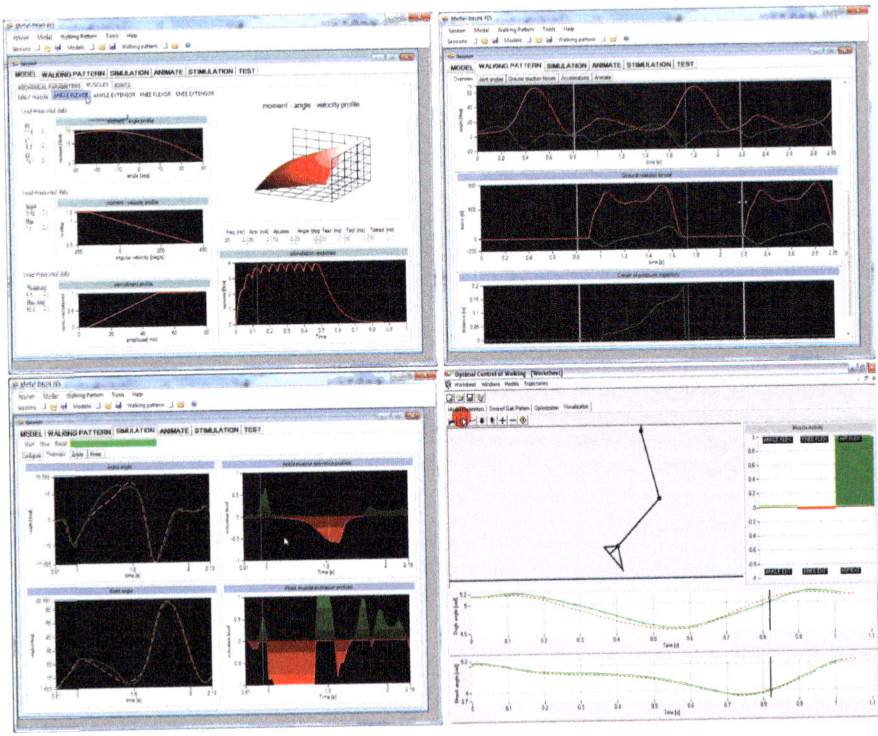

Figure 20.7. Several screenshots from the OptiWalk software tool: model panel for editing and visualizing the parameters for the selected muscle (upper left), trajectory panel showing the desired gait pattern and the part of the trajectory that is to be simulated (upper right), optimization panel with the desired, generated trajectory and muscle activation profiles (lower left), and animation pane with a stick figure model demonstrating the generated gait (lower right).

selected and model parameters are loaded. Biomechanical parameters, muscle profiles, and passive joint properties are presented to the user within the model editor. The model editor can be used to change the existing parameters or to create a new model. The next step is to select a desired gait pattern for the subject. The recorded signals are plotted in the trajectory window, and the user can select the part of the trajectory to be simulated. Simulation is configured and started. The outputs of the simulation, namely, muscle stimulation profiles, and desired and simulation-generated joint angles are plotted. In addition, the generated and desired walking patterns can be visualized in the form of stick figures. If the simulation results are unsatisfactory (e.g., large deviations from the desired trajectories, pulse intensities too high), the simulation can be repeated with another gait pattern and/or with different optimization parameters.

Examples

Here, we provide several illustrative examples demonstrating the use of the OptiWalk software tool. We selected a two-joint model (lower leg) and loaded the parameters from the database of subjects (subject A). For the first simulation, we used a gait pattern recorded from a healthy individual walking at a gait speed of 1.2 m/s as the desired trajectory for the selected subject (see Fig. 20.8). The stimulation frequency was set to 40 pulses per second. Figure 20.9 depicts the results of the simulation. Desired and generated joint angles (left panels) and muscle stimulation profiles (right panels) are shown for one gait stride, starting with a heel strike. The stimulation intensities are normalized in order to compare the results of different simulations (i.e., 0 means stimulation below the motor threshold and 1 denotes maximal intensity).

Note that there is a considerable deviation from the desired joint angles, especially during the stance phase. Also, during the same period, the stimulation pulse train for the knee extensors is saturated, that is, the normalized intensities are equal to one. This means that, although the stimulation is at its maximal level, the generated knee extension moment is still not large enough to extend the knee against the body weight, and ultimately the joint cannot track the reference trajectory. The overall conclusion is that the selected walking pattern is not feasible for the selected subject; the actuators are simply too weak.

Since the simulation demonstrated that the subject was not able to walk at a normal gait speed when assisted by FES, we decided to test if the subject can walk slowly. We selected a trajectory recorded from the same healthy individual but when he was asked to walk at a gait speed of 0.6 m/s. The simulation results for one gait stride are given in Figure 20.10a–d. This time, the deviations from the reference trajectories were very small, and the tracking was almost perfect. Furthermore, the stimulation profiles do not saturate, and the stimulation intensities are generally lower than the ones in Figure 20.9 for all the muscles. To demonstrate the flexibility of our method and OptiWalk,

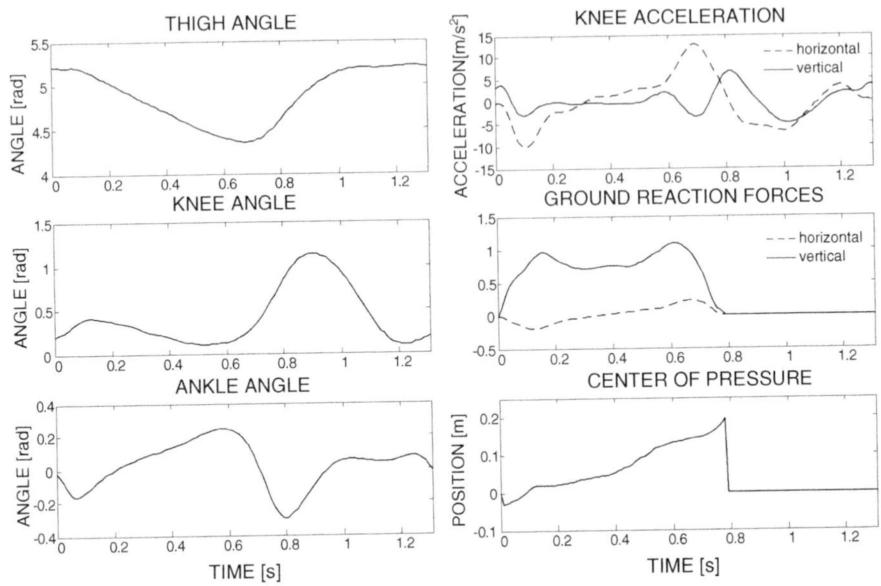

Figure 20.8. Gait pattern recorded from a healthy subject walking at a gait speed of 1.2 m/s used as input for the simulation. One gait stride is shown starting with a heel strike. Knee and ankle angle are the desired trajectories, while the other signals specify the interaction with the rest of the body and the ground. Ground-reaction forces are normalized to body weight.

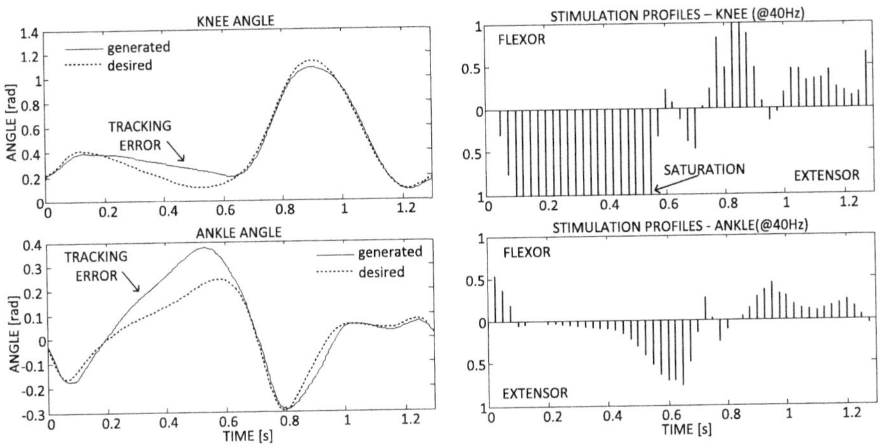

Figure 20.9. Simulation results for subject A and desired trajectory recorded from a healthy individual walking at a gait speed of 1.2 m/s. The panels on the left depict desired (dashed line) and generated (continuous line) knee and ankle joint angles, and the panels on the right show the stimulation profiles for the knee and ankle flexor (top plots) and extensor (bottom plots) muscle groups. The stimulation levels are normalized. Note that the stimulation for the knee extensor saturates during stance, and that this is accompanied by large tracking errors in the joint angle trajectories.

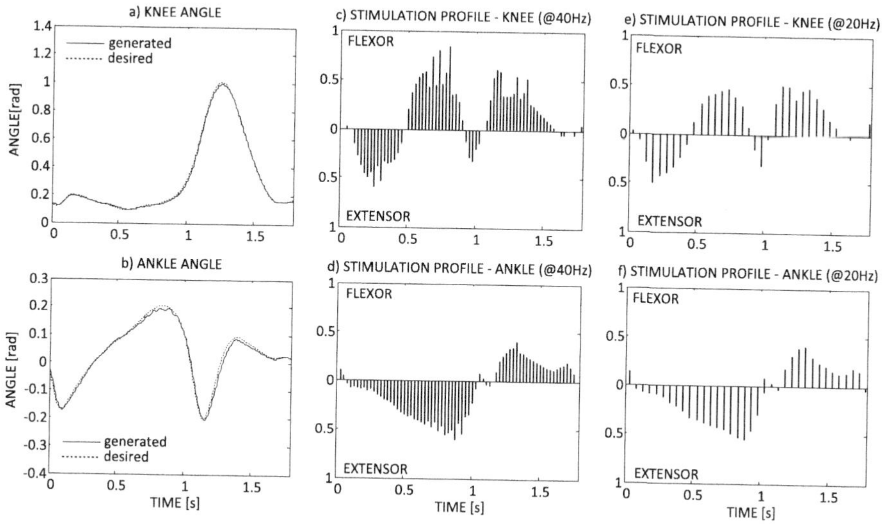

Figure 20.10. Simulation results for subject A and the desired trajectory recorded from a healthy individual walking at a gait speed of 0.6 m/s (slow walk). Panels (a) and (b) show desired (dashed line) and generated (continuous line) knee and ankle joint angles. Panels (c) and (d) are the corresponding stimulation profiles for the knee and ankle flexor (top plots) and extensor (bottom plots) muscle groups for the stimulation frequency of 40 pulses per second. Panels (e) and (f) are the stimulation profiles determined for the stimulation frequency of 20 pulses per second. Note that the stimulation levels do not saturate, and that the tracking errors are very small.

we conducted an additional simulation with the same inputs (subject and walking pattern) but with different stimulation frequency, namely, 20 pulses per second. The obtained stimulation profiles are given in Figure 20.10e,f. Again, we obtained very good tracking (i.e., tracking errors of only few degrees). The overall conclusion is that it is possible to generate the second walking pattern (i.e., slow walking at 0.6 m/s) by applying FES in this subject, and therefore the stimulation profiles in Figure 20.10 can be used to design a sensor-driven control for FES-assisted gait training. The stimulation at 40 Hz might be more comfortable for the subject, while the stimulation at 20 Hz is more desirable if the rate of development of muscle fatigue has to be decreased.

The last simulation demonstrates why the control of FES should be customized to the subject. We used the same walking pattern as previously (Fig. 20.10), but this time we selected a different subject (subject B) from the database of model parameters. Figure 20.11 depicts the results. The quality of tracking is quite similar in both cases, but the stimulation profiles are actually different (e.g., compare the profiles for the ankle muscles). Therefore, to realize exactly the same walking pattern, different subjects require different control signals reflecting the actual status of their impaired motor systems.

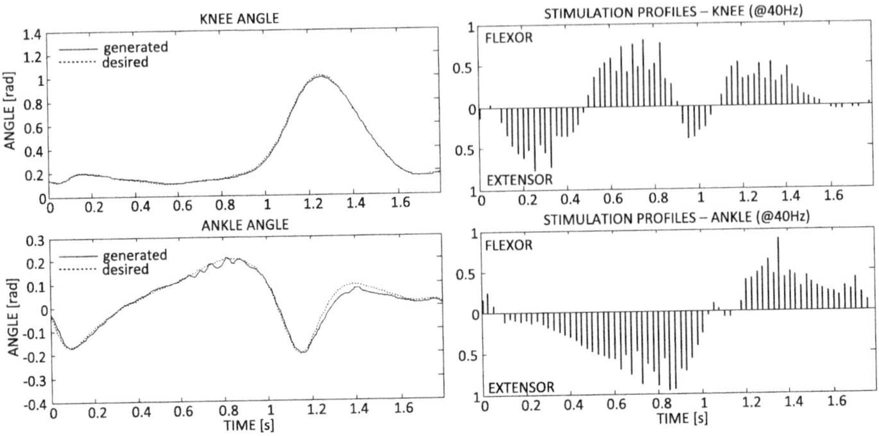

Figure 20.11. Simulation results for subject B and the desired trajectory recorded from a healthy individual walking at a gait speed of 0.6 m/s. The desired trajectory is the same as the one used for the simulation in Figure 20.10 and the quality of tracking is similar, but the profiles for subject A and subject B are still different.

CONCLUSIONS

We presented a method for musculoskeletal modeling and simulations that is designed specifically for the control of surface FES for assisting walking in individuals with hemiplegia. The inputs for the simulation are (1) the model customized with parameters reflecting the individual characteristics of a potential user of FES and (2) a desired walking pattern recorded from a healthy person. The simulation outputs generated joint angles and stimulation profiles for the user's muscles. An important goal was to develop a method that is convenient for practical application. The model is reduced and simplified, so that a minimal set of parameters has to be measured. The simulation based on optimal control decomposes a difficult problem into a series of smaller-scale subproblems that are easier to solve. The simulation lasted for about 6 minutes for one gait stride and a stimulation frequency of 20 Hz, and 13 minutes when the frequency was set to 40 Hz. The modeling and simulation are packaged into a user-friendly software tool with integrated data management and visualization. The software tool can be used for easy off-line testing of different scenarios, in order to select the optimal one that should serve as the basis for online control. The feasibility of generating a selected walking pattern in a specific subject by using FES can be tested, and if the selected gait is shown to be feasible, the obtained stimulation profiles can be used to design a subject-specific sensor-driven FES controller. Using examples, we have illustrated the flexibility of the developed simulation tool as well as the reasons we advocate a subject-specific control design.

The designed stimulation profiles comprise the execution level of the sensor-driven controller (see Fig. 20.1). The next step is to construct a coordination level of control by establishing the mapping between the stimulation levels and gait sensor readings. Importantly, sensors convenient for practical application, such as accelerometers and force-sensing resistors, should be used. The mapping is typically encoded as a set of rules, and these rules can be obtained either by handcrafting or automatically, by using machine learning. Once designed, the rules are downloaded into a programmable stimulator and used to control the stimulation online. This step in the controller design was treated extensively elsewhere (Popović et al., 2003; Jonic et al., 1999; Došen and Popović, 2010).

Our ultimate goal is to provide a framework for safe and effective gait training. The platform for the application of our sensor-driven control is a specially designed walker called Walkaround, developed by D.B. Popović and his group in Belgrade (Veg and Popović, 2008). The walker provides safety against falls and injury, as well as postural assistance, while the sensor-driven FES assists the movement of the affected leg. The wheels of the walker are powered and the walking speed can be controlled, that is, set to the walking speed of the desired trajectory that was used for the simulations. The combined assistance by FES and Walkaround should motivate the user to recruit his or her residual voluntary capacities that were spared by the injury, ultimately resulting in a better, more natural and healthy walking pattern. In that sense, the control we are developing will not act on its own; it will be integrated with the voluntary control of the user that is supported by the walker. The adaptation process will go both ways; that is, the user will learn how to rely on the assistance provided by the system, and the controller will be iteratively adjusted and fine tuned to the user as he or she improves throughout the course of rehabilitation. The latter involves repeated assessments, recordings, and simulations, and the OptiWalk tool is again instrumental for the implementation of this step.

The interactions with the ground and the rest of the body are supplied as inputs for the simulation. However, it is to be expected that, during the actual application of FES, these signals deviate from the prescribed values used for the simulation. The assumption is that the user will be able to compensate for these deviations by using his or her voluntary control and the support provided by the walker. We do not consider muscle fatigue in the simulations, but we assume that the development of muscle fatigue will be monitored and compensated by online control. That is to say that the simulation calculates a pure feedforward control input, which will be modulated during the online operation by the feedback compensating for the disturbances (e.g., modeling uncertainties, muscle fatigue). Our model does not include biarticular muscles explicitly, since this would complicate the model and also the identification process. However, the contribution of biarticular muscles enters the model implicitly, through the properties of the equivalent actuators, since both mono- and bi-articular muscles will be activated in isometric and isokinetic stimulation experiments during the identification of the model parameters.

The next step is an extensive clinical evaluation of the methods presented here. The task is to compare the sensor-driven control using the stimulation profiles obtained through the simulations (subject-specific, gradually modulated intensity) with the sensor-driven control that uses conventional, hand-crafted profiles (ON/OFF stimulation adjusted through trial-and-error). This is an ongoing work and the preliminary results are quite promising.

REFERENCES

Anderson D.E., Madigan M.L., and Nussbaum M.A. (2007). Maximum voluntary joint torque as a function of joint angle and angular velocity: model development and application to the lower limb. *J Biomech* 40:3105–3113.

Audu M.L., Nataraj R., Gartman S.J., and Triolo R.J. (2011). Posture shifting after spinal cord injury using functional neuromuscular stimulation—a computer simulation study. *J Biomech* 44:1639–1645.

Bogataj U., Gros N., Kljajic M., and Acimovic-Janezic R. (1997). Enhanced rehabilitation of gait after stroke: a case report of a therapeutic approach using multichannel functional electrical stimulation. *IEEE Trans Rehabil Eng* 5:221–232.

Bryson A.E. (1999). *Dynamic Optimization*. Menlo Park, CA: Addison Wesley Longman.

Damsgaard M., Rasmussen J., Christensen S.T., Surma E., and de Zee M. (2006). Analysis of musculoskeletal systems in the AnyBody Modeling System. *Simul Model Pract Theory* 14:1100–1111.

Delp S.L., Anderson F.C., Arnold A.S., Loan P., Habib A., John C.T., Guendelman E., and Thelen D.G. (2007). OpenSim: open-source software to create and analyze dynamic simulations of movement. *IEEE Trans Biomed Eng* 54:1940–1950.

Došen S. and Popović D.B. (2008). Accelerometers and force sensing resistors for optimal control of walking of a hemiplegic. *IEEE Trans Biomed Eng* 55:1973–1984.

Došen S. and Popović D.B. (2009). Moving-window dynamic optimization: design of stimulation profiles for walking. *IEEE Trans Biomed Eng* 56:1298–1309.

Došen S. and Popović D.B. (2010). Biomechanical modeling for biologically inspired control of neural prostheses for walking. 7th IFAC Symposium on Modelling and Control in Biomedical Systems, Aalborg, Denmark.

Embrey D.G., Holtz S.L., Alon G., Brandsma B.A., and McCoy S.W. (2010). Functional electrical stimulation to dorsiflexors and plantar flexors during gait to improve walking in adults with chronic hemiplegia. *Arch Phys Med Rehabil* 91:687–696.

Erdemir A., McLean S., Herzog W., and van den Bogert A.J. (2007). Model-based estimation of muscle forces exerted during movements. *Clin Biomech* 22:131–154.

Gartman S.J., Audu M.L., Kirsch R.F., and Triolo R.J. (2008). Selection of optimal muscle set for 16-channel standing neuroprosthesis. *J Rehabil Res Dev* 45:1007–1018.

Ivanenko Y.P., Poppele R.E., and Lacquaniti F. (2006). Motor control programs and walking. *Neuroscientist* 12:339–348.

Jonic S., Jankovic T., Gajic V., and Popović D. (1999). Three machine learning techniques for automatic determination of rules to control locomotion. *IEEE Trans Biomed Eng* 46:300–310.

Khang G. and Zajac F.E. (1989). Paraplegic standing controlled by functional neuromuscular stimulation. I. Computer model and control-system design. *IEEE Trans Biomed Eng* 36:873–884.

Kojović J., Djurić-Jovičić M., Došen S., Popović M.B., and Popović D.B. (2009). Sensor-driven four-channel stimulation of paretic leg: functional electrical walking therapy. *J Neurosci Methods* 181:100–105.

Lyons G.M., Sinkjaer T., Burridge J.H., and Wilcox D.J. (2002). A review of portable FES-based neural orthoses for the correction of drop foot. *IEEE Trans Neural Syst Rehabil Eng* 10:260–279.

Malezic M., Bogataj U., and Gros N. (1987). Evaluation of gait with multichannel electrical stimulation. *Orthopedics* 10:769–772.

Malezic M., Hesse S., Schewe H., and Mauritz K.-H (1994). Restoration of standing, weight-shift and gait by multichannel electrical stimulation in hemiparetic patients. *Int J Rehabil Res* 17:169–179.

Mesci N., Ozdemir F., Kabayel D.D., and Tokuc B. (2009). The effects of neuromuscular electrical stimulation on clinical improvement in hemiplegic lower extremity rehabilitation in chronic stroke: a single-blind, randomised, controlled trial. *Disabil Rehabil* 31:2047–2054.

Musculographics Inc., SIMM software. Available at: http://www.musculographics.com/ (accessed April 15, 2013).

Neptune R.R., Clark D.J., and Kautz S.A. (2009). Modular control of human walking: a simulation study. *J Biomech* 42:1282–1287.

Pandy M.G. (2001). Computer modeling and simulation of human movement. *Annu Rev Biomed Eng* 3:245–273.

Park H. and Durand D.M. (2008). Motion control of musculoskeletal systems with redundancy. *Biol Cybern* 99:503–516.

Popović D., Stein R.B., Oguztoreli M.N., Lebiedowska M., and Jonic S. (1999). Optimal control of walking with functional electrical stimulation: a computer simulation study. *IEEE Trans Rehabil Eng* 7:69–79.

Popović D., Radulović M., Schwirtlich L., and Jauković N. (2003). Automatic vs hand-controlled walking of paraplegics. *Med Eng Phys* 25:63–73.

Riener R. (1999). Model-based development of neuroprostheses for paraplegic patients. *Philos Trans Roy Soc B Biol Sci* 354:877–894.

Riener R. and Edrich T. (1999). Identification of passive elastic joint moments in the lower extremities. *J Biomech* 32:539–544.

Riener R. and Fuhr T. (1998). Patient-driven control of FES-supported standing up: a simulation study. *IEEE Trans Rehabil Eng* 6:113–124.

Shue G., Crago P.E., and Chizeck H.J. (1995). Muscle-joint models incorporating activation dynamics, moment-angle, and moment-velocity properties. *IEEE Trans Biomed Eng* 42:212–223.

Stein R.B., Zehr E.P., Lebiedowska M.K., Popović D.B., Scheiner A., and Chizeck H.J. (1996). Estimating mechanical parameters of leg segments in individuals with and without physical disabilities. *IEEE Trans Rehabil Eng* 4:201–211.

Stevens P. and Hunsaker R.B. (2010). Recent findings regarding the efficacy of functional electrical stimulation in patients with chronic hemiplegia and multiple sclerosis: a narrative literature review. *J Prosthet Orthot* 22:166–171.

Strojnik P., Kralj A., and Ursic I. (1979). Programmed six-channel electrical stimulator for complex stimulation of leg muscles during walking. *IEEE Trans Biomed Eng* BME 26:112–116.

Veg A. and Popović D.B. (2008). Walkaround: mobile balance support for therapy of walking. *IEEE Trans Neural Syst Rehabil Eng* 16:264–269.

Winter D.A. (1990). *Biomechanics and Motor Control of Human Movement*. New York: Wiley Interscience.

Yamaguchi G.T. and Zajac F.E. (1990). Restoring unassisted natural gait to paraplegics via functional neuromuscular stimulation: a computer simulation study. *IEEE Trans Biomed Eng* 37:886–902.

Yan T., Hui-Chan C.W.Y., and Li L.S.W. (2005). Functional electrical stimulation improves motor recovery of the lower extremity and walking ability of subjects with first acute stroke: a randomized placebo-controlled trial. *Stroke* 36:80–85.

Zatsiorsky V.M. and Seluyanov V. (1983). The mass and inertia characteristics of the main segments of human body. In: *Biomechanics VIII-B* (Matsui H. and Kobayashi K., Eds.), pp. 1152. Champaign, IL: Human Kinetics

Zhang D. and Zhu K. (2007). Modeling biological motor control for human locomotion with functional electrical stimulation. *Biol Cybern* 96:79–97.

21

FOCUSED CHAPTER

ACTIGAIT®: A PARTLY IMPLANTABLE DROP-FOOT STIMULATOR SYSTEM

BIRGIT LARSEN AND ANDREI PATRICIU

Neurodan A/S, Aalborg, Denmark

SUMMARY

ActiGait® is a four-channel partly implantable neuroprosthesis to control ankle dorsiflexion during the swing phase of gait. Besides being safe and effective it was the goal of the developers to make a device that was easily handled by users with various levels of disabilities. To ensure optimal application of the ActiGait during market introduction, additional monitoring tools were applied and developed such as preoperative and postoperative magnetic resonance imaging (MRI), intraoperative fluoroscopy, and surface recordings of artifacts generated by the electrical stimulation of the nerve. By employing close monitoring during the establishment of new ActiGait clinical centers, ActiGait has been shown to be well accepted by users in several European countries.

INTRODUCTION AND BACKGROUND

Drop foot is the inability to effectively dorsiflex the ankle during the swing phase of walking. By stimulation of the common peroneal nerve the anterior

Introduction to Neural Engineering for Motor Rehabilitation, First Edition. Edited by Dario Farina, Winnie Jensen, and Metin Akay.

tibial and peronei muscles can be activated, which results in a balanced (eversion–inversion) dorsiflexion during swing phase of walking and at heel-strike in the beginning of stance phase. This technique of applying functional electrical stimulation for correction of drop foot was first proposed by Liberson and co-workers (Liberson et al., 1961) and is applicable when drop foot results from a central nervous system lesion such as hemiparesis following stroke, incomplete lesions of the spinal cord, or multiple sclerosis.

Timing of the stimulation in relation to the phases during gait is most frequently obtained by using a sensor positioned under the foot, but various systems for control of the timing of stimulation have been suggested (Rüterbories et al., 2010).

Electrical stimulation of the peroneal nerve is usually obtained by using surface electrodes applied on the skin over the nerve (Burridge, 2001; Kottink et al., 2004; Stein et al., 2010). However, application of surface electrodes for drop-foot stimulation poses insuperable challenges to some users due to difficulties in correct positioning of the electrodes, discomfort during stimulation, and skin irritation underneath the electrodes (Karsznia et al., 1990; Taylor et al., 1999).

To overcome the problems experienced by some users of surface drop-foot stimulation, the first implantable single-channel peroneal stimulators were developed and successfully implanted in the 1970s (Waters et al., 1975) and 1980s (Strojnik et al., 1987; Kljajic et al., 1992). Waters and co-workers (Waters et al., 1985) reported successful clinical results in 10 out of 16 implanted patients and long-term use in 7 patients (mean time 11.3 years). Several problems were observed using the implantable stimulator system, such as seroma formation and late infections presumably due to the wear of long leads crossing the knee joint, inconvenience and difficulties in operating the external components, and difficulties in obtaining a balanced dorsiflexion during walking. The latest led the authors to recommend dual or multichannel implantable systems in the future to make possible adjustment of eversion–inversion following surgery (for further reading see the review by Lyons et al., 2002).

Kenney and co-workers (Kenney et al., 2002) presented good clinical results of the first two patients who received a two-channel implantable drop-foot stimulator system. Epineural electrodes placed below the knee were used in this device to stimulate the superficial and deep peroneal nerve separately. The system, now known as StimuStep®, became commercially available in 2005 and has shown good clinical results (Kottink et al., 2007).

The first results of the application of a fully implantable closed-loop drop-foot stimulator system, Neurostep®, in one patient were published in 2005 (Hoffer et al., 2005). Two multichambered cuff electrodes were placed on the tibial and common peroneal nerve, respectively, to detect foot contact during gait by the tibial electroneurographic (ENG) signals and thereby time stimulation of the peroneal nerve to obtain ankle dorsiflexors during the swing phase of gait.

The development of the ActiGait system, a partially implantable drop-foot stimulator system, was first started at the Center for Sensory–Motor Interaction (SMI), Aalborg University, by Dr. Morten Haugland and co-workers. The goal was to improve the usability of drop-foot stimulation, in terms of efficient and safe handling of the system. To achieve this, the implant including both stimulator and electrode was positioned above the knee, thereby avoiding leads over the knee region. Further requirements were to minimize the necessity of precise positioning of the external transmission coil above the stimulator, to have wireless transmission from the actuator (heel switch) to the external control unit, and to have a multichannel cuff electrode to make balanced dorsiflexion possible during walking by differentiated stimulation of nerve fascicles within the common peroneal nerve.

In 1994 Dr. Morten Haugland co-founded Neurodan A/S as a spin-off company of Aalborg University, and the system became commercially available under the brand ActiGait. In 2005 Neurodan A/S became a member of the Otto Bock Group.

METHODOLOGY

Description of the System

ActiGait is intended for use by persons suffering from paresis of the ankle dorsiflexor muscles caused by damage to the central nervous system. The largest group of patients is patients with hemiparesis following stroke.

The ActiGait system (Fig. 21.1) consists of an implanted stimulator (4) that has a cuff electrode (5) mounted on the common peroneal nerve, a control unit (1) with an antenna (2) attached to the skin using an antenna fixture (6), and a heel switch (3).

ActiGait activates paretic muscles in the lower leg to lift the foot from the ground during the swing phase of gait. The swing phase is detected by a sensor

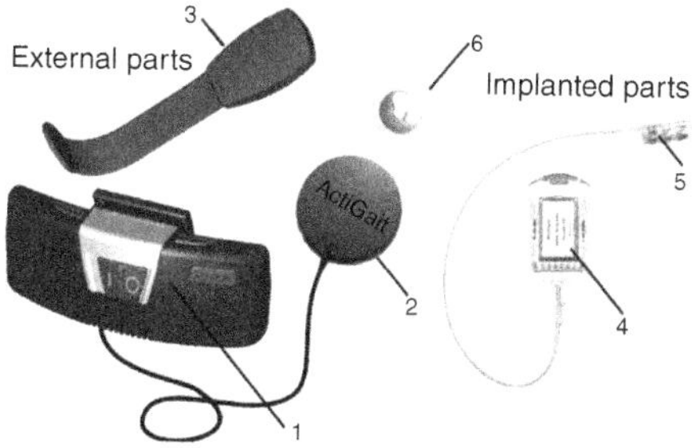

Figure 21.1. ActiGait system parts: 1, control unit; 2, antenna; 3, heel switch; 4, implanted stimulator; 5, cuff electrode; and 6, antenna fixture.

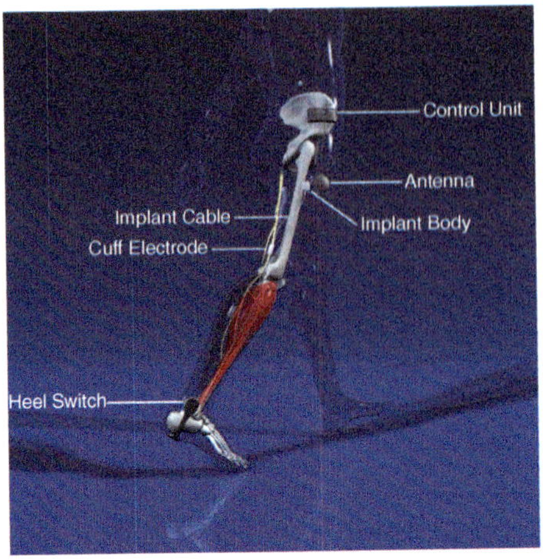

Figure 21.2. Location of the ActiGait system parts.

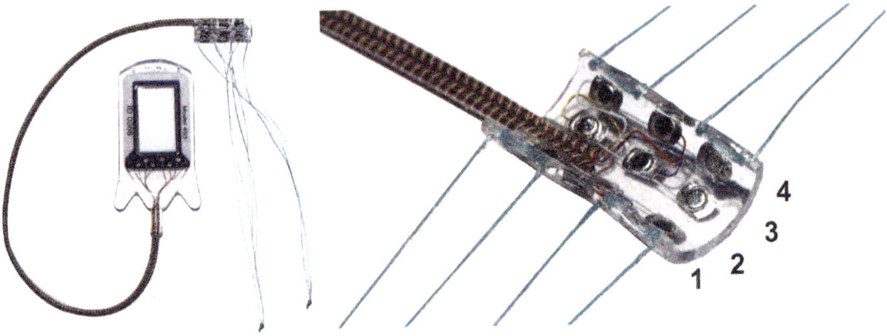

Figure 21.3. The implantable stimulator (left) and nerve cuff electrode held open with the sutures to show the internal platinum–iridium disks (right). The numbers indicate the locations of the stimulator channels.

(heel switch) placed under the heel, and signals are transmitted from the heel switch to a control unit positioned at the waist (Fig. 21.2).

From the control unit the stimulation pulse sequences and the necessary energy are delivered to the implant via a telemetric link by an antenna fixated on the thigh. As shown in Figure 21.3 the implant consists of three components: stimulator, electrode cable, and cuff electrode.

The implanted stimulator activates the common peroneal nerve just above the knee. The stimulator has four independent channels, controlling four separate sets of electrodes situated inside the cuff electrode. Stimulation is applied at different locations around the nerve depending on which channel is active. Thereby different fascicles in the peroneal nerve can be activated, resulting in separate activation of the tibial and peronei muscle groups of the lower leg making a balanced dorsiflexion possible.

The cuff electrode (Fig. 21.3, right) has 12 platinum–iridium disks, spot-welded to the electrode cable and embedded in silicone to form an electrode. To enable firm closing of the cuff around the nerve during implantation, four standard nondegradable multifilament polyester sutures are embedded in the cuff wall, leaving eight ends sticking out of the cuff wall. The sutures are tied pairwise across the opening in the cuff.

The cuff comes in five different sizes, to ensure a correct fit to the nerve of every patient. The cuff diameter should be 25–50% larger than the diameter of the nerve to allow space for any postsurgical swelling.

Surgical Procedure

The ActiGait implant is placed surgically, which requires experience in peripheral nerve operation and handling of active implantable medical devices.

The surgical procedure can be performed under general or spinal anesthesia. An incision about 8–9 cm long is made on the backside of the leg just above the knee fold, and a second incision is made laterally on the middle of the thigh. The common peroneal nerve is located and dissected free before the cuff electrode is placed on the nerve (Fig. 21.4).

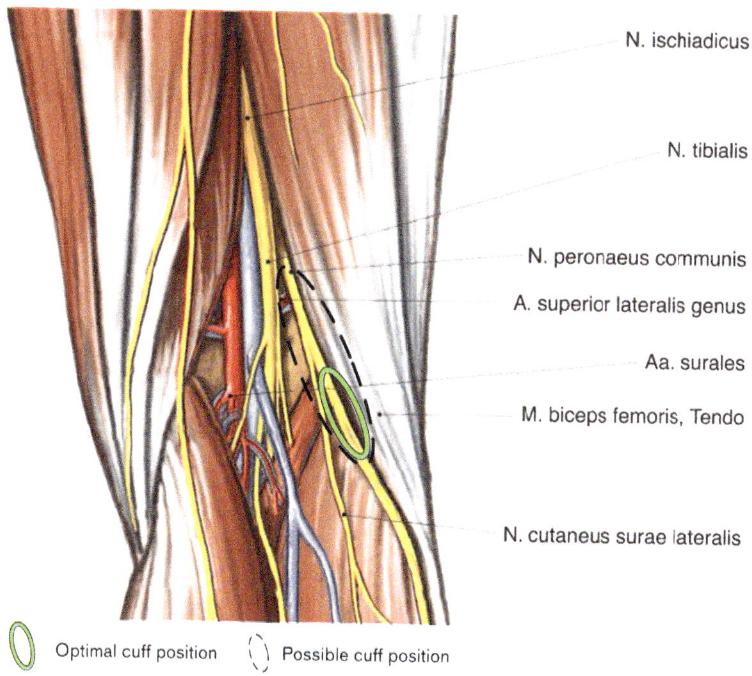

Figure 21.4. Anatomy of the peroneal nerve in the popliteal region. The green oval highlights the suggested region for optimal placement of the cuff electrode. The black oval (dashed line) highlights the possible position of the cuff.

A subcutaneous tunnel from one incision to the other is made under the skin using an insertion tool. This allows the insertion of the implant and passing of cuff and electrode cable through the tunnel. Once the cuff is in place and closed around the peroneal nerve by sutures, a flexion–extension test of the knee is performed to observe the position of the cuff and electrode cable and their movement relative to the nerve. Following an electrical test of the implant function to confirm dorsal flexion of the foot, the implanted stimulator body is fixed in a convenient location on the thigh, and finally both incisions are closed. The last step of the surgical procedure is to verify implant position, by means of fluoroscopic imaging.

Monitoring of the Surgical Procedure and Implant

Various monitoring tools have been used in order to:

1. Avoid inclusion of inappropriate patients (with peripheral nerve damage)
2. Guide surgeons in correctly positioning the cuff electrode during surgery
3. Verify the surgical procedure intraoperatively (to assure correct placement of the cuff electrode and the electrode cable path)
4. Investigate position of the implant postoperatively
5. Investigate correct electrical functioning of the implant

Nerve conduction velocity (NCV) measurements can be performed preoperatively and postoperatively to avoid inclusion of patients with reduced nerve conduction velocity due to peripheral nerve damage and to serve as baseline measure when comparing with the postoperative measurements. The latest is used to verify or disprove negative nerve affection following the implantation.

Magnetic resonance imaging (MRI) is the method of choice to investigate the anatomy of the popliteal region. It is used to find the separation level between the N. peroneus communis and N. cutaneus surae lateralis, which is useful in planning cuff location. MRI measurements can be used to estimate nerve diameter and thus the expected cuff size as well as the thickness of the subcutaneous adipose tissue (guidance in optimizing the stimulator body location). Postoperatively this imaging technique can be used to investigate the position of the implant, the location, rotation, and closure of the cuff electrode, the electrode cable path, and the size, path, and vascularization of the nerves involved. The postoperative imaging sequences are optimized to minimize implant heating and electrical nerve stimulation during scanning.

Fluoroscopic investigation is used to verify the ActiGait surgical procedure. It is performed using a fluoroscopic imaging system (C-arm) intraoperatively, following the implantation. Objectives are to determine the position of the cuff electrode measured from the tibial plateau and to visualize the electrode cable path in the vicinity of the cuff electrode. Images of the knee region are recorded in both antero–posterior and medial–lateral projections to locate the electrode and electrode cable position with extended and 90° flexed knee.

A measurement setup was developed to investigate the correct electrical functioning of the ActiGait implant. It is based on surface recordings, taken above the cuff electrode, of the artifacts generated by the electrical stimulation of the nerve. The measurements allow monitoring the electrical stimulation in various channels of the implant and the cuff–tissue impedance.

RESULTS

According to European Directive 90/385/EEC for Active Implantable Medical Devices (European Directive 90/385/EEC) a clinical evaluation was carried out to demonstrate safety and performance of the system before releasing ActiGait on the European market. Fifteen hemiplegic patients with drop foot following stroke within three Danish clinical centers participated in the study (Burridge et al., 2007). Results showed that the application of ActiGait, including surgical procedure and patients' use, was safe, and no serious device-related adverse events were reported during the study. Ankle dorsiflexion improved during stimulation in all participants and gait was improved. At follow-up approximately one-and-a-half years following implantation gait speed was improved by 19%, corresponding to a mean increase of 0.10 m/s ($p = 0.008$) during stimulation compared with before the operation (long-term effect). At the same time the orthotic effect of stimulation (comparison of gait speed with and without stimulation on the day) improved by 0.7 m/s ($p = 0.011$; Burridge et al., 2007). A users' satisfaction questionnaire posted anonymously approximately one-and-a-half years after implantation revealed that most of the participants used the system all day for the whole week and only one participant reported needing help to put on the external device (Burridge et al., 2008).

Several of the ActiGait users emphasized that the main result of using a drop-foot stimulator system such as ActiGait is the increased confidence during walking. It appears that when they are walking ActiGait users do not focus on their gait to the same extent as before. One ActiGait user expresses it like this: "My nose has got up when I walk in the street." However, increased motivation for physical and social activities are also stressed by several users. Comments such as: "now I don't have to plan if I want to go out and I don't need a wheelchair for longer trips" or "I don't have to ensure the possibility of parking when I go out with my family, I just walk from the parking ground wherever" both indicate a higher degree of freedom and independence of the ActiGait user.

Since clinical evaluation, ActiGait has been spread by controlled introduction to clinical centers in other European countries besides Denmark, and ActiGait is now used in six European countries.

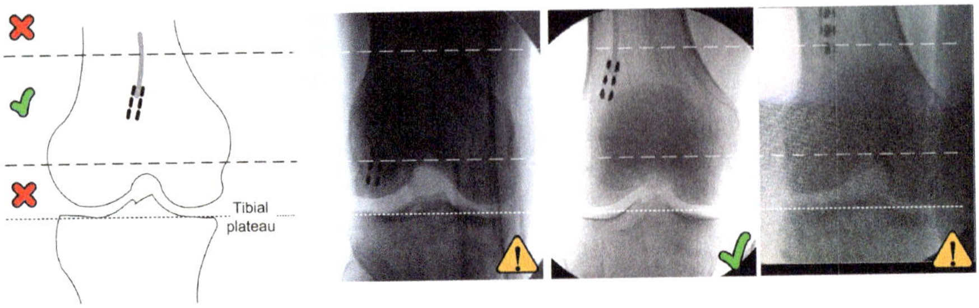

Figure 21.5. Cuff electrode locations in the vicinity of the knee joint: (left) drawing showing the recommended cuff placement; (right) fluoroscopic images showing examples of cuff locations—too low, correct, or too high. The two figures marked with ⚠ recommend close postoperative follow-up.

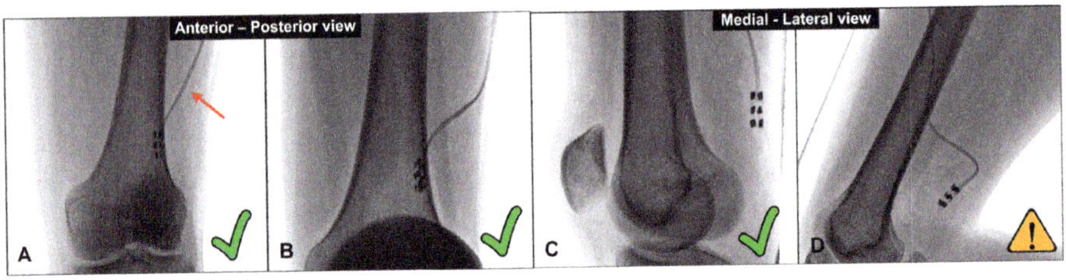

Figure 21.6. Electrode cable path (red arrow) in the vicinity of the cuff electrode as seen on fluoroscopic images. Observe the straight cable path with extended knee (panels A and C) and its bending during a 90° knee flexion (panels B and D). The figure marked with ⚠ recommends close postoperative follow-up.

By distributing the ActiGait system to other countries and clinical cultures, new valuable knowledge has been obtained. The spread of ActiGait outside the more standardized environment of a clinical evaluation has challenged the system and confirmed the importance of thorough knowledge transfer to new clinical ActiGait centers during start-up—and of close monitoring to ensure satisfactory implementation of the treatment.

Close monitoring of the implant was achieved using the methods as described. Fluoroscopic images showing examples of optimal and suboptimal cuff and electrode cable positions are presented in Figure 21.5 and Figure 21.6. The optimal position for the cuff electrode depends on the mechanical properties of the implant (the ActiGait cuff electrode and electrode cable) and electrode cable path in conjunction with the fragile nerve structure and the dynamics of the knee region. In cases of very low or very high cuff placement as well as strong electrode cable bending close to the cuff, it is recommended to have close postoperative follow-up.

The MRIs in Figure 21.7 show the optimal electrode position in the knee region and correct cuff-to-nerve ratio.

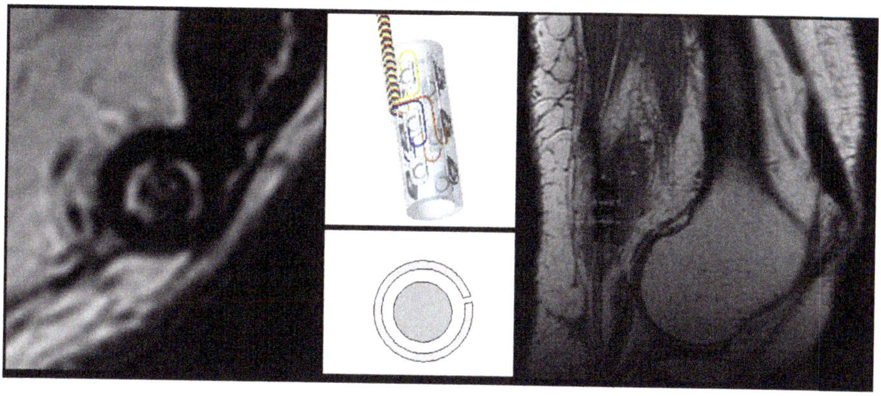

Figure 21.7. T_2 weighted MRIs of the cuff electrode around the peroneal nerve. Left axial MRI and lower drawing show a snugly fitting cuff around the peroneal nerve. The right sagittal MRI shows the position of the cuff electrode in the fossa poplitea region. The ActiGait cuff electrode is illustrated in the upper drawing.

DISCUSSION

The challenges of going to a broader market resulted in a more detailed description of the surgical procedure and in the development of monitoring tools to guide and confirm correct application of ActiGait.

The following steps were identified as being essential to a successful implantation: correct knee incisions allowing a low cuff placement; a longitudinal tunneling to allow the electrode cable to follow the path of the peroneal nerve close to the cuff electrode; and a tension-free electrode cable due to a correct distance between incisions. This highlights the importance of having a flexible electrode cable and a robust cuff electrode to withstand the high mechanical stress in the knee joint region and to ensure long-term reliability and stability. The placement of the cuff close to the knee joint also allows an optimal cuff-to-nerve ratio and in most cases avoids the presence of N. cutaneus surae lateralis inside the cuff. Including the sensory branch in the electrode may lead to decreased effectiveness of the stimulation and discomfort in the foot. However; this goal is sometimes a surgical challenge due to the anatomical variability of the separation level in different patients. A snugly fitting cuff electrode keeps the stimulation low and avoids excessive scar tissue ingrowths and body-fluid accumulation as seen in the MRIs. These are all factors that deteriorate an optimal electrical stimulation and lead to increased charge injection levels.

Presurgical MRI has been shown to be a useful tool in planning knee incision and implantation strategy, especially in the early phase of deployment of the surgical procedure to a new clinical center.

Postsurgical MRI was successfully used to view the cuff electrode position, location, and rotation, and to show how these three correlate with the ActiGait channel setting that is required to obtain the optimal gait pattern.

Intraoperative fluoroscopy has been shown to be a valuable verification tool of ActiGait implantation and is now part of the surgical procedure.

CONCLUSION

The clinical evaluation of ActiGait and subsequent surveillance following market introduction have shown that the ActiGait drop-foot stimulator system is safe and makes a balanced foot lift possible during swing phase of gait in hemiplegic patients with drop foot following stroke. Furthermore, the goal of the developers to achieve a system that is well accepted by users is reached.

Moreover, experiences with the ActiGait have shown that thorough and targeted knowledge transfer to new clinical ActiGait centers, application of suitable monitoring tools, and close monitoring are all decisive in bringing a medical device such as an implantable drop-foot stimulator system successfully on the market.

REFERENCES

Burridge J., Haugland M., Larsen B., Pickering R., Svaneborg N., Iversen H.K., Brøgger Christensen P., Haase J., Brennum J., Sinkjaer T. (2007). A phase II trial to evaluate the ActiGait implanted drop-foot stimulator in established hemiplegia. *J Rehabil Med* 39(3):212–218.

Burridge J., Haugland M., Larsen B., Svaneborg N., Iversen H.K., Brøgger Christensen P., Pickering R., Sinkjaer T. (2008). Patients' perceptions of the benefits and problems of using the ActiGait implanted drop-foot stimulator. *J Rehabil Med* 40:873–875.

Burridge J.H. (2001). Does the drop-foot stimulator improve walking in hemiplegia? *Neuromodulation: Technol Neural Interface* 4(2):77–83.

European Directive 90/385/EEC for Active Implantable Medical Devices. Available at http://ec.europa.eu/enterprise/policies/european-standards/documents/harmonised-standards-legislation/list-references/implantable-medical-devices/index_en.htm.

Hoffer J.A., Baru M., Bedard S., Calderon E., Desmoulin G., Dhawan P., Jenne G., Kerr J., Whittaker M., and Zwimpfer T.J. (2005). Initial results with fully implanted Neurostep™ FES system for foot drop. 10th Annual Conference of the International FES Society Montreal, Canada.

Karsznia A., Dillner S., Ebefors I., and Lundmark P. (1990). Why patients use or reject a peroneal muscle stimulator? In: *Proceedings from the 10th Dubrovnik Meeting 1990. Advances in External Control of Human Extremities (ECHE). Proceedings* Vol. I–X Dubrovnik 1962–1990 (Popovic D., Ed.), pp. 251–260.

Kenney L., Bultstra G., Buschman H., Taylor P., Mann G., Hermens H., Holsheimer J., Nene A., Tenniglo M., van der Aa H., and Hobby J. (2002). An implantable two channel drop foot stimulator: Initial clinical results. *Artif Organs* 26(3):267–270.

Kljajic M., Malezic M., Acimovic R., and Vavken E. (1992). Gait evaluation in hemiparetic patients using subcutaneous peroneal electrical stimulation. *Scand J Rehab Med* 24:121–126.

Kottink A.I., Hermens H., Nene A., Tenniglo M., van der Aa H., Buschman H., and Ijzerman M. (2007). A randomised controlled trial of an implantable 2-channel peroneal nerve stimulator on walking speed and activity in poststroke hemiplegia. *Arch Phys Med Rehabil* 88:971–978.

Kottink A.I.R., Oostendorp L.J.M., Buurke J.H., Nene A.V., Hermens H.J., and Ijzerman M.J. (2004). The orthotic effect of functional electrical stimulation on the improvement of walking in stroke patients with a dropped foot: a systematic review. *Artif Organs* 28:577–586.

Liberson W.T., Holmquest H.J., Scott D., and Dow M. (1961). Functional electrotherapy in stimulation of the peroneal nerve synchronized with the swing phase of gait in hemiparetic patients. *Arch Phys Med Rehabil* 42:202–205.

Lyons G.M., Sinkjær T., Burridge J.H., and Wilcox D.J. (2002). A review of portable FES-based neural orthoses for the correction of drop foot. *IEEE Trans Neural Syst Rehabil Eng* 10(4):260–279.

Rüterbories J., Spaich E.G., Larsen B., and Andersen O.K. (2010). Methods for gait event detection and analysis in ambulatory systems. *Med Eng Phys* 32:545–552.

Stein R.B., Everaert D.G., Thompson A.K., Chong S.L., Whittaker M., Robertson J., and Kuether G. (2010). Long-term therapeutic and orthotic effects of a foot drop stimulator on walking performance in progressive and nonprogressive neurological disorders. *Neurorehabil Neural Repair* 24(2):152–167.

Strojnik P., Acimovic R., Vavken E., Simic V., and Stanic U. (1987). Treatment of drop foot using an implantable peroneal underknee stimulator. *Scand J Rehab Med* 19:37–43.

Taylor P.N., Burridge J.H., Dunkerley A.L., Lamb A., Wood D., and Norton J.A. (1999). Patients' perceptions of the Odstock dropped foot stimulator (ODFS). *Clin Rehabil* 13(5):439–446.

Waters R.L., McNeal D.R., and Perry J. (1975). Experimental correction of footdrop by electrical stimulation of the peroneal nerve. *J Bone Joint Surg Am* 57(8):1047–1054.

Waters R.L., McNeal D.R., Faloon W., and Clifford B. (1985). Functional electrical stimulation of the peroneal nerve for hemiplegia. *J Bone Joint Surg Am* 67:792–793.

22

FOCUSED CHAPTER

SELECTIVITY OF PERIPHERAL NEURAL INTERFACES

WINNIE JENSEN AND KRISTIAN RAUHE HARREBY

Center for Sensory–Motor Interaction, Department of Health Science and Technology, Aalborg University, Aalborg, Denmark

SUMMARY

The aim of a peripheral nerve interface is to provide a connection between the peripheral nerve cells in our bodies and neuroprosthetic devices. Many neural prosthesis applications rely on our ability to interface with individual or specific populations of nerve fibers, typically motor or sensory. It is therefore important to consider and assess how well different types of neural interfaces selectively access specific nerve fibers. This chapter first provides a background for understanding the importance of selectivity by introducing two examples of neural prosthesis applications, and the state of the art of methods for quantitatively assessing selectivity is reviewed. Second, four success criteria for evaluating selectivity are defined, and specific results from an experimental study are provided that compare the selectivity performance of two intraneural peripheral electrodes (tfLIFE and the TIME interfaces) placed in the median nerve in the forelimb of pigs. It is shown that the design and placement of the electrode must be carefully considered before choosing a neural interface for a specific neural prosthesis application.

Introduction to Neural Engineering for Motor Rehabilitation, First Edition. Edited by Dario Farina, Winnie Jensen, and Metin Akay.

INTRODUCTION AND BACKGROUND

Need for Selectivity of Peripheral Neural Interfaces

In normal movement control, motor axons send electrical signals from the central nervous system to the skeletal muscles. The precise control of movement is dependent on sensory feedback and information on the position of the body and limbs is conveyed to the central nervous system from receptors located in the skin, joint capsules, muscle spindles, and tendons. The peripheral nerves are cable-like structures consisting of bundles of axons located within fascicles (see Fig. 22.11 in the Results section). Most peripheral nerves are mixed in such a way that they contain motor, sensory and autonomic axons, and very few peripheral nerves in the body are purely sensory or motor. Individual axons are of varying diameters ranging from 0.5 to 15 μm (according to the Erlanger–Gasser classification of nerve fibers; Dudel et al., 1978). For more detailed information on the anatomy of the peripheral nerves the reader is referred to other chapters within this book.

Injury to the spinal cord may cause permanent loss of voluntary motor function and sensation below the level of the lesion. The neurological impairments are usually permanent and the deficits in motor function and sensation depend on the level and completeness of the lesion. For instance, injury to the thoracic (T1–T12) or the lumbar (L1–L5) regions of the spinal cord may result in loss of motor function in the lower extremities. A peripheral neural interface in a neural prosthesis application can provide a communication link to the peripheral nerves to regain control of body functions. In essence, a neural interface can work as a two-way communication line. It can be used either passively to "listen in" on the neural activity (i.e., the task of sensing or recording from axons) or actively to "deliver information" to the nervous system (i.e., the task of stimulating or activating axons). **Selectivity** is defined as the ability of a peripheral nerve interface to access specific fibers where the fibers of interest will depend on the application. This chapter focuses on the stimulation selectivity of peripheral nerve electrodes. Two examples are briefly introduced in the next section with the targeted fibers being the sensory and the motor axons, respectively.

TWO NEURAL PROSTHESIS APPLICATIONS WHERE STIMULATION SELECTIVITY PLAYS AN IMPORTANT ROLE

Example 1: Providing Sensory Feedback—Activation of Afferent Sensory Fibers

Amputation of a limb may result from trauma or surgical intervention. The amputation of a limb is usually followed by a feeling that the missing body part is still present (Hunter et al., 2003). In 50–80% of amputees, phantom limb pain (PLP) develops in the lost limb (Ephraim et al., 2005). Both periph-

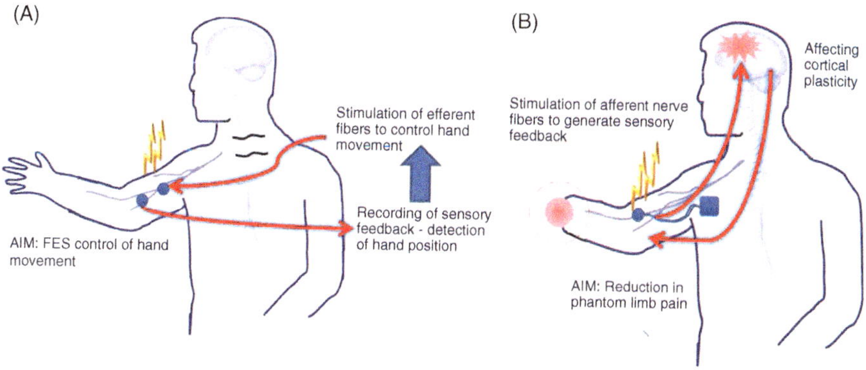

Figure 22.1. Schematic illustration of two applications where detailed control or access to the peripheral nerve is important. (A) Closed-loop functional electrical stimulation (FES) control of paralyzed limb from spinal cord injuries. The hand movement is decoded from activity recorded from the afferent fibers and translated into movement through electrical stimulation of efferent, peripheral nerve fibers. (B) Reduction of phantom limb pain in amputees. Stimulation of afferent fibers in the nerve to generate artificial sensory feedback. The hypothesis is that the stimulation will induce cortical plasticity and assist in the relief of phantom limb pain (see, e.g., Jensen et al., 2010).

eral and central factors have been suggested as the possible cause of PLP (Flor et al., 2006). However, it is still not completely understood why the pain occurs, and there are no fully effective treatments. Several studies have demonstrated the favorable effect of providing sensory feedback related to the missing limb to alleviate PLP. For example, patients who used a myoelectric prosthesis (Lotze et al., 1999; Rossini, 2010) or used daily discrimination training of surface electrical stimuli applied to the stump experienced significant reduction of PLP (Flor et al., 2001). Other studies have shown that intrafascicular, electrical stimulation of the severed peripheral nerves proved to be capable of eliciting tactile or proprioceptive sensations by implanted thin-film longitudinal intrafascicular electrodes (tfLIFE) in human subjects (Dhillon and Horch, 2005; Dhillon et al., 2005; Rossini, 2010). Based on these previous findings, a novel system for alleviating PLP was therefore proposed and is under development by a group of European researchers (www.project-time.eu); see Figure 22.1B. The system is based on providing sensory feedback through the transverse, intrafascicular multichannel electrode (TIME) where the aim is to exploit the favorable effect of sensory feedback to alleviate PLP. However, it is assumed by the group that a much more detailed control of induced sensory modalities (e.g., proprioception), sensation location (where the sensation is induced), or sensation magnitude (how strong the induced sensation is) will be needed compared with the currently available surface stimulation electrodes (see, e.g., Geng et al., 2011). To deliver a wider range of, for example, sensory modalities, it is therefore the goal to be able to selectively activate afferent sensory fibers within the nerve.

Example 2: FES Control of Paralyzed Limbs—Activation of Efferent Motor Fibers

In functional electrical stimulation (FES) the aim is to electrically activate the paralyzed muscles in a controlled way to restore motor function (Prochazka, 1993). FES has proved to be a useful tool for rehabilitation of movement provided that the peripheral nerves and the muscles have remained intact below the lesion (Mortimer, 1981). Functional electrical stimulation can be applied to the subject using an open-loop (feedforward) or a closed-loop (feedback) control strategy. The open-loop systems have been proven to work but muscle fatigue due to excessive stimulation is often observed and the research has turned in the direction of using closed-loop controlled systems. However, the closed-loop controlled FES systems are dependent on feedback from the part of the body being controlled, and the availability of sensors to provide a reliable feedback signal from the controlled limb or organ is therefore essential.

The principle in a neural prosthesis application for closed-loop control of hand function is shown in Figure 22.1A. The recorded afferent activity is fed back to a control box in which the signals are processed and information on, for example, hand position or speed is extracted. Based on this information, efferent nerve fibers are stimulated to activate appropriate muscle groups to achieve, for example, joint flexion or extension. A movement requiring both flexion and extension of the joint will require alternating recruitment and relaxation of an agonist–antagonist muscle pair working around the joint.

To use FES closed-loop control to achieve hand movement, it is therefore necessary to be able to selectively activate specific motor fibers within the nerve.

There are a number of natural sensors whose afferent activity may be appropriate for achieving a feedback signal to be used in closed-loop control of paralyzed muscles. Among these are the following:

1. The sensors receiving information from the body itself. The sensors are located primarily in the muscles (muscle spindles), tendons (Golgi tendon organs), joints, and ligaments and they carry information on movement and orientation of the body in space (proprioception).
2. The sensors receiving information through the body's interaction with the surroundings. These sensors are primarily located in the glaborus skin in humans (e.g., Merkel's cells, Ruffini endings, Meissner corpuscles, and Pacinian corpuscles). The sensors carry information about pressure, touch, vibration, tension, and stretch.

Today sensory information has been extracted from peripheral nerves containing afferent fibers from the skin (Haugland et al., 1999; Inmann et al., 2001; Sinkjær et al., 2003) or muscles (Jensen et al., 2002, 2001, 2000) or bladder (Jezernik et al., 2001, 2000). In the first two cases the recorded afferent activity was used to detect distinct events during movement. For instance, Haugland

and co-workers used afferent activity from the sural nerve to detect "on–off" heel-strike during walking in a human for correction of drop foot (Haugland and Sinkjær, 1995).

ASSESSMENT OF SELECTIVITY

The stimulation selectivity of different designs of peripheral nerve interfaces has been quantified in various ways; see Table 22.1. As described previously, what needs to be evaluated is the ability to interface a distinct population of neural fibers with similar properties, for example, fibers innervating a specific muscle or inducing a specific type of sensation without activating other nearby, nontarget populations. Stimulation selectivity is traditionally assessed by placing the neural interface around or inside the peripheral nerve and delivering electrical stimulation while measuring one or more parameters giving insight into which axons of the nerve are being activated.

In the ideal situation, one would know exactly which axons were activated with a given stimulation and which were not. Such detailed descriptions have been estimated in computer simulations (Raspopovic et al., 2011). However, *in vivo* studies are limited to assessing parameters practical to measure within the framework of the experiment. For example, to assess the ability to evoke motor fiber selectively, the evoked muscle forces of individual muscles have been measured in acute experiments (Veraat et al.,1993; Branner et al., 2001). This approach is especially relevant when the target application for the tested neural interfaces is an FES application where muscle force or movement is the desired output. However, placing sensors in the body to record individual muscle forces is a highly invasive procedure and requires detaching muscles from the bones on which they anchor. Another, and less invasive, method for quantifying evoked muscle activation includes recording of an electromyogram (EMG) from the target muscles (Kundu et al., 2010; Yoo et al., 2004; Deurloo et al., 2000; Badia et al., 2011; Polasek et al., 2007). Measuring EMG is a more indirect measure of muscle activation than the muscle force. However, these measurements do not require detachment of the muscles and may be carried out in chronic animals (Kundu et al., 2010) and acute human experiments (Schiefer et al., 2010). Noninvasive methods for accessing muscle activity have also been described: Branner et al. (2004) used palpation and Grill and Mortimer (1996) measured the torque around the angle joint of a rabbit using a custom-made setup. Palpation requires an experienced examiner with a good understanding of anatomy and physiology and it may be difficult to differentiate activity in closely located and agonist muscles. In addition it is not possible to objectively quantify the level of activation in individual muscles but only the threshold activation. In typical applications, threshold activation will not be sufficient for providing an assessment of the functional activation. Thus, results would likely lead to an overestimation of the selectivity of a tested peripheral nerve interface compared with what

TABLE 22.1. Examples of Studies Evaluating the Performance of Peripheral Neural Interfaces[a]

Author(s)	Interface Design (No. Contacts)	Nerve Model, (Acute, Chronic)	Measured Parameter(s) (No. DOFs)	Assessment of Performance, SI Definition	Comments
Veraat et al. (1993)	Cuff (4 + 4 + 4)	Sciatic nerve in cat (acute)	Muscle force (4)	$SI_j = \frac{\mu_j}{\sum_{i=1}^{N} \mu_i}$	Pure selectivity, general activation level is ignored. When all muscles are activated at the same time, the SI is biased toward $1/N$.
Grill and Mortimer (1996)	Cuff (4 + 4 + 4)	Sciatic nerve in cat (acute)	Ankle joint torque (two axes)	Torque plot separation between activation of different fascicles (current intensity).	Functionally relevant measure. Torque plots are more complex to assess than recruitment curves as muscle recruitment may counteract.
Deurloo et al. (2000)	Cuff (5 or 6)	Sciatic nerve in rabbit (acute)	EMG, V_{pp} (4/2 included in SI)	$SI_j = \frac{\mu_j}{\sum_{i=1}^{N} \mu_i}$	Same type of definition as Veraat et al. (1993).
Branner et al. (2001)	USEA (10 × 10)	Sciatic nerve of cat (acute)	Muscle force, F_{max} and F_{area} (up to 6 muscles at a time)	Recruitment curves were fitted to sigmoid curve. Electrode activation map Excitation overlap between contacts calculated.	Assessment of muscle force is an invasive method only suited for acute studies.

Leventhal and Durand (2004)	FINE cuff (12 – 18)	Sciatic nerve in cat (chronic)	Torque (3 axes)	The selective activation of different fascicles was assessed based on the torque trajectories.	Noninvasive and functionally relevant assessment. However, torque plots are more complex to assess than recruitment curves.
Branner et al. (2004)	USEA (7 × 10 – 10 × 10)	Sciatic nerve of cat (chronic)	Palpation (3 muscle groups)	Electrode activation map	Noninvasive. Only activation thresholds were quantified. May be a poor indicator of selective activation at a functionally relevant activation level.
Yoo et al. (2004)	FINE cuff (12 + 12 + 12)	Hypoglossal nerve in dog (acute)	EMG (4) or ENG (3), V_{pp}	$SI_j = \frac{\mu_j}{\sum_{i=1}^{N} \mu_i}$ When $\mu_j \geq 0.7$ and $\sum_{i=1}^{N\|j} \mu_i \leq 0.3$	ENG and EMG parameters result in similar SI values. Same type of definition as Veraat et al. but with additional restrictions on both target and nontargets.
Polasek et al. (2007)	Cuff (4)	Human upper extremities, acute	EMG, area under curve (max. 4)	$SI_j = \mu_j$ When $\max(\mu_{i,\{1,2\ldots N\|j\}}) < 0.10$	Simple and functional definition

(*Continued*)

TABLE 22.1. (***Continued***)

Author(s)	Interface Design (No. Contacts)	Nerve Model, (Acute, Chronic)	Measured Parameter(s) (No. DOFs)	Assessment of Performance, SI Definition	Comments
Schiefer et al. (2010)	FINE (8)	Femoral nerve of human, acute	EMG, area under curve (6)	$SI_j = \mu_j - \frac{\sum_{i=1, j \neq i}^{N} \mu_{N,i}}{N-1}$ Where max $(\mu_{n,\{1,2...N \setminus j\}}) \leq 0.10$ $SI_d = \frac{\sum_{i=1}^{N} max(SI_j)}{N}$	Same SI as Raspopovic et al. (2011). Same restrictions as Polasek et al. (2007) Cost function of the SI was modified to differentiate between agonists and nonagonists.
Rossini (2010)	tfLIFE (8)	Median and ulnar nerve of human, chronic	Perceived sensation	Map of perceived localization of evoked sensation	Noninvasive assessment. Complex to quantify selectivity and to perform such assessment in an animal model.
Badia et al. (2011)	TIME (10), LIFE (8), and Cuff (4 + 4 + 4)	Sciatic nerve in rat, acute	EMG, $V_{p,negative}$(3)	$SI_j = \frac{\mu_{n,j}}{\sum_{i=1}^{N} \mu_{n,i}}$ $SI_d = \prod_{i=1}^{N} \max(SI_i)$ If $SI_i < 1/N$, then $SI_i = 1/N$	SI is the same type of definition as Veraat et al. (1993). Device selectivity is very punishing for failing to selectively activate single muscles. The more parameters assessed (e.g., channels of EMG), the higher the risk of getting a small SI_d.

Nielsen et al. (2011)	Cuff (1 + 4 + 1)	Sciatic nerve in rabbit (acute)	ENG (3), V_{pp}	$SI_j = \frac{\mu_j}{\sum_{i=1}^{N} \mu_i}$ $SI_d = \frac{\sum_{i=1}^{N} \max\{SI_i\}}{N}$ When $SI_i > 0.70$	Same type of definition as Veraat et al. (1993). When calculating for a whole device target, innervation needed to be > 0.7 to be considered.
Raspopovic et al. 2011	TIME (10)	Computer model of the sciatic nerve of a rat (acute) (no fibrous tissue)	Axon populations (3) Percentage of activated axons	$SI_j = \mu_j - \frac{\sum_{i=1, j \neq i}^{N} \mu_i}{N-1}$	This SI ranges from −1 to +1. Values < 0 indicate than on average other muscles are activated more than the target, j. Obtaining the percentage of axons activated would not be possible in nonsimulated experiments. Selectivity index cannot be higher than the fraction in which μ_j is activated.

[a]All studies sought to describe how well the tested interfaces were able to exclusively activate subpopulations of nerve fibers. The defined selectivity index might be related to, for example, individual muscles (j) or the whole interface including all contacts (d). The following nomenclature has been used: j = target; SI_j = selectivity index for a single target parameter; SI_d = selectivity index for a whole device; N = number of independent parameters quantified; for example, number of EMG channels or muscle forces measured; μ = normalized parameter, for example, EMG, ENG, muscle force.
EMG = electromyogram; ENG = electroneurogram; FINE = flat interface nerve electrode; LIFE = longitudinal intrafascicular electrode; SI = selectivity index; tfLIFE = thin-film longitudinal intrafascicular electrode; TIME = transverse intrafascicular multichannel electrode; USEA = Utah slanted electrode array.

would be the case in a real application. Assessment of the torque produced around a joint (Grill and Mortimer, 1996; Leventhal and Durand, 2004) is a functionally relevant measure, but the method has several drawbacks. For example, the effect of the agonists is lumped together and it is difficult to determine the maximal possible torque because of spillover activation of antagonists.

A more direct way to investigate selectivity is to assess the level of nerve activation after dividing the nerve into several branches (Nielsen et al., 2011; Yoo et al., 2004). Compared with the EMG-based methods the drawback of this approach is that that the electroneurogram (ENG) is an order of magnitude lower than that of the EMG. Also the time between stimulation and the ENG compound action potential is shorter and thus the ENG would be more prone to be affected by stimulation artifacts. The advantage is that the evoked nerve activity is assessed more directly as it is not modified by the biological properties of transferring the nerve activity to muscle contraction. However, when comparing both ENG and EMG-based selectivity assessments, Yoo and co-workers showed that the two methods resulted in similar levels of selectivity when testing a flat interface nerve electrode (FINE) on the hypoglossal nerve in dogs (Yoo et al., 2004).

In humans the selective activation of sensory fibers can be quantified noninvasively by simply asking the subject to describe the sensation type, location, and strength induced (Rossini, 2010; Dhillon and Horch, 2005). The strength of the sensory perception is typically quantified by the subject using the visual analog scale (VAS); location is assessed by marking an area on a drawing; type is quantified by the subject describing the sensations by a set of predefined words.

When a measure has been selected for quantifying nerve activation, the fraction or amount of activation can be quantified in various ways. Typically quantification is assessed by using the maximal peak (e.g., the peak force or the peak-to-peak voltage; Branner et al., 2001; Yoo et al., 2004; Deurloo et al., 2000; Nielsen et al., 2011) or by assessing the information from the whole duration of the evoked responses by calculating the area under the curves of the measured parameter (Branner et al., 2001; Polasek et al., 2007; Schiefer et al., 2010). If recordings are noisy, using the area under the curve may be preferable since noisy spikes may have a great impact on the estimated peak EMG. On the other hand, if there is some overlap between the evoked recording and stimulation artifacts, the peak-to-peak measurement may provide more reliable results as it is more robust to a nonzero baseline.

How to Quantify Selectivity

After implanting the neural interface, electrical stimulation is applied to the electrode's active sites while recording one or more of the parameters described previously. For ease of explaining the various quantification methods, it is assumed in the following that stimulation selectivity is assessed by measuring the muscle EMG activity, although the formulas are valid for any parameter compared.

By varying the stimulation intensity (either the stimulation amplitude or the pulse width), recruitment curves are estimated for the measured parameters. Based on these recruitment curves a **selectivity index** (SI) is defined to quantify how selectively the target is activated compared with the activation of all nontargets. A commonly used index is defined as follows (Deurloo et al., 2000; Badia et al., 2011; Nielsen et al., 2011):

$$SI_{RF,j} = \frac{EMG_j}{\sum_{i=1}^{N} \mathrm{EMG}_i}$$

Here EMG_j refers to the activation level (0, . . . , 1) of target muscle j, and N is the number of muscles from which the EMG is measured. The SI_{RF} (selectivity index, recruitment fraction) represents the fraction of the target muscle activity compared with the total evoked activity in the group of N assessed muscles (including the target and all nontarget muscles). The value of SI_{RF} ranges between 0 and 1 where 0 corresponds to the jth muscle not being selectively activated and 1 corresponds to it being the only muscle active. The SI_{RF} does not take the level of recruitment into account. Thus, a value close to 1 does not indicate whether the muscle was recruited at a functionally relevant level. To compensate for this some authors have implemented restrictions requiring that a minimum recruitment level must be reached. For example, a 70% EMG activation threshold was used by Yoo and co-workers (Yoo et al., 2004). Polasek and co-workers also set activation thresholds for the nontarget muscles (Polasek et al., 2007).

A drawback of the SI_{RF} is that it is biased such that the smaller the number of muscles accessed, the higher the index will tend to be. For example, if two EMGs are assessed and both the target and the nontarget muscles are activated at 50% (i.e., no selectivity), the index will assume a SI_{RF} value of $1 / (1 + 1) = 0.5$, whereas if one target and four nontarget muscles are activated to the same level, it will result in a SI_{RF} value of $1 / (1 + 4) = 0.2$.

An SI definition taking the functional activation level into account is the SI_{RwC} (recruitment with a cost), which is defined as

$$SI_{RwC,j} = EMG_j - \frac{\sum_{i=1, j \neq i}^{N} EMG_i}{N-1}$$

The SI_{RwCj} represents the recruitment of the target muscle activity from which a cost function is subtracted. The **cost function** is defined as the mean activation of all nontarget muscles (Raspopovic et al., 2011; Schiefer et al., 2010). The index can never increase beyond the recruitment of the target muscle. Thus, a high value (close to 1) will indicate a functionally relevant selectivity.

Another index taking the functional level into account is the SI_{RR} (restricted recruitment), which is simply defined as the recruitment of the target (Polasek et al., 2007):

$$SI_{RR,j} = EMG_{n,i}$$

With a limitation on the activation threshold for nontarget muscles:

$$\max(EMG_{n,\{1,2..N|j\}}) < 0.10$$

Thus, this index represents the recruitment of the target with a restriction on the level of activation of nontargets. Dependent on the application of the neural interface being tested, the cost function or limitations on nontargets could be graded or weighted. For example, in a FES application the activation of target antagonists might be given a higher cost or stricter restrictions than agonists (Schiefer et al., 2010).

How to Quantify Selectivity for a Whole Neural Device

The previous selectivity indices describe how well individual targets may be recruited by stimulating individual contact sites. However, when comparing different types of neural interfaces, it is relevant to quantify and compare selectivity indices for the whole device, that is, across all contacts. A "device selectivity index" can be defined based on the SIs previously defined. One intuitive approach is to calculate the average of the maximum obtained SI from each of the muscles assessed (Schiefer et al., 2010):

$$SI_{M,D} = \frac{\sum_{i=1}^{N} \max(SI_i)}{N}$$

Another approach is to emphasize the ability to achieve some selectivity across all parameters assessed (Badia et al., 2011):

$$SI_{P,D} = \prod_{i=1}^{N} \max(SI_i)$$

If $SI_i < 1/N$, then $SI_i = 1/N$.

When taking the product of all the SI_i, poor SIs will reduce the value of the $SI_{P,D}$ significantly. This penalty can, however, be reduced by restricting individual SI_j values to be greater than or equal to $1/N$ (Badia et al., 2011). A drawback with the SI of this device is that the penalty increases greatly when N is large as all the SI_j are by definition less than or equal to 1. Thus, experiments with different N are not comparable using this device SI.

In Table 22.1 we have collected an overview of different methods found in the literature on the assessment of selectivity.

COMPARISON OF THE SELECTIVITY PERFORMANCE OF TWO INTRANEURAL ELECTRODES

The tfLIFE and the TIME are two types of neural interface that were both developed to be placed inside the nerve. However, the two electrodes differ in the way they are orientated inside the nerve. The tfLIFE was developed to be placed along the longitudinal axis of the nerve whereas the TIME was designed to be placed in the transverse direction; see Figure 22.2. For a more detailed description of the historic development of intrafascicular electrodes, see Yoshida et al. (2010). A detailed description of the microtechnology fabrication processes can be found in Hofmann and Koch (2005) and Boretius et al. (2010). The aim of the work described was to compare the stimulation selectivity of the two electrodes. As such the electrodes were placed inside a selected peripheral nerve, and electrical stimulation was delivered through individual contacts to activate motor fibers and cause the muscles in the forelimb to contract. To quantify, evaluate, and compare the ability of the tfLIFE and TIME interfaces to recruit forelimb muscles, the selectivity indices and device selectivity indices were calculated. Four success criteria were defined (based on the state of the art given in the introduction). As such, the best electrode should meet the following requirements:

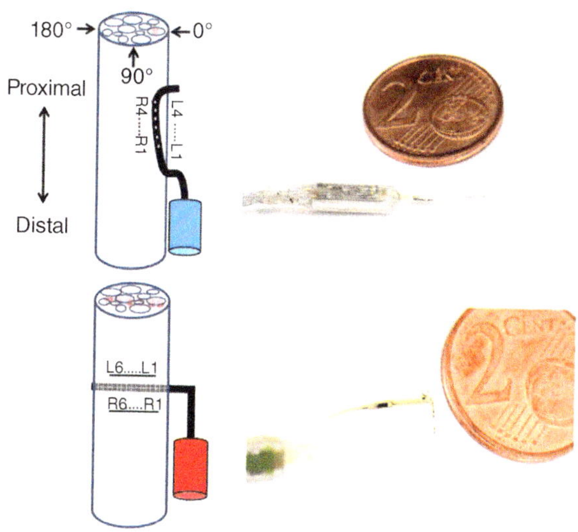

Figure 22.2. Schematic drawing of the different placement strategies and picture of the tfLIFE (top panels) and TIME (bottom panels) electrodes. TIME interfaces were manufactured by T. Stieglitz and T. Boretius, IMTEK, Freiburg, Germany. LIFE interfaces were manufactured by the Fraunhofer Institute, IBMT, Germany.

1. The electrode should be able to recruit as many muscles as possible within the group of muscles innervating the nerve.
2. The electrode should be able to recruit muscles with as high selectivity as possible.
3. The electrode should be able to recruit muscles with the lowest possible activation current.
4. Selective activation of muscles should be reliable and stable; that is, the selectivity should not change significantly over time.

METHODS

Overall, the stimulation selectivity of the tfLIFE and TIME interfaces is based on nine acute animal experiments (farm pigs) and three chronic animal experiments (Göttingen mini-pigs). The technical details of the two electrodes are compared in Table 22.2.

Surgical Procedures: Acute Animals

The acute experiments were carried out using female farm (Landrace-Yorkshire) pigs. The pigs were placed under general anesthesia during the experiments using a combination of anesthetics to ensure hypnosis, analgesia, and muscle relaxation. The animals were intubated and placed on a veterinary ventilator. During the experiment the heart rate and oxygen saturation of the animals were monitored. Animals were placed in a supine position on the surgical table to gain access to peripheral nerves and muscles in the left forelimb. One or more TIME or tfLIFE interfaces were placed in the median nerve

TABLE 22.2. Comparison of the Technical Specifications of the tfLIFE and TIME Interfaces[a]

Parameter	tfLIFE	TIME
Number of active sites	8	12
Indifferent	Yes	Yes
Distance between active sites	1500 μm	440 μm
Size of active sites	10 × 40 μm	Ø = 80 μm
Active site coating	Pt-black	IrOx
Charge injection capacity	—	2.3 mC/cm^2 (120 nC)
Track width and pitch	10 μm	10 μm
Substrate width	160 μm	240 μm

[a]Data on TIME interfaces provided by T. Stieglitz and T. Boretius, IMTEK, Freiburg Germany. Data on tfLIFE interfaces provided by T. Stieglitz, M Schüttler, and S. Kammer at the IBMT, Germany.

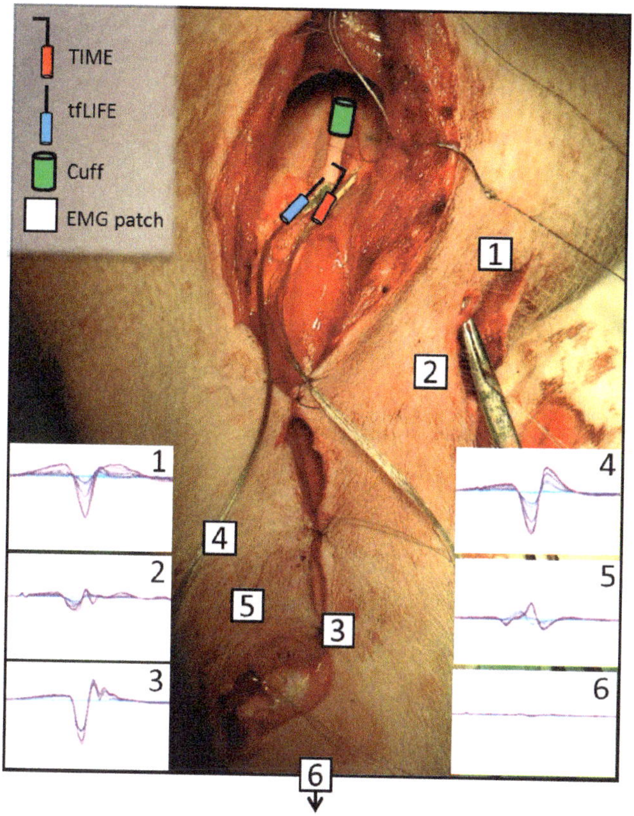

Figure 22.3. Implant sites of TIME, tfLIFE, and cuff in the median nerve of the left forelimb and location of six of the seven EMG patch electrodes. The seventh muscle was located on the lateral side. Examples of the evoked activity during a stimulation sweep are also shown.

approximately 2–3 cm above the elbow joint (see Fig. 22.3). The TIME was placed in the transverse direction of the nerve and the tfLIFE was placed in the longitudinal direction of the nerve. The level of implantation was chosen to resemble the implant site previously used during control of a prosthetic hand in a human amputee subject (Rossini, 2010). The electrodes were placed a few millimeters apart, but such that they both had an entry point at a 135° angle. The tfLIFE had an exit point at the same angle and the TIME had an exit point at a 45° angle. A definition of implant angles in relation to the anatomy and orientation is shown in Figure 22.2. To get a measure of the stimulation-evoked nerve activation of the two intrafasicular electrodes, EMG patch electrodes were implanted on the following seven forearm muscles (see Fig. 22.3): the pronator teres (M1), palmaris longus (M2), flexor carpi radialis

(M3), flexor digit superficialis (M4), flexor digitorum profundus (M5), abductor pollicis brevis (M6), and humeral head of deep digital flexor (M7). After completion of the data collection, the animals were euthanized and the nerves were harvested for histological analysis. For further details on the experimental procedures, see Kurstjens et al. (2008).

Surgical Procedures: Chronic Animals

The chronic experiments were carried out using three female Göttingen minipigs. The implant procedures of the chronic animals were similar to the procedures described for the acute animals with a few exceptions. The implant procedures were performed under sterile conditions. Muscles M1–M5 were instrumented with patch electrodes. In addition to the EMG patch and nerve electrodes, a bipolar cuff electrode was placed around the median nerve proximal to the TIME. The lead-out wires from all electrodes were collected in cables (12 wires in each cable) and tunneled subcutaneously to the back of the pig where they exited the skin (see Fig. 22.4). The animals received a dose of gentamicin antibiotics on the day of surgery to prevent infection developing at the percutaneous exit point. Following surgery, the animals were allowed to recover for 2–4 days before follow-up measurements began. At the end of the implantation period the animals were euthanized and the nerves were harvested for histological analysis.

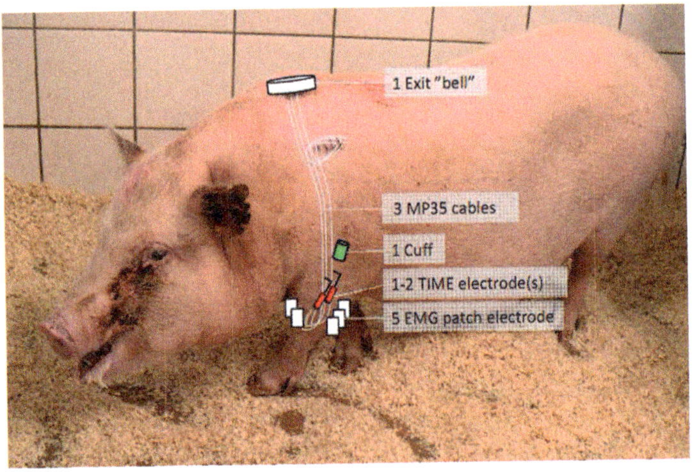

Figure 22.4. The location of the TIME, cuff, and patch electrodes and the subcutaneous route of the lead-out cables are indicated on a Göttingen mini-pig. Access to the electrodes was gained through percutaneous wires and multichannel connectors (Omnetics Nano-circular). Cables and wires were protected at the exit point with a custom-made stainless steel housing sutured to the skin on the back of the animal.

Data Collection and Analysis Procedures

For the acute animals, data were collected for up to 6 hours during the experiment. In the case of the chronic animals, measurements were performed up to three times a week while the animals were placed under general anesthesia (animals were food-deprived 12 hours before going under anesthesia). Typically a follow-up lasted between 1.5 and 2 hours for each of the chronic animals.

A subcutaneous stainless steel needle placed in the skin at the thorax was used as reference electrode for recording and stimulation purposes. Sweeps of 2-Hz stimulation pulses with increasing current were applied (current amplitude ranging from 20 μA up to 1200 μA at a fixed pulse duration of 100 μs). Monopolar stimulation was applied in such a way that each TIME or tfLIFE contact functioned as the cathode. Each stimulation level was repeated three to five times for each contact. To allow comparison of results between animals and recording sessions, the EMG recordings were normalized with the overall, maximum EMG activation as determined by applying supramaximal stimulation (stimulation sweeps up to 3200 μA were applied through the cuff electrode in the chronic animals).

The muscle signals were band-pass filtered at 100–2000 Hz. If further denoising of the EMG signals was necessary, stimulation artifacts were removed in the time window from 0.1 ms prior to onset of the stimulation pulse to 0.9 ms after offset. 50-Hz power-line noise was removed by subtraction of an EMG segment containing no muscle activity.

The level of muscle activation was quantified by calculating the root mean square (RMS) of the segment 3–10 ms after the stimulation onset. All RMS values recorded during a single follow-up session were normalized to the highest RMS value reached for each muscle when stimulating at maximal stimulation intensity through either the cuff or the TIME interfaces. This resulted in a normalized muscle activation parameter going from 0% to 100%; see Fig. 22.5.

The performance of the tfLIFE and TIME interfaces was compared using two different SIs: SI_{RF} and SI_{RwC}. Two response thresholds at 5% and 30% of the maximum activation were implemented for each muscle for the SI_{RF} index. The 5% threshold allows inclusion of muscles activated just above the response threshold (i.e., muscles are activated, but at a very low level) and will likely not produce a functional movement—this threshold was mainly used to block out background noise. Oppositely, the 30% threshold includes only muscles activated above a functional level. To compare the performance of the LIFE and the TIME interfaces, two different device SIs were calculated. The $SI_{M,D}$ and the $SI_{D,RoP}$, defined as

$$SI_{D,RoP} = \sqrt[N]{\prod_{i=1}^{N} \max(SI_{m,i})}$$

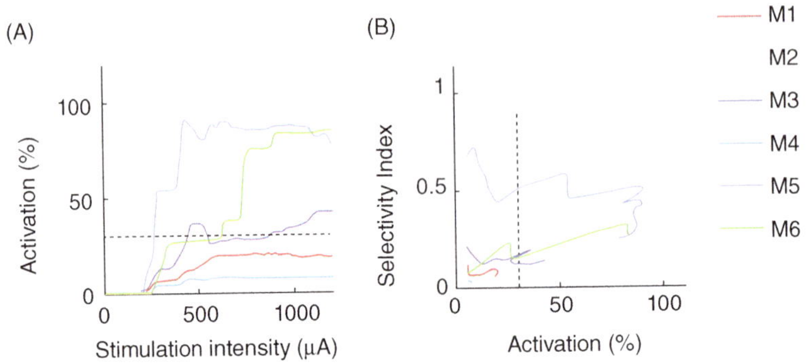

Figure 22.5. (A) Example recruitment curves obtained at day 37 in one chronic animal when stimulating at R2 TIME (T1) contact. The color coding of the muscles is shown at the right. (B) Selectivity plot of the recruitment curves in (A). Note that selectivity is shown on the vertical axis, whereas activation fraction is shown on the horizontal axis. To ignore variations due to noise in the EMG recording, the selectivity is only plotted when muscles are activated more than 0.05%. The dotted lines in both (A) and (B) indicate 30% muscle activation.

The $SI_{D,RoP}$ is a modified version of $SI_{D,M}$ in which the Nth root is added to reduce the cost or penalty of having more muscles.

Histological Evaluation. The thickness of fibrosis formed around the chronically implanted TIME was evaluated. The animals were euthanized and a nerve specimen including the TIME was harvested from the animals. Several samples of 5-μm thickness were cut in the transverse direction of the nerve and processed with hematoxylin and eosin (H&E). To estimate the thickness of the fibrotic scarring, digital pictures were taken through a microscope (Axio Vs40 v. 4.6.30, Carl Zeis Imaging solutions GmbH, Germany). The thickness of the fibrotic scar was estimated as the average of five measured thicknesses (i.e., the distance from the polyimide structure of the electrode to the rim of the fibrotic capsule perpendicularly).

RESULTS

Comparison of TIME and tfLIFE Stimulation Selectivity in the Acute Animal Model

The ability of the two types of electrodes to selectively activate the individual muscles was assessed by comparing the SIs of the individual muscles; see Fig. 22.6. The figure shows three sets of bar plots: the $SI_{RF,7}$ index with a 5% threshold is shown in panel A, the $SI_{RF,7}$ index with a 30% threshold is shown in panel B, and the $SI_{RwC,7}$ index is shown in panel C. The results show that the TIME

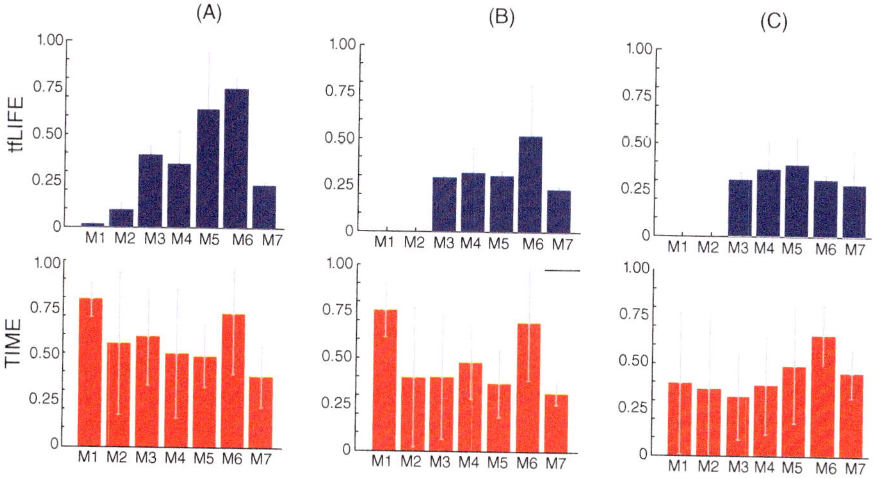

Figure 22.6. Comparison of the selectivity for the seven muscles for the TIME (red, bottom) and tfLIFE (blue, top) electrodes. (A) $SI_{RE,7}$ index with a response threshold of 5%; (B) $SI_{RE,7}$ index with a response threshold of 30%; and (C) $SI_{RwC,7}$ index with a response threshold of 5%.

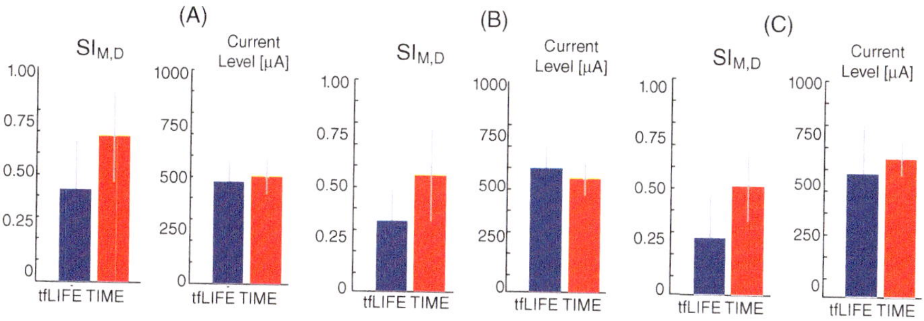

Figure 22.7. Comparison of the device selectivity indices, $SI_{M,D}$ and corresponding activation current, for the TIME (red) and tfLIFE (blue) electrodes. (A) $SI_{M,D}$ based on the $SI_{RE,7}$ index with a response threshold of 5%. (B) $SI_{M,D}$ based on the $SI_{RE,7}$ index with a response threshold of 30%. (C) $SI_{M,D}$ based on the $SI_{RwC,7}$ index with a response threshold of 5%.

recruits all seven muscles in two of three cases, whereas the tfLIFE only recruits five of the seven muscles (panels B and C). The last case (panel A) indicates that the tfLIFE recruits all seven muscles. However, the selectivity is low (recruitment level is just above the 5% threshold), indicating that the recruitment is not functionally useful. The TIME interface shows consistently higher $SI_{RE,7}$ selectivity indices than the tfLIFE.

To compare the overall performance of the TIME and tfLIFE, the device SIs were calculated ($SI_{D,RoP}$ and $SI_{M,D}$); see Figure 22.7, Figure 22.8, Table 22.3,

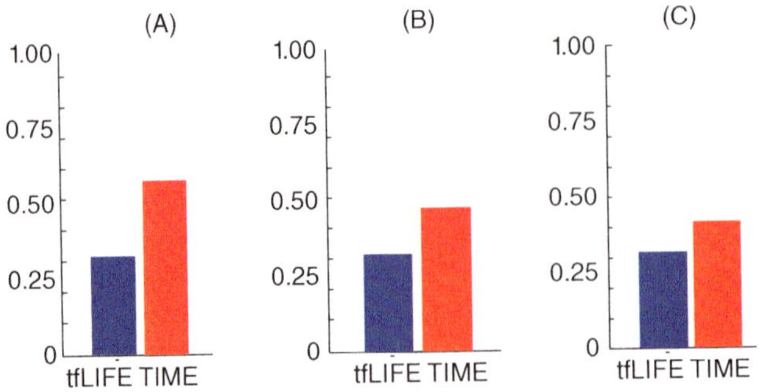

Figure 22.8. Comparison of the device selectivity indices ($SI_{D,RoP}$) for the tfLIFE (blue) and TIME (red) electrodes. (A) $SI_{D,RoP}$ based on the $SI_{RF,7}$ index with a response threshold of 5%. (B) $SI_{D,RoP}$ based on the $SI_{RF,7}$ index with a response threshold of 30%. (C) $SI_{D,RoP}$ based on the $SI_{RwC,7}$ index with a response threshold of 5%.

TABLE 22.3. Comparison of the Device Selectivity Index ($SI_{D,RoP}$) for the TIME and tfLIFE Interfaces[a]

		Device SI ($SI_{D,RoP}$)	
		TIME	tfLIFE
SI Type	Threshold	Mean	Mean
(A) $SI_{RF,7}$	5% threshold	0.56	0.30
(B) $SI_{RF,7}$	30% threshold	0.46	0.25
(C) $SI_{RwC,7}$	5% threshold	0.42	0.25
Overall mean and SD		0.48	0.48

[a]Mean and standard deviation of all electrodes evaluated. The device SI was calculated based on (A) $SI_{RF,7}$ index with a response threshold of 5%, (B) $SI_{RF,7}$ index with a response threshold of 30%, or (C) $SI_{RwC,7}$ index with a response threshold of 5%.
SD = standard deviation

and Table 22.4. Both the bar plots and numbers in the tables show that the device selectivity for the TIME interface is higher than the tfLIFE (statistically significant difference for both SIs, t-test, p-value = 0.30 for $SI_{RF,7}$ and p = 0.02 for $SI_{RwC,7}$). As such the results indicate that, with the given animal model and experimental setup, the TIME interface is able to recruit a larger number of muscles with higher selectivity than the tfLIFE muscle. The stimulation current used at the calculated device SIs was also compared; see Figure 22.7 and Table 22.5. On average, slightly smaller activation currents were needed to activate the muscles with the tfLIFE than with the TIME (505 ± 59 μA and 569 ± 38 μA, respectively). However, the current levels needed for muscle recruitment were comparable between the two interfaces and there were no statistically significant differences (t-test, p = 0.44).

TABLE 22.4. Comparison of the Device Selectivity Index ($SI_{M,D}$) for the TIME and tfLIFE Electrodes[a]

		Device SI ($SI_{M,D}$)			
		TIME		tfLIFE	
SI Type	Threshold	Mean	SD	Mean	SD
(A) $SI_{RF,7}$	5% threshold	0.58	0.14	0.34	0.17
(B) $SI_{RF,7}$	30% threshold	0.48	0.17	0.35	0.26
(C) $SI_{RwC,7}$	5% threshold	0.42	0.11	0.23	0.16
Overall mean and SD		0.49	0.07	0.31	0.05

[a]Mean and SD of all electrodes evaluated. The device SI was calculated based on (A) $SI_{RF,7}$ index with a response threshold of 5%, (B) $SI_{RF,7}$ index with a response threshold of 30%, or (C) $SI_{RwC,7}$ index with a response threshold of 5%.
SD = standard deviation

TABLE 22.5. Comparison of the Activation Current at the Device Selectivity Index (DSI_Mean) for the TIME and tfLIFE Interfaces[a]

		Activation Current at $SI_{M,D}$			
		TIME		tfLIFE	
SI Type	Threshold	Mean	SD	Mean	SD
(A) $SI_{RF,7}$	5% threshold	519 μA	82 μA	556 μA	154 μA
(B) $SI_{RF,7}$	30% threshold	612 μA	64 μA	423 μA	319 μA
(C) $SI_{RwC,7}$	5% threshold	577 μA	100 μA	537 μA	265 μA
Overall mean and SD		569 μA	38 μA	505 μA	59 μA

[a]Mean and SD of all electrodes evaluated. The activation current was determined in the following thee conditions: (A) SI_V index with a response threshold of 5%; (B) SI_V index with a response threshold of 30%; or (C) SI_P index with a response threshold of 5%.
SD = standard deviation.

Stimulation Selectivity of the TIME Interfaces in the Chronic Animal Model

To evaluate the robustness and the stability of the TIME interfaces to selectively activate muscles over a longer period of time, the stimulation selectivity in three chronic mini-pigs was evaluated. In Figure 22.9 a set of recruitment curves from 3–6 muscles obtained from stimulating one active site (LI) in one animal over 12 follow-up sessions are shown (up to 37 days). The x-axis represents the stimulation intensity in microamps and the y-axis represents the amount of recruitment in percent where 100% is maximal activation. The figure illustrates a general trend observed: the mapping between individual TIME contacts and muscle-recruitment curves did not stay constant over the implantation period. It was not possible, however, to monitor the responses

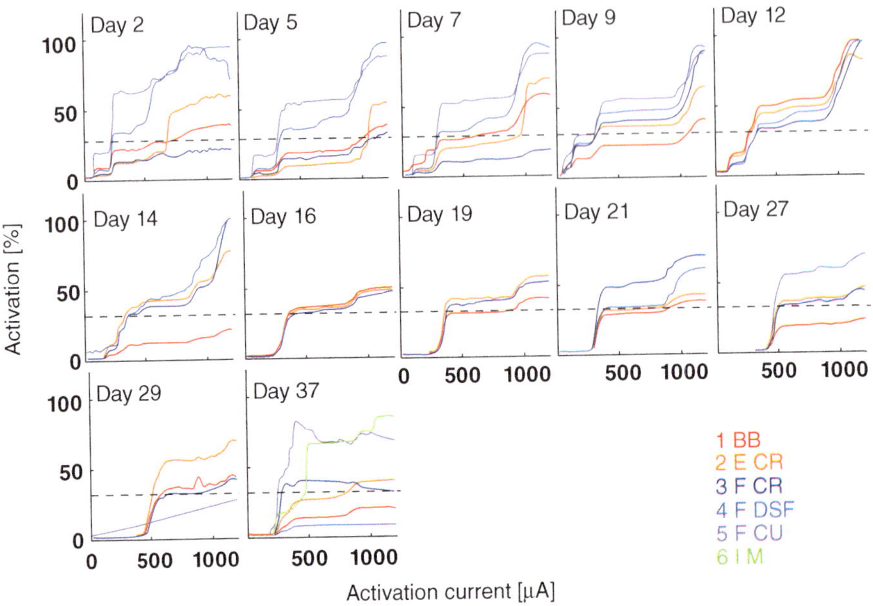

Figure 22.9. Recruitment curves obtained from stimulation of one active site (L1) of a TIME interface in one chronic animal at different follow-up times—from Day 2 (top left panel) to Day 37 (bottom right panel). The *x*-axis represents the stimulation intensity in microamps; the *y*-axis, the amount of recruitment in percent.

from all muscles throughout the implant period since some of the EMG channels broke during the implant period (likely detachment from the muscle surface or broken lead wires).

Figure 22.10 compares the device selectivity and activation current for each implanted electrode over time. The *x*-axis represents two points in time: the first follow-up (FU) and last follow-up. The $SI_{M,D}$ for the individual electrode implants were 0.27 ± 0.07 (day 4), 0.41 ± 0.14 (day 2), and 0.53 ± 0.27 (day 2) at the first follow-up and 0.24 ± 0.07 (day 32), 0.51 ± 0.24 (day 37), and 0.36 ± 0.29 (day 37). There was no statistically significant difference between the first and last follow-up in all three cases ($p = 0.47$, $p = 0.43$, and $p = 0.35$, *t*-tests).

All TIME interfaces were able to activate significant EMG responses with activation currents below the safety limit (Boretius et al., 2010) throughout the implantation period (>30 days). The mean activation current at greater than 30% threshold for the EMG for the individual electrode implants was 225 ± 144 μA (day 4), 430 ± 274 μA (day 2), and 405 ± 214 μA (day 2) at the first follow-up and 435 ± 131 μA (day 32), 497 ± 309 μA (day 37), and 634 ± 317 μA (day 37). There was no statistically significant difference between the first and last follow-up ($p = 0.06$, $p = 0.72$, and $p = 0.20$). However, there was a trend for an increase in the stimulation current needed for achiev-

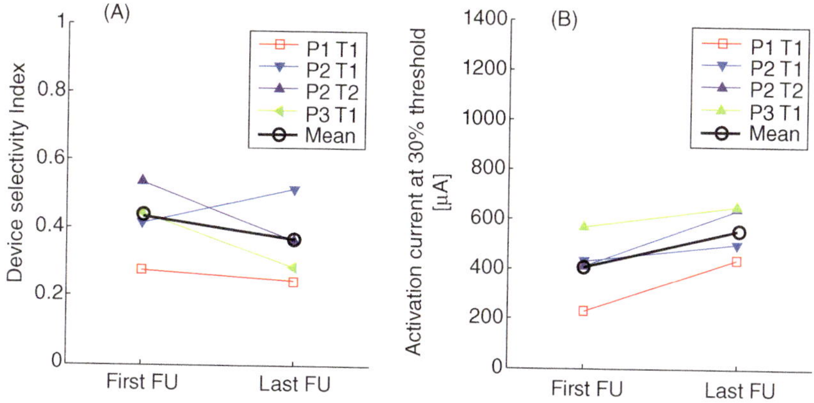

Figure 22.10. Graphical representation of the $SI_{M,D}$ and the recruitment current from the three A-classified TIME interfaces. The points First FU and Last FU on the *x*-axis indicate the first follow-up and last follow-up, respectively. (A) There is no statistical change in $SI_{M,D}$ during the implant period. (B) There is no statistically significant change for the recruitment current needed. However, the required activation current tends to increase from the first to the last follow-up.

ing greater than 30% muscle activation. A majority of the muscles were recruited over time, that is, five out of five (5/5) muscles for all implants at the first follow-up and 6/6, 5/6, and 5/6 muscles at the last follow-up (EMG of additional muscles was measured with percutaneous needles at the last follow-up only).

Histological Evaluation

Figure 22.11 shows the placement of the TIME and tfLIFE interfaces inside the median nerve in the histological specimens. The specimens were harvested in the acute animal experiments and cut in the transverse direction of the nerve. The difference in the orientation and location of the electrodes is clearly visible: the complete TIME structure is visible and the location through the nerve is clear. Conversely, only a short length of the tfLIFE is visible as expected. Two segments of the tfLIFE are visible in the picture because two tfLIFE interfaces were implanted in the same nerve in this case. Histological evaluation was performed in the chronic animals to verify the location of the implanted electrodes after more than 30 days of being implanted and to evaluate the amount of fibrosis formed around the electrode. All four TIME interface implants were verified to be located inside the nerve at the time the experiment was terminated. The thickness of the layer of fibrosis was estimated to be 116.4 ± 9.4 μm, 117.9 ± 38.9 μm, and 100.4 ± 21.1 μm. There was no apparent correlation between the duration of the implant and the thickness of the fibrosis.

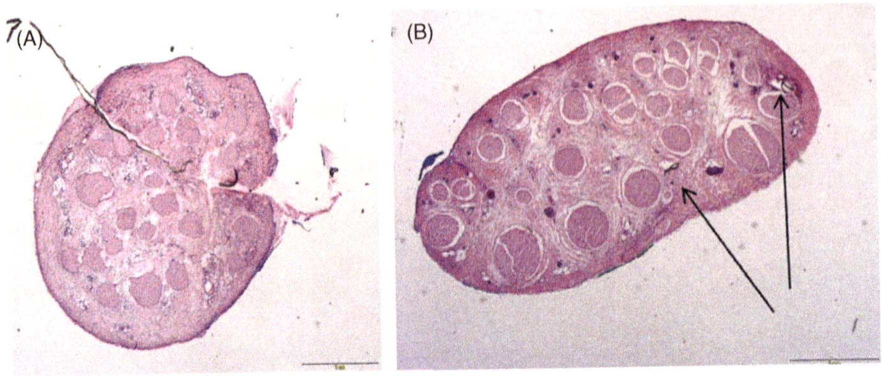

Figure 22.11. H&E stain of tissue sample obtained from the median nerve in one of the acute farm pigs. (A) The TIME interface is seen transversely to the nerve (the broken perineurium to the right is caused by the histological processing procedures). (B) The placement of the tfLIFE is indicated by the black arrows. Both panels show a scale equal to 1 mm in the lower right corner.

DISCUSSION AND CONCLUSIONS

Four success criteria for achieving selectivity were defined and evaluated: (1) the electrode should be able to recruit as many muscles as possible within the group of muscles innervating the nerve; (2) the electrode should be able to recruit muscles with as high selectivity as possible; (3) the electrode should be able to recruit muscles with the lowest possible activation current; and (4) selective activation of muscles should be reliable and stable, that is, the selectivity should not change significantly over time.

Regarding criteria (1) and (2), the experimental results indicated that, with the given animal model and experimental setup, the TIME interface was able to recruit a larger number of muscles with higher selectivity than the tfLIFE.

Regarding criterion (3) slightly smaller activation currents were needed to activate the muscles with the tfLIFE than with the TIME. However, the current levels needed for muscle recruitment were comparable

Regarding criterion (4) our results indicated that the overall device selectivity did not change significantly over time. However, the mapping between individual TIME contacts and muscles did not stay constant. This is likely explained by ongoing changes in the biology (i.e., formation of fibrosis over time) or by micromotions of the electrode. The overall mean of the fibrosis was 108 ± 40 μm, which is comparable to what is found in the literature (see, e.g., Williams and Williams, 1983; Kovacs, 1991).

ACKNOWLEDGMENTS

We thank Aritra Kundu for carrying out the experimental work. We thank Torben Madsen, Ole Sørensen, and Jens Sørensen for assistance during animal experiments. This work was funded by EU Project TIME, project number 224012 (ICT-2-3.6 Micro/Nanosystems).

REFERENCES

Badia J., Boretius T., Andreu D., Azevedo-Coste C., Stieglitz T., and Navarro X. (2011). Comparative analysis of transverse intrafascicular multichannel, longitudinal intrafascicular and multipolar cuff electrodes for the selective stimulation of nerve fascicles. *J Neural Eng* 8:036023.

Boretius T., Badia J., Pascual-Font A., Schuettler M., Navarro X., Yoshida K., and Stieglitz T. (2010). A transverse intrafascicular multichannel electrode (TIME) to interface with the peripheral nerve. *Biosens Bioelectron* 26:62–69.

Branner A., Stein R.B., and Normann R.A. (2001). Selective stimulation of cat sciatic nerve using an array of varying-length microelectrodes. *J Neurophysiol* 85:1585–1594.

Branner A., Stein R.B., Fernandez E., Aoyagi Y., and Normann R.A. (2004). Long-term stimulation and recording with a penetrating microelectrode array in cat sciatic nerve. *IEEE Trans Biomed Eng* 51:146–157.

Deurloo K.E., Holsheimer J., and Bergveld P. (2000). Nerve stimulation with a multi-contact cuff electrode: validation of model predictions. *Arch Physiol Biochem* 108:349–359.

Dhillon G.S. and Horch K. (2005). Direc neural sensory feedback and control of prosthetic arm. *IEEE Trans Neural Syst Rehabil Eng* 13:468–472.

Dhillon G.S., Krüger T.B., Sandhu J.S., and Horch K. (2005). Effects of short-term training on sensory and motor function in severed nerves of long-term human amputees. *J Neurophysiol* 93:2625–2633.

Dudel J., Jänig W., Schmidt R.F., and Zimmermann M. (1978). *Fundamentals of Neurophysiology*. New York: Springer-Verlag.

Ephraim P.L., Wegener S.T., MacKenzie E.J., Dillingham T.R., and Pezzin L.E. (2005). Phantom pain, residual limb pain, and back pain in amputees: results of a national survey. *Arch Phys Med Rehabil* 86:1910–1919.

Flor H., Dencke C., Schaefer M., and Grusser S.M. (2001). Effect of sensory discrimination training on cortical reorganization and phantom limb pain. *Lancet* 357:1763–1764.

Flor H., Nikolajsen L., and Jensen T.S. (2006). Phantom limb pain: a case of maladaptive CNS plasticity. *Neuroscience* 7:873–881.

Geng B., Yoshida K., and Jensen W. (2011). Impacts of selected stimulation patterns on the perception threshold in electrocutaneous stimulation. *J Neuroeng Rehabil* 8:9.

Grill W.M. and Mortimer J.T. (1996). Quantification of recruitment properties of multiple contact cuff electrodes. *IEEE Trans Rehabil Eng* 4:49–62.

Haugland M. and Sinkjær T. (1995). Cutaneous whole nerve recordings used for correction of footdrop in hemiplegic man. *IEEE Trans Rehab Eng* 3:307–317.

Haugland M., Lickel A., Haase J., and Sinkjær T. (1999). Control of FES thumb force using slip information obtained from the cutaneous electroneurogram in quadriplegic man. *IEEE Trans Rehab Eng* 7:215–227.

Hofmann K.P. and Koch K.P. (2005). Final report on design consideration of tfLIFE2. Technical report. Fraunhofer Institute, *Biomedical Engineering*, IBMT, St. Ingbert, Germany.

Hunter J.P., Kztz J., and David K.D. (2003). The effect of tactile and visual sensory inputs on phantom limb awareness. *Brain* 126:579–589.

Inmann A., Haugland M., Haase J., Biering-Sørensen F., and Sinkjær T. (2001). Signals from skin mechanoreceptors used in control of hand grasp neuroprosthesis. *Neuroreport* 12:2817–2820.

Jensen W., Riso R., and Sinkjær T. (2000). Effect of inter-trial delay on whole nerve cuff recordings of muscle afferents in rabbits. *Neuromodulation* 3:39–49.

Jensen W., Lawrence S.M., Riso R., and Sinkjær T. (2001). Effect of initial position on nerve cuff recordings of muscle afferents in rabbits. *IEEE Trans Neural Syst Rehabil Eng* 9:265–273.

Jensen W., Sinkjær T., and Sepulveda F. (2002). Improving signal reliability for on-line joint angle estimation from nerve cuff recordings of muscle afferents. *IEEE Trans Neural Syst Rehabil Eng* 10:133–139.

Jensen W., Micera S., Navarro X., Stieglitz T., Guiraud D., Divoux J.L., Rossini P.M., and Yoshida K. (2010). Development of an implantable transverse intrafascicular multichannel electrode (TIME) system for relieving phantom limb pain. *Conf Proc IEEE Eng Med Biol Soc*, pp. 6214–6217.

Jezernik S., Wen J.G., Rijkhof N.J.M., Djuruus J.C., and Sinkjær T. (2000). Analysis of bladder related nerve cuff electrode recordings from preganglion pelvic nerve and sacrral roots in pigs. *J Urol* 163:1309–1314.

Jezernik S., Grill W.M., and Sinkjær T. (2001). Detection and inhibition of hyperreflexia-like bladder contractions in the cat by sacral nerve root recording and stimulation. *Neurourol Urodyn* 20:215–230.

Kovacs E.J. (1991). Fibrogenic cytokines: the role of immune mediators in the development of scar tissue. *Immunol Today* 12:17–23.

Kundu A., Jensen W., Kurstjens M., Stieglitz T., Boretius T., and Yoshida K. (2010). Dependence of implantation angle of the transverse, intrafascicular electrode (TIME) on selective activation of pig forelimb muscles. *Artif Organs* 34(8):A43, No. 92.

Kurstjens M., Jensen W., and Yoshida K. (2008). Selective activation of pig forearm muscles using thin-film intrafasiscular electrodes implanted in the media nerve. Proceedings 13th IFESS Conference, Freiburg, Germany, Sept 21–25.

Leventhal D.K. and Durand D.M. (2004). Chronic measurement of the stimulation selectivity of the flat interface nerve electrode. *IEEE Trans Biomed Eng* 51: 1649–1658.

Lotze M., Grodd W., Birbaumer N., Erb M., Huse E., and Flor H. (1999). Does use of myoelectric prosthesis reduce cortical reorganization and phantom limb pain? *Nat Neurosci* 2:501–502.

Mortimer J.T. (1981). Motor prostheses. In: *Handbook of Physiology* (xxx, Eds.), pp. 13–45.

Nielsen T.N., Kurstjens G.A., and Struijk J.J. (2011). Transverse versus longitudinal tripolar configuration for selective stimulation with multipolar cuff electrodes. *IEEE Trans Biomed Eng* 58:913–919.

Polasek K.H., Hoyen H.A., Keith M.W., and Tyler D.J. (2007). Human nerve stimulation thresholds and selectivity using a multi-contact nerve cuff electrode. *IEEE Trans Neural Syst Rehabil Eng* 15:76–82.

Prochazka A. (1993). Comparison of natural and artificial control of movement. *IEEE Trans Rehab Eng* 1:7–17.

Raspopovic S., Capogrosso M., and Micera S. (2011). A computational model for the stimulation of rat sciatic nerve using a transverse intrafascicular multichannel electrode. *IEEE Trans Neural Syst Rehabil Eng* 19:333–344.

Rossini P.M. (2010). Double nerve intraneural interface implant on a human amputee for robotic hand control. *Clin Neurophysiol.* To appear.

Schiefer M.A., Polasek K.H., Triolo R.J., Pinault G.C., and Tyler D.J. (2010). Selective stimulation of the human femoral nerve with a flat interface nerve electrode. *J Neural Eng* 7:26006.

Sinkjær T., Haugland M., Inmann A., Hansen M., and Nielsen K. (2003). Biopotentials as command and feedback signals in functional electrical stimulation systems. *Med Eng Phys* 25:29–40.

Veraat C., Grill W.M., and Mortimer J.T. (1993). Selective control of muscle activation with a multipolar nerve cuff electrode. *IEEE Trans Biomed Eng* 40:640–653.

Williams G.T. and Williams W.J. (1983). Granulomatous inflammation—a review. *J Clin Pathol* 36:723–733.

Yoo P.B., Sahin M., and Durand D.D. (2004). Selective stimulation of the canine hypoglossal nerve using a multi-contact cuff electrode. *Ann Biomed Eng* 32:511–519.

Yoshida K., Farina D., Akay M., and Jensen W. (2010). Multichannel intraneural and intramuscular techniques for multiunit recording and use in active prostheses. *Proc IEEE* 98:432–449.

PART V

REHABILITATION THROUGH NEUROMODULATION

As discussed in the General Introduction, the three main approaches of neural engineering to rehabilitation of impaired motor functions are restoration, replacement, and neuromodulation. The focus of Part V is on neuromodulation, which aims to retrain the central nervous system by inducing plasticity through artificial stimulation of afferent pathways and/or by artificial enhancement of efferent neural and muscular signals provided as feedback. This approach is described in six chapters. Some of the interfacing methods described here have also been analyzed in previous chapters, although for different applications. For example, brain–computer interfacing (BCI) methods were introduced in Part III; however, in that section the BCI systems were intended for controlling external devices that would substitute motor functions whereas here the focus is on their use for inducing plastic changes in the central nervous system.

First, in Chapter 23, Daly describes the potential of BCI to facilitate the recovery of impaired motor functions in stroke patients. The following two chapters are both based on electrical stimulation to deliver artificial afferent feedback to the nervous system. In Chapter 24, Popovic and Popovic provide an overview of such an approach, describing the emerging technique of functional electrical therapy. Then, in Chapter 25, Andersen and Spaich describe a more specific approach based on the elicitation of the nociceptive withdrawal

Introduction to Neural Engineering for Motor Rehabilitation, First Edition. Edited by Dario Farina, Winnie Jensen, and Metin Akay.

reflex during gait for stroke rehabilitation. In Chapter 26, Sanguineti and collaborators provide a broad overview of the use of robotics for facilitating the recovery of neuromotor functions. They also analyze the available evidence on the efficacy of robot therapy, which is still controversial. In Chapter 27, Mrachacz-Kersting describes a relatively novel technique for artificially enhancing cortical excitability, referred to as paired associative stimulation. The method is based on the magnetic stimulation of the motor cortex paired with electrical stimulation of peripheral nerves in a precise temporal association. Finally, Chapter 28, by Thompson and Wolpaw, is dedicated to operant conditioning of spinal reflexes and its potential use in motor rehabilitation. The authors show evidence of plastic changes in spinal reflexes following operant conditioning in both animals and humans and preliminary results of the application of this approach in people with incomplete spinal cord injury.

23

FOCUSED CHAPTER

BRAIN–COMPUTER INTERFACE APPLIED TO MOTOR RECOVERY AFTER BRAIN INJURY

JANIS J. DALY

Brain Rehabilitation Research Center of Excellence, MR Gainesville DVA Medical Center, Gainesville, Florida, USA
Department of Neurology, College of Medicine, University of Florida, Gainesville, Florida, USA
Brain Rehabilitation Research Program, McKnight Brain Institute, University of Florida, Gainesville, Florida, USA

SUMMARY

Currently, rehabilitation to treat loss of motor control due to stroke does not return many stroke survivors to normal function. Although brain–computer interfaces (BCIs) were not initially intended to be used for motor relearning, they present a number of advantages compared with other available therapies. Preliminary research indicates that direct brain signal training may contribute to motor control recovery. One means of brain signal training is the use of BCI as a neural feedback system. The use of neural feedback may enhance the recovery of brain function and, thus, motor function. Feasibility studies of noninvasive BCI address the following issues: identification of motor task practice that can produce a brain signal that can be used in rehabilitation, identification of brain signal features that can be used in rehabilitation, and ease and accuracy of the use of BCI training.

Introduction to Neural Engineering for Motor Rehabilitation, First Edition. Edited by Dario Farina, Winnie Jensen, and Metin Akay.

INTRODUCTION

After stroke or head injury, a large percentage of individuals do not recover normal gait or arm function. Some interventions utilize upper and lower limb exercises, both with and without technologies. These peripherally targeted interventions are expected to restore motor control through activity-dependent brain plasticity (Wolf et al., 2006; Alon et al., 2003; Daly et al., 2005). One definition of plasticity is as follows: changes in the strength of synaptic connections in response to either an environmental stimulus or an alteration in synaptic activity in a network (Murphy and Corbett, 2009). These changes can occur in response to motor activity (activity-dependent plasticity). The peripherally targeted interventions are promising, but for many individuals none of these methods restores normal functional motor control. Therefore, it is important to investigate new methods that have potential for success in facilitating the recovery of brain function that will result in the recovery of more normal motor control.

Some researchers are investigating the use of transcranial magnetic stimulation (TMS; Butler and Wolf, 2007), which directs an externally generated electrical or magnetic field into the brain. Others are investigating the use of brain–computer interface (BCI) methods that can be applied to motor recovery training in a neural feedback paradigm; this chapter describes these early investigations of BCI systems applied to the goal of motor recovery after neural injury, as well as some of the difficulties encountered.

BRAIN–COMPUTER INTERFACES

A number of different types of BCI have been developed. There are invasive systems that use cortical recordings (electrocorticography-based BCIs; Daly and Wolpaw, 2008) and others that record neuronal action potentials or local field potentials from within the brain (intracortical BCIs; Daly and Wolpaw, 2008). In contrast, this chapter describes the application of noninvasive BCIs to physical rehabilitation. Noninvasive BCIs include those that use electroencephalographical (EEG) recordings from the surface of the scalp, functional magnetic resonance imaging (fMRI) recordings, and magnetoencephalographic (MEG) recordings.

Noninvasive BCIs can be used to record and analyze brain signals that can then be used to determine a BCI user's intention. Since the late 1990s, a solid body of work has been conducted for developing and testing EEG-BCIs for controlling communication and environmental devices for those who are severely paralyzed (Daly and Wolpaw, 2008). A BCI system typically enacts the following steps: signal feature extraction and brain signal classification are performed to yield control signals; and, with visual graphics or other means, online feedback is delivered to the user so that the user understands whether the brain signal was classified according to the user's intention. Motor recovery was not expected in BCI applications for communication or environmental device control; rather, the purpose of a BCI system use was to improve control of one's environment and communication, for those who were severely paralyzed. There is some basic scientific evidence that suggests BCIs may be applicable to motor learning for those with neural injury.

NEUROSCIENCE EVIDENCE OF BRAIN PLASTICITY AND RELATIONSHIP TO USE OF BCIS FOR MOTOR TRAINING

Central nervous system (CNS) plasticity has been documented by a number of scientists as the adaptation of structural and functional neural mechanisms during the learning of new information and the acquisition of new skills (Nudo, 2006; Wolpaw and Tennissen, 2001; Ziemann et al., 2004; Kempermann et al., 1997; Foster and Dumas, 2001). Normal brain plasticity occurs during normal childhood development and throughout the person's life span (Ziemann et al., 2004; Kempermann et al., 1997; Foster and Dumas, 2001). This plasticity can involve modifications in synaptic strength between neurons in both the brain and the spinal cord (Nudo, 2006; Wolpaw and Tennissen, 2001). In fact, brain plasticity is the basis upon which cognitive and motor skills are acquired. CNS plasticity occurs also in the case of trauma or disease and can involve changes not only in synaptic strength, but also across neural circuits (Ng et al., 1988; Stroemer et al., 1995; Carmichael et al.; Schiene et al., 1996; Carmichael et al., 2001, 2005; Li and Charmichael, 2006; Cramer et al., 2006, 1997; Teasel et al., 2005; Jaillard et al., 2005; Redecker et al., 2000; Frost et al., 2003; Gharbawie et al., 2005; Dancause et al., 2005; Napieralski et al., 1996). After a stroke, extensive plasticity can occur in the cortex and other neural structures, according to both animal models (Nudo, 2006; Chu and Jones, 2000; Nudo et al., 1996; Jones et al., 1999; Nelles et al., 2001; Biernaskie and Corbett, 2001) and human models (Umphred, 1995; Traversa et al., 1997; Carey et al., 2002; Newton et al., 2002; Marshall et al., 2000; Liepert et al., 2001; Johansen-Berg et al., 2002; Neumann-Haefelin et al., 2000). In the case of CNS damage or disease, activity-dependent plasticity can affect the nervous system positively and/or negatively. Plasticity may restore more normal motor function—or, when abnormal movements are repetitively performed, activity-dependent plasticity may solidify or even exacerbate abnormal motor performance.

Using EEG, electrocorticography (ECoG), or single-neuron models, researchers have shown that there are specific brain signal features that are observed during specific coordination tasks, and these features are modified as learning occurs (Leuthardt et al., 2004; Taylor et al., 2002; Carmena et al., 2003; Wolpaw and McFarland, 2004; Fetz, 1969; Daly et al., 2006). One study showed that the learning process, at the neural learning level, can involve a "training neuron array," which then teaches another neuron array (which is at a distance) to activate or deactivate (Jackson et al., 2006). At the pathway level, a study showed that complex pathways are modulated during the learning process of pain perception and management (deCharms, 2007). The findings in these studies, together, support the notion that more direct brain signal training would contribute to the process of motor control recovery and that noninvasive BCIs may have some potential in that regard.

THEORETICAL STRATEGIES FOR BCIS IN MOTOR LEARNING

Successful rehabilitation methods that induce activity-dependent brain plasticity are constructed within the proper framework and are accurately targeted (Daly and Ruff, 2007). Conventional approaches focus the intervention in the upper and lower limbs, that is, the periphery of the body. The thinking is that repetitive movement practice will induce activity-dependent CNS plasticity which will restore more normal function. In contrast, BCI-based approaches could direct measures of brain signal (e.g., EEG-BCIs) to provide feedback to the user, in the expectation of encouraging the brain plasticity that will restore motor control.

Figure 23.1 provides an illustration of two possible uses of BCI-based motor learning. In the first strategy, the concept is to attempt to train more normal brain signal activation (e.g., as measured by specific EEG brain signal features; Fig. 23.1A). The expectation is that more normal brain activation will improve motor function control. The second BCI-based strategy for motor learning utilizes the brain signal, from the intention to move, in order to activate a movement-assistance device (Fig. 23.1B). The expectation for this strategy is that this device-assisted movement will produce more normal afferent signals that may induce the CNS plasticity that can drive recovery of more normal motor function (Fig. 23.1C).

FEASIBILITY STUDIES

Imposed Motor Task

A number of researchers have studied the feasibility of applying noninvasive BCIs to the problem of motor learning and motor recovery after stroke. In this new BCI application, there are a number of issues to address. The first is whether the impaired movement can be used by the stroke survivor, within the BCI environment. In prior BCI applications, this question was not relevant. Rather, the user was screened on a number of cognitive or motor tasks, in order to determine which single task or pair of tasks would produce the most robust brain signal for the user in his or her operation of the BCI. In this case, the brain signal was selected from among an array of motor and imaginary tasks, according to which task produced the most "usable" or "best" brain

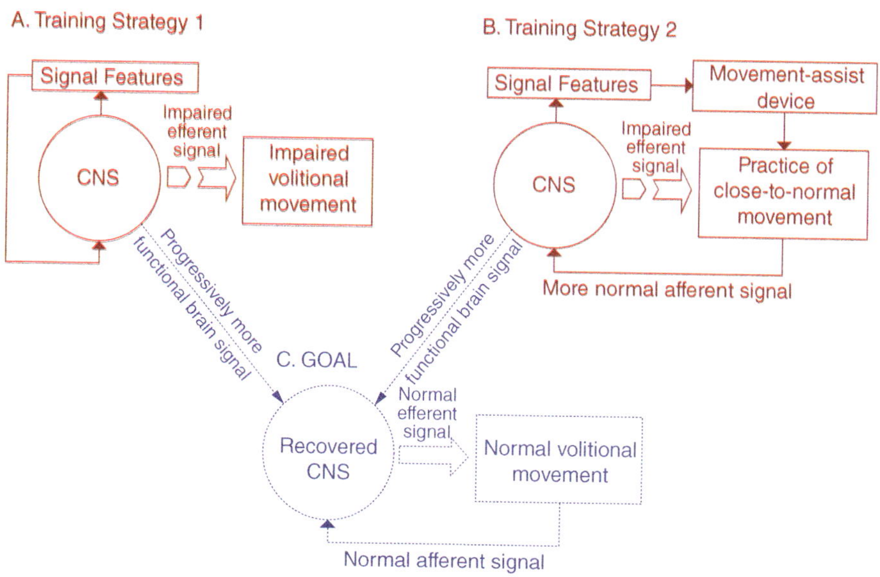

Figure 23.1. Brain–computer interfaces in neurological rehabilitation. With permission from Daly J.J. and Wolpaw J.R., *Lancet Neurology*, 2008;7:1032–1043 (fig. 5).

signal, regardless of whether the task had any bearing on the physical motor task at hand. For example, a patient with locked-in syndrome might use the thought of moving the left foot to activate the light switch in the room. This method of choosing the best brain signal is adequate when one is concerned only with turning on an environmental or communication device. However, in the case of training motor recovery, one is attempting to retrain brain signals for more normal driving of a specific movement or movements; in that case, the cognitive activity that generates the brain signal, ideally, should match the motor function that is being retrained. For example, if finger extension is impaired and retraining is desired, then the cognitive function, used within the BCI environment for motor learning, should be "attempted or imagined finger extension," since that is the very task for which brain control is being retrained.

There are a few studies supporting the concept that an imposed motor task or attempted motor task could generate a usable brain signal for BCI motor training after stroke (Daly et al., 2009; Ang et al., 2009; Buch et al., 2008). First, stroke survivors were able to activate a robotic wrist-movement device using brain signal acquired using magnetoencephalography (MEG; Fig. 23.2; Buch et al., 2008). Using this MEG-based BCI, six of eight stroke survivors were able to activate the robotic wrist movement-assistance device at an accuracy that improved over time to 72.48 ± 18.36% by the final session (20 sessions). Second, stroke survivors were able to use an EEG-BCI, along with a surface

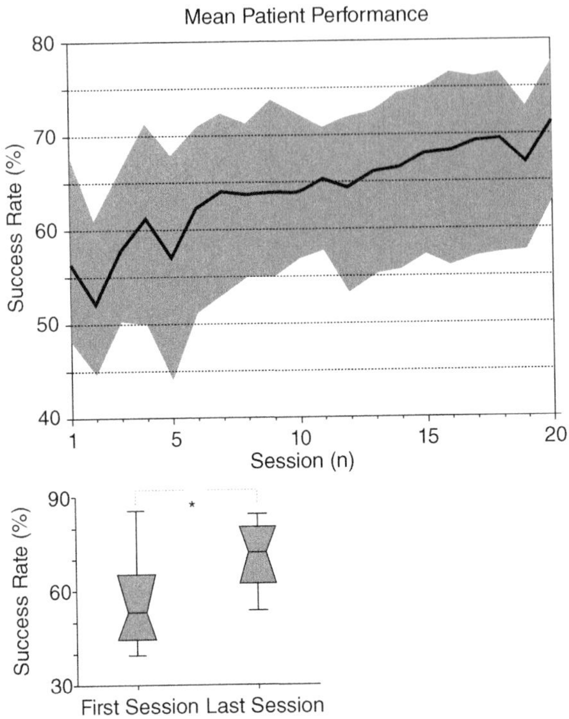

Figure 23.2. Average group success rate as a function of training session. The average success rate for the last training session is 72.48 ± 18.36% (median ± interquartile range). As the total number of training sessions completed by patients was unique, the time-series for each individual was resampled and normalized to 20 sessions (the mode of the session duration across the patient group) using linear interpolation, before being averaged. The gray shaded area represents the 95% confidence interval of the median estimate, which was computed using a bootstrap technique repeated 10,000 times. The boxplot (pre- and post-training median and interquartile range) inset shows a significant group increase in success rate between the first and last training sessions. With permission from Buch E., Weber C., Cohen L., Braun C., Dimyan M.A., Ard T., Mellinger J., Caria A., Soekadar S., Fourkas A., and Birbaumer N. Think to move: a neuromagnetic brain–computer interface (BCI) system for chronic stroke. *Stroke*, 2008;39:910–917 (fig. 2).

functional electrical stimulator (FES), three of three stroke survivors were able to activate the FES device at accuracies ranging from 78% to 98% for imagined movement and from 82% to 98% for attempted movement (Fig. 23.3; Daly et al., 2008) during the last of 9 sessions. Also using an EEG-BCI, but with a shoulder–elbow robot, 48 stroke survivors (89% of the group) could operate the BCI–robot system at greater than chance accuracy, with accuracy

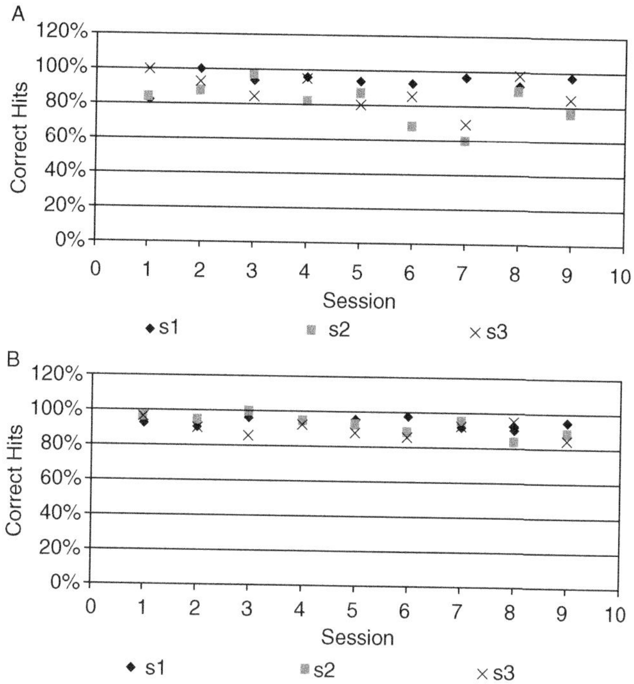

Figure 23.3. (A) Brain signal accuracy for three different subjects is shown across 9 BCI + FES intervention sessions that were held 3 times per week for 3 weeks. The subjects were 3 stroke survivors with severe impairment of wrist/finger movement. Brain signal accuracy is shown for imagined wrist movement (s2, s3) or finger movement (s1), as percentage correct. (B) Brain signal accuracy for three different subjects is shown across 9 BCI + FES intervention sessions that were held 3 times per week for 3 weeks. The subjects were 3 stroke survivors with severe impairment of wrist/finger movement. Brain signal accuracy is shown for attempted wrist movement (s2, s3) or finger movement (s1), as percentage correct. With permission from Daly J.J., Cheng R.C., Hrovat K., Litinas K.H., Rogers J.M., and Dohring M.E. Development and testing of non-invasive BCI + FES/robot system for use in motor re-learning after stroke. International Functional Electrical Stimulation Society, Freiburg, Germany, September 2008. *Biomedizinische Technik*, vol. 53, suppl. 1, pp. 200-202, figs. 4 and 5.

ranging across subjects from 60% to 99% (Ang et al., 2010). These results, together, provide evidence that BCIs could be used in the learning strategy described in Figure 23.1B, in conjunction with a movement-assistance device. This combination of BCI and movement-assistance technology could prove to be advantageous versus movement-assistance technologies alone; however, there is little support for that possibility, to date.

Initial Session Accuracy

It is important for a motor learning tool to be easily used by the learner. For example, if a cane requires 10 sessions before one is able to begin using it to learn to walk, it would be considered impractical as a gait training tool. In the same manner, a motor learning BCI application must be accurately usable at the beginning of its application in motor learning. There is some preliminary information on initial session accuracy. For example, for the MEG-BCI combined with a wrist–hand robot, initial median accuracy was 52.84 ± 20.59% (Fig. 23.2; Buch et al., 2008). For the EEG-BCI combined with FES, initial session accuracy ranged from 82% to 99% for imagined movement and from 96% to 99% accuracy in the initial session for attempted movement (Fig. 23.3; Daly et al., 2008). Finally, for EEG-BCI combined with robotics, initial session accuracy ranged from approximately 70% to 90% (Fig. 23.4; Ang et al., 2010).

Motor Recovery in Response to BCI Use

To date, there is no solid evidence that BCIs can induce motor recovery beyond what is achievable with other less complex motor learning methods.

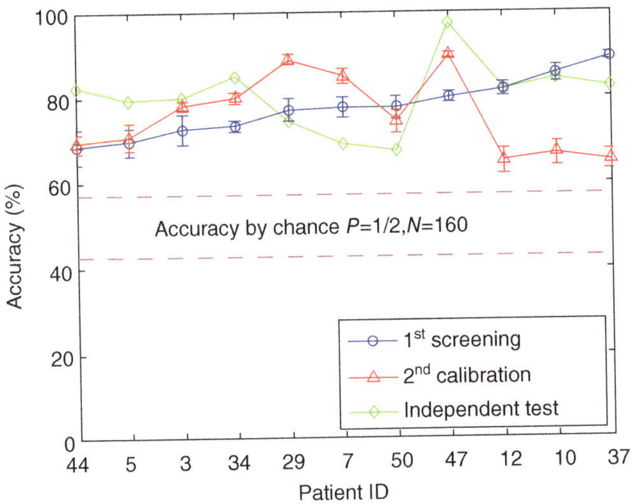

Figure 23.4. Plots of the accuracies of the screening session from 11 patients, accuracies of second calibration session of the patient in performing motor imagery of the stroke-affected hand versus background rest, and accuracies of session-to-session transfer of the calibration to another to independent test session. With permission from Ang K.K., Guan C., Chua K.S.G., Ang B.T., Kuah C., Wang C., Phua K.S., Chin Z.Y., and Zhang H. Clinical study of neurorehabilitation in stroke using EEG-based motor imagery brain–computer interface with robotic feedback. 32nd Annual International Conference of the IEEE EMBS, Buenos Aires, Argentina, 2010 (fig. 3, p. 5552).

One study reported no recovery of motor function in 8 stroke survivors, after 13–22 sessions of MEG-BCI training combined with wrist–hand robot movement-assistance (Buch et al., 2008). Two case studies reported recovery of motor control in response to either MEG-BCI or EEG-BCI, both combined with another technology, goal-directed physical therapy (Broetz et al., 2010) or with FES-assisted movement practice (Daly et al., 2009), respectively. In the first study, the participant was 14 months poststroke and received treatment over a year's time with both MEG-BCI and 1 hour per day of physical therapy (two weeks) and EEG-BCI and 1 hour per day of physical therapy), as well as 1hour per day of home exercise assigned by the physical therapist. At the end of the year's time, hand function had improved according to objective measure (Broetz et al., 2010). In the second study, the participant was 6 months poststroke and received 9 sessions of combined EEG-BCI and surface FES training for index finger extension of isolated movement control (i.e., no other fingers moving at the same time). At the end of the 3 weeks of treatment, the participant had recovered 26 degrees of isolated movement control of index finger extension (Daly et al., 2009). Both of these studies are promising, though case studies are not generalizable; further, since combined interventions were used, it is not possible to determine, from these two studies, whether the motor recovery could have occurred in response to the interventions other than BCI training. To address those concerns, we can note the only available randomized controlled trial of BCI (Ang et al., 2009), which utilized an EEG-BCI in conjunction with a shoulder–elbow movement-assistance device. For the 18 chronic-phase participants, randomized to BCI–robotics or robotics alone, there was no significant additive advantage for BCI training, according to the Fugl–Meyer upper limb coordination scale ($p = 0.475$; Ang et al., 2009).

CONCLUSIONS

Brain–computer interfaces hold some potential to assist in restoring motor function after stroke or in other types of brain injury or disease. One problem is that there is little information available regarding the relationship between motor control and EEG signal features that are available in online analysis, at the time of brain signal generation when learner feedback is critical. With more study, greater clarity may be reached regarding the usefulness or importance that particular noninvasively acquired brain signal features may have during the learning process. In that case, BCIs may prove efficacious in providing useful feedback to the learner during the motor learning process. In addition, BCIs may also prove useful in combination therapies such as BCI and robotics or BCI and FES.

Though BCIs were not initially intended for application in motor learning, there are a number of advantages afforded by BCIs during motor learning compared with other available therapies. First, BCIs demand that the user focus attention and engage his or her own brain neural cells. Second, in using noninvasive BCIs, there is no imposition into the brain of an externally generated electrical or magnetic field, as there is in the use of TMS. Third, conventional therapies direct attention to the periphery of the body, that is the limbs; in contrast, BCIs direct the user's attention to generation of brain signals. Fourth, one of the critical motor learning principles required for brain plasticity is to focus attention purposefully on the task (Singer et al., 1993).

Preliminary studies of BCI applications to motor learning in the presence of brain injury have produced mixed results. Some stroke survivors have been able to gain control of brain signals with accuracies that would be reasonable to use in a motor learning paradigm. Some users have shown feasibility in gaining control of brain signals during the initial session(s) of BCI training. To date, there have been no results reported that would support the use of BCI as a motor learning therapy, having any greater benefit than less complex therapies such as motor therapy with no technologies or with technologies such as robotics or functional electrical stimulation. Investigations into the usefulness of BCIs are currently in the preliminary stage.

REFERENCES

Alon G., Sunnerhagen K.S., Geurts A.C., and Ohry A. (2003). A home-based, self-administered stimulation program to improve selected hand functions of chronic stroke. *NeuroRehabilitation* 18:215–225.

Ang K.K., Guan C., Chua K.S., Ang B.T., Kuah C., Wang C., Phua K.S., Chin Z.Y., and Zhang H. (2009). A clinical study of motor imagery-based brain–computer interface for upper limb robotic rehabilitation. *Conf Proc IEEE Eng Med Biol Soc* 2009: 5981–5984.

Ang K.K., Guan C., Chua K.S.G., Ang B.T., Kuah C., Wang C., Phua K.S., Chin Z.Y., and Zhang H. (2010). Clinical study of neurorehabilitation in stroke using EEG-based motor imagery brain–computer interface with robotic feedback. In: 32nd Annual International Conference of the IEEE EMBS, Buenos Aires, Argentina.

Biernaskie J. and Corbett D. (2001). Enriched rehabilitative training promotes improved forelimb motor function and enhanced dendritic growth after focal ischemic injury. *J Neurosci* 21(14):5272–5280.

Broetz D., Braun C., Weber C., Soekadar S.R., Caria A., and Birbaumer N. (2010). Combination of brain–computer interface training and goal-directed physical therapy in chronic stroke: a case report. *Neurorehabil Neural Repair* 24(7):674–679. Epub.

Buch E., Weber C., Cohen L., Braun C., Dimyan M.A., Ard T., Mellinger J., Caria A., Soekadar S., Fourkas A., and Birbaumer N. (2008). Think to move: a neuromagnetic brain–computer interface (BCI) system for chronic stroke. *Stroke* 39:910–917.

Butler A.J. and Wolf S.L. (2007). Putting the brain on the map: use of transcranial magnetic stimulation to assess and induce cortical plasticity of upper-extremity movement. *Phys Ther* 87:719–736.

Carey J.R., Kimberley T.J., Lewis S.M., et al. (2002). Analysis of fMRI and finger tracking training in subjects with chronic stroke. *Brain* 125(Pt 4):773–788.

Carmena J.M., Lebedev M.A., Crist R.E., et al. (2003). Learning to control a brain–machine interface for reaching and grasping by primates. *PLoS Biol* 1(2):E42.

Carmichael S.T., Wei L., Rovainen C.M., and Woolsey T.A. (2001). New patterns of intracortical projections after focal cortical stroke. *Neurobiol Dis* 8(5):910–922.

Carmichael S.T., Archibeque I., Luke L., Nolan T., Momiy J., and Li S. (2005). Growth-associated gene expression after stroke: evidence for a growth-promoting region in peri-infarct cortex. *Exp Neurol* 193(2):291–311.

Chu C.J. and Jones T.A. (2000). Experience-dependent structural plasticity in cortex heterotopic to focal sensorimotor cortical damage. *Exp Neurol* 166(2):403–414.

Cramer S.C., Nelles G., Benson R.R., et al. (1997). A functional MRI study of subjects recovered from hemiparetic stroke. *Stroke* 28(12):2518–2527.

Cramer S.C., Shah R., Juranek J., Crafton K.R., and Le V. (2006). Activity in the peri-infarct rim in relation to recovery from stroke. *Stroke* 37(1):111–115.

Daly J.J. and Ruff R.L. (2007). Construction of efficacious gait and upper limb functional interventions based on brain plasticity evidence and model-based measures for stroke patients. *Scientific World J* 20(7):2031–2045.

Daly J.J. and Wolpaw J.R. (2008). Brain–computer interfaces in neurological rehabilitation. *Lancet Neurol* 7:1032–1043.

Daly J.J., Hogan N., Perepezko E., et al. (2005). Response to upper-limb robotics and functional neuromuscular stimulation following stroke. *J Rehabil Res Dev* 42: 723–736.

Daly J.J., Fang Y., Perepezko E., and Yue G. (2006). Prolonged brain motor planning time and elevated cognitive effort during a linear movement task following stroke. *IEEE Trans Neural Syst Eng* 14(2):168–171.

Daly J.J., Cheng R.C., Hrovat K., Litinas K.H., Rogers J.M., and Dohring M.E. (2008). Development and testing of non-invasive BCI + FES/robot system for use in motor re-learning after stroke. International Functional Electrical Stimulation Society, Freiburg, Germany.

Daly J.J., Cheng R., Rogers J., Litinas K., Hrovat K., and Dohring M. (2009). Feasibility of a new application of non-invasive brain computer interface: a case study of training for recovery of volitional motor control after stroke. *J Neurol Phys Ther* 33(4):203–211.

Dancause N., Barbay S., Frost S.B., et al. (2005). Extensive cortical rewiring after brain injury. *J Neurosci* 2;25(44):10167–10179.

deCharms R.C. (2007). Reading and controlling human brain activation using real-time functional magnetic resonance imaging. *Trends Cogn Sci* 11(11):473–481.

Fetz E.E. (1969). Operant conditioning of cortical unit activity. *Science* 163(870): 955–958.

Foster T.C. and Dumas T.C. (2001). Mechanism for increased hippocampal synaptic strength following differential experience. *J Neurophysiol* 85(4):1377–1383.

Frost S.B., Barbay S., Friel K.M., Plautz E.J., and Nudo R.J. (2003). Reorganization of remote cortical regions after ischemic brain injury: a potential substrate for stroke recovery. *J Neurophysiol* 89(6):3205–3214.

Gharbawie O.A., Gonzalez C.L., Williams P.T., Kleim J.A., and Whishaw I.Q. (2005). Middle cerebral artery (MCA) stroke produces dysfunction in adjacent motor cortex as detected by intracortical microstimulation in rats. *Neuroscience* 130(3): 601–610.

Jackson A., Mavoori J., and Fetz E.E. (2006). Long-term motor cortex plasticity induced by an electronic neural implant. *Nature* 444(7115):56–60.

Jaillard A., Martin C.D., Garambois K., Lebas J.F., and Hommel M. (2005) Vicarious function within the human primary motor cortex? A longitudinal fMRI stroke study. *Brain* 128(Pt 5):1122–1138.

Johansen-Berg H., Dawes H., Guy C., Smith S.M., Wade D.T., and Matthews P.M. (2002). Correlation between motor improvements and altered fMRI activity after rehabilitative therapy. *Brain* 125(Pt 12):2731–2742.

Jones T.A., Chu C.J., Grande L.A., and Gregory A.D. (1999) Motor skills training enhances lesion-induced structural plasticity in the motor cortex of adult rats. *J Neurosci* 19(22):10153–10163.

Kempermann G., Kuhn H.G., and Gage F.H. (1997). Genetic influence on neurogenesis in the dentate gyrus of adult mice. *Proc Natl Acad Sci U S A* 94(19):10409–10414.

Leuthardt E.C., Schalk G., Wolpaw J.R., Ojemann J.G., and Moran D.W. (2004). A brain–computer interface using electrocorticographic signals in humans. *J Neural Eng* 1(2):63–71.

Li S. and Carmichael S.T. (2006). Growth-associated gene and protein expression in the region of axonal sprouting in the aged brain after stroke. *Neurobiol Dis* 23(2):362–373.

Liepert J., Uhde I., Graf S., Leidner O., and Weiller C. (2001). Motor cortex plasticity during forced-use therapy in stroke patients: a preliminary study. *J Neurol* 248(4):315–321.

Marshall R.S., Perera G.M., Lazar R.M., Krakauer J.W., Constantine R.C., and DeLaPaz R.L. (2000). Evolution of cortical activation during recovery from corticospinal tract infarction. *Stroke* 31(3):656–661.

Murphy T.H. and Corbett D. (2009). Plasticity during stroke recovery: from synapse to behavior. *Nat Rev Neurosci* 10:861–872.

Napieralski J.A., Butler A.K., and Chesselet M.F. (1996). Anatomical and functional evidence for lesion-specific sprouting of corticostriatal input in the adult rat. *J Comp Neurol* 373(4):484–497.

Nelles G., Jentzen W., Jueptner M., Muller S., and Diener H.C. (2001). Arm training induced brain plasticity in stroke studied with serial positron emission tomography. *Neuroimage* 13(6 Pt 1):1146–1154.

Neumann-Haefelin T., Moseley M.E., and Albers G.W. (2000). New magnetic resonance imaging methods for cerebrovascular disease: emerging clinical applications. *Ann Neurol* 47(5):559–570.

Newton J., Sunderland A., Butterworth S.E., Peters A.M., Peck K.K., and Gowland P.A. (2002). A pilot study of event-related functional magnetic resonance imaging of monitored wrist movements in patients with partial recovery. *Stroke* 33(12): 2881–2887.

Ng S.C., de la Monte S.M., Conboy G.L., Karns L.R., and Fishman M.C. (1988). Cloning of human GAP-43: growth association and ischemic resurgence. *Neuron* 1(2): 133–139.

Nudo R.J. (2006). Mechanisms for recovery of motor function following cortical damage. *Curr Opin Neurobiol* 16(6):638–644.

Nudo R.J., Wise B.M., SiFuentes F., and Milliken G.W. (1996). Neural substrates for the effects of rehabilitative training on motor recovery after ischemic infarct. *Science* 272(5269):1791–1794.

Redecker C., Luhmann H.J., Hagemann G., Fritschy J.M., and Witte O.W. (2000). Differential downregulation of GABAA receptor subunits in widespread brain regions in the freeze-lesion model of focal cortical malformations. *J Neurosci* 20(13): 5045–5053.

Schiene K., Bruehl C., Zilles K., et al. (1996). Neuronal hyperexcitability and reduction of GABAA-receptor expression in the surround of cerebral photothrombosis. *J Cereb Blood Flow Metab* 16(5):906–914.

Singer R., Lidor R., and Cauraugh J.H. (1993). To be aware or not aware? What to think about while learning and performing a motor skill. *Sport Psychologist* 7:19–30.

Stroemer R.P., Kent T.A., and Hulsebosch C.E. (1995). Neocortical neural sprouting, synaptogenesis, and behavioral recovery after neocortical infarction in rats. *Stroke* 26(11):2135–2144.

Taylor D.M., Tillery S.I., and Schwartz A.B. (2002). Direct cortical control of 3D neuroprosthetic devices. *Science* 296(5574):1829–1832.

Teasell R., Bayona N.A., and Bitensky J. (2005). Plasticity and reorganization of the brain post stroke. *Top Stroke Rehabil* 12(3):11–26.

Traversa R., Cicinelli P., Bassi A., Rossini P.M., and Bernardi G. (1997). Mapping of motor cortical reorganization after stroke. A brain stimulation study with focal magnetic pulses. *Stroke* 28(1):110–117.

Umphred D.A. (1995). *Neurological Rehabilitation*. St. Louis, MO: Mosby.

Wolf S.L., Winstein C.J., Miller J.P., et al. (2006). Effect of constraint-induced movement therapy on upper extremity function 3 to 9 months after stroke: the EXCITE randomized clinical trial. *JAMA* 296:2095–2104.

Wolpaw J.R. and McFarland D.J. (2004). Control of a two-dimensional movement signal by a noninvasive brain–computer interface in humans. *Proc Natl Acad Sci U S A* 101(51):17849–17854.

Wolpaw J.R. and Tennissen A.M. (2001). Activity-dependent spinal cord plasticity in health and disease. *Annu Rev Neurosci* 24:807–843.

Ziemann U., Ilic T.V., Pauli C., Meintzschel F., and Ruge D. (2004). Learning modifies subsequent induction of long-term potentiation-like and long-term depression-like plasticity in human motor cortex. *J Neurosci* 24(7):1666–1672.

24

REVIEW CHAPTER

FUNCTIONAL ELECTRICAL THERAPY OF UPPER EXTREMITIES

MIRJANA B. POPOVIĆ[1,2,3] AND DEJAN B. POPOVIĆ[1,3]

[1]*School of Electrical Engineering, University of Belgrade, Belgrade, Serbia*
[2]*Institute for Multidisciplinary Research, Belgrade, Serbia*
[3]*Center for Sensory–Motor Interaction, Aalborg University, Aalborg, Denmark*

SUMMARY

The functional electrical therapy (FET) systems described here relate to the methods used in recovery of upper limb functioning in humans with a sensory–motor lesion in the central nervous system (CNS) that has impacted reaching and grasping. FET is a noninvasive neuromuscular stimulation modality that integrates task-oriented intensive exercise with electrical stimulation. This ideal modality for activity-dependent neurorehabilitation consists of a multichannel stimulator, a set of surface electrodes, and the controller that mimics spatial and timing synergies of hand muscle activation found in able-bodied subjects when performing functional tasks. FET functionally assists the user while the sensory feedback associated with the performance of the activities contributes to the relearning process of the brain. This functional sensory feedback should be maximized both in intensity and duration and should not be limited to short clinical training sessions. The neurophysiological basis of these therapeutic effects is still insufficiently understood. Therefore, it is as yet unclear how the therapy should be optimized. Better results may be expected with selective functional electrical stimulation.

Introduction to Neural Engineering for Motor Rehabilitation, First Edition. Edited by Dario Farina, Winnie Jensen, and Metin Akay.

INTRODUCTION AND BACKGROUND

Motor Recovery

Motor impairment after stroke is the major cause of permanent disability. Recovery of hand motor function is crucial in order to perform activities of daily living. Currently, neuroscience-based rehabilitation seeks to promote spontaneous functional motor recovery by capitalizing on the inherent potential of the brain for plastic reorganization after stroke. Rehabilitative interventions aimed at functional motor recovery in stroke patients are based mainly on active movement training. Clinical trials have provided new insights into the methods used to assist motor recovery after stroke. A common conclusion is that interventions emphasizing intense, active, repetitive task-oriented movements have high value in this regard.

To promote the effects of training and practice, biomedical engineers, neuroscientists, and clinicians have started an intense joint collaboration since the early 1990s. This technological approach holds the promise of enhancing traditional poststroke recovery in different ways: as intensive exercise and augmented feedback (Sunderland et al., 1992), constraint-induced movement therapy (Taub et al., 1999), and exercise in virtual environments could provide feedback to aid skills learning (Merians et al., 2006); robotic assistive devices with sensory feedback for repetitive practice could provide therapy for long periods of time, consistently and measurably (Takahashi et al., 2008; Volpe et al., 2008, 2009); functional electrical stimulation (FES) of muscles might enable movements not otherwise possible during the practice of tasks such as reaching to grasp an object (Alon et al., 2007; D.B. Popovic and M.B. Popovic, 2010). These are only a part of the increasing number of technological developments that have been applied in samples of stroke patients and have been shown to provide a clear incremental reduction of motor impairments. Thus they offer the opportunity to build a better outcome for patients. These treatments are based on the ability of the patients to perform actions with the affected hand or arm and therefore require residual motor ability. Many patients, however, are prevented from engaging in training based on these treatments due to having no residual hand motor functions.

Electrotherapy

Electrotherapy was introduced by Liberson et al. (1961) in order to reduce foot drop. Merletti and co-workers (Merletti et al., 1975) applied two or three independently controlled channels of electrical stimulation (ES) to augment elbow extension and fingers-and-wrist extension in hemiplegic patients. With stimulation, patients were able to move an object from one location to another, which they could not do without the stimulation. The conclusions from this small clinical trial were that an improvement was obtained in elbow extension without stimulation over a period of three months, but not of hand opening movement.

After initial positive results, ES systems for functional support of the upper extremities were developed for spinal cord injury and stroke, both implanted and with electrodes attached to the skin. In many cases, these systems were developed to provide immediate functional support, although therapeutic effects, resulting in improved voluntary movement control, have been reported in several studies.

More recently, studies have been performed that investigate the therapeutic effect of ES primarily. Their shared conclusion is that ES of peripheral sensory–motor systems works as a therapy (Lin et al., 2010; Hara, 2010; Chen et al., 2010; Thrasher et al., 2008; M.R. Popovic et al., 2006; Hughes et al., 2009), although there is no conclusive guidance on which technique works the best for a given indication (Chae et al., 1998; Francisco et al., 1998; Dimitrijević et al., 1996; Sonde et al., 2000).

Functional Electrical Therapy

Functional electrical therapy (FET) is a noninvasive neuromuscular stimulation modality that assists motor activities intended by the user, thus enabling a paretic user to perform functional activities during a period when he or she is not yet able to perform these functions self-sustained, thus promoting normal functioning.

In parallel, FET activates sensory channels that provide maximal afferent inflow to the brain, consistent with the performance of the activities, thus training the neuromuscular system to perform the function independently. The FET approach promotes functional reorganization of the brain and, consequently, functional relearning. The functional sensory information thus generated is hypothesized to result in intensive functional brain training of the activities performed (Fig. 24.1). The functional support and maximal sensory functional feedback to the spinal cord and the brain make FET very different and, preliminary data indicates, superior to alternative forms of traditional therapies (neuromuscular facilitation, therapeutic neuromuscular stimulation, below-motor-threshold stimulation, intensive exercise, etc.).

During the therapy period, a FET system can be provided to each stroke patient for use during his or her daily activities, thus resulting in intensive training. Recent studies suggest that FET is a promising modality for promoting the relearning of reaching and grasping in paretic persons (M.B. Popović et al., 2002; D.B. Popović et al., 2009b).

Functional electrical therapy (FET) may be applied with stimulation systems that use several surface electrodes, a multichannel stimulator, and a preprogrammed sequence of stimulation that can be triggered by a switch, several switches, or signals from a sensor (Nathan, 1993; Saxena et al., 1995; Prochazka et al., 2001; D.B. Popović and M.B. Popović, 1998; M.B. Popović and D.B. Popović, 2001; Fiseković and Popović, 2001). Various stimulation systems have been evaluated: with implantable electrodes, or fully implantable systems that use a preprogrammed stimulation that is controlled by a switch or sensory

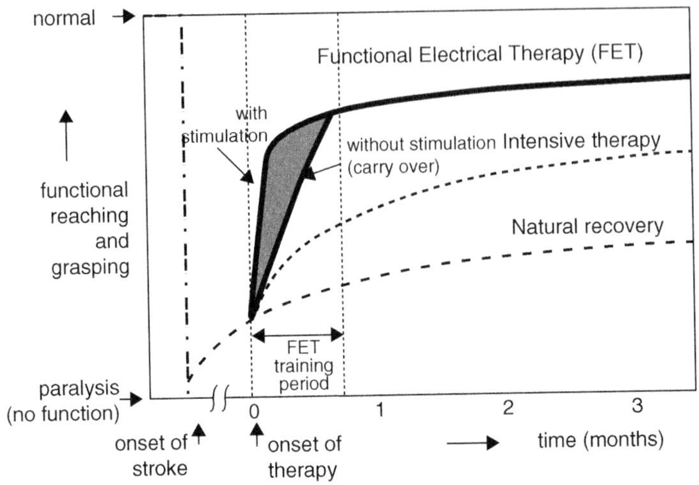

Figure 24.1. Hypothesis for recovery of functioning promoted by functional electrical therapy (FET).

signal in an open-loop or closed-loop control scheme. Several systems based on surface stimulation were designed to operate in humans with chronic paralysis or paresis caused by a cerebrovascular accident or spinal cord injury (e.g., NESS, Bionic Glove, COMPEX, Belgrade Grasping System (BGS), UNAFET, ACTIGRIP) (Prochazka et al., 1997; D.B. Popović et al., 1999; D.B. Popović and M.B. Popović, 1998; Ijzerman et al., 1996; Knutson et al., 2006).

These stimulators use the well-known principles described basically in the work of Liberson et al. (1961). Most of the stimulation systems possess the flexibility to be programmed for various applications other than the one that they were originally designed for; however, the best stimulation protocol is unknown.

METHODOLOGY

The FET methods described here comprise the use of a novel apparatus that provides missing or diminished functions of the upper limb during intensive daily exercise performing typical daily activities that require these functions. The FET system consists of a multichannel stimulator, a set of surface electrodes, and the novel controller that mimics spatial and timing synergies (M.B. Popović, 2003; Iftime et al., 2005; Micera et al., 2005) of muscle activation found in able-bodied subjects when performing functional tasks. The local ethics committees approved the experimental procedure for the FET in conformity with the Declaration of Helsinki.

Patient Selection

Patients are included in a study if they have had a stroke more than two weeks and less than six months earlier, if they have no hemineglect, no other severe cognitive impairments, no severe sensory impairments, no frozen shoulder, and no contractures at the elbow and wrist, are aged over 18, and have cognitive status sufficient to learn how to use FET. Patients are excluded for the following reasons: if, prior to stroke, they were dependent on care for activities of daily living; if they have a severe medical condition in either arm that precludes participation in the study; if they have a previous disease or injury, a pre-existing neurological disease or injury, a severe cognitive disability, or severe aphasia; or if they need electrical life support devices. Informed consent must be obtained from each patient.

Patients are characterized as belonging to the higher or lower functioning group based on their ability to actively control the wrist and fingers when entering the study. The patients in the higher functioning group (HFG) could actively extend the wrist more than 20°, and the metacarpophalangeal (MCP) and interphalangeal (IP) joints of all digits 10°. Patients in the lower functioning group (LFG) could extend the paretic wrist 10° only, and the MP and IP joints of the thumb and a minimum of two other digits 10°.

Functional Electrical Therapy

Functional electrical therapy is applied daily for three consecutive weeks. This is an average length of time that patients spend in a rehabilitation institution. The FET session consists of 30-minute electrically assisted grasping while functionally using various objects (e.g., can, pen, comb, toothbrush, finger food). The objects are selected to force the subjects to practice palmar (Fig. 24.2, bottom panels), lateral (Fig. 24.2, top panels), and precision grasps. A task-oriented use of an object consists of the following phases: reach, grasp, manipulate, bring back to the original post, and release the object. The control for FET is a slowed down replica of the grasping recorded in able-bodied subjects. The hand functioning is externally assisted with a specially designed stimulator: the modified Belgrade Grasping System (BGS) or prototype of ActiGrip or UNAFET, or other suitable neuroprosthesis. The typical stimulation parameters are $f = 50$ Hz, $T = 200$ μs, and $I = 20$–45 mA.

Four channels of ES are applied via self-adhesive surface electrodes positioned over the following muscle groups: finger flexors (flexor digitorum profundus m. and flexor digitorum superficialis m.), finger extensors (extensors communis digitorum m.), thumb extensor (extensor pollicis longus m.), and the thenar muscle group (abductor pollicis m. and m. opponens); Figure 24.2. The positions of electrodes are carefully marked on the skin; this helps the patients later to position the electrodes on their own. Patients use a simple toggle switch: first push triggers “the open to grasp synergy” and

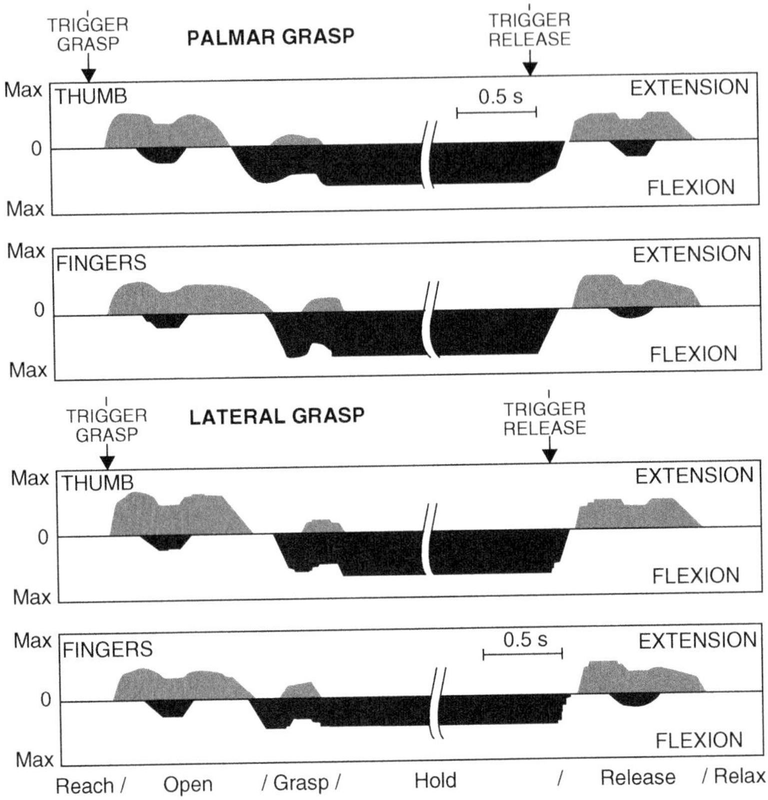

Figure 24.2. Stimulation pattern for the two most common grasps: lateral (top two panels) and palmar (bottom two panels).

second triggers "the release sequence" that assists hand functioning during FET sessions.

Assessment

The outcome measures of FET include the following: Upper Extremity Functioning Test (UEFT) to assess the performance of activities of daily living (ADLs); Drawing Test (DT) to assess coordination of the arm joint movements; Modified Ashworth Scale (MAS) to assess spasticity of key arm muscles; and Reduced Upper Extremity Motor Activity Log (RUE/MAL) questionnaire to assess personal satisfaction derived from daily functional arm use.

The first three outcome measures are evaluated before treatment onset (0 weeks), upon completion of treatment (3 weeks), and 6, 13, and 26 weeks after

Date	23.12.2008 Before FET		22.01.2009 After FET		27.07.2009 6 mo follow-up	
ACTIVITY	YES/ NO	NUMBER OF REPETITIONS	YES/ NO	NUMBER OF REPETITIONS	YES/ NO	NUMBER OF REPETITIONS
combing hair	NO	0	YES	3	YES	6
using a fork	NO	0	YES	15	YES	17
picking up a VHS tape	YES	4	YES	12	YES	11
picking up a full beer can	NO	0	YES	16	YES	16
picking up a full small (pop/soda) can	NO	0	YES	21	YES	26
writing with a pen	NO	0	NO	0	YES	4
handling the telephone receiver	YES	2	NO	0	YES	13
brushing teeth	NO	0	YES	9	YES	9
pouring from a one litter juice box	NO	0	NO	0	NO	0
drinking feom a mug	NO	0	YES	15	YES	13
handling finger food	NO	0	YES	1	YES	5

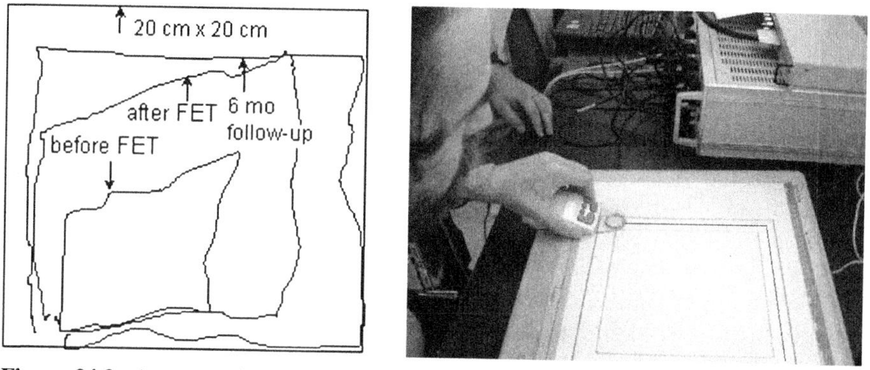

Figure 24.3. An example of assessment performed before and after 3 weeks of FET, and 6 months later: (top) UEFT; (bottom) DT—right, patient performs DT on digitizing board; left, results of DT for one patient who enrolled in the clinical trial in acute phase.

the treatment. The assessment of user needs is made at the beginning and at the end of follow-up (0 and 26 weeks) by the RUE/MAL questionnaire.

The Upper Extremity Function Test (UEFT; M.B. Popovic et al., 2004) determines the number of successfully accomplished trials selected from ADLs during 2-minute intervals (Fig. 24.3, top table). The following 11 tasks are tested: (1) combing hair, (2) using a fork, (3) picking up a VHS tape, (4) drinking from a full juice can, (5) drinking from a small bottle, (6) writing with a pen, (7) using the telephone receiver, (8) brushing teeth, (9) pouring from a full one-liter box, (10) drinking from a mug, and (11) eating finger food. The activities in UEFT are selected to include palmar (tasks 4, 5, 7, and 9), lateral grasps (1, 2, 3, 8, and 10), and precision grasps (6 and 11). The UEFT also evaluates the ability to handle small objects (1, 2, 5, 6, 8, 10, and 11) and large objects (3, 4, 7, and 9), as well as light objects (1, 2, 3, 6, 8, and 11) and heavy objects (4, 5, 7, and 9). UEFT determines the differences in the performance of selected ADL sbefore and after the therapy.

The Drawing Test (DT) was validated as a tool (Eder et al., 2005) to assess coordination capacity between the shoulder and elbow. For this purpose we have developed a simple method that comprises PC, digitizing board with cordless mouse or pen, and custom-made software. In DT subjects are required to track a square (20 cm × 20 cm) on the digitizing board (Fig. 24.3, bottom left panel). The movements are self-paced. The DT score is the ratio between two surfaces: the achieved area and the area of the desired square.

The Modified Ashworth Scale (MAS) is an ordinal scale of tone intensity (0, 1, 1+, 2, 3, and 4) of key muscles of the hemiplegic arm. It is reproducible; it is proven reliable only at the elbow; variability in the patients' condition and the test conditions contributes to unreliability.

The Reduced Upper Extremity Motor Activity Log (RUE/MAL) is a structured interview examining "how much" and "how well" the subjects use their more affected arm outside of the laboratory setting. The subjects rate the amount of use of their hemiplegic arm ("Amount" scale) and the quality of their movement during the functional activities indicated ("How Well" scale). The scores in the "Amount" and "How Well" scales range from 60 to 0. The results are expressed as a percentage of the maximum score. The RUE/MAL questionnaire includes 12 activities: pick up the phone; open a door; eat finger foods; control the bathroom faucet; pick up a glass, a bottle, or a can; brush teeth; use a key to unlock the door; write on a paper; use a CD or floppy disk; use utensils for eating; pick up a cup by the handle; and carry an object in the hand.

RESULTS

Functional electrical therapy was evaluated in two clinical trials (M.B. Popovic et al., 2003, 2004).

In the first clinical study (M.B. Popovic et al., 2003) we described a clinical randomized single-blinded study of the effects of FET on the paretic arms of subjects with acute hemiplegia caused by strokes. FET consisted of a 30-min exercise for 3 consecutive weeks in addition to conventional therapy. Twenty-eight acute hemiplegic subjects participated in a 6-month study. The subjects were divided into LFGs and HFGs and were randomly assigned to controls or FET groups. Controls performed the same amount of exercise, yet without ES. The outcome measures included the UEFT, DT, MAS, and RUE/MAL.

The FET and control groups showed a recovery trend in all outcome measures. The gains in FET groups were much larger compared with the gains in control groups. The speed of recovery in FET groups was substantially faster compared with the recovery rate in control groups during the first 3 weeks (treatment). The LFG subjects showed less improvement than the HFG subjects in both the FET and control groups.

In the second study (M.B. Popovic et al., 2004) we reported that there are indications that both intensive exercise and ES have a beneficial effect on arm function in poststroke hemiplegic patients. FET was applied 30 min daily for 3 weeks. Forty-one acute hemiplegic patients volunteered in the 18-month single-blinded crossover study. Nineteen patients (Group A) participated in FET during the acute phase of their hemiplegia, and 22 patients (Group B) participated in FET during the chronic phase of their hemiplegia. Group B patients were controls during FET in acute hemiplegia, and Group A patients were controls during the FET in chronic hemiplegia. Thirty-two patients completed the study. The outcome measures included use of UEFT, DT, MAS, and RUE/MAL.

Patients who participated in Group A reached functionality of the paretic arm, on average, in less than 6 weeks, and maintained this near-normal use of the arm and hand throughout the follow-up. The gains in all outcome scores were significantly larger in Group A after FET and at all follow-ups compared with the scores before the treatment. The gains in patients who participated in Group B were measurable, yet not significant. The speed of recovery was higher during the FET period compared with the follow-up period. The gains in Group A were significantly larger compared with the gains in Group B. The conclusion was that FET greatly promotes the recovery of the paretic arm if applied during the acute phase of hemiplegia.

The results just described were included in a FES Topic Cluster List (FES-TCL, 1983–2010) and the 2004 National Clinical Guidelines (RCPL, 2004) for stroke rehabilitation.

CONCLUSIONS

Functional electrical therapy provides a method for integrating ES of only hand muscles and the whole upper extremity exercise in order to improve the physical ability of persons having a disabled upper extremity due to sensory–motor impairment of the CNS. FET allows patients to train performing functional movements and learn new strategies of optimal use of preserved sensory–motor mechanism. This training could be understood as the process of recalibration of the natural control system (Iftime Nielsen et al., 2010). FET is the ideal modality for activity dependent neurorehabilitation.

The FET method comprises the following steps:

1. Impose functional ES of neural pathways from the FET apparatus, based on a voluntary command, using a stimulation pattern that mimics the natural flow of neural activities to the impaired upper extremity.
2. Generate the missing components of a functional movement in parallel with the voluntary exercise of the same functional movement based on

stimulation of the efferent neural pathways synchronized with volitional movement.
3. Enhance the afferent input by ES in synchrony with the biological afferent activity caused by the functional movement of the limb.

The fact that user-driven electrical muscle stimulation—but not machine-paced electrical muscle stimulation—improves the motor function of the hemiparetic arm and hand (Hara, 2008) may justify the encouraging results of clinical studies with FET.

Functional electrical therapy (Barsi et al., 2008; Eder et al., 2006), as well as most of the techniques mentioned in the Introduction (Liepert, 2010), considers to only a peripheral degree the manner in which the different treatments benefit from the now very well accepted effects of neural plasticity. Namely, the human brain has a power to change its structure and function in order to optimize the performance of frequently performed activities. The brain learns from its own actions. The afferent sensory information, giving feedback about the consequences of voluntarily controlled actions, is essential for this learning process. There is increasing evidence that it is the way adult persons relearn their motor activities after a stroke.

Discomfort and Pain

Electrical stimulation activates not only motor fibers, but also sensory fibers and pain receptors too (Bastos Conforto et al., 2010), thus producing discomfort and pain. It is expected that selective stimulation (Popovic-Bijelic et al., 2005) not only minimizes discomfort (Kuhn et al., 2010), but also provides better performance (D.B. Popovic and M.B. Popovic, 2009). Discomfort may be minimized by optimizing stimulation parameters, electrode location, and electrode size. For example, small electrodes (0.8×0.8 cm^2) are more comfortable for thin fat layers and superficial nerves, and large electrodes (4.1×4.1 cm^2) are more comfortable for thicker fat layers and deeper nerves at a constant recruitment (Kuhn et al., 2010).

Other FET Applications

Functional electrical therapy methodology has been translated to therapy for gait (Kojovic et al., 2009) and lower back pain (D.B. Popović et al., 2009a). More recent technology may apply FET in conjunction with telerehabilitation (Hermann et al., 2010).

FUTURE DIRECTIONS

Application of ES in stroke patients is still limited to a few clinical groups that provide dedicated clinical services. Therefore, three directions for future research are recommended by Ijzerman et al. (2009). Our knowledge of brain plasticity and the use of different ES strategies to modulate the neural system should be expanded. We foresee therapies combining different training principles, for example, FES and robotics or neuromodulating drugs. Finally, it is expected that clinical and economic evidence will influence transferring these interventions to a wider community.

To improve the quality of stroke rehabilitation, quality initiatives (Dowla and Chan, 2010) must be scientifically driven, continuing the trend of evidence-based practice (Ottawa Panel, 2006; EBRSR, 2005) in medicine.

A promising approach is to deploy BCI technology as a tool for functional motor recovery of the upper extremities after stroke (Millán et al., 2010). Despite there being no literature available on the use of BCI-controlled FES orthosis, some studies on the topic of BCI in conjunction with FES have emerged.

ACKNOWLEDGMENTS

We thank Thomas Sinkjaer from SMI, Aalborg University, Denmark; Laslo Švirtlih, Aleksandra Stefanović, Stevan Jović, and Andjelka Pjanović from Rehab Center "M Zotovic," Belgrade, Serbia; Peter Broger Christensen and Ulle Sonnenborg from Hummel Center, Aarhus, Denmark;

Nikola Jorgovanović, Strahinja Došen, and Goran Bijelić from UNA Systems (previously UNA Consulting), Belgrade, Serbia, for their valuable and versatile contribution in developing FET methods and systems.

Funding

The Danish National Research Foundation and the Ministry for Science and Technology, Serbia (projects #1747, #145041, #175016, #6117, and # 11019), partly supported the research.

REFERENCES

Alon G., Levitt A.F., and McCarthy P.A. (2007). Functional electrical stimulation enhancement of upper extremity functional recovery during stroke rehabilitation: a pilot study. *Neurorehabil Neural Repair* 21:207–215.

Barsi G.I., Popovic D.B., Tarkka I.M., Sinkjaer T., and Grey M.J. (2008). Cortical excitability changes following grasping exercise augmented with electrical stimulation. *Exp Brain Res* 191(1):57–66.

Bastos Conforto A., Nocelo Ferreiro K., Tomasi C., et al. (2010). Effects of somatosensory stimulation on motor function after subacute stroke. *Neurorehabil Neural Repair* 24(3):263–272.

Chae J., Bethoux F., Bohine T., Dobos L., Davis T., and Friedl A. (1998). Neuromuscular stimulation for upper extremity motor and functional recovery in acute hemiplegia. *Stroke* 29(5):975–979.

Chen W.L., Chen S.C., Chen C.C., Chou C.H., Shih Y.Y., Chen Y.L., and Kuo T.S. (2010). Patient-driven loop control for ambulation function restoration in a non-invasive functional electrical stimulation system. *Disabil Rehabil* 32(1):65–71.

Dimitrijević M.M., Stokić D.S., Wawro A.W., and Wun C.C. (1996). Modification of motor control of wrist extension by mesh-glove electrical afferent stimulation in stroke patients. *Arch Phys Med Rehabil* 77(3):252–258.

Dowla N. and Chan L. (2010). Improving quality in stroke rehabilitation. *Top Stroke Rehabil* 17(4):230–238.

Eder C., Popovic M.B., Popovic D.B., Stefanovic A., Schwirtlich L., and Jovic S. (2005). The drawing test: a tool for assessing motor coordination in post-stroke hemiplegic subjects. *Arch Phys Med Rehabil* 86(2):289–295.

Eder C.F., Sokić D., Čovočković-Šternić N., Mijajlović M., Savić M., Sinkjær T., and Popović D.B. (2006). Symmetry of post-movement beta-ERS and motor recovery from stroke: a low-resolution EEG pilot study. *Eur J Neurol* 13:1312–1323.

EBRSR (2005). *Upper Extremity Interventions*, 8th ed. London, ON, Canada: Evidence Based Review of Stroke Rehabilitation (EBRSR). http://www.ebrsr.com/uploads/Module-10_upper-extremity.pdf (accessed April 9, 2013).

FES-TCL (1983–2010). Functional Electrical Stimulation Topic Cluster List of JRehResDev (1983–2010). Available at http://www.rehab.research.va.gov/cluster/fes.html (accessed April 9, 2013).

Fišeković N. and Popović D.B. (2001). New controller for functional electrical stimulation systems. *Med Eng Phys* 23:391–399.

Francisco G., Chae J., Chawla H., et al. (1998). Electromyogram-triggered neuromuscular stimulation for improving the arm function of acute stroke survivors: a randomised pilot study. *Arch Phys Med Rehabil* 79(5):570–575.

Hara Y. (2008). Neurorehabilitation with new functional electrical stimulation for hemiparetic upper extremity in stroke patients. *J Nippon Med Sch* 75:4–14.

Hara Y. (2010). Novel functional electrical stimulation for neurorehabilitation. *Brain Nerve* 62(2):113–124.

Hermann V.H., Herzog M., Jordan R., Hofherr M., Levine P., and Page S.J. (2010). Telerehabilitation and electrical stimulation: an occupation-based, client-centered stroke intervention. *Am J Occupational Therapy* 64(1):73–81.

Hughes A.M., Freeman C.T., Burridge J.H., Chappell P.H., Lewin P.L., and Rogers E. (2009). Feasibility of iterative learning control mediated by functional electrical stimulation for reaching after stroke. *Neurorehabil Neural Repair* 23(6):559–568.

Iftime Nielsen S.D., Došen S., Popovic M., and Popovic D. (2010). Learning arm/hand coordination with an altered visual input. *Computational Intelligence Neurosci* Article ID 520781, 12 pages, DOI:10.1155/2010/520781.

Iftime S.D., Egsgaard L.L., and Popovic M.B. (2005). Automatic determination of synergies by radial basis function artificial neural networks for control of a neural prosthesis. *IEEE Trans Neur Syst Rehab Eng* 13(4):482–489.

Ijzerman M.J., Stoffers T.S., Klatte M.A., Snoeck G., Vorsteveld J.H., and Nathan R.H. (1996). The NESS Handmaster orthosis: restoration of hand function in C5 and stroke patients by means of electrical stimulation. *J Rehabil Sci* 9:86–89.

Ijzerman M.J., Renzenbrink G.J., and Geurts A.C.H. (2009). Neuromuscular stimulation after stroke: from technology to clinical deployment. *Experts Rev Neurotherapies* 9(4):541–552.

Knutson J., Audu M., and Triolo R. (2006). Interventions for mobility and manipulation after spinal cord injury: a review of orthotic and neuroprosthetic options. *Top Spinal Cord Injury Rehabil* 11(4, March):61.

Kojovic J., Djuric-Jovcic M., Dosen S., Popovic M.B., and Popovic D.B. (2009). Sensor-driven four-channel stimulation of paretic leg: functional electrical walking therapy. *J Neurosci Methods* 181(1):100–105.

Kuhn A., Keller T., Lawrence M., and Morari M. (2010). The influence of electrode size on selectivity and comfort in transcutaneous electrical stimulation of the forearm. *IEEE Neural Syst Rehabil Eng* 18(3):255–262.

Liberson W.T., Holmquest H.J., Scott D., et al. (1961). Functional electrotherapy, stimulation of the peroneal nerve synchronized with the swing phase of the gait of hemiplegic patients. *Arch Phys Med* 42:101–105.

Liepert J. (2010). Evidence-based methods in motor rehabilitation. *J Neurologie Neurochirurgie Psychiatrie* 11(1):5–10.

Lin Z., Chen L., and Yan T. (2010). Effectiveness of functional electrical stimulation on functional recovery of the affected upper extremities in subjects with stroke in randomized controlled trial. *Chinese J Rehabil Med* 25(2):152–155.

Merians A.S., Poizner H., Boian R., Burdea G., and Adamovich S. (2006). Sensorimotor training in a virtual reality environment: does it improve functional recovery post-stroke? *Neurorehabil Neural Repair* 20:252–267.

Merletti R., Aćimović R., Grobelnik S., and Cvilak G. (1975). Electrophysiological orthosis for the upper extremity in hemiplegia: feasibility study. *Arch Phys Med Rehabil* 56(12):507–513.

Micera S., Carpaneto J., Posteraro F., Cenciotti L., Popovic M.B., and Dario P. (2005). Characterization of upper arm synergies during reaching tasks in subjects affected by neurological disorders. *Clin Biomech* 20(9):939–946.

Millán J.d.R., Rupp R., and Müller-Putz G.R. (2010). Combining brain–computer interfaces and assistive technologies: state-of-the-art and challenges. *Frontiers Neuroprosthet* 4(161):1–15.

Nathan R.H. (1993). Control strategies in FNS systems for the upper extremities. *Crit Rev Biomed Eng* 21(6):485–568.

Ottawa Panel (2006). Ottawa Panel evidence-based clinical practice guidelines for post-stroke rehabilitation. Topics in Rehabilitation, Clinical Practice Guidelines. *Top Stroke Rehabil* 13(1):1–116. Available at http://www.thomasland.com/strokeguide.html (accessed MM DD, YYYY).

Popović D.B and Popović M.B. (1998). *Belgrade Grasping System, Electronics, Faculty of Electrical Engineering, University of BanjaLuka*, 2(1):25–32.

Popovic D.B. and Popovic M.B. (2009). Automatic determination of the optimal shape of a surface electrode: selective stimulation. *J Neurosci Methods* 178(1): 174–181.

Popovic D.B. and Popovic M.B. (2010). New trends in neurorehabilitation of subjects with central nervous system lesions. *Zdravniski Vestnik* 79:296–301.

Popović D.B., Stojanović A., Vulović D., Pjanović A., Radosavljević S., and Popović M.B. (1999). Clinical evaluation of the Bionic glove. *Arch Phys Med Rehabil* 80: 299–304.

Popović D.B., Popović M.B., and Sinkjær T. (2002). Neurorehabilitation of upper extremities in humans with sensory-motor impairment. *Neuromodulation* 5(1):54–67.

Popović D.B., Popović M.B., Sinkjær T., Stefanovic A., and Schwitlich L. (2004). Neuroprosthesis for therapy of paretic arm in hemiplegic subjects: a cross-over study. *Can J Physiol Pharmacol* 82:749–756.

Popović D.B., Bijelić G., Miler V., Došen S., Popović M.B., and Schwirtlich L. (2009a). Lumbar stimulation belt for therapy of low-back pain. *Artif Organs* 33(1):54–60.

Popović D.B., Sinkjær T., and Popović M.B. (2009b). Electrical stimulation as a means for achieving recovery of function in stroke patients. *J Neurorehabil* 25:45–58.

Popović M.B. (2003). Control of neural prostheses for grasping and reaching. *Med Eng Phys* 25(1):41–50.

Popović M.B. and Popović D.B. (2001). Cloning biological synergies improves control of elbow neuroprosthesis. *IEEE Eng Med Biol* 20(1):74–81.

Popovic M.B., Popovic D.B., and Sinkjær T. (2002). Restitution of reaching and grasping promoted by functional electrical therapy. *Artif Organs* 26(3):271–275.

Popović M.B., Popović D.B., Sinkjær T., Stefanovic A., and Schwirtlich L. (2003). Clinical evaluation of functional electrical therapy in acute hemiplegic subjects. *J Rehabil Res* Dev 40(5):443–454.

Popovic M.B., Popovic D.B., Schwirtlich L., and Sinkjær T. (2004). Clinical evaluation of functional electrical therapy (FET) in chronic hemiplegic subjects. *Neuromodulation* 7(2):133–140.

Popovic M.R., Thrasher T.A., Adams M.E., Takes V., Zivanovic V., and Tonack M.I. (2006). Functional electrical therapy: retraining grasping in spinal cord injury. *Spinal Cord* 44(3):143–151.

Popovic-Bijelic A., Bijelic G., Jorgovanovic N., Bojanic D., Popovic D.B., and Popovic M.B. (2005). Multi-field surface electrode for selective electrical stimulation. *Artif Organs* 29(6):448–452.

Prochazka A., Gauthier M., Wieler M., and Kenwell Z. (1997). The Bionic glove: an electrical stimulator garment that provides controlled grasp and hand opening in quadriplegia. *Arch Phys Med Rehabil* 78(6):608–614.

Prochazka A., Mushahwar V.K., and McCreery D.B. (2001). Neural prostheses. *J Physiol* 533(1):99–109.

RCPL (2004). *National Clinical Guidelines for Stroke*, 2nd ed. London, U.K.: Royal College of Physicians of London (RCPL), U.K. Available at http://www.rcplondon.ac.uk/publications/national-clinical-guidelines-stroke (accessed April 9, 2013).

Saxena S., Nikolić S., and Popović D.B. (1995). An EMG-controlled grasping system for tetraplegics. *J Rehabil Res Dev* 32:17–24.

Sonde L., Kalimo H., Fernaeus S.E., and Viitanen M. (2000). Low TENS treatment on post-stroke paretic arm: a three-year follow-up. *Clin Rehabil* 14:114–119.

Sunderland A., Tinson D.J., Bradley E.L., et al. (1992). Enhanced physical therapy improves recovery of arm function after stroke—a randomised controlled trial. *J Neurol Neurosurg Psych* 5(7):530–535.

Takahashi C.D., Der-Yeghiaian L., Le V., Motiwala R.R., and Cramer S.C. (2008). Robot-based hand motor therapy after stroke. *Brain* 131:425–437.

Taub E., Uswatte G., and Pidikiti R. (1999). Constraint-induced movement therapy: a new family of techniques with broad application to physical rehabilitation—a clinical review. *J Rehabil Res Dev* 36(3):237–251.

Thrasher T.A., Zivanovic V., McIlroy W., and Popovic M.R. (2008). Rehabilitation of reaching and grasping function in severe hemiplegic patients using functional electrical stimulation therapy. *Neurorehabil Neural Repair* 22(6):706–714.

Volpe B.T., Lynch D., Rykman-Berland A., Ferraro M., Galgano M., Hogan N., and Krebs H.I. (2008). Intensive sensorimotor arm training mediated by therapist or robot improves hemiparesis in patients with chronic stroke. *Neurorehabil Neural Repair* 22(3):305–310.

Volpe B.T., Huerta P.T., Zipse J.L., Rykman A., Edwards D., Dipietro L., Hogan N., and Krebs H.I. (2009). Robotic devices as therapeutic and diagnostic tools for stroke recovery. *Arch Neurol* 66:1086–1090.

25

FOCUSED CHAPTER

GAIT REHABILITATION USING NOCICEPTIVE WITHDRAWAL REFLEX–BASED FUNCTIONAL ELECTRICAL THERAPY IN STROKE PATIENTS

OLE K. ANDERSEN AND ERIKA G. SPAICH

Integrative Neuroscience Group, Center for Sensory–Motor Interaction, Aalborg University, Aalborg, Denmark

SUMMARY

The nociceptive withdrawal reflex is the motor reaction initiated to protect the body from a potentially harmful stimulus. It involves a spinal reflex pathway and activates multiple muscles in both limbs to ensure adequate withdrawal while preserving balance and maintaining the ongoing motor program. This integrated movement is highly relevant for supporting the swing phase during gait therapy in subacute hemiparetic patients as it typically involves flexion of the joints. This chapter describes the modulation of the reflex size depending on the phase of the gait cycle and the stimulation site on the foot. Flexion of the hip and knee was observed irrespective of stimulation site while the largest reflexes were detected when stimulation was delivered at heel-off. For the ankle joint, dorsiflexion was prevalent when stimulation was applied at the medial, distal sole of the foot. Applying repetitive stimulation at 15 Hz (four bursts) increased the

Introduction to Neural Engineering for Motor Rehabilitation, First Edition. Edited by Dario Farina, Winnie Jensen, and Metin Akay.

mechanical response substantially. This information is needed in order to utilize the reflex response optimally for gait support. In 15 subacute stroke patients, the swing phase was supported by the reflex evoked by four repetitive stimuli applied in the arch of the foot at heel-off during daily gait therapy. This reflex-based gait therapy was continued for one month. When tested one month after therapy, a significantly higher preferred gait velocity (average of 27%) was found compared with a control group receiving the same gait therapy but without reflex support. Future studies into more advanced stimulation paradigms are needed to explore the full potential of functional electrical therapy supported by the withdrawal reflex.

INTRODUCTION

The lower limb nociceptive withdrawal reflex (NWR) can be elicited using stimulation of the sole of the foot. It involves coordinated activation of multiple muscles in the ipsilateral and contralateral limb leading to adequate withdrawal while maintaining balance and ensuring continuation of the ongoing motor programs (Andersen 2007; Spaich et al., 2004b). The NWR has been shown to present a modular organization where the size of the reflex response and the pattern of unloading depend on the stimulation site (Andersen et al. 1999; Sonnenborg et al., 2000). The motor context may modulate the site dependency; for instance, upright standing alters the balance between flexor and extensor muscles but the intrinsic modular organization is still present (Andersen et al., 2005). The modular organization is also present during gait, where the nociceptive withdrawal reflex elicited by a single electrical stimulus at the sole of the foot elicits site-modulated responses (Spaich et al., 2004b). Moreover, the reflex response is modulated during the phases of the gait cycle as detected in healthy subjects with the strongest flexor reflexes during the swing phase (Spaich et al., 2004b).

The site- and gait-phase modulation of the reflex response indicates an interesting application of the reflex in the rehabilitation of the hemiplegic gait, where the reflex involving flexion of the ipsilateral joints may support the swing phase during gait. Furthermore, the most appropriate swing trajectory could be achieved by selecting the optimal stimulus location and stimulus timing during the gait cycle whereby the evoked response would improve ground clearance, resulting in a more functional gait. Using the NWR for supporting gait has been investigated in detail in relation to people with spinal cord injuries (Granat et al., 1993; Bajd et al., 1999), where the strong afferent input may even trigger spinal stepping (Nicol et al. 1995). Reflex habituation, that is, gradually reduced reflex size despite constant stimulus intensity, has hindered the wider use of this technique (Granat et al., 1991; Dimitrijevic et al., 1972). However, alternating the stimulus location from step to step may have a dishabituating effect. It is expected that site-specific responses will, in addition allow a better control of the evoked movement and consequently a more functional gait will be achieved and relearned.

For improving the control of the evoked response, repetitive stimulation may be used. Repetitive stimulation triggers the spinal temporal summation mechanisms (Arendt-Nielsen et al., 2000). The number of pulses and stimulus frequency could therefore be part of the control scheme. Controlling the reflex response is nevertheless complicated because the delay from stimulus onset until the mechanical reflex response is present is substantial (several hundred milliseconds), preventing traditional feedback control and, moreover, the relationship between the stimulus train and the mechanical output is partly unknown and with large individual variation.

Eliciting the withdrawal reflex by electrical stimulation allows the best control of the evoked response. However, electrical stimulation may be unpleasant, which hinders the use of this technique as part of a permanent device for patients with partly intact perception. Instead, the NWR may be used during subacute gait therapy. The impact of such training is most likely largest if reflex-supported gait is combined with intensive gait training. This may trigger constructive cortical plasticity on the basis of the sensory–proprioceptive afferent activity associated with functionally sound gait obtained via support from the reflex.

The dependence of the withdrawal reflex response on stimulation site and gait phase was tested in a group of hemiplegic patientss. Furthermore, the effects of changing the onset of a train of repetitive stimuli during the gait cycle and the use of the reflex for gait support during daily gait therapy were investigated.

MATERIALS AND METHODS

Eliciting the Nociceptive Withdrawal Reflex

The NWR was elicited by transcutaneous electrical stimulation delivered to the sole of the foot (2.63 cm^2 surface area electrodes, AMBU, Denmark) with reference to a common anode placed on the dorsum of the foot (7 × 10 cm electrode, Pals, Axelgaard Ltd., United States). The method is described in full detail in Spaich et al. (2004b) and Andersen et al. (1999). The stimulus site was computer controlled from step to step. Each stimulus consisted of a constant-current square pulse train of five 1-ms unipolar pulses delivered at 200 Hz, which was perceived as a single burst. For normalizing the stimulation intensity, the pain threshold for a single stimulus was assessed at each stimulation site in order to obtain as equal an afferent input as possible, and the stimulation intensity was set as a percentage of the pain threshold at each site.

Repetitive Stimulation

For supporting hip flexion, the interstimulus interval has to be appropriate to provoke adequate hip flexion at swing initiation. Four bursts were delivered as a fixed train at a frequency of 15 Hz.

Stimulus Timing in the Gait Cycle

Stimulus timing was controlled by a foot switch placed under the heel. A set of control steps was acquired before collecting reflex data. From these control steps, the mean duration of the gait cycle was estimated. Reflexes were subsequently elicited at fixed time intervals. To minimize habituation, an interstimulus interval of 10–15 steps was chosen.

For detecting stimulus site and gait phase dependences in hemiplegic patients, four phases of the gait cycle were selected (two time points within the stance phase and two within the swing phase).

For testing temporal summation, focus was on the period immediately after heel off-to ensure the reflex response occurred while the ipsilateral foot was in the swing phase. Four moments were selected between heel-off and approximately toe-off with a gait speed of 3 km/h on a treadmill (0 ms, around 90 ms, 180 ms, and 270 ms after heel-off). Further, to reduce interstep variability, an auditory cue was generated by the computer at a period equal to the mean control step. Using the auditory cue the subjects attempted to maintain a constant gait cycle duration and thereby also a constant walking speed.

For studying the use of reflex-supported therapy, stimulation was delivered immediately after heel-off.

Functional Electrical Therapy Protocol

Two groups of acute stroke individuals received 20 days of gait training, 30 minutes per day, with at least 15 minutes of overground walking. In one group, gait was supported by electrical stimulation delivered to the arch of the foot and timed manually at heel-off (hence there were no alternations between stimulation sites). The other group received no electrical stimulation. Evaluations of walking quality were performed at inclusion, immediately after completion of training, and 1 month after completion of training.

Subjects

For detecting reflex modulation during the hemiplegic gait, seven chronic stroke patients aged 55–65 years participated in the experiments (full details are given in Spaich et al., 2006).

Fourteen healthy subjects aged 22–32 years participated in the experiments to study the optimal stimulation time and temporal summation when using repetitive stimulation (for full details see Spaich et al., 2009).

Thirty stroke patients (36–83 years old) participated in a single blinded experiment to study the use of reflex support during daily gait therapy. Only patients experiencing their first ever cerebrovascular accident (CVA) or their second CVA with asymptomatic first CVA were included. The patients, who were included no more than six weeks after the stroke and were able to walk a maximum of 10 meters without help from a physiotherapist, were allocated to one of two groups balanced at inclusion regarding walking performance assessed by the Functional Ambulation Category (FAC) score (Holden et al., 1984).

Outcome Measures

Surface electromyograms (EMGs) were obtained from main lower leg muscles of the ipsilateral leg of healthy and chronic stroke individuals. EMG was acquired using a bipolar configuration, amplified and band-pass filtered (5–500 Hz, 2nd order) before sampling at 2 kHz. Furthermore, leg kinematics was assessed in the sagittal plane by goniometers on the ipsilateral ankle, knee, and hip joints (Biometrics Ltd., Gwent, United Kingdom). The FAC score and walking velocity were assessed in the group who received daily gait therapy while the subjects walked between 1 and 4 times along a 9-m-long line.

RESULTS

Stimulus Site and Gait Phase Modulation in Chronic Stroke Patients

Gait phase modulation was observed in the EMG recorded from the tibialis anterior (TA) muscle (see Fig. 25.1). Hence, the largest EMG response was

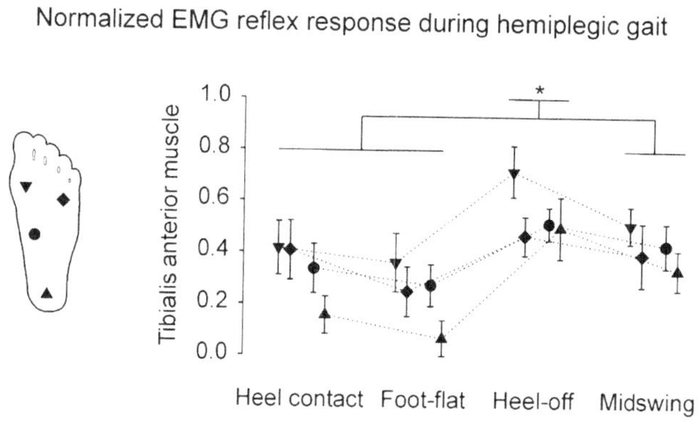

Figure 25.1. Tibialis anterior EMG reflex response to electrical stimulation at the specified stimulation sites (see symbols in the inset) during hemiplegic gait. The tibialis anterior reflex response was assessed in the interval 60–300 ms after stimulus onset as the difference between the mean poststimulation EMG and the corresponding mean EMG activity in a control step recorded prior to stimulation. All values are normalized to the peak EMG activity detected during unperturbed gait (grand mean across subjects is $72.4 \pm 17.3\ \mu V$) and presented as mean ± standard error of the mean; the asterisk (*) indicates that the reflexes detected after stimulation at heel-off were larger than those following stimulation at other moments during the gait cycle.

observed when stimulation was applied at heel-off (post-hoc analysis, $P < 0.001$). Moreover, the largest reflexes were detected when stimulation was applied to the median forefoot (post-hoc analysis, $P < 0.02$) which clearly indicates a modular organization with a receptive field for the main dorsiflexor muscle covering the medial, distal part of the sole of the foot in agreement with observations in healthy volunteers (Grimby, 1963a; Spaich et al., 2004b). A perspective of this organization is the possibility to steer the evoked movement by selecting appropriate stimulation site and timing.

For the ankle joint, however, there was no stimulation site dependency in the kinematic response as dorsiflexion was evoked at all four sites, in contrast to observations in young and middle-aged healthy volunteers (Spaich et al., 2004a, 2006). Using a single stimulus an average of 40% higher than the pain threshold, the mean dorsiflexion response was 2.2 ± 0.2 degrees compared with the mean trajectories in steps with no stimulation. For both the knee and hip joints, flexion was also prevalent independent of stimulation site and gait phase, opposite to observations in young healthy volunteers (Spaich et al., 2004a). For the hip joint 1.8 ± 0.1 degrees of additional hip flexion were observed while the knee was additionally flexed 2.9 ± 0.2 degrees. Similar loss of stimulation site dependency has been observed in patients during rest (Grimby, 1963b).

Repetitive Stimulation Effect on Hip, Knee, and Ankle Flexion

Four repetitive stimuli at 15 Hz delivered to the sole of the foot in healthy volunteers evoked significant hip flexion (see Fig. 25.2). Though the response was independent of stimulation site (forefoot, arch of the foot, and heel), it was dependent on the phase of the gait cycle as the smallest reflexes were detected close to toe-off (272 ± 30 ms after heel-off, post-hoc analysis $P < 0.05$). Up to 12° of hip flexion was obtained for a stimulus intensity 30% above the pain threshold. The missing stimulus site dependency suggests that the reflex receptive field for the hip flexor muscles covers the entire sole of the foot, in agreement with observations in chronic stroke patients for a single stimulus (Spaich et al., 2006). Knee flexion independent of the stimulation site was evoked by the repetitive stimuli as also observed in chronic stroke patients for a single stimulus (Spaich et al., 2006); the response was, however, phase modulated as smaller reflexes were detected with increasing stimulation onsets after heel-off (post-hoc analysis $P < 0.05$). At the ankle joint, largest dorsiflexion was achieved when stimulating the forefoot and arch of the foot (post-hoc analysis $P < 0.001$) and phase modulation resulted in increased responses for increased stimulation onsets after heel-off (post-hoc analysis $P < 0.02$). In relation to rehabilitation, these results indicate that functionally appropriate joint kinematics can be obtained by repetitively stimulating the sole of the foot with the right timing.

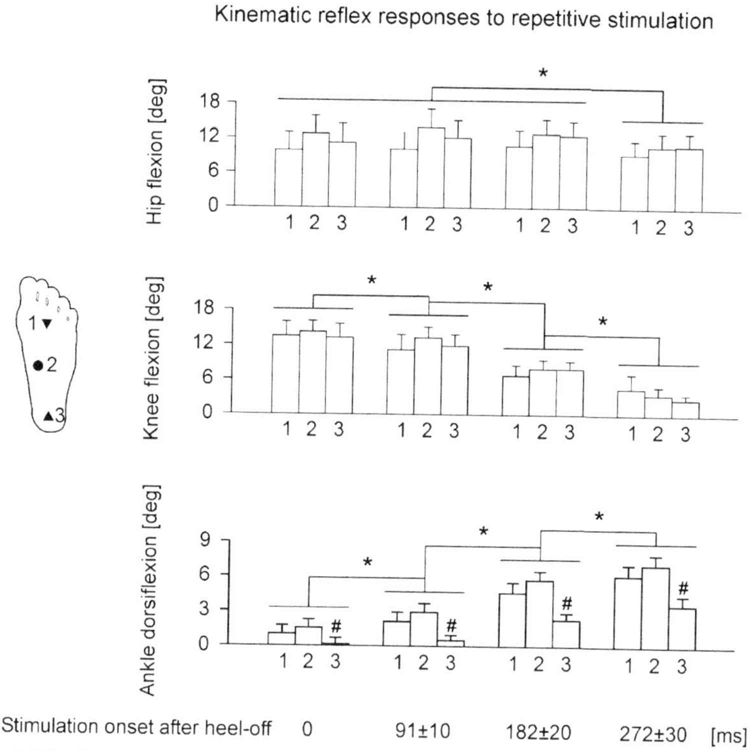

Figure 25.2. Average kinematic reflex responses evoked by repetitive stimulation at 15 Hz across 14 healthy subjects. The stimulation onsets measured from the moment of heel-off are indicated. Values are mean plus standard error of the mean; # indicates site modulation, and * indicates phase modulation. The insert illustrates the three stimulation sites on the sole of the foot.

Withdrawal Reflexes for Functional Electrical Therapy

Supporting the gait function in the acute phase poststroke with electrical stimulations in the arch of the foot at heel-off to evoke the withdrawal reflex had a positive effect on walking velocity (see Fig. 25.3). Statistically, the preferred walking velocity was faster in the patients who received electrical stimulation during therapy, but only 1 month after termination of therapy, whereas no differences were detected at inclusion and immediately after finishing therapy (post-hoc analysis $P < 0.01$). A tendency for group differences in walking velocities was detected at inclusion; however, the groups were matched based on FAC scores (1 ± 0.2, range 0–2, for both groups at inclusion). Finally, the individuals who received withdrawal reflex stimulation had a tendency to score better in the FAC test at completion of treatment and one month later.

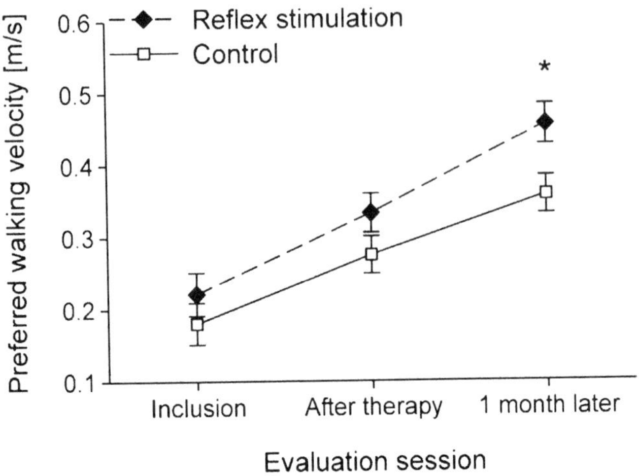

Figure 25.3. Average preferred walking velocity (mean ± standard error of the mean) for the two groups of hemiparetic patients ($N = 15$ in each group) before daily therapy, immediately after completing daily therapy, and again one month after completing therapy. The asterisk (*) indicates $P < 0.01$ in post-hoc analysis.

CONCLUSIONS

The withdrawal reflex is an integrated movement involving flexion of the ipsilateral hip, knee, and ankle joints when elicited by a painful stimulus at the initiation of the swing phase in a group of chronic stroke patients. Repetitive stimulation provides a strong afferent input that results in significant and phase-modulated reflex responses during gait. Activation of sensory nerve fibers at one or a few skin sites on the foot of the hemiparetic limb leads to natural activation of multiple muscles whereby support to achieve sufficient ground clearance during the swing phase was generated. The reflex involves all joints in the ipsilateral leg, but balance is also preserved by reactions in the pelvis and contralateral leg as shown in healthy individuals (Rietdyk and Patla, 1998; Andersen et al., 2003; Spaich et al., 2004a). This has interesting perspectives for supporting the initiation and execution of the swing phase during subacute gait therapy in hemiparetic patients despite the discomfort associated with the painful stimulus. Proper timing is critical to ensure the provoked flexion also leads to functionally optimal forward propulsion. This can be obtained by acknowledging the substantial delay from stimulus onset until peak mechanical reflex response.

This chapter documents a positive effect from utilizing the reflex during daily gait therapy in subacute stroke patients. This effect most likely reflects cortical plasticity associated with completion of functionally appropriate movements during gait training. Hence, the improvement is a result of the voluntary effort by the patients to move the paretic leg forward combined with the electrical stimuli that ensured proper completion of the swing phase. This motor training paradigm could therefore be labeled as reflex-based functional electrical therapy (Popovic et al., 2003). This paradigm might also prevent patients from learning compensatory movements (e.g., circumduction of the leg) since ground clearance is better due to hip and ankle dorsiflexion and thereby more symmetric gait might be achieved.

Future studies of reflex-based functional electrical therapy should explore more advanced stimulation paradigms. For instance, the stimulation site could continuously be changed to reduce reflex habituation (Dimitrijevic and Nathan, 1971; Carstens and Ansley, 1993) and to obtain the most functionally appropriate movement trajectory. As an example, stimulation at places other than the sole of the foot might be relevant to generate a forward movement of the lower leg by eliciting extension reflexes (targeting the reflex-receptive fields of extensor muscles). Closed-loop control should also be tested—as the present results reflect manual triggering of the stimulation—as well as fixed stimulation intensities and sites, whereas continuous control of the stimulation parameters (site, intensity, number of pulses, etc.) might result in more optimal support for the individual patient during daily therapy (Emborg et al., 2009).

ACKNOWLEDGMENTS

The Svend Andersen Foundation and the Danish Research Council for Technology and Production have supported the authors' studies

REFERENCES

Andersen O.K. (2007). Studies of the organization of the human nociceptive withdrawal reflex. Focus on sensory convergence and stimulation site dependency. *Acta Physiol* 189(Suppl 654):1–35.

Andersen O.K., Sonnenborg F.A., and Arendt-Nielsen L. (1999). Modular organization of human leg withdrawal reflexes elicited by electrical stimulation of the foot sole. *Muscle Nerve* 22:1520–1530.

Andersen O.K., Sonnenborg F., Matjacic Z., and Arendt-Nielsen L. (2003). Foot-sole reflex receptive fields for human withdrawal reflexes in symmetrical standing position. *Exp Brain Res* 152:434–443.

Andersen O.K., Spaich E.G., Madeleine P., and Arendt-Nielsen L. (2005). Gradual enlargement of human withdrawal reflex receptive fields following repetitive painful stimulation. *Brain Res* 1042:194–204.

Arendt-Nielsen L., Sonnenborg F.A., and Andersen O.K. (2000). Facilitation of the withdrawal reflex by repeated transcutaneous electrical stimulation: an experimental study on central integration in humans. *Eur J Appl Physiol* 81:165–173.

Bajd T., Kralj A., Stefancic M., and Lavrac N. (1999). Use of functional electrical stimulation in the lower extremities of incomplete spinal cord injured patients. *Artif Organs* 23:403–409.

Carstens E. and Ansley D. (1993). Hindlimb flexion withdrawal evoked by noxious heat in conscious rats: magnitude measurement of stimulus–response function, suppression by morphine and habituation. *J Neurophysiol* 70:621–629.

Dimitrijevic M.R. and Nathan P.W. (1971). Studies of spasticity in man. 5. Dishabituation of the flexion reflex in spinal man. *Brain* 94:77–90.

Dimitrijevic M.R., Faganel J., Gregoric M., Nathan P.W., and Trontelj J.K. (1972). Habituation: effects of regular and stochastic stimulation. *J Neurol Neurosurg Psychiat* 35:234–242.

Emborg J., Bendtsen J.D., Spaich E.G., and Andersen O.K. (2009). A novel method exploiting the nociceptive withdrawal reflexes in rehabilitation of hemiplegic gait. *IFMBE Proc* 25:84–87.

Granat M.H., Nicol D.J., Baxendale R.H., and Andrews B.J. (1991). Dishabituation of the flexion reflex in spinal cord–injured man and its application in the restoration of gait. *Brain Res* 559:344–346.

Granat M.H., Heller B.W., Nicol D.J., Baxendale R.H., and Andrews B.J. (1993). Improving limb flexion in FES gait using the flexion withdrawal response for the spinal cord injured person. *J Biomed Eng* 15:51–56.

Grimby L. (1963a). Normal plantar response: integration of flexor and extensor reflex components. *J Neurol Neurosurg Psychiat* 26:39–50.

Grimby L. (1963b). Pathological plantar response: disturbance of the normal integration of flexor and extensor reflex components. *J Neurol Neurosurg Psychiat* 26: 314–321.

Holden M.K., Gill K.M., Magliozzi M.R., Nathan J., and Piehl-Baker L. (1984). Clinical gait assessment in the neurologically impaired. Reliability and meaningfulness. *Phys Ther* 64:35–40.

Nicol D.J., Granat M.H., Baxendale R.H., and Tuson S.J.M. (1995). Evidence for a human spinal stepping generator. *Brain Res* 684:230–232.

Popovic M.B., Popovic D.B., Sinkjaer T., Stefanovic A., and Schwirtlich L. (2003). Clinical evaluation of functional electrical therapy in acute hemiplegic subjects. *J Rehabil Res Dev* 40:443–453.

Rietdyk S. and Patla A.E. (1998). Context-dependent reflex control: some insights into the role of balance. *Exp Brain Res* 119:251–259.

Sonnenborg F.A., Andersen O.K., and Arendt-Nielsen L. (2000). Modular organization of excitatory and inhibitory reflex receptive fields elicited by electrical stimulation of the foot sole in man. *Clin Neurophysiol* 111:2160–2169.

Spaich E.G, Andersen O.K., and Arendt-Nielsen L. (2004a). Tibialis anterior and soleus withdrawal reflexes elicited by electrical stimulation of the foot sole during gait. *Neuromodulation* 7:126–132.

Spaich E.G., Arendt-Nielsen L., and Andersen O.K. (2004b). Modulation of lower limb withdrawal reflexes during gait: a topographical study. *J Neurophysiol* 91:258–266.

Spaich E.G., Hinge H.H., Arendt-Nielsen L., and Andersen O.K. (2006). Modulation of the withdrawal reflex during hemiplegic gait: effect of stimulation site and gait phase. *Clin Neurophysiol* 117:2482–2495.

Spaich E.G., Emborg J., Collet T., Arendt-Nielsen L., and Andersen O.K. (2009). Withdrawal reflex responses evoked by repetitive painful stimulation delivered on the sole of the foot during late stance: site, phase, and frequency modulation. *Exp Brain Res* 194:359–368.

26

FOCUSED CHAPTER

ROBOT–ASSISTED NEUROREHABILITATION

VITTORIO SANGUINETI,[1,2] MAURA CASADIO,[1,3] LORENZO MASIA,[2] VALENTINA SQUERI,[2] AND PIETRO G. MORASSO[2]

[1]*Department of Informatics, Systems and Telematics, University of Genoa, Genoa, Italy*
[2]*Department of Robotics, Brain and Cognitive Sciences, Italian Institute of Technology, Genoa, Italy*
[3]*Sensory Motor Performance Program, Rehabilitation Institute of Chicago and Department of Physiology, Northwestern University Medical School, Chicago, Illinois, USA*

SUMMARY

This chapter addresses the use of robots to facilitate the recovery of neuromotor functions. While early applications of robot therapy were based on commercially available industrial robots, there is now a large variety of devices that are specifically designed for human–robot interaction and rehabilitation applications. We discuss the underlying design criteria and address the related problem of controlling these devices.

Early approaches to robot-based therapy were mostly empirical and limited by the available hardware, designed for industrial applications. More recent applications build on the current understanding of the physiology of the reorganization of the nervous system following a lesion. The chapter is focused on rehabilitation of the upper limb, but similar principles are applicable to the lower limb.

Introduction to Neural Engineering for Motor Rehabilitation, First Edition. Edited by Dario Farina, Winnie Jensen, and Metin Akay.

We address the neuromotor recovery of chronic stroke survivors and examine two case studies in more detail. Then we discuss the available evidence on the efficacy of robot therapy. Finally, we suggest that robots may be potentially useful in the context of other neurological diseases, for example, multiple sclerosis and cerebral palsy.

INTRODUCTION AND BACKGROUND

The use of robots as aids to the treatment of persons with neuromotor disabilities has a relatively long history, dating back to the early 1990s. The large variety of existing applications may be classified as (i) assistive, (ii) cognitive and communication, and (iii) therapeutic. Assistive robots are programmed to perform physical tasks for the well-being of a person with a disability that would otherwise have to be performed by an attendant. These robots have a certain degree of autonomy but are directly controlled by the disabled person. Mobility and manipulation aids (e.g., "intelligent" wheelchairs) are examples of assistive robots. Cognitive and communication aids aim at promoting attention, motivation, and social interaction capabilities of persons with cognitive impairment (e.g., educational tools for promoting development in children with autism). Therapy robots are designed to carry out physical interaction exercises with upper or lower extremities, aimed at promoting neuromotor recovery, for instance in stroke survivors. In this chapter, we specifically focus on therapy robots aimed at the recovery of neuromotor functions. We mostly discuss rehabilitation of the upper limb, but similar principles and technologies apply to the lower limb.

Existing robot therapy devices differ widely in terms of mechanical design, number of degrees of freedom, and control architectures. One common feature is that they all incorporate sensors of different types, including movement sensors; therefore, they integrate both therapeutic and measuring functionalities. As therapeutic devices, they may be programmed to implement a variety of highly reproducible, repetitive exercise protocols and/or interaction modalities. As measuring devices, they are capable of detecting and quantifying many aspects of both movement and physical interaction with users (movements, forces, and possibly physiological signals).

Therapy robots are specifically designed to deliver and/or support rehabilitation exercises aimed at neurological patients, in particular, stroke survivors. In the following sections we discuss the technological aspects (mechanical and control design) and their applications to therapy. Then we examine two case studies in detail, and summarize the overall outcomes of robot therapy. We initially focus on the neuromotor recovery of stroke survivors, for which several randomized clinical trials are available. Finally, we discuss the potential use of robots in the context of other neurological diseases, for example, multiple sclerosis and cerebral palsy.

METHODOLOGY

Therapeutic Exercises

Different approaches have been proposed for using robots to promote neuromotor recovery. Early approaches were mostly heuristic and limited by the available hardware, designed for industrial applications. More recent applications build on the current, improving understanding of the physiology of reorganization of the nervous system after a lesion. Following the taxonomy proposed in Marchal-Crespo and Reinkensmeyer (2009), existing approaches to robot-assisted exercise can be summarized into three broad scenarios:

1. *Haptic Simulation.* Robots can be used for haptic rendering in virtual environments, in which subjects practice with a variety of interaction tasks, generally inspired by activities of daily living (ADLs). Robots and visual displays allow the joint visual and haptic interaction with virtual objects. The advantages of a virtual environment over practice in a real-life context include greater safety and flexibility, including adaptation to the subjects and the possibility to quantify performance.
2. *Challenge-Based.* The robot provides disturbances and/or perturbations that make a task more difficult or challenging with respect to their performance without the robot. Several approaches are possible. For instance, during exercise the robot may generate perturbations that oppose subjects' movements or require greater forces. Another possibility is to amplify motor errors, so that displayed movements appear to be less accurate (e.g., more curved) than they actually are.
3. *Assistive.* The robot provides forces that facilitate task performance or task completion. The goal is to help subjects to move the weakened limbs in a desired way, in specific motor tasks such as grasping, reaching, walking.

These categories are not mutually exclusive, in the sense that they can be combined in a specific application. Furthermore, there are no immediate associations between the different approaches and specific robots. Rather, most robot devices can be programmed to work according to each of the three categories. For a comprehensive review of robot therapy scenarios, the interested reader can refer to Marchal-Crespo and Reinkensmeyer (2009).

Mechanical Design

Early applications of robot therapy were based on commercially available industrial robots, whose design did not consider the safety requirements inherent in any human–robot interaction. More recent devices are specifically designed for human–robot interaction and rehabilitation applications, and they are direct descendants of the force-reflecting controllers used for teleoperation. One core requirement of these "haptic" devices is an ability to provide a realistic force–position interface between the human operator and the virtual environment. To do so, the device controller must be programmed to render the desired stiffness, viscosity, and mass properties of the virtual environment, as well as the contact forces with the virtual solid objects.

The view of therapy robots as haptic devices is general enough to encompass all three application scenarios. In particular, the assistive, resistive, and perturbing forces used in challenge-based and assistive scenarios can be seen as special types of haptic effects, not associated with specific virtual objects.

While the preservation of stability is a primary requirement, the quality of haptic rendering can be characterized in terms of three criteria (Massie and Salisbury, 1994), which must drive both robot (hardware) and controller (software) design: (i) free space must feel free; (ii) solid objects must feel stiff; and (iii) virtual constraints must not be easily saturated. Criterion (i) implies that when there is no contact with virtual objects, the mechanics of the robot must be transparent—"backdrivable"—to the human operator. The latter perceived robot dynamics—in particular, the apparent friction and mass at the point (or points) of interaction—as a disturbance to haptic rendering. The apparent mass is mostly determined by the mass of the device and by its distribution. Friction is mostly determined by mechanical transmission (gearboxes, cables, linkages). Presence of gears also increases the magnitude of the perceived inertia. Criterion (ii) relates to the fact that collisions and contacts with simulated objects must exhibit a high degree of realism. This requires that the magnitude of the impedance that the robot can generate at its point(s) of interaction should be as large as possible. The greater the impedance the robot can generate, the wider the range of the exercises and forms of assistance that subjects can experience. The maximum impedance is determined by actuators, but the controller also plays a crucial role. Criterion (iii) implies that the haptic device must be capable of producing enough force. Again, this capability is mostly determined by the strength of actuators. The three criteria are clearly conflicting. More powerful actuators tend to be heavier. Mechanical design may help reduce some aspects of the dynamics, but even in this case conflicts may arise. Motors can be kept proximal with respect to the point of interaction, thus reducing the apparent mass. However, this requires additional parts (linkages, cables) to provide force transmission, which may increase friction. Gearboxes allow the use of smaller motors but increase the apparent inertia of the device and may be an additional source of friction and/or backlash. In fact, many haptic devices use direct-drive (no gearboxes) transmission mechanisms. Similarly, the effect of gravity can be compensated by using counterweights, but the latter further increase the overall inertia.

In terms of mechanical structure, there are two classes of therapy robots: (i) wearable devices or exoskeletons, which have multiple points of interaction with the user's limb, and (ii) endpoint robots or manipulanda, in which user and robot have only one point of interaction (end effector). Examples

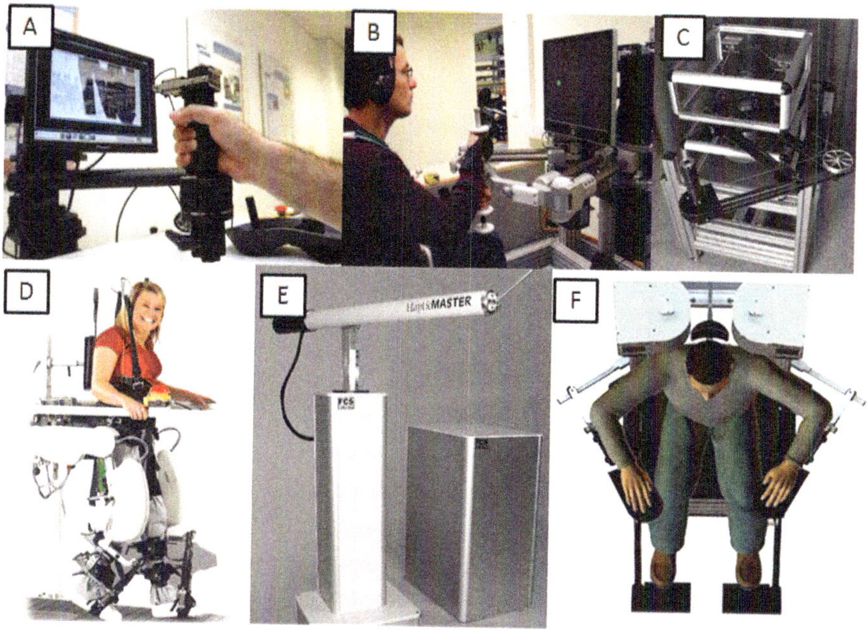

Figure 26.1. Robots used for therapy. (A) Inmotion2 (IMT, USA); (B) Physioassistant (Celin, Italy); (C) wBot planar manipulandum; (D) Lokomat (Hocoma, Switzerland); (E) HapticMASTER (Moog, USA); (F) KinARM (B.KIN Technologies, Canada).

of exoskeleton robots include KINARM (B.Kin, Canada) (Scott, 1999) and MULOS (Johnson et al. 2001), for the upper limbs, and Lokomat (Hocoma, Switzerland; Jezernik et al., 2003), for the lower limbs. Examples of manipulandum robots include Inmotion2 (IMT, USA), formerly known as MIT-Manus (Aisen et al., 1997), PhysioAssistant (Celin, Italy), formerly known as Braccio di Ferro (upper limb), and the GT I gait trainer (Reha-Stim, Germany; lower limb). See Figure 26.1 for examples of robots used for therapy (Werner et al., 2002).

Control Architectures

Technological advances have provided more lightweight materials and actuators, with greater torque-to-weight ratios. Larger and more powerful actuators, drive mechanisms, and links can support greater forces, but they may end up with a greater perceived friction and inertia. Low-power motors coupled to lightweight mechanisms have lower dynamics but are unable to achieve large force (and therefore impedance) levels. As a consequence, further improvements are more likely to come from improvement in control rather than physical means. In robotic rehabilitation, two main classes of control schemes are widely used: impedance control and admittance control.

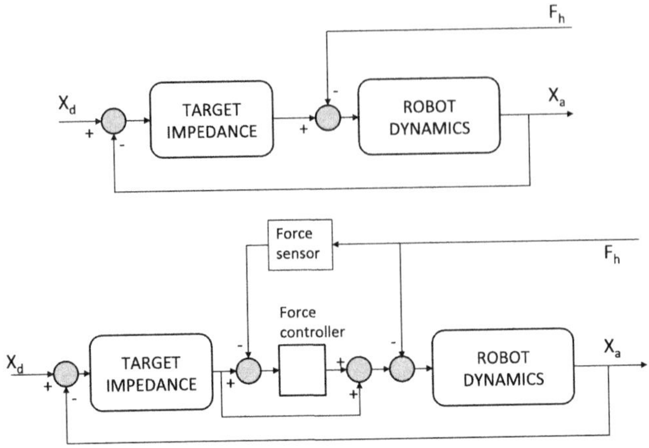

Figure 26.2. Open-loop (top) and closed-loop (bottom) impedance controller.

An **impedance controller** detects the movements of the interacting human and restitutes a force at the point(s) of interaction. An **admittance controller** detects the force exerted by the human and translates it into robot movements. In some applications, force is used as an additional input to the impedance controller. Likewise, the admittance controller may use displacement as an additional input. We briefly review both control modalities, with examples of devices that use these control schemes. More details can be found in Carignan and Cleary (2000).

Impedance Controllers. The open-loop impedance control scheme is schematically depicted in Figure 26.2 (top). The "target impedance" module specifies the impedance Z_d of the simulated environment, which determines the control effort F_d in terms of the difference $\Delta X_d = X_d - X_a$, where X_d and X_a are, respectively, the desired and actual position of the device. The overall impedance felt by the user, that is, the transfer function between the robot motion X_a and the hand force $F_{h,}$, is a combination of the programmed impedance Z_d and the dynamics of the robot. Therefore, this control scheme is only applied when robot dynamics is negligible compared with the target impedance. Open-loop impedance controllers are easy to implement and are widely used in robotic rehabilitation. Devices like MIT-Manus, Braccio di Ferro, ATR's Parallel direct-drive air-magnet Floating Manipulandum (PFM; Gomi and Kawato, 1997), and vBot (Howard et al., 2009) all use this control scheme.

In some applications, for gravity compensation or even manipulation of bulky anatomical parts such as the lower limb, the actuators may need to deliver high torques, and it may be difficult to decrease the intrinsic dynamics by physical means. In these situations, open-loop controllers have a poor performance. To improve it, a force feedback loop can be added as in Figure 26.2 (bottom), which has the effect of (approximately) transforming the robot into

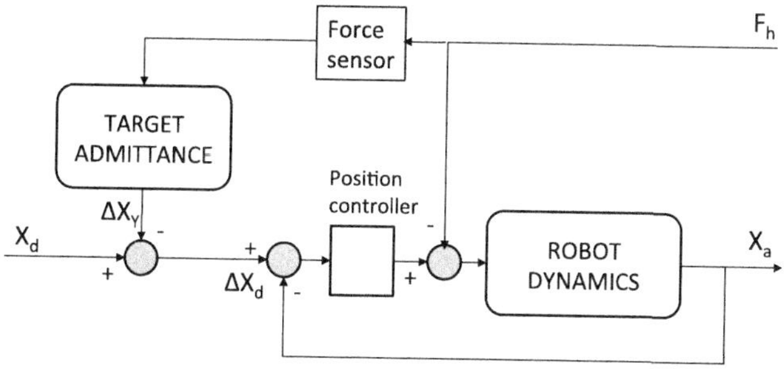

Figure 26.3. Admittance control scheme.

a pure force generator, so that the controller can be more accurate in generating the desired impedance. Devices using this control scheme include the Lokomat system (Jezernik et al., 2003).

Admittance Controllers. Admittance control schemes (see Fig. 26.3) usually consist of two nested control loops. In the outer loop the human-generated force, provided by a force–torque sensor, is translated into a robot movement through a "target admittance" block, Y_d, which specifies the desired behavior of the manipulator at the interface with the subject. In other words, Y_d reflects the desired haptic rendering behavior and generates a desired device position. The inner loop is a position controller and is used to compensate for the intrinsic robot dynamics. In other words, the robot is transformed into a position (displacement) generator. The inner loop is generally implemented by the servos of the motor at a very high speed (>3 kHz). The outer loop can be updated at lower speeds to limit the computational burden.

Admittance control is used in devices with low levels of backdrivability; an example is the HapticMASTER (Moog, USA; Van der Linde et al., 2002), which is a widely used device in the field of neurorehabilitation and neuroscience; see, for instance, the ACT^{3D} and GENTLE/S systems. It is a three-dimensional design that operates using admittance control. When the user exerts a force on the device, the device reacts by generating the appropriate displacement. This kind of active compensation is effective even when robots are very stiff and powerful, but has the disadvantage of being intrinsically less safe in case of failure.

To summarize, impedance and admittance controls have complementary stability and performance properties. Impedance control does not explicitly compensate for dynamics; therefore it works better when dealing with stiff environments, in which the robot dynamics can be neglected. However, robot dynamics and the update rate of the controller may limit the maximum amount of stiffness that the robot can deliver. In contrast, admittance control is better at compensating for large robot dynamics. Therefore, it can accurately render free space. However, large robots using admittance control may usually deliver greater target impedances.

In conclusion, no control scheme is clearly superior. Rather, preference in a particular application depends on the intrinsic dynamics of the robot and on the target impedance that has to be rendered.

Assistance Modalities

In assistive scenarios, robot devices have been frequently used to enforce passive movements (passive mobilization). Repetitive passive training may improve recovery, at least in specific clinical conditions (Fazekas et al., 2006; Hesse et al., 2003; P. Lum et al., 2002), as it counteracts the deterioration of the mechanical properties of tendon and muscle tissues which are an indirect consequence of the reduced mobility associated with limb paresis. However, better results are obtained when using techniques that take the adaptive nature of the nervous system into consideration. These techniques include active-assisted exercises, in which the robot guides the movement along a desired path. Variants are represented by active-constrained techniques, in which the robot only allows movement when voluntary motor commands are appropriately directed toward the target, or active-triggered techniques, in which the robot only provides assistance if subjects are unable to complete the task by themselves.

Challenge-based scenarios include active-resisted exercises, in which the robot resists the desired movement, and adaptive techniques, in which the robot provides an unfamiliar dynamic environment, to which subjects are required to adapt.

Like haptic rendering, assistance (or resistance) may take the form of a mechanical impedance, implemented through either impedance or admittance control schemes. Alternatively, the robot may simply counteract the effects of body dynamics. For instance, the T-WREX (Housman et al., 2009), commercialized as ARMEO (Hocoma, Switzerland), is a passive orthosis that uses springs to counteract the effect of gravity in three-dimensional arm movements. In a similar way, all gait robots are equipped with weight support devices that allow them to compensate for body weight during robot-assisted gait movements.

Bilateral trainers represent a special category of therapy robots. The first bilateral device (Burgar et al., 2000) focused on shoulder and elbow movements (P.S. Lum et al., 2002). Control is based on the mirror image movement enabler (MIME) concept: a modified industrial robot applies forces by using position control on the impaired arm, with the goal of replicating the movements of the other arm mirror-symmetrically. Based on a similar bilateral approach, the BiManuTrack device (Hesse et al., 2003) is used for training pronosupination of the forearm and flexion–extension of the wrist. The control of the impaired side can be either passive or active, and the movement may be mirrorlike or parallel. As regards the lower limb, the bilateral approach is followed by the LOPES gait training robot, where the unimpaired leg determines the state of other leg by using a method called complementary limb motion estimation (Vallery et al., 2009). Other studies were performed with just two unilateral robots, such as two planar manipulanda or two HapticMaster robots (Lewis and Perreault, 2009). To date, in the majority of bimanual control schemes the two sides are not requested to cooperate but rather interact in a master–slave fashion, with few exceptions (Squeri et al., 2009).

Regulation of Assistance

A study focusing on locomotion (Emken et al., 2007) suggested that the motor system behaves as a "greedy" optimizer that quickly incorporates the assistive forces generated by the robot into the motor plan in order to reduce the degree of voluntary control (and therefore muscle activation) while keeping the position error small. This phenomenon is known as the **slacking effect**, and is likely to occur during active-assisted exercises (and, even more, during passive training): a constant assistive force may induce a reduction of voluntary control, which may have adverse effects on recovery. To prevent slacking, assistance should be provided "as needed," that is, subjects should exercise their task with the minimal amount of manual assistance that is compatible with task completion.

As subjects improve and therefore their performance increases, the amount of assistance needs to be continuously regulated and adapted to their improvements. Based on this principle, several robot control strategies have been designed, for both the upper limb and gait training; see Marchal-Crespo and Reinkensmeyer (2009) for a review. In fact, Ferraro et al. (2003) and Hogan et al. (2006) found that a treatment protocol that adapts to the subject's motor ability achieves a better recovery compared with a training protocol in which assistive forces are not adapted.

Another point that can contribute to motor skill learning is the regulation of task difficulty (or, equivalently, the "desired" or target performance). Computational models of motor learning further suggest that large initial errors may prevent learning (Sanger, 2004). The challenge point theory (Guadagnoli and Lee, 2004) states that "optimal" learning is achieved when the difficulty of the task is appropriate for the participant's level of expertise (i.e., when the challenge point is reached). This would predict that providing a difficult task to a less skilled participant would result in less learning with a similar amount of practice, compared with training when the task difficulty is adjusted to the skill level. This theory was supported by a more recent study (Marchal-Crespo et al., 2010) that demonstrated that a force field that guided subjects through a desired movement was more beneficial for less skilled participants. Similar effects have been observed in the neurorehabilitation field. Overall, these considerations suggest that a controller that (i) maximally promotes subject involvement, (ii) provides enough assistance so that subjects complete desired movements, and (iii) adapts to the subjects' skill level and to his or her improvements will maximize motor-skill learning as well as neuromotor recovery.

However, translating it into actual mechanisms of regulation is far from easy. The problem of "optimally" regulating assistance is currently an open research area, and only heuristic, ad-hoc solutions are currently available. The task of deriving an optimal (i.e., assist-as-needed) controller would be relatively straightforward if the dynamics of the learning or recovery process were completely known. Although there have been attempts to model this process (Casadio et al., 2009b; Han et al., 2008), these models are currently not accurate enough to allow the design of robust controllers.

The general goal of regulation of assistance is to decrease assistance as performance improves. This is often achieved through simple linear control models (Krebs et al., 2003; Riener et al., 2005). Controller parameters are usually set heuristically. However, stability of the (closed-loop) recovery process is critically dependent on these parameters, which may be a problem as very little prior information is available on the dynamics of the learning–relearning process. Moreover, the latter may be highly subject-dependent.

Model-based controllers rely on the exact knowledge of the parameters of the trainee learning model. These parameters can be obtained by observing how the dynamics of recovery is affected by varying amounts of assistance. On the other hand, effective training requires a fully functioning controller. This is a dual control problem (exploration vs. exploitation dilemma). Moreover, the recovery process is nonlinear and inherently noisy (it includes a random component accounting for exploration of action space), which makes this approach quite problematic. Another possibility is to use adaptive controllers, which do not require a detailed knowledge of the learning process and automatically adapt to it.

Assessing Recovery

Motor learning obtained by repeated movements is not a unitary phenomenon. Rather, it affects many different components of sensory and motor processing. In normal subjects, the repetition of a task usually improves motor performance in terms of accuracy and speed of movement. In neurological rehabilitation, the assessment of motor recovery should include, in addition to smoothness, the efficacy and efficiency of the movement.

All these aspects are usually influenced by the quality of force control. Taken together, these aspects could provide a more complete picture of the mechanisms underlying motor recovery following rehabilitation.

Usually, therapists compare the subject's performance during a specific task with the normal kinesiology associated with that task. Therefore, learning performance can be quantified by the discrepancy between the system's actual and target outcomes. In this perspective, the components of recovery may provide a clearer understanding of the learning mechanisms and, if necessary, may be used to modify the rehabilitation strategies.

As stated before, robots are not only powerful tools used to deliver a rehabilitation therapy, but are also useful for functional evaluation and to monitor the effects of the therapy in real time.

However, it is important to rely not only on kinematic indicators (i.e., Colombo et al., 2008) provided by the robot, but also to assess the recovery by using clinical scales; see www.rehabmeasures.org for a general overview. The Fugl–Meyer score, a 0–66 scale limited to the arm section (FMA), is the most widely used test to evaluate arm mobility. The Ashworth score, a 1–4 scale, is also extensively used for evaluating spasticity; the Medical Research Council's Power Grading Scale is used to quantify strength and range of motion.

For evaluating the influence of robot therapy on daily life activities, the most common scale is the Functional Independence Measure (FIM). Tests that directly measure the functional recovery of the arm are the Wolf Motor Function Test (WMFT) and the Action Research Arm Test (ARAT). The ABILHAND test provides information on the ability to deal with bimanual tasks.

Finally, it is also important to evaluate the psychological aspect involved in robot therapy, because depression, motivation, and degree of participation (Colombo et al., 2007) can influence its outcome to a great extent.

RESULTS

How does robot therapy work in practice? Is it effective? Two applications of robots to promote the recovery of arm movements in stroke survivors are discussed in some detail. Then we summarize the outcomes of randomized clinical trials addressing robot therapy.

Both applications focus on chronic stroke survivors who were initially unable to complete the required movements with their affected arm without assistance. All studies use the same robot system, a planar manipulandum with two degrees of freedom (Casadio et al., 2006). Hand trajectory is measured through optical encoders and an open-loop impedance controller modulates the force transmitted to the hand. Therefore, hand motion is not imposed but results from the interaction between the forces generated by the robot and the forces generated by the subject.

In both cases, the subjects' inclusion criteria were chronic (at least one year after stroke) and stable clinical conditions for at least one month before entering the study. The exclusion criteria were the inability to understand instructions about the exercise protocol, and the presence of other neurocognitive problems.

Case Study 1: Arm Extension

This study (details are given in Casadio et al., 2009a) specifically focused on facilitating the active execution of arm extension movements. This was motivated by the observation that many stroke subjects are unable to actively perform these movements, particularly in specific directions. In contrast, wide inward movements are dominated by the flexion pattern that characterizes this pathology.

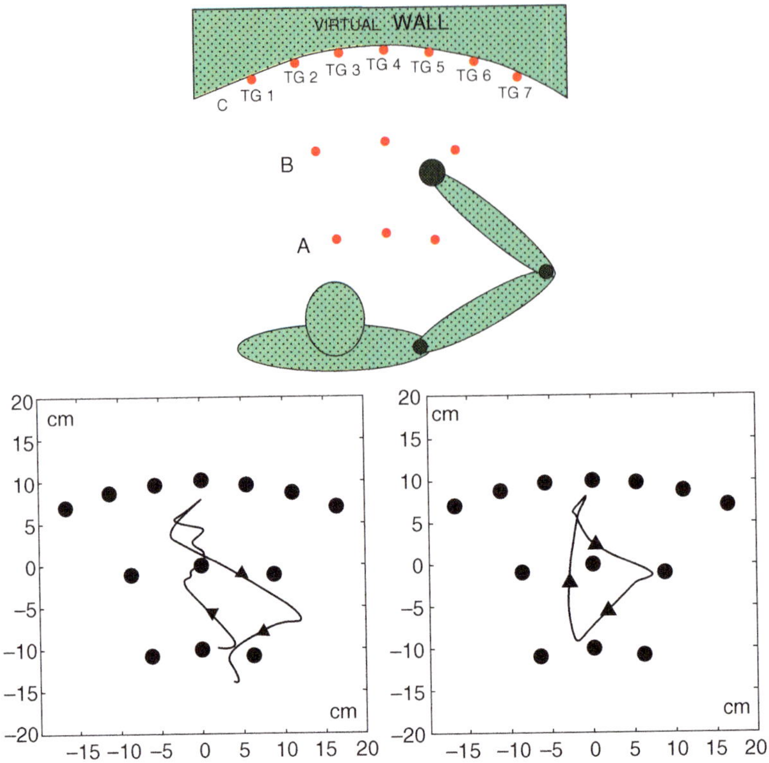

Figure 26.4. Arm extension task. The targets are arranged in three layers: A, B, C. The C layer is just in front of a virtual wall. The distance between adjacent layers was 10 cm. A target was considered as reached when its distance from the hand was less than 2 cm.

The task consists of hitting a set of targets, arranged in the horizontal plane (Fig. 26.4, top) according to three layers: inner (A), middle (B), and outer (C). Reaching the outer targets requires nearly full extension of the arm. Target sequences were generated according to the following scheme: A→C→B→A. In this way, outward movements had to be performed in one step (A→C), whereas inward movements were performed in two steps (C→B and B→A). When a target was presented to the subject, the robot generated a constant-magnitude assistive force directed toward the target. The assistive force was developed gradually with a ramp-and-hold profile (rise time, 1 s), and switched off as soon as the subject hit the target. The next target was presented after a pause of 1 s. Assistance also had a speed-dependent component, aimed at improving the interaction between the subject and the robot. A virtual wall provided additional haptic feedback.

The exercise protocol started with a test phase, during which individual subjects became familiar with the apparatus and in which a physical therapist selected the minimum force level that evoked a functional response, that is, a (possibly incomplete) movement in the intended direction. One block of trials included repetitions of the A→C→B→A sequence with different targets in random order, for a total of 63 movements.

To specifically train proprioceptive functions, we alternated blocks of trials with and without vision. In the latter case, the subjects were blindfolded, but could still feel target direction through the assistive force. The first training session initiated with two blocks of trials (vision, no-vision), using the same level of force determined in the test session. After a little rest, the therapist considered the level of performance and asked the subject about fatigue. The decision could be (1) to terminate the session, (2) to continue with the same force level, (3) to continue with a reduced force (10–20% less than the initial magnitude). The procedure was iterated until the decision to stop was agreed by the subject and the therapist. In the following sessions, training always started with the initial force level; then, if possible, the level of assistance was decreased. If subjects reached a level of assistance with a force below 4 N, the no-vision blocks were eliminated. The whole training protocol consisted of 10 sessions (1–2 sessions per week, about 1 hour each), plus the initial test session. The protocol was "assist-as-needed" in the sense that the therapist kept force magnitude to a minimum. Assistance allowed subjects to initiate the movements, but in no way did it impose the trajectory, the reaching time, or the speed profile.

Nine stroke survivors (2 males, 7 females, age 52 ± 14) participated in this study. Disease duration was 34 ± 19 months (range 12–76), in which the majority were ischemic in nature (7/9). Impairment was evaluated by means of the Fugl–Meyer score. The average FMA score was 15 ± 13 (range 5–41). The average Ashworth score of muscle spasticity was 1.9 ± 0.9 (range 1–3).

An example of a trial in an early and a late phase of training is depicted in Figure 26.4 (bottom left), which shows (middle) the A→C→B→A trajectories. In early sessions, the outward movement (A→C) is segmented into a sequence of submovements. The motor performance in late training sessions (Fig. 26.4, bottom right) suggests a clear improvement. At the same time, the level of robot assistance could be reduced from 12 N to 6 N; movement duration was shorter, and the number of submovements was reduced. The residual error after the first submovement decreases as well. In the overall population of subjects, the initial level of assistance ranged between 25 N and 5 N, and was generally greater for subjects who initially had lower FMA scores. Over sessions, we found a statistically significant change in the FMA score, from 15± 13 to 20 ± 13, corresponding to an average 4.8 ± 2.4 improvement. Evaluation of the FMA at follow-up (3 months) resulted in a substantial preservation of the improvement (FMA = 20 ± 13, no significant difference from that assessed at the end of treatment). Four subjects even displayed an improvement in their FMA score. No change was observed in the subjects' Ashworth score.

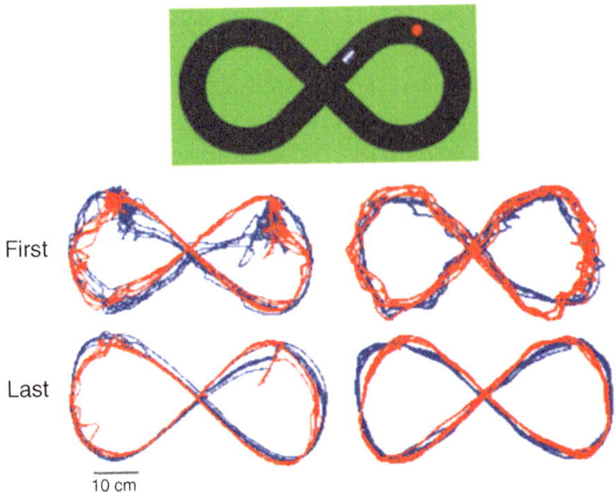

Figure 26.5. Visuomotor tracking task. (top) Visual tracking task. (bottom) Trajectories at early and late training, for vision (left) and no-vision (right) trials.

Case Study 2: Visuomotor Tracking

In this task (Vergaro et al., 2010a), subjects had to continuously track a moving visual target, moving on a figure-of-eight trajectory (length = 90 cm, time period = 15 s); see Figure 26.5 (top). The target was represented visually as a small red circle and haptically as an attractive force field. The current position of the hand was continuously displayed (as the picture of a small car). For each subject, the force field magnitude was initially selected as the minimum level capable of inducing the initiation of movement; the range of the assistive force was 3–30 N (from the least to the greatest impairment). The moving target stopped if the distance from the cursor was greater than 2 cm. The experimental protocol was organized into blocks of 10 trials each, which included 10 repetitions of the figure-of-eight. Similar to Case Study 1, within each training session, two blocks of trials were alternated, with eyes open and eyes closed.

Within each block, half of the trials were clockwise and half were counterclockwise. One session lasted approximately 45 minutes. At the end of each block, the robot estimated a performance score, based on the number of stops and the overall movement duration. If the score exceeded a threshold, the level of assistance was reduced. Unlike the previous exercise, assistance here is automatically adapted to the observed performance.

The therapy cycle included up to 10 sessions (2–3 sessions per week). Improvements were evaluated with clinical scales (FMA, Ashworth) and movement indicators (average speed, duration, tracking error, stop time).

Ten chronic hemiparetic subjects participated in this study (3 males, 7 females, age = 53 ± 15 years, disease duration = 4 ± 2 years, FMA = 23 ± 14). Figure 26.5 (bottom) displays changes in the tracking trajectories in a typical stroke survivor between the first and the last session. Statistical analysis resulted in highly significant effects of session for the mean speed. As regards assistance, we found no significant effects. This is no surprise, as the expected outcome of assistance regulation is that performance is relatively insensitive to assistance.

We found significant vision and session × vision effects. The presence of vision did not have a systematic effect, likely due to the fact that subjects vary widely in their level of sensory impairment.

Subjects with an initially more severe impairment with eyes closed resulted in a greater improvement in eyes-closed trials, and vice versa. Improved performance is also reflected in the increased FMA score (from 23 ± 14 to 27 ± 15, corresponding to an average 3.4 ± 1.9 increase). The level of assistance was reduced on average by 28%.

As in the previous experiment, subjects consistently improved their performance. Moreover, proprioceptive problems—revealed by a discrepancy between initial performance with eyes open and closed—tended to reduce over training.

Clinical Trials Involving Robots

Many robots have been designed for upper limb rehabilitation following stroke. However, in spite of the rising number of publications on robot-assisted therapy, the number of high-quality clinical trials is still low. A 2008 review (Kwakkel et al., 2008) pointed out that only four such devices were tested in randomized controlled trials (RCTs): MIT-Manus, ARM Guide, MIME, and BiManuTrack. This review reported a total of 10 RCTs (for a total of 218 stroke survivors) that assessed both motor and functional recovery of the upper limb following robot therapy. These studies confirmed an improvement on motor function after stroke. An earlier review (Prange et al., 2006) reported an average improvement of 3.7 ± 0.5 points of the Fugl–Meyer score. This is consistent with the outcomes of the two trials illustrated before.

In spite of earlier suggestions (P.S. Lum et al., 2002) that robot-mediated therapy can be beneficial for activities of daily living (ADLs), convincing evidence for significant changes in the FIM or other ADL indicators is still lacking. The previously mentioned studies suggested that improvement was limited to the body regions under test—namely, shoulder and elbow. In contrast, performance in ADLs is highly affected by hand functionality. Likewise, a Cochrane review (Mehrholz et al., 2008) investigated eleven RCTs (328 subjects in total) that compared electromechanical and robot-mediated therapy to other rehabilitation techniques or no training. No evidence was found for an effect of robot-assisted arm training on ADLs, though both arm motor function and motor strength were found to improve significantly.

The impact of the bimanual robot training on recovery has not been extensively studied yet. P.S. Lum et al. (2006) suggested that unimanual training is indeed more beneficial than bimanual training. Another recent Cochrane review (Coupar et al., 2010) confirmed this finding, and suggested that simultaneous bilateral training—using conventional rehabilitation methods—has no effects on arm or hand recovery. However, to date there is no sufficient evidence to support any solid conclusions (Kwakkel et al., 2008) and whether "two hands are better than one" (Rose and Winstein, 2004) still remains an open and debated topic.

How does robot therapy compare with conventional therapy or other technology-based approaches? A large multicenter RCT confronting robot therapy, intensive physical therapy and usual care (Lo et al., 2010) confirmed that robots are indeed effective but found no significant advantage over conventional physical therapy. Nevertheless, a systematic review (Langhorne et al., 2009) comparing different approaches to stroke rehabilitation pointed out that robot therapy is among the most effective techniques for the rehabilitation of both upper and lower limbs. A recent statement from the American Heart Association (Miller et al., 2010) indicates that varied repetitive task practice, including robot therapy, has a Class I treatment effect (treatment should be administered) with a Level A evidence (multiple populations evaluated). However, no recommendations are made on whether one approach should be preferred.

As regards the use of robots for gait training, the field is still in its infancy and not many RCTs are available. The available evidence (Langhorne et al., 2009; Moseley et al., 2005) suggests that robot-assisted gait training is no better than treadmill training, with or without body weight support.

CONCLUSIONS

What Determines Recovery?

We have shown that robot therapy is effective in promoting neuromotor recovery, but what about its mechanisms of action? Several studies have demonstrated that exercise facilitates the recovery of motor functions following stroke (Nudo, 2006). Intensity and frequency of practice are major determinants of recovery (Kwakkel et al., 2006); however, mere repetitive task training resulted in modest increases in lower limb function and no improvement in the upper limb (French et al., 2007).

However, robot-assisted therapy requires activities that are not only repetitive, but also task-oriented and attention-demanding. It has been argued (Adkins et al., 2006; Krakauer, 2006) that robot therapy relies on some form of motor learning or skill acquisition (Kleim et al., 1998, 2004; Remple et al., 2001). In humans, motor learning is a set of cognitive processes associated with practice, training, or experience that results in relatively permanent changes in motor behavior (Schmidt and Wrisberg, 2000) and that induces reorganization in both the lesioned and the nonlesioned brain areas (Nudo, 2003). The training schedule affects motor learning as well: practice of a variety of tasks leads to a better retention than practice of a single task (Kantak et al., 2010).

Robot assistive therapy may be beneficial (Marchal-Crespo and Reinkensmeyer, 2009) because (i) assistive forces may help subjects to complete the motor task even in the early phases of the recovery process, which in turn may increase motivation; (ii) it elicits the "right" afferent signals (proprioceptive, tactile), thus promoting the emergence of the appropriate associations in sensory and motor cortical areas; (iii) it may induce a sensation of greater stability of the external environment, a necessary condition for long-term, more-stable adaptation to occur; (iv) it interleaves effort by the participant (essential to provoke motor plasticity) with stretching of muscles and connective tissue, which is helpful to prevent stiffening of soft tissues; (v) it may help inducing brain plasticity through moving the limb in a manner that self-generated effort cannot achieve; (vi) it may help to perform more movements in a shorter amount of time, potentially allowing more intense practice; (vii) it allows subjects to practice a task more intensively by simply making it safer. All these factors may contribute to the recovery in robot-assisted exercise, but none of them has been tested empirically.

Wrist and Hand Rehabilitation

Given the results obtained in the recovery of arm and gait functions, it is natural to extend robot therapy to the rehabilitation to the wrist and the hand. The field is still in its infancy due to the technical difficulty of designing robots that both are lightweight and have multiple degrees of freedom. Many different designs have been proposed for wrist and hand rehabilitation, ranging from 1 to 18 degrees of freedom; see Balasubramanian et al. (2010) for a comprehensive review. However, very few have been tested with stroke subjects, and even fewer have evolved into commercially available devices. As regards the wrist, for instance Masia et al. (2009) reported improved functions after training with a wrist exoskeleton through a tracking task with progressive dynamic splinting.

Hand rehabilitation is even more challenging. Existing devices range from cable-driven mechanisms (Dovat et al., 2008) able to act on each individual finger to haptic knobs (Lambercy, et al., 2007) focusing on coordinated action with the wrist and the forearm. Wrist (Krebs et al., 2007) and hand (Masia et al., 2007) modules have been added to the Inmotion2 device (IMT, USA). Other commercially available devices include the AMADEO (Tyromotion, Austria), for hand and finger movements, and the BiManuTrack, aimed at arm and wrist functions.

Using Robots for Proprioceptive Training

In stroke survivors, motor impairment is frequently associated with degraded proprioceptive and/or somatosensory functions (Tyson et al., 2007). These subjects may have difficulties with estimating the position of their arm in the absence of vision, and/or they may be unable to integrate visual and proprioceptive information. During assistive training, they may not be capable of detecting the presence, magnitude, and direction of assistive forces.

Impaired proprioception may affect motor recovery (Carey et al., 1993). Like motor deficits, proprioceptive deficits may decrease through repeated exercise (Dechaumont-Palacin et al., 2007). When both visual and kinesthetic information of a limb are available, vision is usually the dominant source of information. As a consequence, proprioceptive impairment may be masked by vision if the latter is available. In subjects with both proprioceptive and motor impairments, assistive exercise might be more effective if at least part of the training were performed without vision. In this context, the contribution of robotic devices to neuromotor rehabilitation may turn out to be crucial. Moreover, different training conditions—either presence or absence of vision—may help facilitate the recovery of both sensory and motor functions (Casadio et al., 2009a; Vergaro et al., 2010a).

Other Than Stroke

Multiple Sclerosis. Multiple sclerosis (MS) is associated with a variety of symptoms and functional deficits, in proportions that vary widely from patient to patient. About 30% of subjects show functionally relevant cerebellar deficits, and the associated symptoms (ataxia and kinetic tremor) are highly disabling. Exercise therapy may be beneficial for subjects with MS (Rietberg et al., 2005) and multidisciplinary rehabilitation programs may improve their experience in terms of activity and participation but cannot change the level of impairment (Khan et al., 2007). Few studies have addressed the use of robots for the treatment of MS subjects. In MS subjects with severe gait impairment, Beer et al. (2008) compared robot-assisted gait training with conventional walking therapy. Similarly, Lo and Triche (2008) compared body weight supported treadmill gait training with and without robot assistance. Both studies found a significant effect of treatment but no significant advantage of robots over conventional therapy. Due to the different types and degrees of impairment, the timing and mode of rehabilitation treatment should be set individually. As regards the upper limb, in MS subjects with cerebellar symptoms, Vergaro and co-workers (Vergaro et al., 2010b) found that a subject-specific sensorimotor adaptation task resulted in an improved coordination. Although preliminary, these results suggest that incoordination symptoms may benefit from robot-mediated training that challenge the ability to adapt to novel dynamics.

Cerebral Palsy. Cerebral palsy (CP) is a group of nonprogressive, but often changing, motor impairment syndromes, secondary to lesions or anomalies of the brain arising in the early stages of development. Although motor impairment is the leading factor in CP, sensory disorders have been described; sensorimotor cognitive functions are likely affected as well. Brain lesions that cause CP occur early in neurodevelopment or are due to birth accidents, and there is ample evidence that plasticity is greater in the developing brain. For this reason, and since limb control capabilities are age-related, the recovery mechanisms of children may differ from those observed in adults (Masia et al., 2011). Only a few pilot studies have addressed the use of robots for the treatment of CP, based on haptic simulation (Fluet et al., 2010) or assistive scenarios (Frascarelli et al., 2009) similar to those used for stroke. Significant improvements were reported, which point to the need for more systematic investigations. However, CP may require different, more specific approaches. Furthermore, working with children may require specific devices (smaller, lightweight, with better compensation of dynamics) and more engaging exercises.

Final Remarks

The use of robots to promote neuromotor recovery is clearly a hot topic in the rehabilitation community, as witnessed by the increasing number of publications in the medical and engineering literature (Brochard et al., 2010). Although most of the research still focuses on the technology (devices, control modalities), several devices have evolved into commercial products, approved for clinical use. During the last five years, a number of randomized controlled trials have been published, confirming that these devices are indeed effective, even though it is unclear whether they are better than conventional high-quality therapy.

This is remarkable, as very little is known about the mechanisms of action of recovery. A better understanding of the way robot therapy works and, more generally, how physical assistance might facilitate motor skill learning and relearning may lead to novel, more principled approaches that may lead to a wider range of applications and to even more effective recovery.

ACKNOWLEGMENT

This work is partly supported by EU Grant FP7-ICT-231724 (HUMOUR).

REFERENCES

Adkins D.L., Boychuk J., Remple M.S., and Kleim J.A. (2006). Motor training induces experience-specific patterns of plasticity across motor cortex and spinal cord. *J Appl Physiol* 101(6):1776–1782.

Aisen M.L., Krebs H.I., Hogan N., McDowell F., and Volpe B.T. (1997). The effect of robot-assisted therapy and rehabilitative training on motor recovery following stroke. *Arch Neurol* 54(4):443–446.

Balasubramanian S., Klein J., and Burdet E. (2010). Robot-assisted rehabilitation of hand function. *Curr Opin Neurol* 23(6):661–670.

Beer S., Aschbacher B., Manoglou D., Gamper E., Kool J., and Kesselring J. (2008). Robot-assisted gait training in multiple sclerosis: a pilot randomized trial. *Mult Scler* 14(2):231–236.

Brochard S., Robertson J., Médée B., and Rémy-Néris O. (2010). What's new in new technologies for upper extremity rehabilitation? *Curr Opin Neurol* 23(6):683–687.

Burgar C.G., Lum P.S., Shor P.C., and Van der Loos H.F.M. (2000). Development of robots for rehabilitation therapy: the Palo Alto VA/Stanford experience. *J Rehab Res Dev* 36:663–673.

Carey L.M., Matyas T.A., and Oke L.E. (1993). Sensory loss in stroke patients: effective training of tactile and proprioceptive discrimination. *Arch Phys Med Rehabil* 74(6):602–611.

Carignan C.R. and Cleary K.R. (2000). Closed-loop force control for haptic simulation of virtual environments. *Haptics-e: Electronic J Haptics Res* 1(2):1–14.

Casadio M., Morasso P.G., Sanguineti V., and Arrichiello V. (2006). Braccio di Ferro: a new haptic workstation for neuromotor rehabilitation. *Technol Health Care* 13:1–20.

Casadio M., Morasso P., Sanguineti V., and Giannoni P. (2009a). Minimally assistive robot training for proprioception enhancement. *Exp Brain Res* 194(2):219–231.

Casadio M., Novakovic V., Morasso P.G., and Sanguineti V. (2009b). Modeling the dynamics of the recovery process in robot therapy. Paper presented at the 2009 Virtual Rehabilitation International Conference, Haifa, Israel.

Colombo R., Pisano F., Mazzone A., et al. (2007). Design strategies to improve patient motivation during robot-aided rehabilitation. *J Neuroeng Rehabil* 4:3.

Colombo R., Pisano F., Micera S., et al. (2008). Assessing mechanisms of recovery during robot-aided neurorehabilitation of the upper limb. *Neurorehabil Neural Repair* 22(1):50–63.

Coupar F., Pollock A., van Wijck F., Morris J., and Langhorne P. (2010). Simultaneous bilateral training for improving arm function after stroke. *Cochrane Database Syst Rev* (4):CD006432.

Dechaumont-Palacin S., Marque P., De Boissezon X., et al. (2007). Neural correlates of proprioceptive integration in the contralesional hemisphere of very impaired patients shortly after a subcortical stroke: an fMRI study. *Neurorehabil Neural Repair*.

Dovat L., Lambercy O., Gassert R., et al. (2008). HandCARE: a cable-actuated rehabilitation system to train hand function after stroke. *IEEE Trans Neural Syst Rehabil Eng* 16(6):582–591.

Emken J.L., Benitez R., Sideris A., Bobrow J.E., and Reinkensmeyer D.J. (2007). Motor adaptation as a greedy optimization of error and effort. *J Neurophysiol* 97(6): 3997–4006.

Fazekas G., Horvath M., and Toth A. (2006). A novel robot training system designed to supplement upper limb physiotherapy of patients with spastic hemiparesis. *Int J Rehabil Res* 29(3):251–254.

Ferraro M., Palazzolo J.J., Krol J., Krebs H.I., Hogan N., and Volpe B.T. (2003). Robot-aided sensorimotor arm training improves outcome in patients with chronic stroke. *Neurology* 61(11):1604–1607.

Fluet G.G., Qiu Q., Kelly D., et al. (2010). Interfacing a haptic robotic system with complex virtual environments to treat impaired upper extremity motor function in children with cerebral palsy. *Dev Neurorehabil* 13(5):335–345.

Frascarelli F., Masia L., Di Rosa G., et al. (2009). The impact of robotic rehabilitation in children with acquired or congenital movement disorders. *Eur J Phys Rehabil Med* 45(1):135–141.

French B., Thomas L.H., Leathley M.J., et al. (2007). Repetitive task training for improving functional ability after stroke. *Cochrane Database Syst Rev* (4):CD006073.

Gomi H. and Kawato M. (1997). Human arm stiffness and equilibrium-point trajectory during multi-joint movement. *Biol Cybern* 76(3):163–171.

Guadagnoli M.A. and Lee T.D. (2004). Challenge point: a framework for conceptualizing the effects of various practice conditions in motor learning. *J Mot Behav* 36(2):212–224.

Han C.E., Arbib M.A., and Schweighofer N. (2008). Stroke rehabilitation reaches a threshold. *PLoS Comput Biol* 4(8):e1000133.

Hesse S., Schulte-Tigges G., Konrad M., Bardeleben A., and Werner C. (2003). Robot-assisted arm trainer for the passive and active practice of bilateral forearm and wrist movements in hemiparetic subjects. *Arch Phys Med Rehabil* 84(6):915–920.

Hogan N., Krebs H.I., Rohrer B., et al. (2006). Motions or muscles? Some behavioral factors underlying robotic assistance of motor recovery. *J Rehabil Res Dev* 43(5): 605–618.

Housman S.J., Scott K.M., and Reinkensmeyer D.J. (2009). A randomized controlled trial of gravity-supported, computer-enhanced arm exercise for individuals with severe hemiparesis. *Neurorehabil Neural Repair* 23(5):505–514.

Howard I.S., Ingram J.N., and Wolpert D.M. (2009). A modular planar robotic manipulandum with end-point torque control. *J Neurosci Methods* 181(2):199–211.

Jezernik S., Colombo G., Keller T., Frueh H., and Morari M. (2003). Robotic orthosis Lokomat: a rehabilitation and research tool. *Neuromodulation* 6(2):108–115.

Johnson G.R., Carus D.A., Parrini G., Scattareggia Marchese S., and Valeggi R. (2001). The design of a five-degree-of-freedom powered orthosis for the upper limb. *Proc Inst Mech Eng (H)* 215(3):275–284.

Kantak S.S., Sullivan K.J., Fisher B.E., Knowlton B.J., and Winstein C.J. (2010). Neural substrates of motor memory consolidation depend on practice structure. *Nat Neurosci* 13(8):923–925.

Khan F., Turner-Stokes L., Ng L., and Kilpatrick T. (2007). Multidisciplinary rehabilitation for adults with multiple sclerosis. *Cochrane Database Syst Rev* (2):CD006036.

Kleim J.A., Barbay S., and Nudo R.J. (1998). Functional reorganization of the rat motor cortex following motor skill learning. *J Neurophysiol* 80(6):3321–3325.

Kleim J.A., Hogg T.M., VandenBerg P.M., Cooper N.R., Bruneau R., and Remple M. (2004). Cortical synaptogenesis and motor map reorganization occur during late, but not early, phase of motor skill learning. *J Neurosci* 24(3):628–633.

Krakauer J.W. (2006). Motor learning: its relevance to stroke recovery and neurorehabilitation. *Curr Opin Neurol* 19(1):84–90.

Krebs H.I., Palazzolo J.J., Dipietro L., et al. (2003). Rehabilitation robotics: performance-based progressive robot-assisted therapy. *Autonomous Robots* 15:7–20.

Krebs H.I., Volpe B.T., Williams D., et al. (2007). Robot-aided neurorehabilitation: a robot for wrist rehabilitation. *IEEE Trans Neural Syst Rehabil Eng* 15(3):327–335.

Kwakkel G., Kollen B., and Twisk J. (2006). Impact of time on improvement of outcome after stroke. *Stroke* 37(9):2348–2353.

Kwakkel G., Kollen B.J., and Krebs H.I. (2008). Effects of robot-assisted therapy on upper limb recovery after stroke: a systematic review. *Neurorehabil Neural Repair* 22(2):111–121.

Lambercy O., Dovat L., Gassert R., Burdet E., Teo C.L., and Milner T. (2007). A haptic knob for rehabilitation of hand function. *IEEE Trans Neural Syst Rehabil Eng* 15(3):356–366.

Langhorne P., Coupar F., and Pollock A. (2009). Motor recovery after stroke: a systematic review. *Lancet Neurol* 8(8):741–754.

Lewis G.N. and Perreault E.J. (2009). An assessment of robot-assisted bimanual movements on upper limb motor coordination following stroke. *IEEE Trans Neural Syst Rehabil Eng* 17(6):595–604.

Lo A.C. and Triche E.W. (2008). Improving gait in multiple sclerosis using robot-assisted, body weight supported treadmill training. *Neurorehabil Neural Repair* 22(6):661–671.

Lo A.C., Guarino P.D., Richards L.G., et al. (2010). Robot-assisted therapy for long-term upper-limb impairment after stroke. *New Engl J Med* 362(19):1772–1783.

Lum P., Reinkensmeyer D., Mahoney R., Rymer W.Z., and Burgar C. (2002). Robotic devices for movement therapy after stroke: current status and challenges to clinical acceptance. *Top Stroke Rehabil* 8(4):40–53.

Lum P.S., Burgar C.G., Shor P.C., Majmundar M., and Van der Loos M. (2002). Robot-assisted movement training compared with conventional therapy techniques for the rehabilitation of upper-limb motor function after stroke. *Arch Phys Med Rehabil* 83(7):952–959.

Lum P.S., Burgar C.G., Van der Loos M., Shor P.C., Majmundar M., and Yap R. (2006). MIME robotic device for upper-limb neurorehabilitation in subacute stroke subjects: a follow-up study. *J Rehabil Res Dev* 43(5):631–642.

Marchal-Crespo L. and Reinkensmeyer D.J. (2009). Review of control strategies for robotic movement training after neurologic injury. *J Neuroeng Rehabil* 6:20.

Marchal-Crespo L., McHughen S., Cramer S.C., and Reinkensmeyer D.J. (2010). The effect of haptic guidance, aging, and initial skill level on motor learning of a steering task. *Exp Brain Res* 201(2):209–220.

Masia L., Krebs H.I., Cappa P., and Hogan N. (2007). Design and characterization of hand module for whole-arm rehabilitation following stroke. *IEEE–ASME Trans Mechatronics* 12(4):399–407.

Masia L., Casadio M., Giannoni P., Sandini G., and Morasso P. (2009). Performance adaptive training control strategy for recovering wrist movements in stroke patients: a preliminary, feasibility study. *J Neuroeng Rehabil* 6:44.

Masia L., Frascarelli F., Morasso P., et al. (2011). Reduced short term adaptation to robot generated dynamic environment in children affected by cerebral palsy. *J Neuroeng Rehabil* 8:28.

Massie T. and Salisbury J.K. (1994). The PHANTOM haptic interface: a device for probing virtual objects. Paper presented at the ASME Winter Annual Meeting: Symposium on Haptic Interfaces for Virtual Environment and Teleoperator Systems.

Mehrholz J., Platz T., Kugler J., and Pohl M. (2008). Electromechanical and robot-assisted arm training for improving arm function and activities of daily living after stroke. *Cochrane Database Syst Rev* (4):CD006876.

Miller E.L., Murray L., Richards L., et al. (2010). Comprehensive overview of nursing and interdisciplinary rehabilitation care of the stroke patient: a scientific statement from the American Heart Association. *Stroke* 41(10):2402–2448.

Moseley A.M., Stark A., Cameron I.D., and Pollock A. (2005). Treadmill training and body weight support for walking after stroke. *Cochrane Database Syst Rev* (4):CD002840.

Nudo R.J. (2003). Adaptive plasticity in motor cortex: implications for rehabilitation after brain injury. *J Rehabil Med* 41(Suppl):7–10.

Nudo R.J. (2006). Mechanisms for recovery of motor function following cortical damage. *Curr Opin Neurobiol* 16(6):638–644.

Prange G.B., Jannink M.J., Groothuis-Oudshoorn C.G., Hermens H.J., and Ijzerman M.J. (2006). Systematic review of the effect of robot-aided therapy on recovery of the hemiparetic arm after stroke. *J Rehabil Res Dev* 43(2):171–184.

Remple M.S., Bruneau R.M., VandenBerg P.M., Goertzen C., and Kleim J.A. (2001). Sensitivity of cortical movement representations to motor experience: evidence that skill learning but not strength training induces cortical reorganization. *Behav Brain Res* 123(2):133–141.

Riener R., Lunenburger L., Jezernik S., Anderschitz M., Colombo G., and Dietz V. (2005). Patient-cooperative strategies for robot-aided treadmill training: first experimental results. *IEEE Trans Neural Syst Rehabil Eng* 13(3):380–394.

Rietberg M.B., Brooks D., Uitdehaag B.M., and Kwakkel G. (2005). Exercise therapy for multiple sclerosis. *Cochrane Database Syst Rev* (1):CD003980.

Rose D.K. and Winstein C.J. (2004). Bimanual training after stroke: are two hands better than one? *Top Stroke Rehabil* 11(4):20–30.

Sanger T.D. (2004). Failure of motor learning for large initial errors. *Neural Comput* 16(9):1873–1886.

Schmidt R.A. and Wrisberg C.A. (2000). *Motor Learning and Performance*, 2nd ed. Champaign, IL: Human Kinetics.

Scott S.H. (1999). Apparatus for measuring and perturbing shoulder and elbow joint positions and torques during reaching. *J Neurosci Methods* 89(2):119–127.

Squeri V., Casadio M., Vergaro E., Giannoni P., Morasso P., and Sanguineti V. (2009). Bilateral robot therapy based on haptics and reinforcement learning: feasibility study of a new concept for treatment of patients after stroke. *J Rehabil Med* 41(12):961–965.

Tyson S., Hanley M., Chillala J., Selley A.B., and Tallis R.C. (2007). Sensory loss in hospital-admitted people with stroke: characteristics, associated factors and relationship with function. *Neurorehabil Neural Repair*.

Vallery H., van Asseldonk E.H., Buss M., and van der Kooij H. (2009). Reference trajectory generation for rehabilitation robots: complementary limb motion estimation. *IEEE Trans Neural Syst Rehabil Eng* 17(1):23–30.

Van der Linde R.Q., Lammertse P., Frederiksen E., and Ruiter B. (2002). The HapticMaster, a new high-performance haptic interface. Paper presented at the EuroHaptics.

Vergaro E., Casadio M., Squeri V., Giannoni P., Morasso P., and Sanguineti V. (2010a). Self-adaptive robot training of stroke survivors for continuous tracking movements. *J Neuroeng Rehabil* 7(1):13.

Vergaro E., Squeri V., Brichetto G., et al. (2010b). Adaptive robot training for the treatment of incoordination in multiple sclerosis. *J Neuroeng Rehabil* 7:37.

Werner C., Von Frankenberg S., Treig T., Konrad M., and Hesse S. (2002). Treadmill training with partial body weight support and an electromechanical gait trainer for restoration of gait in subacute stroke patients: a randomized crossover study. *Stroke* 33(12):2895–2901.

27

FOCUSED CHAPTER

PAIRED ASSOCIATIVE STIMULATION

NATALIE MRACHACZ-KERSTING

Center for Sensory–Motor Interaction (SMI), Department of Health Science and Technology, Aalborg University, Aalborg, Denmark

SUMMARY

Paired associative stimulation (PAS) is a stimulation protocol designed to alter noninvasively the excitability of cortical projections to the target muscle. It is based on the concept of spike timing dependent plasticity, where a presynaptic volley arrives at the postsynaptic neuron at the same time as it is being depolarized. PAS exhibits many characteristics of associative long-term potentiation (LTP), which is a candidate mechanism for learning and memory formation and storage. PAS thus has the potential to increase our understanding of learning in the intact human cortex. In addition, studies are emerging where PAS is applied in a rehabilitation setting to alter the excitability of projections weakened by insult. Studies where PAS targets lower limb muscles, however, are limited. This chapter focuses on the possible use of PAS in altering the control of lower limb muscles and in particular the tibialis anterior (TA) muscle. The motor evoked potential (MEP) elicited by transcranial magnetic stimulation (TMS) is enhanced when PAS is delivered in synchrony with rhythmic voluntary contraction. However, clinical application of such an intervention focuses on populations with

Introduction to Neural Engineering for Motor Rehabilitation, First Edition. Edited by Dario Farina, Winnie Jensen, and Metin Akay.

motor dysfunction. The study presented here investigated whether an artificially induced muscular contraction could result in a similar effect as a voluntary contraction. Furthermore, the possible functional implications following the application of PAS are reported.

Results reveal that PAS delivered to the resting TA results in significant facilitation of respective motor cortical projections without a concomitant change in basic functional measures. Further, the effects on TA MEP amplitude were similar for PAS and PAS delivered concomitantly with an artificially induced muscular contraction. Possible reasons are discussed.

INTRODUCTION AND BACKGROUND

Paired associative stimulation (PAS) consists of the repetitive pairing of a peripheral electrical and a central magnetic stimulus at low frequency. The first stimulus frequently is a single electrical pulse applied with surface electrodes to the peripheral nerve that innervates the target muscle. This is followed a set time later by a second stimulus applied using transcranial magnetic stimulation (TMS) over that area of the motor cortex that has direct corticospinal projections to the target muscle. PAS was first introduced by Stefan et al. (2000), who based it on studies of associative long-term potentiation (LTP) and long-term depression (LTD) in animal models (Bi and Poo, 1998). In such studies, correlated activation of presynaptic and postsynaptic neurons is artificially induced and the continued pairing of these two events leads to a strengthening of the synapse that outlasts the period of stimulation. Depending on the relative timing between the two stimuli, the direction of the synaptic change is one of either potentiation or depression. LTP is one mechanism for inducing synaptic plasticity thought to underlie processes of memory storage and learning (Letzkus et al., 2007).

The PAS protocol developed by Stefan and colleagues for the first time demonstrated that pairing of two separate inputs noninvasively and at a systems level can lead to changes in excitability of the postsynaptic neuron that are fast evolving, long lasting yet reversible (Stefan et al., 2000)—the postsynaptic site being in this case the direct corticospinal tract linking the motor cortex with hand muscles. Later studies revealed that the direction of the synaptic change could be altered to LTD by timing the TMS stimulus to depolarize the postsynaptic neuron after the afferent volley had reached the presynaptic neuron (Wolters et al., 2003). Further, the effects are dependent on activation of NMDA-receptors and involvement of (L-type voltage-gated) Ca^{2+} channels (Stefan et al., 2002; Wolters et al., 2003). In this way, Stefan and colleagues were able to demonstrate that indeed many components of PAS resemble those of LTP.

In the following years, a number of research labs throughout the world have continued to investigate the mechanisms involved that lead to the changes in excitability of the cortical projections to the target muscle following the application of PAS (Ridding and Taylor, 2001; K. Kujirai et al., 2006; Ridding and Flavel, 2006; Roy et al., 2007; Quartarone et al., 2006; Rosenkranz and Rothwell, 2006; Ridding and Uy, 2003; Fratello et al., 2006). This has extended to reports on abnormal effects in patient populations and functional benefits following application in rehabilitation settings (Tsuji et al., 2005; Castle-Lacanal et al., 2007; Muller et al., 2007; Tecchio et al., 2008; Pellicciari et al., 2009; Quartarone et al., 2009; Tamura et al., 2009; Kang et al., 2011). Many of these studies have implemented differing protocols for the induction of PAS, which resulted in the publication of a consensus paper on motor cortex plasticity protocols (Ziemann et al., 2008). This document demonstrated the interest in the area of neuroplasticity protocols and their importance in providing further information on mechanisms of memory formation and learning in the intact human. It also underlined the use of protocols such as PAS as possible new rehabilitation strategies for the improvement of motor function in impaired populations.

The report highlighted a lack in the literature on the effects following PAS applied to lower limb muscles. In fact at that time only two papers were cited (Mrachacz-Kersting et al., 2007; Roy et al., 2007). Currently the application of PAS to human upper limb and specifically hand muscles continues to dominate the literature. Of those papers where the target muscles have been in the lower limb, there is a wide range of protocols that have been implemented. Some of these have been modeled directly after Stefan et al. (2000). It is well known that the organization of neural pathways of lower limb muscles differs from that of the upper limb muscles. For example, the afferent volley arriving at the sensory cortex following the median nerve stimulus transiently inhibits the MEPs in hand muscles evoked by TMS (Tokimura et al., 2000) whereas the afferent volley from the tibial nerve facilitates both the TA and soleus MEPs (Roy and Gorassini, 2008). Furthermore, the I-waves that are thought to be involved in the effects following PAS are predominantly I3-waves (K. Kujirai et al., 2006). These are readily elicited in upper limb muscles at low TMS intensities when the current induced in the brain flows from anterior to posterior (Sakai et al., 1997). For lower limb muscles this is not the case. In fact it is not possible to preferentially recruit I3-waves when lower limb muscles are targeted (Di Lazzaro et al., 2001).

Mrachacz-Kersting et al. (2007) demonstrated that, in order to attain similar effects following PAS in the lower limb muscle TA compared with upper limb muscles as investigated by Stefan et al. (2000), PAS had to be applied at a higher frequency and with a combined voluntary dorsiflexion movement. However, the authors also reported that if the timing between the afferent volley and the TMS stimulus was optimized for each subject by recording the exact arrival time of the afferent volley, then PAS resulted in increases even with the muscle at rest. Roy and Gorassini (2008) further explored the time frame in relation to the TMS stimulus, in which the afferent volley would still act on a depolarized postsynaptic neuron. Although that study implies that the timing is not as crucial as thought, the maximum effects were attained when the afferent volley was elicited 40 ms prior to the TMS stimulus.

As has been stated, PAS was originally applied to artificially induce a memory formation at a systems level and, more important, in an intact system. However, since its development, PAS has also found applicability in rehabilitation. For upper limb muscles, studies are more abundant and differences of effects following PAS are found between controls and dystonic patients (Tamura et al., 2009; Quartarone et al., 2009; Kang et al., 2011), stroke patients (Tsuji et al., 2005) as well as between controls and healthy elderly subjects (Tecchio et al., 2008; Mueller-Dahlhaus et al., 2008; Pellicciari et al., 2009). For lower limb muscles Roy et al. (2010) have shown that PAS can increase the TA MEP amplitude in 7 out of 13 patients. These patients appeared to have a stronger functional ascending sensory pathway.

Few of these studies have also investigated the functional consequences following PAS in terms of improved performance even though the primary motor cortex (M1) is most closely related to simple aspects of movements such as force production (Ashe, 1997; Georgopoulos et al., 1992). It is thus conceivable that alterations in the excitability of the cortical projections from M1 to the target muscle may lead to functional changes. Indeed, Uy and Ridding (2003) reported increased TA MEP amplitudes in five out of nine stroke subjects following a high-frequency stimulation of the motor point of TA, paired with single-pulse TMS over a four-week period. This was accompanied by an improvement in functional measures such as cadence, stride length, and stride time. However, it is not known if the facilitation induced by this type of dual stimulation protocol is greater than that resulting from the same peripheral stimulus when applied alone.

Aim 1 of this study was to investigate whether the PAS-targeted intact TA muscle in humans can lead to functional changes as assessed by measures of maximal force and rate of force development. Aim 2 was to establish if a combination of repetitive electrical stimulation (rES) of a mixed nerve and PAS would produce an effect above and beyond the two applied separately. The reasoning behind this latter aim is that PAS-induced changes when targeting the TA are more readily induced in an active motor neuron pool (Mrachacz-Kersting et al., 2007). However, in rehabilitation settings, many patients are unable to produce a voluntary contraction of their dorsiflexors. Since similar brain areas are activated following active, passive, and electrically stimulated ankle dorsiflexion (Francis et al., 2009), rES was implemented to induce an artificial dorsiflexion movement at the ankle joint.

METHODOLOGY

Twenty participants (13 males, 7 females) aged between 20 and 46 years with a mean age of 27 ± 7 years undertook this study. All subjects were classified as having their right side as their dominant side, in accordance to the Edinburgh Handedness Index (Oldfield, 1971). At the time of the study none of the subjects had any neurological disorders or other health issues that may have prevented them from participating in the study. The study was approved by the University of Auckland Human Participants Ethics Committee (Reference number: 2006/039).

Electromyographic Recording

Surface electromyogram (EMG) activity was recorded from the target muscle, the right TA, in accordance with Cram and Kasman (1998). Recording was achieved with disposable bipolar surface electrodes (20-mm Blue Sensor Ag–AgCl, AMBU A/S, Denmark). Data was collected at a sampling rate of 4000 Hz using a custom-made Labview v6.1 based computer program (Mr. Kick v1.42e, Knud Larsen, Center for Sensory–Motor Interaction, SMI, University of Aalborg, Denmark). Following data collection, data was band-pass filtered from 20 Hz to 2 kHz and stored on the laboratory computer for further off-line analysis.

Peripheral Nerve Stimulation

Electrical mixed peripheral nerve stimulation was achieved by placing stimulating electrodes (32 mm, PALS Platinum, Patented Conductive Fabric Neurostimulation Electrodes, Axelgaard Manufacturing Co., Ltd., United States) over the deep branch of the right common peroneal nerve [L4 and L5]. The cathode was placed slightly distal to the superior aspect of the muscular origin of the TA. The anode was placed approximately two centimeters distal and one centimeter medial to the cathode. Stimulation, at the position of the electrodes, gave rise to palpable responses of the distal right TA tendon. The location of palpation was standardized to be over the ankle, at the level of the superior extensor retinaculum. This electrode site could also elicit M-waves in the TA upon stimulation. To ensure specificity of stimulation, palpation of synergistic peroneal muscles (innervated by the superficial branch of the peroneal nerve [L5, S1, and S2]) and antagonistic soleus (SOL) muscle (innervated by the tibial nerve [S1–S2]) was performed to ensure negligible activation. Stimulation was generated with a Digitimer DS7A isolated high-voltage electrical stimulator (maximum current of 10 mA).

Motor threshold (MT) was defined as the minimum amount of current required to generate a visible M-wave in the EMG trace. Interventional peripheral nerve stimulation was delivered at $1.1 \times$ MT for the right TA (stimulus width 1 ms).

Somatosensory Evoked Potentials

Surface electrodes (7-mm Ag–AgCl Hydrospot disposable adhesive electrodes from Physiometrix) were placed at scalp sites CPz and CP_2 referenced to Fz.

Site definitions were in accordance with the 10/20 system of EEG electrode placement (Yamada et al., 1996; Yamada, 2000). Sampling rate was 10 kHz, the band-pass filter of 3–1000 Hz. Peripheral electrical stimuli were delivered at 0.2–0.22 Hz to the same TA site as outlined previously. The stimulation intensity was set to $1 \times$ MT with a stimulus width of 1 ms. The latency of the maximal activation of the somatosensory cortex was determined by averaging a minimum of 1000 (maximum 3000) trials.

Experimental Procedures

Participants were seated in a comfortable chair so that the hip was flexed to 90°, the knee to 130°, and the ankle to 90°. The right foot was strapped into a custom-made footplate to ensure that the stated joint angles were maintained. The left foot was unrestrained and permitted to relax comfortably on a footrest in a similar position to the right foot.

Maximum Voluntary Contraction. The subjects were asked to perform an isometric maximal voluntary contraction (MVC) of the TA while their foot was restrained by the footplate. The subjects were instructed to "pull their foot up toward them as hard and fast as possible by rotating at their ankle joint" for two seconds. At least one minute rest phase was given between each of three trials. The EMG activity during each trial was recorded. The trial with the largest EMG activity, as quantified by the root mean square (RMS) value, was deemed the maximum.

Functional Measures. The maximal rate of force generation associated with dorsiflexion was measured. During recording of MVC, the force associated with each MVC was measured using a force transducer (ELHM-T3M load cell manufactured by ENTRAN, Hampton, VA; range 0–10 kN) attached to the plantar aspect of the footplate. The position of the footplate could be altered so that the axis of rotation of a participant's ankle was aligned to the axis of rotation of the cradle. Data from the force transducer was collected using Chart (Chart v5.4.1, Copyright 1994–2006 ADInstruments) and stored on the laboratory computer for off-line analysis.

Transcranial Magnetic Stimulation. Focal TMS was delivered using a figure-of-eight double cone coil (110-mm diameter), connected to a Magstim 200 magnetic stimulator (Magstim Company, Dyfed, U.K.). The coil was held firmly on top of the participant's head with the direction of the induced current flowing in a posterior-to-anterior (PA) direction. The best site for stimulation (the hot spot) for the right TA was found by delivering a series of suprathreshold stimuli (at approximately 50% of maximum stimulator output). The hot spot was determined by the coordinate that elicited the most consistent MEPs compared with adjacent coordinates. This site was approximately two centimeters rostral and one centimeter contralateral to the vertex. Once the hot spot was determined, the coil position was maintained by markings on the cap.

The resting threshold (RTh) was defined as the minimum stimulus intensity required to generate five MEPs, with a peak-to-peak amplitude greater than 50 μV, when 10 consecutive TMS stimuli were administered. During this process the target muscle was at rest.

Motor evoked potentials were elicited with single TMS pulses prior to, immediately following, and 30 min after the cessation of the respective interventions. The intensity of these pulses was set to a percentage of stimulator output that evoked a peak-to-peak MEP of approximately 0.5 mV (range, 0.129–1.184 mV; average, 0.44 ± 0.21 mV). The high variability in preintervention MEP reflects the individual variation in orientation of TA representation of the motor cortex. In some participants a mean preintervention MEP of 0.5 mV could not be achieved, irrespective of stimulator intensity. Sixteen single pulses were taken and these pulses were randomly separated by five to seven seconds to remove expectation effects and refractoriness.

Intracortical Inhibition. Short interval intracortical inhibition (SICI) was measured in nine subjects using a paired-pulse TMS protocol (T. Kujirai et al., 1993). Two magnetic stimulators were connected through an external timer delay. The conditioning stimulus (CS) was set to 90% of active threshold (ATh). Active threshold was defined as the minimum stimulator intensity required to generate p–p MEPs of at least 100 μV in at least 5 out of 10 consecutive trials while the subject held a constant level of isometric contraction in their right TA (Kischka et al., 1993; Rossini et al., 1994). This level of contraction was set as having an electromyographic root mean square (EMG RMS) equivalent to 5–10% of the EMG RMS associated with the MVC. Increasing levels of precontraction increase the p–p MEP amplitude so the 5–10% of MVC precontraction was strictly maintained in response to verbal feedback from the experimenter.

The test stimulus was delivered 2.5 ms following the conditioning stimulus in accordance with T. Kujirai et al. (1993) and Roshan et al. (2003). The strength of the original test stimulus (as a percentage of maximum stimulator output, MSO) was the same as that used to elicit the original MEPs for each subject. For post-test measures, the strength of the test stimuli was adjusted so that the mean p–p MEP amplitude was matched to the p–p MEP prior to the intervention.

Interventions

Participants were required to undertake three interventional sessions. Consecutive sessions were separated by a minimum of seven days. In these sessions one of three interventions, PAS, rES, or rES + PAS, were delivered in a pseudorandomized order for 30 minutes and their effects assessed.

Paired Associative Stimulation. Paired associative stimulation consisted of a single peripheral electrical stimulus followed by a single TMS stimulus. The two stimuli were separated by the optimized interstimulus interval (ISI) for that individual. As described by Mrachacz-Kersting et al. (2007), this ISI consisted of the subject's somatosensory evoked potential (SEP) peak latency plus six milliseconds of central processing delay. Three hundred and sixty pairs of stimuli were delivered at 0.2 Hz over a 30-minute intervention period. A depiction of the stimulation protocol is outlined in Figure 27.1A.

Repetitive Electrical Stimulation of a Mixed Nerve. An artificial muscle contraction was evoked by a rES protocol delivered at the site of peripheral electrical stimulation as described previously. A 30-Hz train of interventional peripheral electrical stimulation (stimulus width 1 ms, intensity 1.1 × MT) was applied for one second, which was sufficient to elicit a visible contraction of the TA. Each train was followed by a 1.5-s rest phase where no stimulation was applied and this pairing was applied consistently for the 30-minute intervention (Fig. 27.1B).

Combined PAS and rES. The rES was applied as outlined previously. A TMS pulse was delivered at the optimized ISI following the last electrical stimulus of the electrical train delivered in rES. This was done every two trains so that PAS was delivered at the rate of 0.2 Hz as achieved when PAS was delivered alone (Fig. 27.1C).

Attention was standardized across interventions by delivering a set auditory reminder every five minutes. The auditory reminder consisted of asking about the subject's well-being and comfort then instructing them "remember you need to think about your TA."

Statistical Analysis

The effects of the three interventions on the pre and post MEP amplitude and latency were analyzed using a one-way repeated-measures analysis of variance (ANOVA). The effects of the interventions on the time course (pre, post, and 30 min post) of MEP sizes or on MEP threshold were analyzed using a two-way repeated-measures ANOVA. Significance was taken if the p-value was less than or equal to the critical α value of 0.05. Data has been expressed as mean ± standard deviation (SD).

RESULTS

Corticospinal Excitability

In Figure 27.2 the averaged (16 sweeps) raw TA MEP data prior to and after PAS (Fig. 27.2A), rES (Fig. 27.2B), and rES + PAS (Fig. 27.2C) are presented

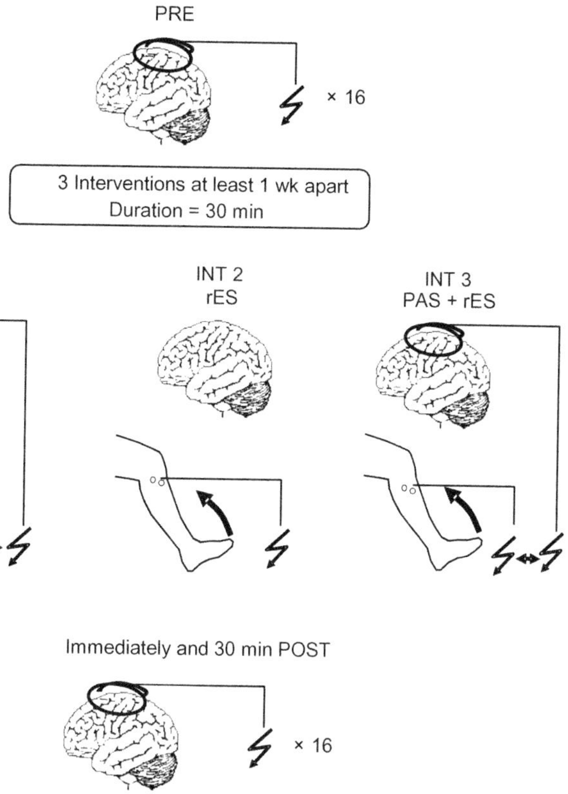

Figure 27.1. The experimental protocol. The figure shows a basic experimental protocol. MEPs were taken before, immediately after, and thirty minutes after the application of one of the three interventions. (A) PAS stimulus protocol. The single peripheral electrical stimulus is followed by a TMS stimulus when activation of the motor cortex occurs. This latency is termed the interstimulus interval (ISI) and is made up of the latency of a subject's somatosensory evoked potential (SEP) peak and six milliseconds of central processing time. This pairing of stimuli is applied once every five seconds for thirty minutes. (B) rES stimulus protocol. A one-second 30-Hz pulse was followed by a one-and-a-half-second rest phase. This was applied for thirty minutes. (C) A novel stimulus protocol, rES + PAS. First, rES was elicited as per Figure 27.1B, but every other burst of electrical stimuli was associated with a TMS pulse. This TMS pulse followed the application of the last peripheral electrical stimulus in a stimulus block by the ISI used in PAS. The number of electrical bursts was consistent with a rES-only protocol and the frequency of PAS application was maintained. This protocol was applied for thirty minutes.

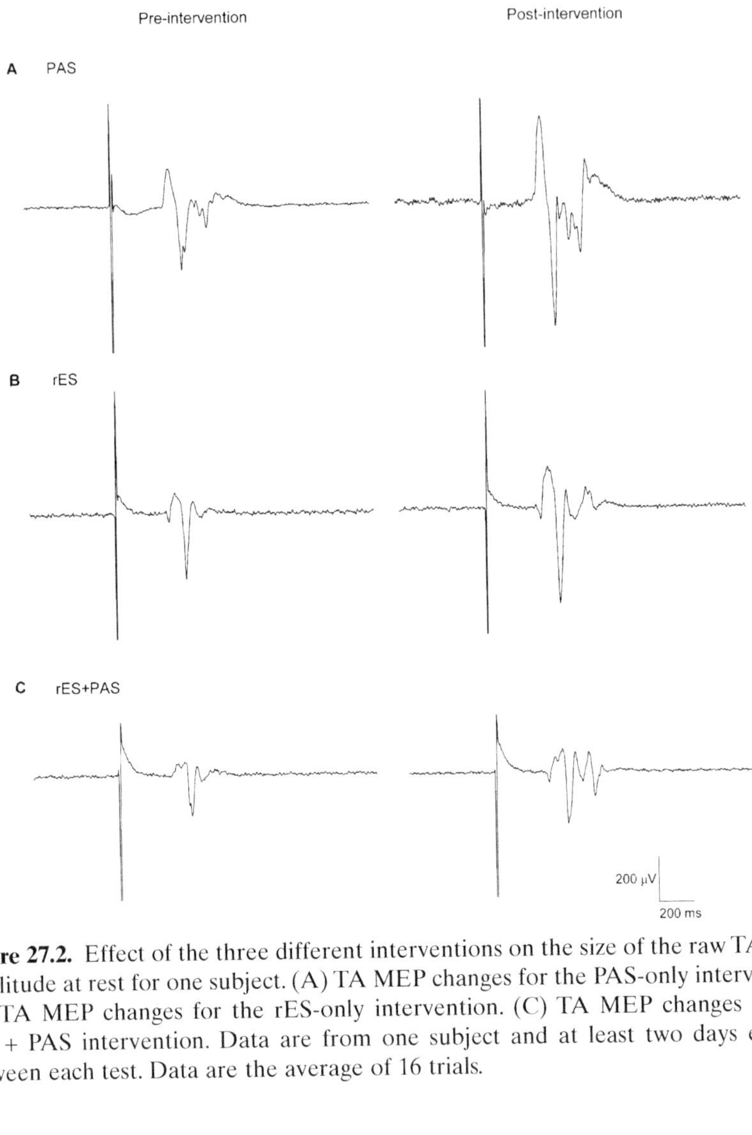

Figure 27.2. Effect of the three different interventions on the size of the raw TA MEP amplitude at rest for one subject. (A) TA MEP changes for the PAS-only intervention. (B) TA MEP changes for the rES-only intervention. (C) TA MEP changes for the rES + PAS intervention. Data are from one subject and at least two days elapsed between each test. Data are the average of 16 trials.

for one subject. For this subject, the average values of the MEPs prior to the interventions were 0.54 mV, 0.43 mV, and 0.37 mV for PAS, rES, and rES + PAS, respectively. Immediately following the interventions the values were 1.20 mV, 0.70 mV, and 0.45 mV.

Across all subjects, the preintervention average TA MEP size was 0.37 ± 0.23 mV for PAS, 0.43 ± 0.23 mV for rES, and 0.39 ± 0.21 mV for rES + PAS. This was not statistically different ($p = 0.60$)

Immediately following the intervention, these values were 0.73 ± 0.19 mV, 0.66 ± 0.18 mV, and 0.67 ± 0.28 mV PAS, rES, and rES + PAS respectively

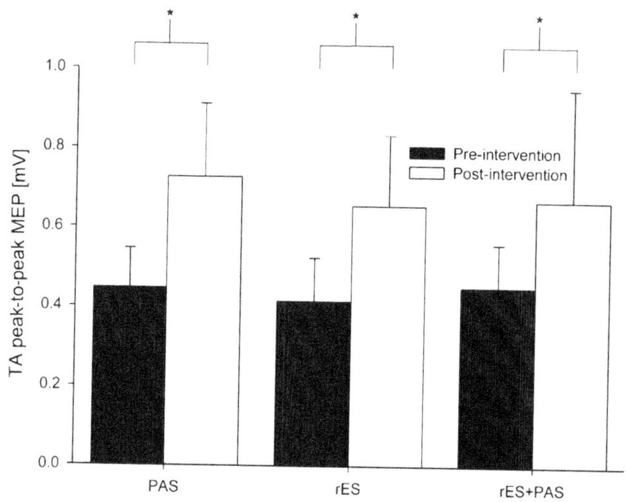

Figure 27.3. Effect of the three different interventions on the size of the TA MEP amplitude across all subjects. (A) For 20 subjects, recordings were made prior to and immediately following the cessation of the intervention. (B) For 16 subjects recordings were made prior to, immediately after, and 30 minutes following the cessation of the intervention. The MEP amplitudes are expressed in millivolts. Also shown are standard deviations.

(Fig. 27.3A). When the change is expressed for each subject individually as a percentage of preintervention, then the average TA MEP amplitude increased by 81 ± 48%, 57 ± 36%, and 44 ± 43% immediately following the cessation of the PAS, rES, and rES + PAS interventions.

Two-way repeated-measures ANOVA revealed no significant difference between the three interventions ($p = 0.88$); however, there were significant differences across the factor time ($p = 0.01$).

Resting Threshold

In 16 subjects the RTh was assessed prior to, immediately following, and 30 minutes following the cessation of the respective interventions. Prior to the interventions there was no significant difference in the mean RTh across interventions (PAS, 47.94 ± 4.26% MSO; rES, 45.94 ± 4.62% MSO; rES + PAS, 47.38 ± 4.43% MSO; $p = 0.2014$).

Resting threshold did not change significantly immediately after PAS (−1.00 ± 4.24% MSO), or thirty minutes after PAS (−1.00 ± 4.24% MSO; $p = 0.1347$). The mean RTh immediately following rES did not change (0.00 ± 4.96% MSO), and changes remained nonsignificant after 30 minutes (−0.04 ± 4.67% MSO; $p = 0.9060$). However, changes in RTh after rES + PAS were significant; immediately after (−2.56 ± 4.04% MSO) and 30 minutes after (−2.00 ± 3.89% MSO; $p = 0.002$) the cessation of the intervention.

Short Interval Intracortical Inhibition

In nine subjects short interval intracortical inhibition (SICI) was assessed prior to and immediately following the interventions. The test MEP used for postintervention measures of SICI was adjusted to produce a MEP of similar size prior to the intervention. SICI was expressed as a factor of the test MEP amplitude such that a value of one would signify no SICI and 0.5 would signify a depression of the test MEP by 50%. Values so expressed for SICI prior to the interventions were 0.51 ± 0.27 (rES), 0.49 ± 0.23 (PAS), and 0.51 ± 0.21 (rES + PAS). After intervention there was a significant change in SICI for the rES intervention (0.82 ± 0.36, $t = 2.827$, $p = 0.02$). However, SICI did not change for PAS (0.40 ± 0.19, $t = 0.542$, $p = 0.6$) or rES + PAS (0.6 ± 0.49, $t = 0.163$, $p = 0.87$).

Functional Measures

There was no significant change in the maximum rate of force generation immediately following PAS, rES, and rES + PAS. The change expressed as a percentage of preintervention value was −9 ± 10.63% ($p = 0.1540$), −10 ± 10.79% ($p = 0.1007$), and 0.0 ± 9.58 ($p = 0.9449$) for PAS, rES, and rES + PAS, respectively.

The correlation between changes in the maximal rate of force generation and changes in the TA MEP amplitude, immediately following the respective intervention, were assessed. Changes in maximal rate of force production were not correlated with changes in TA MEP amplitude induced by PAS alone (correlation coefficient 0.0318, r^2 value 0.0201, $p = 0.6141$), rES alone (correlation coefficient −0.0536, r^2 value 0.0396, $p = 0.4602$), or PAS + rES (correlation coefficient 0.0744, r^2 value 0.1195, $p = 0.2070$).

CONCLUSIONS

The study has confirmed the findings from previous studies that both rES and PAS when applied alone can enhance the excitability of the cortical projections to the TA (Khaslavskaia et al., 2002; Khaslavskaia and Sinkjaer, 2005; Knash et al., 2003; Thompson and Stein, 2004; Mrachacz-Kersting et al., 2007). The results of the study discussed here extend these findings by demonstrating the following: (i) a combination of these two protocols does not further enhance the induced facilitation, suggesting that these two protocols may interact; (ii) there is an effect on both resting motor threshold and MEP following the rES + PAS protocol, suggesting a dual mechanism for the induced effects in this intervention; (iii) basic functional measures are unaltered, whereas the motor representation of the muscle involved in these measures is significantly increased.

Interaction between PAS and rES

Evidence presented here suggests that changes in excitability following all three interventional protocols were similar in magnitude and mostly restricted to the cortical projections of the target muscle. Further, the effects lasted for at least half an hour after the cessation of the interventions. Several past studies have also indicated a cortical site for changes following either rES alone (Khaslavskaia et al., 2002; Khaslavskaia and Sinkjaer, 2005; Thompson et al., 2006) or PAS (Mrachacz-Kersting et al., 2007; Roy et al., 2007) when targeting the TA. In addition, although PAS does not result in any significant changes in SICI, such changes are reported when rES is applied alone. The main idea in this study was thus that the rES would prime the motor cortex and allow an increased effect of the subsequently applied PAS. Stefan et al. (2002) have suggested that a transient disinhibition of the primary motor cortex at a time when the TMS is applied contributes to the facilitatory changes observed following the application of PAS. Several studies have implied that a reduction in SICI or a disinhibited motor cortex is favorable for the induction of training-induced changes in cortical excitability (Perez et al., 2004) in healthy subjects. In stroke populations, a bout of electrical stimuli applied prior to motor training designed to induce use-dependent plasticity, increases the effects of motor learning significantly (Sawaki, 2005).

In this study, SICI was significantly reduced following rES, providing support for this idea. However, Kaelin-Lang et al. (2002) reported a significant increase in MEPs of abductor digiti minimi (ADM) without an associated release of SICI after a two-hour period of ulnar nerve stimulation. However, SICI did show a tendency to decrease; the authors speculate that the power of their study only allowed changes in SICI of 35% or greater and that changes may have been more subtle.

It was an unexpected finding that the combination of the rES and PAS protocols did not induce a further facilitation of the MEP above and beyond their algebraic summation when applied separately. There are several possibilities why rES + PAS was unable to enhance the effects of the two protocols when applied separately. First, it is possible that both rES and PAS act on the same population of cortical neurons and thus the same neural circuitry. In past studies, the effects of rES versus active dorsiflexion (DF) have been shown to activate similar areas in the brain when compared with passive movement (Francis et al., 2009). Motor learning has been shown to interfere with induction of plastic changes by PAS (Stefan et al., 2006; Ziemann et al., 2004). It is conceivable that rES similar to learning a simple motor task affects the induction of PAS. This interference may be the result of the synapses having attained a limit in their ability to change. This is reminiscent of metaplasticity, which has been inferred to be responsible for the fact that protocols designed to induce LTP affect the subsequent induction of LTP through either a different or the same protocol (Ziemann et al., 2004; Siebner et al., 2004). Bienenstock et al. (1982) proposed a dynamic adaptation of a modification threshold (often referred to as the Bienenstock–Cooper–Munro, BCM, rule of synaptic modification); thus synaptic plasticity is regulated according to the previous history of the neural activity across that synapse. A train of rES may cause overpotentiation of selected synapses that are also targeted by the PAS protocol.

A second possibility is that a transient reduction in SICI following a train of rES removes inhibition from circuits that under normal conditions of SICI would be inaccessible to PAS, that is, those neurons with the highest threshold to TMS. SICI is reduced following a single electrical stimulus to a mixed nerve (Stefan et al., 2002; Poon et al., 2008) though it is not known if a train of such stimuli induces the same effect. However, the observations that rES reduces SICI after application for two hours (Kaelin-Lang et al., 2002) as well as our report of a reduction in SICI following the rES protocol favor such a hypothesis. At first sight, a disinhibition of the primary motor cortex at a time when the TMS pulse arrives would be in favor of the TMS pulse being able to exert an effect over more cortical cells. However, the current data does not yield definite conclusions on this. Rather, several other possibilities exist. For example, following rES some of the cells normally targeted by the PAS protocol may be in a hyperexcited state and thus unable to respond to the next TMS stimulus; however, then we would expect the facilitation to be much less following the combined protocol and this is not supported by the presented data. Alternatively, since the rES + PAS protocol resulted in similar facilitation, this may imply that the subsequent TMS pulse was not only acting on a transiently disinhibited cortex, but also was able to access and act on a different set of neurons that may or may not be part of a different circuitry. The fact that the RTh decreased only after rES + PAS lends support to this hypothesis. Changes in RTh are indicative of an alteration in membrane excitability (Ziemann et al., 2004). Since the onset of the MEPs remained unchanged following all three interventions, it suggests that it is not those neurons (circuits) with the lowest threshold to TMS, but more likely the high-threshold cells that are changed. However, a third possibility must be addressed. In a more recent study, Peurala et al. (2008) provided convincing evidence that SICI tested at interstimulus intervals of 2–3 ms between the subthreshold first and suprathreshold second stimuli may be contaminated by short interval intracortical facilitation (SICF). Indeed changes in MEP size following PAS have been associated with concomitant increases in SICF (Stefan et al., 2002). Thus in this study we cannot rule out that the reduced SICI was at least in part due to an increase in SICF. If this is the case, then it may explain further why the RTh was decreased while there were no changes in SICI for the rES + PAS protocol.

Functional Measures

The assessment of changes in motor function in normal healthy individuals is challenging. The motor performance of the healthy normal population is arguably near the maximal potential and less easily increased. In this study we did not find significant changes in our behavioral measure of rate of peak acceleration. Muellbacher et al. (2000) failed to see functional measures following an rTMS intervention that caused significant increases in MEP amplitude. However, changes in excitability of the circuitry within the motor cortex have been shown to precede behavioral recovery (Aggelopoulos et al., 2008). Hamdy et al. (1998) applied repetitive electrical stimulation to the pharynx at a frequency of 10 Hz that increased MEP size but did not alter the swallowing function. However, when this protocol was applied to stroke patients (with dysphagia) there was both an increase in MEP and an improvement in function, and these two were significantly correlated (Fraser et al., 2002).

The motor cortex most probably requires extended periods of facilitation, such as those that can be observed upon multiple exposures to interventions, to be associated with greater function. The greatest potential for protocols that increase the excitability of the motor cortex still likely resides in the ongoing application of protocols in conjunction with conventional long-term rehabilitation techniques. As PAS is based on associative LTP, a candidate mechanism for learning as well as memory formation and storage, it warrants further investigation into how this can be used in a rehabilitative setting.

REFERENCES

Aggelopoulos N., Chakrabarty S., and Edgley S. (2008). Presynaptic control of transmission through group II muscle afferents in the midlumbar and sacral segments of the spinal cord is independent of corticospinal control. *Exp Brain Res* 187(1):61–70.

Ashe J. (1997). Force and the motor cortex. *Behavioural Brain Res* 86:1–15.

Bi G. and Poo M. (1998). Synaptic modifications in cultured hippocampal neurons: dependence on spike timing, synaptic strength, and postsynaptic cell type. *J Neurosci* 18(24):10464–10472.

Bienenstock E.L., Cooper L.N., and Munro P.W. (1982). Theory for the development of neuron selectivity: orientation specificity and binocular interaction in visual cortex. *J Neurosci* 2(1):32–48.

Castle-Lacanal E., Gerdelat-Mas A., Marque P., Loubinoux I., and Simonetta-Moreau M. (2007). Induction of cortical plastic changes in wrist muscles by paired associative stimulation in healthy subjects and post-stroke patients. *Exp Brain Res* 180(1):113–122.

Cram J. and Kasman G. (1998). *Introduction to Surface Electromyography*, 1st ed. Jones and Bartlett.

Di Lazzaro V., Oliviero A., Profice P., Meglio M., Cioni B., Tonali P., and Rothwell J.C. (2001). Descending spinal cord volleys evoked by transcranial magnetic and electrical stimulation of the motor cortex leg area in conscious humans. *J Physiol* 537(3):1047–1058.

Francis S., Lin X., Aboushoushah S., White T.P., Phillips M., Bowtell R., and Constantinescu C.S. (2009). fMRI analysis of active, passive and electrically stimulated ankle dorsiflexion. *NeuroImage* 44(2):469–479.

Fraser C., Power M., Hamdy S., Rothwell J., Hobday D., Hollander I., Tyrell P., Hobson A., Williams S., and Thompson D. (2002). Driving plasticity in human adult motor cortex is associated with improved motor function after brain injury. *Neuron* 34(5):831–840.

Fratello F., Veniero D., Curcio G., Ferrara M., Marzano C., Moroni F., Pellicciari M.C., Bertini M., Rossini P.M., and De Gennaro L. (2006). Modulation of corticospinal excitability by paired associative stimulation: reproducibility of effects and intraindividual reliability. *Clin Neurophysiol* 117(12):2667–2674.

Georgopoulos A.P., Ashe J., Smyrnis N., and Taira M. (1992). The motor cortex and the coding of force. *Science* 256:1692–1695.

Hamdy S., Rothwell J.C., Aziz Q., Singh K.D., and Thompson D.G. (1998). Long-term reorganization of human motor cortex driven by short-term sensory stimulation. *Nature Neurosci* 1(1):64–68.

Kaelin-Lang A., Luft A.R., Sawaki L., Burstein A.H., Sohn Y.H., and Cohen L.G (2002). Modulation of human corticomotor excitability by somatosensory input. *J Physiol (London)* 540(2):623–633.

Kang J.S., Terranova C., Hilker R., Quartarone A., and Ziemann U. (2011). Deficient homeostatic regulation of practice-dependent plasticity in writer's cramp. *Cereb Cortex* 21:1203–1212.

Khaslavskaia S. and Sinkjaer T. (2005). Motor cortex excitability following repetitive electrical stimulation of the common peroneal nerve depends on the voluntary drive. *Exp Brain Res* 162(4):497–502.

Khaslavskaia S., Ladouceur M., and Sinkjaer T. (2002). Increase in tibialis anterior motor cortex excitability following repetitive electrical stimulation of the common peroneal nerve. *Exp Brain Res* 145(3):309–315.

Kischka U., Fajfr R., Fellenberg T., and Hess C.W. (1993). Facilitation of motor evoked potentials from magnetic brain stimulation in man: a comparative study of different target muscles. *J Clin Neurophysiol* 10(4):505–512.

Knash M., Kido A., Gorassini M., Chan K.M., and Stein R. (2003). Electrical stimulation of the human common peroneal nerve elicits lasting facilitation of cortical motor-evoked potentials. *Exp Brain Res* 153(3):366–377.

Kujirai K., Kujirai T., Sinkjaer T., and Rothwell J.C. (2006). Associative plasticity in human motor cortex during voluntary muscle contraction. *J Neurophysiol* 96(3):1337–1346.

Kujirai T., Caramia M.D., Rothwell J.C., Day B.L., Thompson P.D., Ferbert A., Wroe S., Asselman P., and Marsden C.D. (1993). Corticocortical inhibition in human motor cortex. *J Physiol (London)* 471(1):501–519.

Letzkus J.J., Kampa B.M., and Stuart G.J. (2007). *Does Spike Timing-Dependent Synaptic Plasticity Underlie Memory Formation?* 34th ed., pp. 1070–1076. Hoboken, NJ: Wiley-Blackwell.

Muellbacher W., Ziemann U., Boroojerdi B., and Hallett M. (2000). Effects of low-frequency transcranial magnetic stimulation on motor excitability and basic motor behavior. *Clin Neurophysiol* 111(6):1002–1007.

Mueller-Dahlhaus J., Orekhov Y., Liu Y., and Ziemann U. (2008). Interindividual variability and age-dependency of motor cortical plasticity induced by paired associative stimulation. *Exp Brain Res* 187(3):467–475.

Muller J.F., Orekhov Y., Liu Y., and Ziemann U. (2007). Homeostatic plasticity in human motor cortex demonstrated by two consecutive sessions of paired associative stimulation. *Eur J Neurosci* 25(11):3461–3468.

Mrachacz-Kersting N., Fong M., Murphy B.A., and Sinkjaer T. (2007). Changes in excitability of the cortical projections to the human tibialis anterior following paired associative stimulation. *J Neurophysiol* 97:1951–1958.

Oldfield R.C. (1971). The assessment and analysis of handedness: the Edinburgh inventory. *Neuropsychologia* 9(1):97–113.

Pellicciari M.C., Miniussi C., Rossini P.M., and De Gennaro L. (2009). Increased cortical plasticity in the elderly: changes in the somatosensory cortex after paired associative stimulation. *Neuroscience* 163(1):266–276.

Perez M.A., Lungholt B.K.S., Nyborg K., and Nielsen J.B. (2004). Motor skill training induces changes in the excitability of the leg cortical area in healthy humans. *Exp Brain Res* 159(2):197–205.

Peurala S.H., Mueller-Dahlhaus M., Arai N., and Ziemann U. (2008). Interference of short-interval intracortical inhibition (SICI) and short-interval intracortical facilitation (SICF). *Clin Neurophysiol* 119(10):2291–2297.

Poon D., Roy F., Gorassini M., and Stein R. (2008). Interaction of paired cortical and peripheral nerve stimulation on human motor neurons. *Exp Brain Res* 188(1):13–21.

Quartarone A., Rizzo V., Bagnato S., Morgante F., Sant'Angelo A., Girlanda P., and Roman Siebner H. (2006). Rapid-rate paired associative stimulation of the median nerve and motor cortex can produce long-lasting changes in motor cortical excitability in humans. *J Physiol (London)* 575(2):657–670.

Quartarone A., Rizzo V., Terranova C., Morgante F., Schneider S., Ibrahim N., Girlanda P., Bhatia K.P., and Rothwell J.C. (2009). Abnormal sensorimotor plasticity in organic but not in psychogenic dystonia. *Brain* 132(10):2871–2877.

Ridding M. and Flavel S. (2006). Induction of plasticity in the dominant and non-dominant motor cortices of humans. *Exp Brain Res* 171(4):551–557.

Ridding M.C. and Taylor J.L. (2001). Mechanisms of motor-evoked potential facilitation following prolonged dual peripheral and central stimulation in humans. *J Physiol (London)* 537(2):623–631.

Ridding M.C. and Uy J. (2003). Changes in motor cortical excitability induced by paired associative stimulation. *Clin Neurophysiol* 114(8):1437–1444.

Rosenkranz K. and Rothwell J.C. (2006). Differences between the effects of three plasticity inducing protocols on the organization of the human motor cortex. *European J Neurosci* 23(3):822–829.

Roshan L., Paradiso G.O., and Chen R. (2003). Two phases of short-interval intracortical inhibition. *Exp Brain Res* 151(3):330–337.

Rossini P.M., Barker A.T., Berardelli A., Caramia M.D., Caruso G., Cracco R.Q., Dimitrijevic M.R., Hallett M., Katayama Y., Lucking C.H., Maertens de Noordhout A.L., Marsden C.D., Murray N.M.F., Rothwell J.C., Swash M., and Tomberg C. (1994). Non-invasive electrical and magnetic stimulation of the brain, spinal cord and roots: basic principles and procedures for routine clinical application. Report of an IFCN committee. *Electroencephalogr Clin Neurophysiol* 91(2):79–92.

Roy F.D. and Gorassini M.A. (2008). Peripheral sensory activation of cortical circuits in the leg motor cortex of man. *J Physiol (London)* 586(17):4091–4105.

Roy F.D., Norton J.A., and Gorassini M.A. (2007). Role of sustained excitability of the leg motor cortex after transcranial magnetic stimulation in associative plasticity. *J Neurophysiol* 98(2):657–667.

Roy F.D., Yang J.F., and Gorassini M.A. (2010). Afferent regulation of leg motor cortex excitability after incomplete spinal cord injury. *J Neurophysiol* 103(4):2222–2233.

Sakai K., Ugawa Y., Terao Y., Hanajima R., Furubayashi T., and Kanazawa I. (1997). Preferential activation of different I waves by transcranial magnetic stimulation with a figure-of-eight-shaped coil. *Exp Brain Res* 113(1):24–32.

Sawaki L. (2005). Use-dependent plasticity of the human motor cortex in health and disease. *IEEE Eng Med Biol Magazine* 24(1):36–39.

Siebner H.R., Lang N., Rizzo V., Nitsche M.A., Paulus W., Lemon R.N., and Rothwell J.C. (2004). Preconditioning of low-frequency repetitive transcranial magnetic stimulation with transcranial direct current stimulation: evidence for homeostatic plasticity in the human motor cortex. *J Neurosci* 24(13):3379–3385.

Stefan K., Kunesch E., Cohen L.G., Benecke R., and Classen J. (2000). Induction of plasticity in the human motor cortex by paired associative stimulation. *Brain* 123(3):572–584.

Stefan K., Kunesch E., Benecke R., Cohen L.G., and Classen J. (2002). Mechanisms of enhancement of human motor cortex excitability induced by interventional paired associative stimulation. *J Physiol* 543(Pt 2):699–708.

Stefan K., Wycislo M., Gentner R., Schramm A., Naumann M., Reiners K., and Classen J. (2006). Temporary occlusion of associative motor cortical plasticity by prior dynamic motor training. *Cereb Cortex* 16(3):376–385.

Tamura Y., Ueki Y., Lin P., Vorbach S., Mima T., Kakigi R., and Hallett M. (2009). Disordered plasticity in the primary somatosensory cortex in focal hand dystonia. *Brain* 132(3):749–755.

Tecchio F., Zappasodi F., Pasqualetti P., Gennaro L.D., Pellicciari M.C., Ercolani M., Squitti R., and Rossini P.M. (2008). Age dependence of primary motor cortex plasticity induced by paired associative stimulation. *Clin Neurophysiol* 119(3):675–682.

Thompson A.K. and Stein R.B. (2004). Short-term effects of functional electrical stimulation on motor-evoked potentials in ankle flexor and extensor muscles. *Exp Brain Res* 159:491–500.

Thompson A., Doran B., and Stein R. (2006). Short-term effects of functional electrical stimulation on spinal excitatory and inhibitory reflexes in ankle extensor and flexor muscles. *Exp Brain* Res 170(2):216–226.

Tokimura H., Di L., Tokimura V.Y., Oliviero A., Profice P., Insola A., Mazzone P., Tonali P., and Rothwell J.C. (2000). Short latency inhibition of human hand motor cortex by somatosensory input from the hand. *J Physiol* 523(Pt 2):503–513. [Erratum. *J Physiol (London)* 2000 May 1; 524(Pt 3):942.]

Tsuji T., Suzuki K., Masakado Y., Ota T., Kimura A., Liu M., and Chino N. (2005). Long-lasting effects of paired associative stimulation in hemiparetic stroke patients. Int Congress Ser 1278:280–283.

Uy J. and Ridding M.C. (2003). Increased cortical excitability induced by transcranial DC and peripheral nerve stimulation. *J Neurosci Methods* 127(2):193–197.

Wolters A., Sandbrink F., Schlottmann A., Kunesch E., Stefan K., Cohen L.G., Benecke R., and Classen J. (2003). A temporally asymmetric Hebbian rule governing plasticity in the human motor cortex. *J Neurophysiol* 89(5):2339–2345.

Yamada T. (2000). Neuroanatomic substrates of lower extremity somatosensory evoked potentials. *J Clin Neurophysiol* 17(3):269–279.

Yamada T., Matsubara M., Shiraishi G., Yeh M., and Kawasaki M. (1996). Topographic analyses of somatosensory evoked potentials following stimulation of tibial, sural and lateral femoral cutaneous nerves. *Electroencephalogr Clin Neurophysiol/Evoked Potentials Sect* 100(1):33–43.

Ziemann U., Iliac T.V., Pauli C., Meintzschel F., and Ruge D. (2004). Learning modifies subsequent induction of long-term potentiation-like and long-term depression-like plasticity in human motor cortex. *J Neurosci* 24(7):1666–1672.

Ziemann U., Paulus W., Nitsche M.A., Pascual-Leone A., Byblow W.D., Berardelli A., Siebner H.R., Classen J., Cohen L.G., and Rothwell J.C. (2008). Consensus: motor cortex plasticity protocols. *Brain Stimulation* 1(3):164–182.

28

FOCUSED CHAPTER

OPERANT CONDITIONING OF SPINAL REFLEXES FOR MOTOR REHABILITATION AFTER CNS DAMAGE

AIKO K. THOMPSON AND JONATHAN R. WOLPAW

Program for Translational Neurological Research, Helen Hayes Hospital and the Wadsworth Center, New York State Department of Health, New York, USA

SUMMARY

Spinal reflex behaviors can be changed through operant conditioning, in which modification of a behavior is brought about by the consequence of that behavior. Previous work in animals shows that successful operant conditioning of a spinal reflex changes not only the reflex itself but also other behaviors (e.g., locomotion) that involve the same pathway. Therefore, it may be possible to use spinal reflex conditioning to change a spinal reflex pathway so as to ameliorate movement disabilities due to central nervous system (CNS) damage. Indeed, in rats with partial spinal cord injury (SCI), appropriate reflex conditioning can restore more normal locomotion. We hypothesize that appropriate reflex conditioning will improve movement recovery in people with CNS damage.

Spasticity, a common problem after CNS damage, is seen as exaggerated stretch reflexes and abnormal reflex modulation during movements. The normal

Introduction to Neural Engineering for Motor Rehabilitation, First Edition. Edited by Dario Farina, Winnie Jensen, and Metin Akay.

modulation of the soleus H-reflex, an electrical analog of the stretch reflex, is often greatly diminished or entirely eliminated during walking (i.e., the reflex remains large even in the early stance or swing phase, where the H-reflex is normally very small or absent). We demonstrated that operant conditioning can increase or decrease the soleus H-reflex in neurologically intact people, and we are now investigating whether down-conditioning can reduce abnormal soleus H-reflex excitability in people with incomplete SCI and can thereby alleviate spastic gait. Our preliminary results in people with incomplete SCI indicate that soleus H-reflex conditioning is possible is this population and may improve gait.

INTRODUCTION AND BACKGROUND

Neural circuits in the spinal cord are capable of generating movement without direct input from the brain (i.e., reflexes and rhythmic movements) and are plastic throughout life (Stein, 2008). Because these circuits participate in essentially all motor behaviors, methods for inducing and guiding spinal cord plasticity could help to restore motor functions after chronic central nervous system (CNS) damage or disease (Wolpaw, 2006). Operant conditioning that changes specific spinal reflex pathways might, therefore, become a useful approach for improving recovery of motor function.

Activity-Dependent Plasticity in the Spinal Cord

The spinal cord receives descending and peripheral inputs continually throughout life. In the short term, these inputs elicit appropriate movements—for example, through voluntary muscle activation (Capaday et al., 1999; Schubert et al., 1997) and reflex modulation (E.P. Zehr and Stein, 1999; Stein, 1995; Brooke et al., 1997). Furthermore, in the long term, they gradually establish and preserve spinal cord pathways in a condition that supports the full repertoire of motor behaviors (Nielsen et al., 1993; Schneider and Capaday, 2003). Gradual activity-dependent plasticity, induced and guided by descending and associated peripheral inputs, shapes spinal cord function in early development and throughout life (Wolpaw and Tennissen, 2001).

Both descending and peripheral inputs have crucial roles in shaping, producing, and maintaining normally functioning adult spinal cord pathways that support motor skills such as posture control, locomotion, dancing, and playing musical instruments. Early-life disturbances to the crucial descending activity lead to an inappropriately functioning adult spinal cord. During early life, the corticospinal tract (CST), which is important for motor control and skill learning (Wolpaw, 2007; Adkins et al., 2006), develops its normal pattern of mainly contralateral innervation of motoneurons (Eyre, 2003). However, perinatal CST damage (e.g., cerebral palsy) prevents the normal development of these pathways, resulting in the abnormal persistence of bilateral CST projections to the adult spinal cord. Perinatal supraspinal damage also prevents proper maturation of spinal proprioceptive reflexes (Myklebust et al., 1986; O'Sullivan et al., 1998; Hodapp et al., 2007) that contribute to normal motor behaviors (Dietz, 2002; Dietz et al., 2002) and results in lifelong motor disabilities. Similarly, appropriate descending and peripheral inputs during early development are essential for normal development of flexor withdrawal reflexes to noxious stimuli (Waldenstrom et al., 2003; Levinsson et al., 1999) and urinary function (de Groat, 2002).

Peripheral and descending inputs continue to modify spinal cord pathways during early development and throughout adult life. Thus, CNS damage that disturbs descending and peripheral activity produces reflex abnormalities and motor dysfunction in the adult.

Abnormally Functioning Spinal Reflexes in Motor Dysfunction

Normally, spinal reflexes are modulated in functionally appropriate ways to enable the motor task being performed. For instance, soleus H-reflex gain decreases from standing to walking, and even further from walking to running (Stein and Capaday, 1988). This task-dependent adjustment of reflex gain prevents saturation of the motor output and the reflex feedback loop. Nonreciprocal inhibition, which comes mainly from Golgi tendon organs and is present during standing, disappears or changes to excitation in some phases of walking (Pearson and Collins, 1993; Stephens and Yang, 1996; Pearson, 1995). In walking, the extensor muscles need to remain active to support the body as long as the limb is loaded, and positive feedback from Golgi tendon organs helps to maintain extensor activity (Pearson, 1995; Dietz and Duysens, 2000; Duysens et al., 2000; Hiebert et al., 1996; Whelan et al., 1995). In general, reflex modulation across motor tasks appears to contribute to effective task performance (Pearson, 1995; Dietz and Duysens, 2000; Duysens et al., 2000).

Nervous system injuries often lead to abnormal spinal reflexes. Normal task-dependent modulation of the soleus H-reflex is diminished or absent (or even reversed) in some people with incomplete spinal cord injury (SCI; Stein et al., 1993; Yang et al., 1991). After SCI, normal nonreciprocal Ib inhibition of the soleus by stimulation of the medial gastrocnemius nerve disappears (Morita et al., 2006) and recurrent inhibition of the soleus is exaggerated (Shefner et al., 1992). These losses of appropriate reflex modulation are likely to contribute to motor dysfunction.

Altered spinal reflexes may interfere with the already weakened supraspinal control of locomotion in people with SCI (Stein et al., 1993; Yang et al., 1991; Crone et al., 2003; Hiersemenzel et al., 2000; Thompson et al., 2009c). Foot drop (i.e., drop and drag of the foot in the swing phase of the step cycle), one of the most common problems after CNS damage, probably results from changes in both spinal and supraspinal pathways. Ankle dorsiflexion is reduced after SCI due to disruption of the CST activity that normally contributes to ankle dorsiflexion during the swing-phase (Capaday et al., 1999; Schubert et al., 1997; Petersen et al., 2001). Ankle dorsiflexion is often further diminished by exaggerated stretch reflex responses from extensor triceps surae muscles (including soleus; Yang et al., 1991; Fung et al., 1994). These reductions in dorsiflexion cause foot drop.

During normal walking in humans, reflexes are modulated depending on the phase of the step cycle (Stein and Capaday, 1988; Schneider et al., 2000; Sinkjaer et al., 1996). However, after SCI or other CNS damage, both task-dependent modulation (see the preceding discussion; Thompson et al., 2009c; Boorman et al., 1996) and phase-dependent reflex modulation may be greatly diminished or entirely lost (Stein et al., 1993; Yang et al., 1991; Fung and Barbeau, 1994). Furthermore, abnormal reciprocal inhibition between ankle plantar- and dorsiflexors may contribute to exaggerated stretch reflexes and/or foot drop (Crone et al., 2003; Thompson et al., 2009c; Boorman et al., 1996; Ashby and Wiens, 1989). (Other mechanisms can also contribute to these spastic movement abnormalities, e.g., Aymard et al., 2000.) Figure 28.1 shows soleus H-reflexes, and soleus and tibialis anterior (TA) EMG, during walking in a normal person and in two people with incomplete SCI. In the normal person (Fig. 28.1A), H-reflex size is greatly modulated over the step-cycle: it gradually increases from the start of stance, peaks at the end of stance (i.e., about the same time as the peak soleus EMG activity), then rapidly falls and remains low or absent through the swing phase. In contrast, in the first person with SCI (Fig. 28.1B), the soleus H-reflex falls after push-off (as it should) but recovers too soon (i.e., in the middle of the swing phase) and remains high through midstance. This abnormal modulation probably contributes to the diminished second TA burst from the end of swing to early stance, which causes foot drop at the end of swing. In addition, the abnormally high H-reflex gain during early stance probably underlies the clonic soleus EMG activity in this person (and many people with SCI) during early-middle stance. Such abnormal muscle activity causes instability during the stance–swing transition and in initial foot placement in early stance (Yang et al., 1991; Fung and Barbeau, 1994). Another frequent locomotor reflex abnormality is loss of reflex modulation throughout the step cycle, as shown in the second person with SCI (Fig. 28.1C; Stein et al., 1993). This loss of modulation may reflect saturation of the reflex loop (Yang et al., 1991).

Possible Therapeutic Benefits of Reflex Operant Conditioning

As described previously, H-reflex modulation is often impaired in people with chronic CNS damage, and this abnormality can affect locomotor EMG activity (Stein et al., 1993; Burne et al., 2005; Dietz and Sinkjaer, 2007). Unsuppressed extensor reflex activity from swing to early stance can counteract ankle dorsiflexion and contribute to foot drop, and high reflex gain in the swing–stance transition can cause clonus that makes the ankle unstable (see Fig. 28.1B). Furthermore, lack of reflex modulation during walking may lead to the tonic activation of the soleus throughout the step cycle and result in little or no TA activity in the swing phase (Fig. 28.1C). Thus, decreasing extensor reflex gain

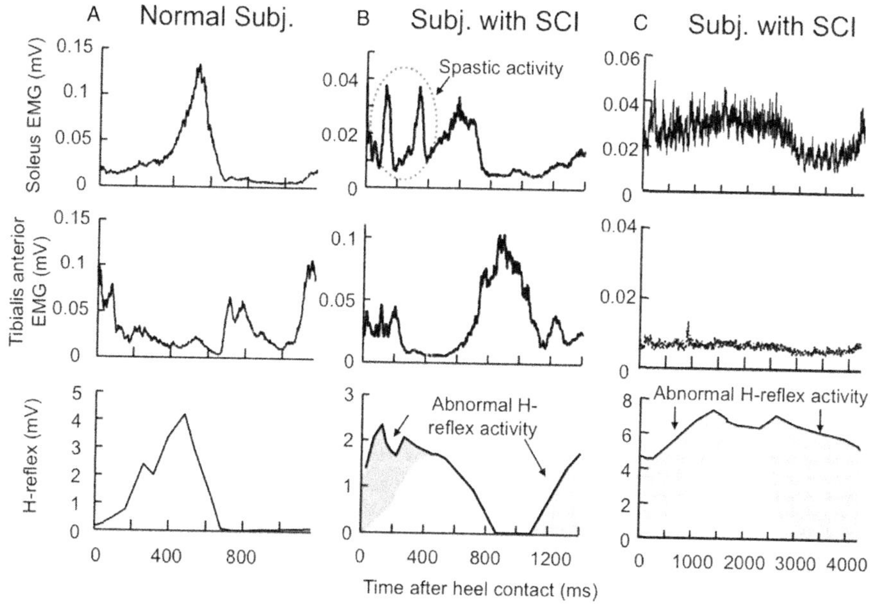

Figure 28.1. Soleus and tibialis anterior (TA) EMG activity and soleus H-reflex modulation during walking. (A) In a normal subject, soleus EMG activity gradually increases from heel-contact to push-off, then rapidly decreases to almost no activity, and stays quiet for the entire swing phase. TA EMG activity typically occurs in two bursts: one from the end of stance to early swing and another during the swing–stance transition. The soleus H-reflex modulation pattern is similar to the soleus EMG pattern. (B) In a subject with chronic incomplete SCI, the stretch of the soleus by the ankle angle change at heel-contact triggers clonic soleus EMG activity in the early- to middle-stance phase. This abnormal reflex activity leads to the early stance instability. In addition, H-reflex modulation is impaired. The H-reflex is suppressed after push-off (i.e., 750 ms after heel-contact), but it recovers too early (i.e., in the middle of the swing phase) and stays high through the midstance phase. This probably contributes to the abnormally low TA activity during late swing to early stance. (C) Another subject with chronic incomplete SCI shows little reflex modulation throughout the step cycle. The soleus is active throughout the step cycle and TA activity is almost absent. In panels B and C, shaded areas indicate abnormal, excessive reflex activity during walking.

and/or restoring its normal phase-dependent modulation might improve locomotion. When treadmill training improves locomotion in SCI-injured cats, the excitatory and inhibitory effects of group I pathways also changes appropriately (Cote et al., 2003). This provides evidence for the connection between restoring more normal spinal reflexes and improving locomotion. Thus, operant conditioning of spinal reflexes could offer a promising new approach to achieving therapeutic goals. The next sections review reflex conditioning in laboratory animals. These animal studies have led to the recent and ongoing studies in humans with or without partial spinal cord injuries.

Operant Conditioning of Spinal Reflexes in Laboratory Animals

The standard protocol for operant conditioning of spinal reflexes was first developed in monkeys (Wolpaw, 1987; Wolpaw and O'Keefe, 1984), adapted and used extensively in rats (X.Y. Chen and Wolpaw, 1995), and more recently tested in mice (Carp et al., 2006). Although the protocol was first applied to the spinal stretch reflex (Wolpaw and O'Keefe, 1984), subsequent work has focused on the H-reflex (Wolpaw, 1987; X.Y. Chen and Wolpaw, 1995; Wolpaw, 1997). The rat H-reflex conditioning protocol is summarized in Figure 28.2B.

Figure 28.2C shows the results of operant conditioning in rats, monkeys, and mice. In each species, exposure to the up-conditioning (▲) or down-conditioning (▼) mode gradually changes the size of the reflex in the correct direction. Successful conditioning, defined as a change of more than 20% in the correct direction (X.Y. Chen and Wolpaw, 1995; Wolpaw et al., 1993), occurs in 75–80% of the animals. In the remaining animals, the reflex remains within 20% of its control value.

Chen and co-workers (Y. Chen et al., 2006) showed that up-conditioning of the soleus H-reflex can improve locomotion in rats with incomplete SCI. Mid-thoracic hemisection of the right lateral column shortened the right stance phase and thereby produced an asymmetry in locomotion. As Figure 28.3 illustrates, H-reflex up-conditioning, which enhanced the right soleus burst during the stance phase (Y. Chen et al., 2005), eliminated this asymmetry. This study suggested that reflex conditioning protocols might improve motor function in people with partial SCI.

Current Understanding of the Mechanisms of Reflex Conditioning

An ongoing series of animal studies has begun to reveal the complex pattern of spinal and supraspinal plasticity underlying H-reflex conditioning (Wolpaw and Tennissen, 2001; Wolpaw, 2007; Wolpaw, 1997). A positive shift in motoneuron firing threshold (possibly resulting from a change in the activation voltage of Na^+ channels) can largely account for H-reflex down-conditioning (reviewed in Wolpaw, 1997). Down-conditioning is also accompanied by marked increases in identifiable GABAergic interneurons in the ventral horn and GABAergic terminals on the soleus motoneuron (Wolpaw, 2010). There is also evidence for changes in several other synaptic populations on the motoneuron, in motor unit properties, in other spinal interneurons, and even on the contralateral side of the spinal cord (Wolpaw, 1997). Up-conditioning and down-conditioning appear to have different mechanisms; they are not mirror images of each other. Up-conditioning may result from plasticity in spinal interneurons (Wolpaw, 2010).

The CST is the only major descending tract essential for H-reflex conditioning (X.Y. Chen et al., 2006c; X.Y. Chen et al., 2002; X.Y. Chen and Wolpaw, 2002). Thus, it is presumably CST activity that induces the spinal cord plasticity

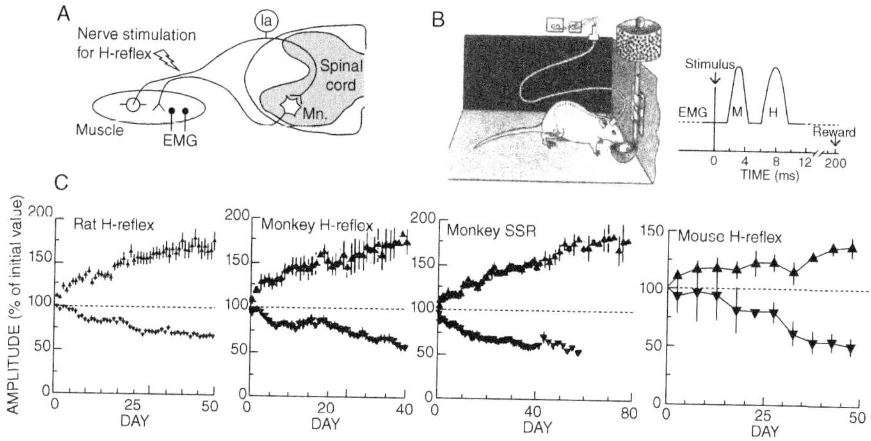

Figure 28.2. (A) Main pathway of the spinal stretch reflex (SSR) and its electrical analog, the H-reflex. Excitation of the Ia afferents (and possibly large group II afferents) (Zehr, 2002; Magladery et al., 1951) activates the motoneurons innervating the same muscle and its synergists. If the afferents are excited by muscle stretch, the response is the SSR. If the afferents are excited by electrical stimulation, the response is the H-reflex. Although the pathway is entirely spinal, it is strongly influenced by supraspinal activity. (B) The rat H-reflex conditioning protocol. Rats are chronically implanted with fine-wire EMG electrodes in the soleus muscle and a stimulating cuff on the posterior tibial nerve. The implanted wires connect through a headmount and a flexible tether and commutator to EMG amplifiers and a nerve-cuff simulator. Soleus EMG is monitored continuously (24 h/day) in the freely moving animal. Whenever the absolute value of soleus EMG remains within a specified range for a random varying 2.3–2.7 s period, a stimulus through the nerve cuff that is kept just above M-wave threshold elicits the M-wave and the H-reflex. In the course of its normal activity, the animal usually provides 2500–8000 of these H-reflex trials per day. For the first 10 days, the animal is exposed to the control mode, in which no reward occurs and the H-reflex is simply measured to determine its baseline (i.e., control) value. For the next 50 days, the rat is exposed to the up-conditioning (HRup) or down-conditioning (HRdown) mode, in which a food reward occurs if the H-reflex is above (HRup) or below (HRdown) a criterion value. Background EMG and M-wave amplitude remain constant throughout. (C) Operant up-conditioning and down-conditioning of a spinal reflex in different animal models. From left, soleus H-reflex in rats, triceps surae H-reflex in monkeys, biceps brachii SSR in monkeys, and triceps surae H-reflex in mice. Generally, the time courses and magnitudes of change are similar across different species and muscles. (From Chen et al., 2010; Wolpaw and Chen, 2009.)

underlying H-reflex change. Furthermore, cerebellar–cortical connections appear to be essential for establishing and maintaining supraspinal plasticity that in turn establishes and maintains the spinal cord plasticity directly responsible for H-reflex change (Wolpaw and Chen, 2006; Chen and Wolpaw, 2005). In sum, H-reflex conditioning appears to depend on a hierarchy in which plasticity in the brain induces plasticity in the spinal cord (Chen et al., 2002;

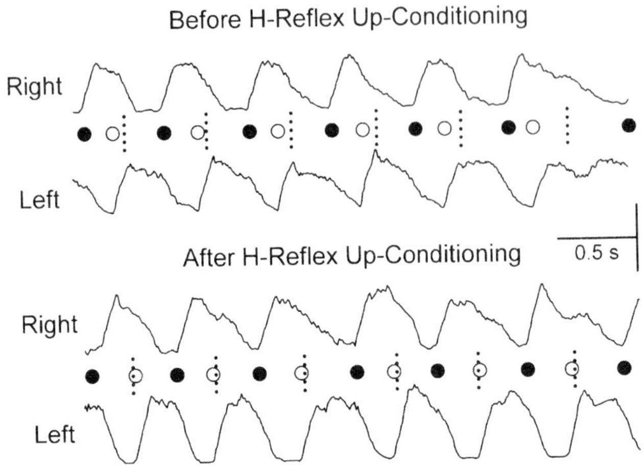

Figure 28.3. The effects of soleus H-reflex up-conditioning on locomotion in a rat with midthoracic right lateral column transection. The traces show the EMG bursts from right and left soleus muscles during treadmill walking before (top) and after (bottom) H-reflex up-conditioning has increased the right soleus H-reflex size. The presumed onsets of the right (•) and left (○) stance phases of the gait cycle are indicated in the middle of each set of traces. The short vertical dashed lines indicate the midpoints between right burst onsets, where the left burst onsets should occur. Before H-reflex up-conditioning, the left burst onset occurs too early, and the gait is asymmetrical. H-Reflex up-conditioning strengthens the right soleus EMG burst and corrects the left burst onset timing—and thereby restores the gait symmetry. Horizontal scale bar, 0.5 s; vertical scale bar, 100 and 150 μV for the right and left EMG bursts, respectively (From Y. Chen et al., 2006.)

Chen and Wolpaw, 2002; Wolpaw and Chen, 2006; Chen and Wolpaw, 2005; Chen et al., 2006a; Thompson et al., 2009a). Operant conditioning of a spinal reflex produces multisite plasticity that extends considerably beyond the reflex pathway conditioned, and can thereby have complex effects on motor function.

METHODS AND RESULTS

As reviewed previously, spinal reflexes can be changed through operant conditioning. Animal studies show that this conditioning changes not only the reflex itself but also other behaviors, such as locomotion, that use the same pathway. Thus, it should be possible to use spinal reflex conditioning to change a spinal reflex pathway so as to reduce motor abnormalities resulting from CNS damage. Indeed, in rats with abnormal locomotion due to partial SCI, appropriate reflex conditioning can restore more normal locomotion (Y. Chen et al., 2006). We now hypothesize that appropriate operant conditioning of a spinal reflex will improve motor function recovery after CNS damage in people. To test this hypothesis, we developed an H-reflex operant conditioning protocol for human subjects and are using it to evaluate spinal reflex conditioning as a new therapeutic approach.

Operant Conditioning of the Soleus H-Reflex in Normal Humans

Reflex conditioning was first applied in humans to the biceps brachii stretch reflex (Wolf and Segal, 1996; Segal and Wolf, 1994) and more recently has been applied to the human soleus H-reflex (Thompson et al., 2009a). Spinal reflex conditioning in humans uses a protocol comparable to that developed in animals. The only significant difference from the animal protocol is that humans perform many fewer trials (i.e., only 2–5% as many) and these trials are confined to three 1-hour sessions per week. The newly developed soleus H-reflex protocol comprises six baseline sessions, 24 conditioning sessions at a rate of 3 sessions per week, and four follow-up sessions over the next three months. Sessions are always at the same time of day (to control for diurnal variation in the reflex; Lagerquist et al., 2006). In each session, the soleus H-reflex is elicited while the subject maintains a natural standing posture and a defined stable level of soleus background EMG. The size of the M-wave is kept constant within and across sessions. In each baseline session, three blocks of 75 control H-reflexes (i.e., 225 H-reflexes) are elicited. In each conditioning or follow-up session, 20 control H-reflexes are elicited as in the baseline sessions and then three blocks of 75 (i.e., 225) conditioned H-reflexes are elicited. In these conditioned H-reflex trials, the subject is encouraged to increase (HRup mode) or decrease (HRdown mode) the H-reflex and is given visual feedback after each stimulus that indicates whether the H-reflex was larger (HRup) or smaller (HRdown) than a criterion value. Good performance in changing reflex size appropriately earns an additional monetary reward. Background EMG and M-wave size are kept stable throughout data collection.

Figure 28.4 summarizes the results of soleus H-reflex conditioning in normal humans. Over the 24 conditioning sessions, H-reflex size gradually increased in 6 of 8 HRup subjects and decreased in 8 of 9 HRdown subjects, resulting in final sizes of 140(±12SEM)% and 69(±6)% of baseline size, respectively. In these people, the final H-reflex change was the sum of within-session change (i.e., task-dependent adaptation) and across-session (i.e., long-term) change. Task-dependent adaptation arose within 4–6 sessions and persisted thereafter, while long-term change began after 10–12 sessions and increased gradually from then on. (See Thompson et al., 2009a, for complete presentation and discussion of task-dependent adaptation and long-term change.)

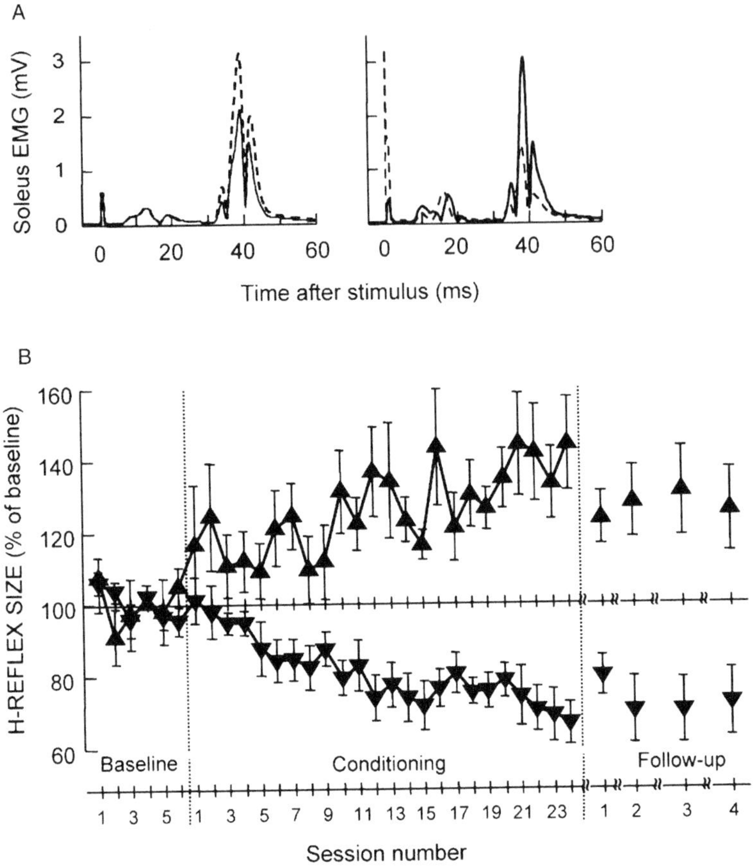

Figure 28.4. Humans can learn to change H-reflex size in response to an operant conditioning protocol. (A) Average H-reflexes from two representative people for a baseline session (solid) and for the last conditioning session (dashed). After the 24 conditioning sessions, the H-reflex is larger in the up-conditioning (HRup) subject (left) and smaller in the down-conditioning (HRdown) subject (right). (From Thompson et al., 2009a.) (B) Average H-reflexes (±SEM) for 6 successful HRup and 8 successful HRdown subjects for baseline, conditioning, and follow-up (12, 30, 60, and 90 days after the end of conditioning) sessions. As in animals, H-reflex size gradually increases in the HRup subjects (upward triangles) and decreases in the HRdown subjects (downward triangles) over the course of the conditioning period. (From Thompson et al., 2009a.)

This study showed that people performing only 225 reflex conditioning trials a day, 3 days a week, displayed gradual reflex change similar in course and nearly equal in magnitude to that of animals that performed 20–50 times as many trials. This finding shows that H-reflex conditioning is possible in humans, and that it does not require the several thousand trials per day typically completed by animals. (Animals probably do not need that many trials either, but that remains to be determined.) The success rate of 82% (14 of 17 subjects changed H-reflex size significantly in the correct direction) was also similar to that of animals (Wolpaw, 1987; Chen and Wolpaw, 1995; Wolpaw, 1997; Thompson et al., 2009a). Furthermore, this study showed that exposure to the reflex operant conditioning protocol over several months produced both short-term adaptation and long-term plasticity in the spinal reflex pathway. The long-term plasticity appeared to be a lasting change in the reflex pathway that persisted outside of the reflex conditioning protocol and lasted for at least several months after reflex conditioning ceased (see Thompson et al., 2009a, for full explication). This suggests that it should be possible to use reflex conditioning protocols to guide such long-term plasticity so as to essentially reeducate abnormally functioning spinal reflex pathways (Wolpaw, 2006, 2010) and thereby alleviate motor disabilities associated with incomplete SCI (e.g., Fig. 28.1). Animal studies (Y. Chen et al., 2006) (see Fig. 28.3) support this possibility.

Operant Conditioning of Soleus H-Reflex in Incomplete Spinal Cord Injury

Operant conditioning is a powerful method for inducing changes in specific spinal pathways. Because abnormally functioning spinal reflexes contribute to movement disabilities (e.g., Fig. 28.1), methods for reducing reflex abnormalities may help to reduce motor disabilities. Segal and Wolf (1994) showed that it is possible to operantly condition the biceps brachii stretch reflex in humans with incomplete SCI (Fig. 28.5A). In this initial study, they did not address the question of whether such conditioning can produce therapeutic benefits. Thus, to test the hypothesis that appropriately applied operant conditioning of a spinal reflex can improve motor function recovery, we investigated whether reflex conditioning could improve locomotor function in people with spasticity after incomplete SCI. In people with chronic incomplete SCI, spasticity is seen as exaggerated stretch reflexes and abnormal reflex modulation in the ankle extensor muscles. As described above and illustrated in Figure 28.1, normal soleus H-reflex modulation during walking is often absent or greatly diminished (i.e., the reflex amplitude remains high even in the early stance or swing phase, where the H-reflex is normally very small or absent). Thus, we applied operant down-conditioning to reduce soleus H-reflex excitability, and hopefully thereby alleviate spastic gait in this population (Thompson et al., 2013).

The subjects were adults with chronic (0.7–49 years duration) incomplete spinal cord damage who suffered from ankle extensor spasticity and foot drop. All were medically stable and ambulatory. The protocol was the same as the one used in normal subjects (Thompson et al., 2009a), except that the number of conditioning sessions was increased from 24 to 30. The 6 baseline and 30 down-conditioning sessions occurred at the rate of 3 sessions per week for 12

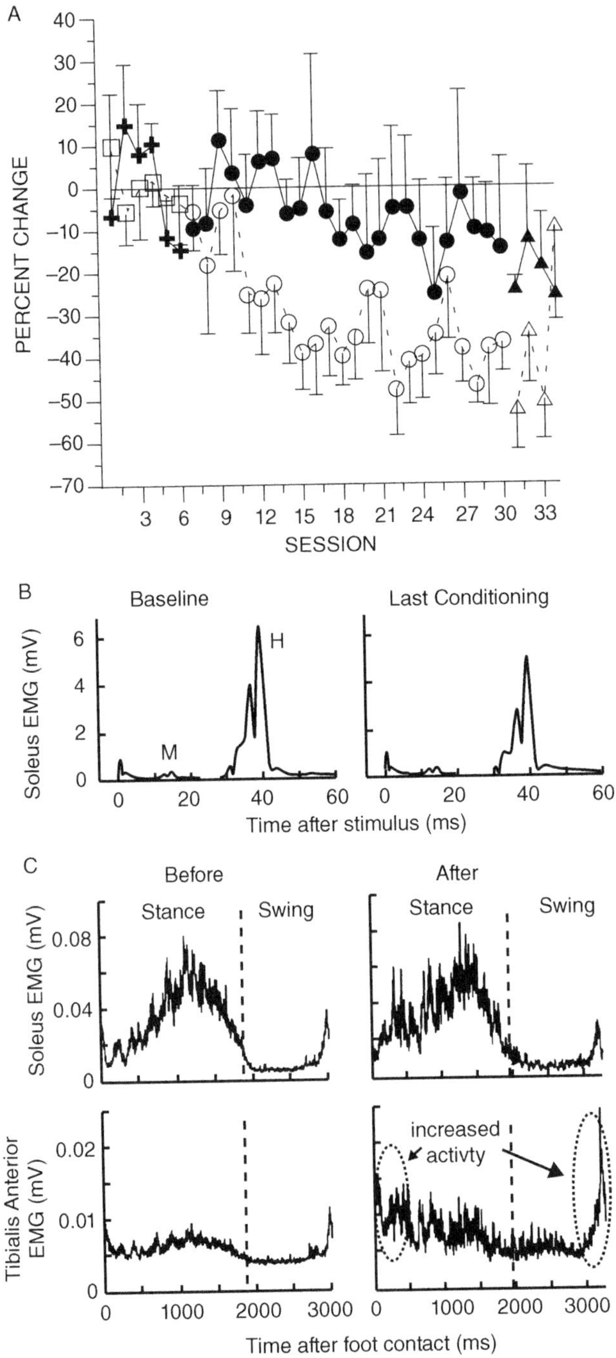
A
PERCENT CHANGE
40
30
20
10
0
−10
−20
−30
−40
−50
−60
−70
3 6 9 12 15 18 21 24 27 30 33
SESSION
B
Baseline
Last Conditioning
Soleus EMG (mV)
6
4
2
0
M
H
0 20 40 60
Time after stimulus (ms)
C
Before
After
Stance
Swing
Soleus EMG (mV)
0.08
0.04
Tibialis Anterior EMG (mV)
0.02
0.01
increased activty
0 1000 2000 3000
Time after foot contact (ms)

weeks. Soleus and tibialis anterior (TA) background EMG and soleus M-wave size are kept at constant levels throughout the study.

The results indicate that operant down-conditioning of the soleus H-reflex is possible in people after incomplete SCI. Six of the nine down-conditioning subjects had significantly smaller H-reflexes after down-conditioning. Figure 28.5B shows the before and after results for one of these subjects. Chen and co-workers (Chen et al., 1999; Chen et al., 1996) found that the success rate for reflex conditioning in rats with SCI was inversely correlated with the severity of the injury. Thus, incomplete SCIs may impair reflex conditioning. Nevertheless, the study of Segal and Wolf (1994) and these new results indicate that conditioning is possible in many people with SCI, although it may take longer.

In all six successful subjects, a timed 10-m walking test revealed improved walking speed after conditioning, and gait symmetry also improved significantly. Figure 28.5C shows soleus and TA locomotor EMG in a spastic subject before and after successful down-conditioning of the soleus H-reflex. In this subject there was little change in the soleus EMG, while the TA EMG was increased, especially during the swing–stance transition phase. This alleviated foot drop, although not enough to eliminate it. This improved TA activation may have been due to reductions in soleus stretch reflex gain and/or reciprocal inhibition of the TA, both of which tend to reduce TA activity during the swing phase. In another subject (not shown) after successful H-reflex conditioning, ankle extensor spasticity was significantly reduced, and as an apparent result, the soleus EMG pattern changed from tonic activity throughout the step cycle (e.g., Fig. 28.1C) to a clearly phase-modulated pattern (i.e., appropriately reduced soleus activity during the swing phase). Furthermore, these subjects reported that they were walking faster and farther in their daily lives. In contrast, locomotion

Figure 28.5. (A) Average biceps brachii spinal stretch reflexes (SSR) over 34 sessions (i.e., 6 baseline, 24 conditioning, and 4 follow-up sessions) in people with incomplete SCI who were (□/○) or were not (+/●) exposed (i.e., the control group) to the down-conditioning protocol. In the down-conditioning group, SSR size declines steadily over the conditioning sessions and remains low in the follow-up sessions. In contrast, SSR size shows only a small insignificant decrease in the control group. (Figure from Segal and Wolf, 1994.) (B) Average H-reflexes (all 225 trials averaged for each trace) in a representative subject with SCI for a baseline session (left) and for the final conditioning session (right). The final H-reflex is substantially smaller than the baseline H-reflex, while the background EMG and M-wave size have not changed. (C) Soleus and TA EMG activity during walking before and after soleus H-reflex down-conditioning in the same subject. Before H-reflex conditioning, there is almost no TA EMG activity throughout the step cycle. After down-conditioning, there is little change in soleus EMG, while TA EMG increases, especially in the period from late swing to early stance. Increased TA activity in this period reduces foot drop.

did not improve significantly in down-conditioning subjects in whom the H-reflex did not decrease or in unconditioned control subjects.

This initial study (Thompson et al., 2013) suggests that reflex-conditioning protocols can enhance recovery of function after incomplete spinal cord injuries and possibly in other disorders also. Because they can target specific spinal pathways, these protocols could address each person's particular deficits, and thus might complement other rehabilitation methods.

DISCUSSION

Important Issues for Future Therapeutic Applications of Reflex Conditioning

Spinal reflex conditioning protocols may provide a unique new approach to rehabilitation. Because these protocols can target specific reflex pathways, they could be customized to address the specific motor deficits of each person. It should be possible to design conditioning protocols that complement existing therapeutic methods, such as treadmill training (Edgerton et al., 2008; Harkema et al., 1997; Maegele et al., 2002; Wernig et al., 2000) and constraint-induced movement therapy (Wolf et al., 2006; Taub and Uswatte, 2003; Taub et al., 1999), in order to maximize the recovery of useful motor function. Furthermore, when CNS regeneration becomes possible, methods such as spinal reflex conditioning are likely to be needed for reeducating newly regenerated connections to support effective function (Wolpaw, 2006, 2010). Without the activity-dependent plasticity that could be induced and guided by appropriate reflex conditioning protocols, regenerated connections are likely to display diffuse infantile-like responses and dysfunctional movement control. At the same time, animal studies indicate that these protocols produce complex patterns of plasticity that can extend considerably beyond the targeted reflex pathway (Wolpaw, 2007, 2010; Wolpaw and Carp, 2006; Chen et al., 2006b). As a result, the principles of designing reflex conditioning protocols that serve the needs of individual patients will require careful study.

This new therapeutic approach has limitations. Current spinal reflex conditioning protocols are successful in 75–80%, not 100%, of the subjects. This success rate could be lower, and the associated therapeutic effects may vary, in people with CST damage (e.g., incomplete SCI), since studies to date indicate that the CST is needed for successful conditioning (Chen et al., 2002; Chen and Wolpaw, 2002). In addition, effective application of the current protocols requires a well-trained investigator or therapist with knowledge of the relevant neurophysiology. However, this requirement may become less critical in the future if a semiautomated hardware–software system for reflex conditioning can be developed.

Development of Other Operant Conditioning Protocols

While the operant conditioning protocol described here has been applied mainly to the H-reflex and the spinal stretch reflex, comparable protocols might be applied to other reflexes (e.g., reciprocal inhibition; Chen et al., 2006b). Extension to other reflex pathways (e.g., other proprioceptive or cutaneous reflexes, reflexes underlying bladder or bowel function) could offer further therapeutic possibilities. In addition, it may be possible to sharpen the focus of reflex conditioning by incorporating a reflex conditioning protocol into complex behaviors such as locomotion. Thus, for example, it should be possible to place H-reflex conditioning trials in specific phases of the step-cycle. By targeting the exact time when reflex activity is abnormal, such a protocol might help restore appropriate phase-dependent reflex modulation during walking.

Operant conditioning protocols that target CST connections might also be therapeutically useful. The motor evoked potential (MEP) to transcranial magnetic stimulation (TMS) is often reduced after SCI (Davey et al., 1999), and more recent studies suggest that functional recovery is associated with increased MEP size (Thomas and Gorassini, 2005; Everaert et al., 2010). Our initial results suggest that operant up-conditioning of the TA MEP to TMS can strengthen CST connectivity after SCI and may thereby improve the control of ankle dorsiflexor muscles (Thompson et al., 2009b; Abel et al., 2010). Finally, operant conditioning protocols that focus on modifying EEG activity over sensorimotor cortex might conceivably provide an additional approach to improving cortical control of motor function (see Daly and Wolpaw, 2008, for discussion).

CONCLUSIONS

Damage to the CNS often produces abnormal spinal reflexes (Stein et al., 1993; Yang et al., 1991; Crone et al., 2003; Hiersemenzel et al., 2000; Thompson et al., 2009c) that contribute to motor disabilities (Burne et al., 2005; Dietz and Sinkjaer, 2007; Hultborn, 2003; Nielsen et al., 2007). Thus, methods for appropriately modifying reflex pathways might help to restore more effective motor function. Operant conditioning protocols can change specific spinal reflex pathways both in normal people and in people with partial SCIs (Thompson et al., 2009a; Wolf and Segal, 1996; Segal and Wolf, 1994). Encouraged by results in animals, studies exploring the therapeutic potential of this new rehabilitation methodology in people with partial SCIs have begun. The preliminary results are promising. In the future, conditioning protocols might be developed for modifying other spinal reflexes, CST connections, or cortical activity, and might thereby also promote functional recovery in people after CNS damage.

ACKNOWLEDGMENTS

Work in the authors' laboratories is supported by the National Institutes of Health (NS69551, NS22189, HD36020, NS061823) and the New York State Spinal Cord Injury Research Trust (C023685).

REFERENCES

Abel B.M., Pomerantz F., DeFrancesco E., Lichtman S.W., and Thompson A.K. (2010). Operant up-conditioning of the ankle dorsiflexor motor evoked potential in people with chronic incomplete spinal cord injury. In: Society for Neuroscience 40th Annual Meeting. San Diego, CA, Program No. 82.16.

Adkins D.L., Boychuk J., Remple M.S., and Kleim J.A. (2006). Motor training induces experience-specific patterns of plasticity across motor cortex and spinal cord. *J Appl Physiol* 101:1776–1782.

Ashby P. and Wiens M. (1989). Reciprocal inhibition following lesions of the spinal cord in man. *J Physiol* 414:145–157.

Aymard C., Katz R., Lafitte C., et al. (2000). Presynaptic inhibition and homosynaptic depression: a comparison between lower and upper limbs in normal human subjects and patients with hemiplegia. *Brain* 123:1688–1702.

Boorman G.I., Lee R.G., Becker W.J., and Windhorst U.R. (1996). Impaired "natural reciprocal inhibition" in patients with spasticity due to incomplete spinal cord injury. *Electroencephalogr Clin Neurophysiol* 101:84–92.

Brooke J.D., Cheng J., Collins D.F., McIlroy W.E., Misiaszek J.E., and Staines W.R. (1997). Sensori-sensory afferent conditioning with leg movement: gain control in spinal reflex and ascending paths. *Prog Neurobiol* 51:393–421.

Burne J.A., Carleton V.L., and O'Dwyer N.J. (2005). The spasticity paradox: movement disorder or disorder of resting limbs? *J Neurol Neurosurg Psychiatry* 76:47–54.

Capaday C., Lavoie B.A., Barbeau H., Schneider C., and Bonnard M. (1999). Studies on the corticospinal control of human walking. I. Responses to focal transcranial magnetic stimulation of the motor cortex. *J Neurophysiol* 81:129–139.

Carp J.S., Tennissen A.M., Chen X.Y., and Wolpaw J.R. (2006). H-reflex operant conditioning in mice. *J Neurophysiol* 96:1718–1727.

Chen X.Y. and Wolpaw J.R. (1995). Operant conditioning of H-reflex in freely moving rats. *J Neurophysiol* 73:411–415.

Chen X.Y. and Wolpaw J.R. (2002). Probable corticospinal tract control of spinal cord plasticity in the rat. *J Neurophysiol* 87:645–652.

Chen X.Y. and Wolpaw J.R. (2005). Ablation of cerebellar nuclei prevents H-reflex down-conditioning in rats. *Learn Mem* 12:248–254.

Chen X.Y., Wolpaw J.R., Jakeman L.B., and Stokes B.T. (1996). Operant conditioning of H-reflex in spinal cord–injured rats. *J Neurotrauma* 13:755–766.

Chen X.Y., Wolpaw J.R., Jakeman L.B., and Stokes B.T. (1999). Operant conditioning of H-reflex increase in spinal cord–injured rats. *J Neurotrauma* 16:175–186.

Chen X.Y., Carp J.S., Chen L., and Wolpaw J.R. (2002). Corticospinal tract transection prevents operantly conditioned H-reflex increase in rats. *Exp Brain Res* 144:88–94.

Chen X.Y., Carp J.S., Chen L., and Wolpaw J.R. (2006a). Sensorimotor cortex ablation prevents H-reflex up-conditioning and causes a paradoxical response to down-conditioning in rats. *J Neurophysiol* 96:119–127.

Chen X.Y., Chen L., Chen Y., and Wolpaw J.R. (2006b). Operant conditioning of reciprocal inhibition in rat soleus muscle. *J Neurophysiol* 96:2144–2150.

Chen X.Y., Chen Y., Chen L., Tennissen A.M., and Wolpaw J.R. (2006c). Corticospinal tract transection permanently abolishes H-reflex down-conditioning in rats. *J Neurotrauma* 23:1705–1712.

Chen X.Y., Chen Y., Wang Y., et al. (2010). Reflex conditioning: a new strategy for improving motor function after spinal cord injury. *Ann N Y Acad Sci* 1198(Suppl 1):E12–E21.

Chen Y., Chen X.Y., Jakeman L.B., Schalk G., Stokes B.T., and Wolpaw J.R. (2005). The interaction of a new motor skill and an old one: H-reflex conditioning and locomotion in rats. *J Neurosci* 25:6898–6906.

Chen Y., Chen X.Y., Jakeman L.B., Chen L., Stokes B.T., and Wolpaw J.R. (2006). Operant conditioning of H-reflex can correct a locomotor abnormality after spinal cord injury in rats. *J Neurosci* 26:12537–12543.

Cote M.P., Menard A., and Gossard J.P. (2003). Spinal cats on the treadmill: changes in load pathways. *J Neurosci* 23:2789–2796.

Crone C., Johnsen L.L., Biering-Sorensen F., and Nielsen J.B. (2003). Appearance of reciprocal facilitation of ankle extensors from ankle flexors in patients with stroke or spinal cord injury. *Brain* 126:495–507.

Daly J.J. and Wolpaw J.R. (2008). Brain–computer interfaces in neurological rehabilitation. *Lancet Neurol* 7:1032–1043.

Davey N.J., Smith H.C., Savic G., Maskill D.W., Ellaway P.H., and Frankel H.L. (1999). Comparison of input–output patterns in the corticospinal system of normal subjects and incomplete spinal cord injured patients. *Exp Brain Res* 127:382–390.

de Groat W.C. (2002). Plasticity of bladder reflex pathways during postnatal development. *Physiol Behav* 77:689–692.

Dietz V. (2002). Proprioception and locomotor disorders. *Nat Rev Neurosci* 3:781–790.

Dietz V. and Duysens J. (2000). Significance of load receptor input during locomotion: a review. *Gait and Posture* 11:102–110.

Dietz V. and Sinkjaer T. (2007). Spastic movement disorder: impaired reflex function and altered muscle mechanics. *Lancet Neurol* 6:725–733.

Dietz V., Muller R., and Colombo G. (2002). Locomotor activity in spinal man: significance of afferent input from joint and load receptors. *Brain* 125:2626–2634.

Duysens J., Clarac F., and Cruse H. (2000). Load-regulating mechanisms in gait and posture: comparative aspects. *Physiol Rev* 80:83–133.

Edgerton V.R., Courtine G., Gerasimenko Y.P., et al. (2008). Training locomotor networks. *Brain Res Rev* 57:241–254.

Everaert D.G., Thompson A.K., Chong S.L., and Stein R.B. (2010). Does functional electrical stimulation for foot drop strengthen corticospinal connections? *Neurorehabil Neural Repair* 24:168–177.

Eyre J.A. (2003). Developmental plasticity of the corticospinal system. In: *Plasticity in the Human Nervous System Investigations with Transcranial Magnetic Stimulation* (Boniface S. and Ziemann U., Eds.), pp. 62–89. Cambridge, UK: Cambridge University Press.

Fung J. and Barbeau H. (1994). Effects of conditioning cutaneomuscular stimulation on the soleus H-reflex in normal and spastic paretic subjects during walking and standing. *J Neurophysiol* 72:2090–2104.

Harkema S.J., Hurley S.L., Patel U.K., Requejo P.S., Dobkin B.H., and Edgerton V.R. (1997). Human lumbosacral spinal cord interprets loading during stepping. *J Neurophysiol* 77:797–811.

Hiebert G.W., Whelan P.J., Prochazka A., and Pearson K.G. (1996). Contribution of hind limb flexor muscle afferents to the timing of phase transitions in the cat step cycle. *J Neurophysiol* 75:1126–1137.

Hiersemenzel L.P., Curt A., and Dietz V. (2000). From spinal shock to spasticity: neuronal adaptations to a spinal cord injury. *Neurology* 54:1574–1582.

Hodapp M., Klisch C., Mall V., Vry J., Berger W., and Faist M. (2007). Modulation of soleus H-reflexes during gait in children with cerebral palsy. *J Neurophysiol* 98:3263–3268.

Hultborn H. (2003). Changes in neuronal properties and spinal reflexes during development of spasticity following spinal cord lesions and stroke: studies in animal models and patients. *J Rehabil Med* 41 Suppl:46–55.

Lagerquist O., Zehr E.P., Baldwin E.R., Klakowicz P.M., and Collins D.F. (2006). Diurnal changes in the amplitude of the Hoffmann reflex in the human soleus but not in the flexor carpi radialis muscle. *Exp Brain Res* 170:1–6.

Levinsson A., Luo X.L., Holmberg H., and Schouenborg J. (1999). Developmental tuning in a spinal nociceptive system: effects of neonatal spinalization. *J Neurosci* 19:10397–10403.

Maegele M., Muller S., Wernig A., Edgerton V.R., and Harkema S.J. (2002). Recruitment of spinal motor pools during voluntary movements versus stepping after human spinal cord injury. *J Neurotrauma* 19:1217–1229.

Magladery J.W., Porter W.E., Park A.M., and Teasdall R.D. (1951). Electrophysiological studies of nerve and reflex activity in normal man. IV. The two-neurone reflex and identification of certain action potentials from spinal roots and cord. *Bull Johns Hopkins Hosp* 88:499–519.

Morita H., Shindo M., Momoi H., Yanagawa S., Ikeda S., and Yanagisawa N. (2006). Lack of modulation of Ib inhibition during antagonist contraction in spasticity. *Neurology* 67:52–56.

Myklebust B.M., Gottlieb G.L., and Agarwal G.C. (1986). Stretch reflexes of the normal infant. *Dev Med Child Neurol* 28:440–449.

Nielsen J., Crone C., and Hultborn H. (1993). H-reflexes are smaller in dancers from The Royal Danish Ballet than in well-trained athletes. *European J Appl Physiol Occupational Physiol* 66:116–121.

Nielsen J.B., Crone C., and Hultborn H. (2007). The spinal pathophysiology of spasticity—from a basic science point of view. *Acta Physiol (Oxford)* 189:171–180.

O'Sullivan M.C., Miller S., Ramesh V., et al. (1998). Abnormal development of biceps brachii phasic stretch reflex and persistence of short latency heteronymous reflexes from biceps to triceps brachii in spastic cerebral palsy. *Brain* 121(Pt 12):2381–2395.

Pearson K.G. (1995). Proprioceptive regulation of locomotion. *Curr Opin Neurobiol* 5:786–791.

Pearson K.G. and Collins D.F. (1993). Reversal of the influence of group Ib afferents from plantaris on activity in medial gastrocnemius muscle during locomotor activity. *J Neurophysiol* 70:1009–1017.

Petersen N.T., Butler J.E., Marchand-Pauvert V., et al. (2001). Suppression of EMG activity by transcranial magnetic stimulation in human subjects during walking. *J Physiol* 537:651–656.

Schneider C. and Capaday C. (2003). Progressive adaptation of the soleus H-reflex with daily training at walking backward. *J Neurophysiol* 89:648–656.

Schneider C., Lavoie B.A., and Capaday C. (2000). On the origin of the soleus H-reflex modulation pattern during human walking and its task-dependent differences. *J Neurophysiol* 83:2881–2890.

Schubert M., Curt A., Jensen L., and Dietz V. (1997). Corticospinal input in human gait: modulation of magnetically evoked motor responses. *Exp Brain Res* 115: 234–246.

Segal R.L. and Wolf S.L. (1994). Operant conditioning of spinal stretch reflexes in patients with spinal cord injuries. *Exp Neurol* 130:202–213.

Shefner J.M., Berman S.A., Sarkarati M., and Young R.R. (1992). Recurrent inhibition is increased in patients with spinal cord injury. *Neurology* 42:2162–2168.

Sinkjaer T., Andersen J.B., and Larsen B. (1996). Soleus stretch reflex modulation during gait in humans. *J Neurophysiol* 76:1112–1120.

Stein R.B. (1995). Presynaptic inhibition in humans. *Prog Neurobiol* 47:533–544.

Stein R.B. (2008). The plasticity of the adult spinal cord continues to surprise. *J Physiol* 586:2823.

Stein R.B. and Capaday C. (1988). The modulation of human reflexes during functional motor tasks. *Trends Neurosci* 11:328–332.

Stein R.B., Yang J.F., Belanger M., and Pearson K.G. (1993). Modification of reflexes in normal and abnormal movements. *Prog Brain Res* 97:189–196.

Stephens M.J. and Yang J.F. (1996). Short latency, non-reciprocal group I inhibition is reduced during the stance phase of walking in humans. *Brain Res* 743:24–31.

Taub E. and Uswatte G. (2003). Constraint-induced movement therapy: bridging from the primate laboratory to the stroke rehabilitation laboratory. *J Rehabil Med* 41 Suppl:34–40.

Taub E., Uswatte G., and Pidikiti R. (1999). Constraint-induced movement therapy: a new family of techniques with broad application to physical rehabilitation—a clinical review. *J Rehabil Res Dev* 36:237–251.

Thomas S.L. and Gorassini M.A. (2005). Increases in corticospinal tract function by treadmill training after incomplete spinal cord injury. *J Neurophysiol* 94:2844–2855.

Thompson A.K., Chen X.Y., and Wolpaw J.R. (2009a). Acquisition of a simple motor skill: task-dependent adaptation plus long-term change in the human soleus H-reflex. *J Neurosci* 29:5784–5792.

Thompson A.K., DeFrancesco E., Lichtman S.W., and Pomerantz F. (2009b). Operant conditioning of motor evoked potentials to transcranial magnetic stimulation in people with chronic incomplete spinal cord injury. In: Society for Neuroscience 39th Annual Meeting; Chicago, IL, 2009: 79.10.

Thompson A.K., Estabrooks K.L., Chong S., and Stein R.B. (2009c). Spinal reflexes in ankle flexor and extensor muscles after chronic central nervous system lesions and functional electrical stimulation. *Neurorehabil Neural Repair* 23:133–142.

Thompson A.K., Pomerantz F.R., and Wolpaw J.R. (2013). Operant conditioning of a spinal reflex can improve locomotion after spinal cord injury in humans. *J Neurosci* 33:2365–2375.

Waldenstrom A., Thelin J., Thimansson E., Levinsson A., and Schouenborg J. (2003). Developmental learning in a pain-related system: evidence for a cross-modality mechanism. *J Neurosci* 23:7719–7725.

Wernig A., Nanassy A., and Muller S. (2000) Laufband (LB) therapy in spinal cord lesioned persons. *Prog Brain Res* 128:89–97.

Whelan P.J., Hiebert G.W., and Pearson K.G. (1995). Stimulation of the group I extensor afferents prolongs the stance phase in walking cats. *Exp Brain Res* 103:20–30.

Wolf S.L. and Segal R.L. (1996). Reducing human biceps brachii spinal stretch reflex magnitude. *J Neurophysiol* 75:1637–1646.

Wolf S.L., Winstein C.J., Miller J.P., et al. (2006). Effect of constraint-induced movement therapy on upper extremity function 3 to 9 months after stroke: the EXCITE randomized clinical trial. *JAMA* 296:2095–2104.

Wolpaw J.R. (1987) Operant conditioning of primate spinal reflexes: the H-reflex. *J Neurophysiol* 57:443–459.

Wolpaw J.R. (1997). The complex structure of a simple memory. *Trends Neurosci* 20:588–594.

Wolpaw J.R. (2006). The education and re-education of the spinal cord. *Prog Brain Res* 157:261–280.

Wolpaw J.R. (2007) Spinal cord plasticity in acquisition and maintenance of motor skills. *Acta Physiol (Oxford)* 189:155–169.

Wolpaw J.R. (2010). What can the spinal cord teach us about learning and memory? *The Neuroscientist* 16:532–549.

Wolpaw J.R. and Carp J.S. (2006). Plasticity from muscle to brain. *Prog Neurobiol* 78:233–263.

Wolpaw J.R. and Chen X.Y. (2006). The cerebellum in maintenance of a motor skill: a hierarchy of brain and spinal cord plasticity underlies H-reflex conditioning. *Learn Mem* 13:208–215.

Wolpaw J.R. and Chen X.Y. (2009). Operant conditioning of reflexes. In: *Encyclopedia of Neuroscience* (Squire L.R., Ed.), pp. 225–233. Oxford, U.K.: Academic Press.

Wolpaw J.R. and O'Keefe J.A. (1984). Adaptive plasticity in the primate spinal stretch reflex: evidence for a two-phase process. *J Neurosci* 4:2718–2724.

Wolpaw J.R. and Tennissen A.M. (2001). Activity-dependent spinal cord plasticity in health and disease. *Annu Rev Neurosci* 24:807–843.

Wolpaw J.R., Herchenroder P.A., and Carp J.S. (1993). Operant conditioning of the primate H-reflex: factors affecting the magnitude of change. *Exp Brain Res* 97:31–39.

Yang J.F., Fung J., Edamura M., Blunt R., Stein R.B., and Barbeau H. (1991). H-reflex modulation during walking in spastic paretic subjects. *Can J Neurol Sci* 18:443–452.

Zehr E.P. (2002). Considerations for use of the Hoffmann reflex in exercise studies. *Eur J Appl Physiol* 86:455–468.

Zehr E.P. and Stein R.B. (1999). What functions do reflexes serve during human locomotion? *Prog Neurobiol* 58:185–205.

INDEX

Introduction to Neural Engineering for Motor Rehabilitation, First Edition. Edited by Dario Farina, Winnie Jensen, and Metin Akay.

IEEE Press Series in Biomedical Engineering

The focus of our series is to introduce current and emerging technologies to biomedical and electrical engineering practitioners, researchers, and students. This series seeks to foster interdisciplinary biomedical engineering education to satisfy the needs of the industrial and academic areas. This requires an innovative approach that overcomes the difficulties associated with the traditional textbooks and edited collections.

Series Editor: Metin Akay, University of Houston, Houston, Texas

1. *Time Frequency and Wavelets in Biomedical Signal Processing*
Metin Akay

2. *Neural Networks and Artificial Intelligence for Biomedical Engineering*
Donna L. Hudson, Maurice E. Cohen

3. *Physiological Control Systems: Analysis, Simulation, and Estimation*
Michael C. K. Khoo

4. *Principles of Magnetic Resonance Imaging: A Signal Processing Perspective*
Zhi-Pei Liang, Paul C. Lauterbur

5. *Nonlinear Biomedical Signal Processing, Volume 1, Fuzzy Logic, Neural Networks, and New Algorithms*
Metin Akay

6. *Fuzzy Control and Modeling: Analytical Foundations and Applications*
Hao Ying

7. *Nonlinear Biomedical Signal Processing, Volume 2, Dynamic Analysis and Modeling*
Metin Akay

8. *Biomedical Signal Analysis: A Case-Study Approach*
Rangaraj M. Rangayyan

9. *System Theory and Practical Applications of Biomedical Signals*
Gail D. Baura

10. *Introduction to Biomedical Imaging*
Andrew G. Webb

11. *Medical Image Analysis*
Atam P. Dhawan

12. *Identification of Nonlinear Physiological Systems*
David T. Westwick, Robert E. Kearney

13. *Electromyography: Physiology, Engineering, and Non-Invasive Applications*
Roberto Merletti, Philip Parker

14. *Nonlinear Dynamic Modeling of Physiological Systems*
Vasilis Z. Marmarelis

15. *Genomics and Proteomics Engineering in Medicine and Biology*
Metin Akay

16. *Handbook of Neural Engineering*
Edited by Metin Akay

17. *Medical Image Analysis, Second Edition*
Atam P. Dhawan

18. *Advanced Methods of Biomedical Signal Processing*
Edited by Sergio Cerutti, Carlo Marchesi

19. *Epistemology of the Cell: A Systems Perspective on Biological Knowledge*
Edward R. Dougherty, Michael L. Bittner

20. *Micro and Nanotechnologies for Engineering Stem Cells and Tissues*
Murugan Ramalingam, Esmaiel Jabbari, Seeram Ramakrishna,
Ali Khademhosseini

21. *Introduction to Neural Engineering for Motor Rehabilitation*
Dario Farina, Winnie Jensen, Metin Akay

IEEE Press Series in Biomedical Engineering

Series Editor: Metin Akay, University of Houston, Houston, Texas

A complete list of the titles in this series appears at the end of this volume.

IEEE Press Series in Biomedical Engineering

The focus of our series is to introduce current and emerging technologies to biomedical and electrical engineering practitioners, researchers, and students. This series seeks to foster interdisciplinary biomedical engineering education to satisfy the needs of the industrial and academic areas. This requires an innovative approach that overcomes the difficulties associated with the traditional textbooks and edited collections.

Series Editor: Metin Akay, University of Houston, Houston, Texas

Time Frequency and Wavelets in Biomedical Signal Processing
Metin Akay

Neural Networks and Artificial Intelligence for Biomedical Engineering
Donna L. Hudson, Maurice E. Cohen

Physiological Control Systems: Analysis, Simulation, and Estimation
Michael C. K. Khoo

Principles of Magnetic Resonance Imaging: A Signal Processing Perspective
Zhi-Pei Liang, Paul C. Lauterbur

Nonlinear Biomedical Signal Processing, Volume 1, Fuzzy Logic, Neural Networks, and New Algorithms
Metin Akay

Fuzzy Control and Modeling: Analytical Foundations and Applications
Hao Ying

Nonlinear Biomedical Signal Processing, Volume 2, Dynamic Analysis and Modeling
Metin Akay

Biomedical Signal Analysis: A Case-Study Approach
Rangaraj M. Rangayyan

System Theory and Practical Applications of Biomedical Signals
Gail D. Baura

Introduction to Biomedical Imaging
Andrew G. Webb

Medical Image Analysis
Atam P. Dhawan

Identification of Nonlinear Physiological Systems
David T. Westwick, Robert E. Kearney

Electromyography: Physiology, Engineering, and Non-Invasive Applications
Roberto Merletti, Philip Parker

Nonlinear Dynamic Modeling of Physiological Systems
Vasilis Z. Marmarelis

Genomics and Proteomics Engineering in Medicine and Biology
Metin Akay

Handbook of Neural Engineering
Edited by Metin Akay

Medical Image Analysis, Second Edition
Atam P. Dhawan

Advanced Methods of Biomedical Signal Processing
Edited by Sergio Cerutti, Carlo Marchesi

Epistemology of the Cell: A Systems Perspective on Biological Knowledge
Edward R. Dougherty, Michael L. Bittner

Micro and Nanotechnologies for Engineering Stem Cells and Tissues
Murugan Ramalingam, Esmaiel Jabbari, Seeram Ramakrishna,
Ali Khademhosseini

Introduction to Neural Engineering for Motor Rehabilitation
Dario Farina, Winnie Jensen, Metin Akay

Printed and bound by CPI Group (UK) Ltd, Croydon, CR0 4YY
16/09/2026
14963410-0001